Natural Products for Antibacterial Drug Development: Recent Advancement of Computational Approach

Tripti Sharma • Chita Ranjan Sahoo
Debdutta Bhattacharya • Sanghamitra Pati
Editors

Natural Products for Antibacterial Drug Development: Recent Advancement of Computational Approach

Springer

Editors
Tripti Sharma
School of Pharmaceutical Sciences and Research
Chhatrapati Shivaji Maharaj University
Panvel, Navi Mumbai, Maharashtra, India

Debdutta Bhattacharya
Department of Health Research, Ministry of Health and Family Welfare, Government of India
ICMR-Regional Medical Research Centre
Bhubaneswar, Odisha, India

Chita Ranjan Sahoo
Department of Health Research, Ministry of Health and Family Welfare, Government of India
ICMR-Regional Medical Research Centre
Bhubaneswar, Odisha, India

Sanghamitra Pati
Department of Health Research, Ministry of Health and Family Welfare, Government of India
ICMR-Regional Medical Research Centre
Bhubaneswar, Odisha, India

ISBN 978-981-97-9633-5 ISBN 978-981-97-9634-2 (eBook)
https://doi.org/10.1007/978-981-97-9634-2

This Springer imprint is published by the registered company Springer Nature Singapore Pte Ltd.
The registered company address is: 152 Beach Road, #21-01/04 Gateway East, Singapore 189721, Singapore

Preface

In the ongoing battle against bacterial infections, the development of new and effective antibacterial drugs is of paramount importance. With the rise of antibiotic resistance and the dwindling pipeline of novel antibiotics, it has become crucial to explore alternative approaches for drug discovery. This book focuses on the recent advancements in computational methods for the discovery and development of natural products as potential antibacterial agents.

The use of natural products in medicine has a long and rich history, with many of the current antibiotics being derived from natural sources. However, the process of discovering and developing new antibacterial drugs from natural sources can be time-consuming and challenging. This is where computational approaches come into play, offering efficient and innovative solutions to streamline the drug discovery process.

This book aims to provide a comprehensive overview of the latest computational methods and techniques used in the discovery and development of natural products as antibacterial agents. It covers a wide range of topics, including virtual screening, molecular docking, machine learning, and structure-based drug design. The readers will gain valuable insights into how these computational tools can be utilized to identify promising lead compounds, optimize their structures, and predict their antibacterial activity.

By bringing together contributions from leading experts in the field, this book offers a unique perspective on the integration of computational approaches with natural product research. It highlights the potential of these methods to accelerate the discovery of novel antibacterial drugs and addresses the challenges and future directions in this rapidly evolving field.

This book is intended for researchers, scientists, and students in the fields of medicinal chemistry, pharmacology, and computational biology who are interested in exploring the potential of natural products for antibacterial drug development. By providing a detailed understanding of the latest computational advancements, this book aims to inspire and guide future research efforts in this important area of drug discovery.

We hope that this book will serve as a valuable resource, fostering innovation and collaboration in the ongoing quest to discover new antibacterial drugs and address the pressing global health challenge of antibiotic resistance.

Navi Mumbai, Maharashtra, India — Tripti Sharma
Bhubaneswar, Odisha, India — Chita Ranjan Sahoo
Bhubaneswar, Odisha, India — Debdutta Bhattacharya
Bhubaneswar, Odisha, India — Sanghamitra Pati

Contents

Editors and Contributors

About the Editors

Tripti Sharma currently serves as Professor and Principal at School of Pharmaceutical Sciences and Research, Chhatrapati Shivaji Maharaj University, Panvel, Navi Mumbai, Maharashtra, India. She has 19 years of teaching experience in Pharmaceutical Chemistry at undergraduate and postgraduate levels. She has 14 years of research experience in computational drug design and drug development. During this period, she has been involved in teaching and research. She is a life member of the Association of Pharmaceutical Teachers of India (APTI), Orissa Chemical Society. To her credit, she has more than 45 publications in national and international journals of repute and 9 book chapters. She is serving as a reviewer in various national/international Journals as well.

Chita Ranjan Sahoo is presently availing ICMR-Centenary Post-Doctoral Fellowship (PDF) from ICMR, Department of Health Research, Ministry of Health and Family Welfare, Government of India, and he received PhD at Central Research Laboratory, Institute of Medical Sciences and SUM Hospital, and Department of Pharmaceutical Chemistry, School of Pharmaceutical Sciences, Siksha 'O' Anusandhan (Deemed to be University), Bhubaneswar, Odisha. He obtained his master's degree in biotechnology, Utkal University, Odisha, India. He has published several research articles, review articles, abstracts, book chapter, and books with the leading houses. His contribution toward problem to key solution in research achieving Chief Minister Award- for the year 2016-17 as youth innovation, Young Scientist Award in Science-2018, Rajiv Gandhi Prativa Puraskar Odisha Forum and a few more. He has been associated with various professional societies.

Debdutta Bhattacharya presently works as Scientist-E at ICMR-Regional Medical Research Centre (Department of Health Research, Ministry of Health and Family Welfare, Government of India), Bhubaneswar, India. He completed his PhD in Microbiology under Pondicherry University from ICMR-RMRC, Port Blair. He is working with ICMR for last 16 years in various capacities. He is trained in Public Health Microbiology and Infectious Disease Biology. His areas of research interest are One Health, Antimicrobial resistance, Infectious Disease Epidemiology, and zoonotic disease. He has been part of various outbreak investigation team deputed by Emergency Medical Relief team deputed by Ministry of Health and Family Welfare, Government of India, and by the institute. He was involved as a lead scientist from the center in the development of the 1st Odisha-based Antigen detection kit (ImCOV-Ag) in collaboration with IMGENEX India Pvt. Ltd. He has published more than 90 papers in national and international journals of repute on various aspects of infectious diseases. He serves as editorial panel member of several journals. He is also the recipient of ICMR Gold Medal and Bill and Melinda Gates Foundation Professional Development Award 2022 for his work on Infectious Diseases.

Sanghamitra Pati is presently working as Scientist-G and Director at ICMR-Regional Medical Research Centre (Department of Health Research, Ministry of Health and Family Welfare, Government of India), Bhubaneswar, India. She is a physician turned laboratory scientist and public health researcher. Dr. Pati completed her MBBS (1992) and MD (1998) at MKCG Medical College, Berhampur, Odisha, India. She completed her PhD on Multimorbidity in Primary care on June 28, 2023, at Maastricht University, The Netherlands. In 2000, she received Joint Japan/World Bank Graduate Scholarship (JJ/WBGS) to pursue Master of Public Health (MPH) from the University of Maastricht, The Netherlands. Since August 2016, Dr. Pati has been serving as the Director of ICMR-Regional Medical Research Centre, Bhubaneswar, for advancing research programs aimed at improvement of public health through addressing different causes of both communicable and non-communicable diseases. Her Wellcome Trust-funded research is the first in India and any Low and Middle Income Country to assess the burden and impact of multimorbidity (coexistence of multiple chronic illness in a person) in patients attending healthcare settings. The MAQ-PC tool developed by her is now being used by many researchers across the LMICs to measure multimorbidity. She has co-authored 280+ scientific publications in peer-reviewed journals. She has an h-index of 39 and i10-index of 122. Dr. Pati is a fellow of UICC (Union for International Cancer Control) and Indian Public Health Association (IPHA) and a Distinguished Lown Scholar, at Harvard School of Public Health, Boston. She is also the recipient of APACT Young Investigator Award (2007, Taiwan), Early Career Award from International Society of Behavioral Medicine (2010, 2012), and International Engagement Award, by The Wellcome Trust UK (2015–16). In 2021, she was elected as the Fellow of National Academy of Medical Sciences (FAMS), India, and Fellow of Royal Society of Public Health (FRSPH), UK. She is also Fellow Member of Royal Society of Public Health, London; Royal Academy of Medicine, Ireland; International Society of Global Health, Edinburgh; and Royal Society of Tropical Medicine and Hygiene, London.

Contributors

Diptikanta Acharya Department of Biotechnology, School of Sciences, GIET University, Gunupur, Odisha, India

Biswajit Banerjee Department of Pharmaceutical Chemistry, School of Pharmaceutical Sciences, Siksha 'O'Anusandhan (Deemed to be University), Bhubaneswar, Odisha, India

Rashmi Ranjan Behera Department of Life Science, School of Sciences, Gandhi Institute of Engineering and Technology, Gunupur, Odisha, India

Debdutta Bhattacharya Department of Health Research, Ministry of Health and Family Welfare, Government of India, ICMR-Regional Medical Research Centre, Bhubaneswar, Odisha, India

Subha Laxmi Bhushan Department of Life Science, School of Sciences, Gandhi Institute of Engineering and Technology, Gunupur, Odisha, India

Anshuman Chandra School of Physical Sciences, Jawaharlal Nehru University, New Delhi, India

Smriti Dewangan MATS School of Pharmacy, MATS University, Raipur, Chhattisgarh, India

Debasmita Dubey Department of Medical Research, IMS & SUM Hospital, Siksha 'O' Anusandhan Deemed to be University, Bhubaneswar, Odisha, India

T. Gayatri Department of Biotechnology, School of Sciences, OP Jindal University, Raigarh, Chhattisgarh, India

Vijay Kumar Goel School of Physical Sciences, Jawaharlal Nehru University, New Delhi, India

Nainee Goyal School of Biotechnology, Gautam Buddha University, Greater Noida, Uttar Pradesh, India

Jublee Jasmine Department of Biosciences, College of Pharmaceutical Sciences, Berhampur, Odisha, India

Goutam Jha Department of Biotechnology, School of Sciences, GIET University, Gunupur, Odisha, India

Smarita Lenka Department of Medical Research, IMS & SUM Hospital, Siksha 'O' Anusandhan Deemed to be University, Bhubaneswar, Odisha, India

Gyanranjan Mahalik Department of Botany, School of Applied Sciences, Centurion University of Technology and Management, Brahmapur, Odisha, India

Ramchander Merugu Department of Biochemistry, Mahatma Gandhi University, Nalgonda, Telangana, India

Monali Priyadarsini Mishra School of Paramedics and Allied Health Sciences, Centurion University of Technology and Management, Bhubaneswar, Odisha, India

Saswat S. Mohapatra Molecular Microbiology Lab, P.G. Department of Biotechnology, Berhampur University, Berhampur, Odisha, India

Ipsa Padhy Department of Pharmaceutical Chemistry, School of Pharmaceutical Sciences, Siksha 'O'Anusandhan (Deemed to be University), Bhubaneswar, Odisha, India

Rabindra Nath Padhy Central Research Laboratory, IMS and SUM Hospital, Sikhsa 'O' Anusandhan Deemed to be University, Bhubaneswar, Odisha, India

Madhulita Panda College of Pharmaceutical Sciences, Berhampur, Odisha, India

Somya Ranjan Panda Department of Life Science, School of Sciences, Gandhi Institute of Engineering and Technology, Gunupur, Odisha, India

Girish Kumar Pati Department of Gastroenterology and Hepatobiliary Sciences, IMS and SUM Hospital, Sikhsa 'O' Anusandhan Deemed to be University, Bhubaneswar, Odisha, India

Sanghamitra Pati Department of Health Research, Ministry of Health and Family Welfare, Government of India, ICMR-Regional Medical Research Centre, Bhubaneswar, Odisha, India

Swarup Kumar Patnaik Department of Gastroenterology and Hepatobiliary Sciences, IMS and SUM Hospital, Sikhsa 'O' Anusandhan Deemed to be University, Bhubaneswar, Odisha, India

Sandhyarani Patra Department of Biochemistry, School of Sciences, GIET University, Gunupur, Odisha, India

Ganesh Patro College of Pharmaceutical Sciences, Berhampur, Odisha, India

Ghanishtha Prusty Department of Life Science, School of Sciences, Gandhi Institute of Engineering and Technology, Gunupur, Odisha, India

Paresh Kumar Prusty Department of Microbiology, Aravali Veterinary College, Sikar, Rajasthan, India

Anjali Rath Department of Life Science, School of Sciences, Gandhi Institute of Engineering and Technology, Gunupur, Odisha, India

Varsha Rawat MATS School of Pharmacy, MATS University, Raipur, Chhattisgarh, India

Sumedha Saha Advanced Laboratory for Plant Genetic Engineering, Advanced Technology Development Centre, IIT, Kharagpur, West Bengal, India

Chita Ranjan Sahoo Department of Health Research, Ministry of Health and Family Welfare, Government of India, ICMR-Regional Medical Research Centre, Bhubaneswar, Odisha, India

Mamalisa Sahoo School of Paramedics and Allied Health Sciences, Centurion University of Technology and Management, Bhubaneswar, Odisha, India

Somanath Sahoo Department of Biotechnology, School of Sciences, GIET University, Gunupur, Odisha, India

Sagarika Satapathy Department of Biotechnology, School of Sciences, GIET University, Gunupur, Odisha, India

Tripti Sharma School of Pharmaceutical Sciences and Research, Chhatrapati Shivaji Maharaj University, Panvel, Navi Mumbai, Maharashtra, India

Shivangi Department of Biotechnology, School of Sciences, GIET University, Gunupur, Odisha, India

Kenguruka Sidoine Central Research Laboratory, IMS and SUM Hospital, Sikhsa 'O' Anusandhan Deemed to be University, Bhubaneswar, Odisha, India

Anand Prakash Singh IIMT College of Engineering, Greater Noida, Uttar Pradesh, India

Nagendra Singh School of Biotechnology, Gautam Buddha University, Greater Noida, Uttar Pradesh, India

Santosh Kumar Swain Department of Otorhinolaryngology and Head and Neck Surgery, All India Institute of Medical Sciences, Bhubaneswar, Odisha, India

Kanishka Uthansingh Department of Gastroenterology and Hepatobiliary Sciences, IMS and SUM Hospital, Sikhsa 'O' Anusandhan Deemed to be University, Bhubaneswar, Odisha, India
Central Research Laboratory, IMS and SUM Hospital, Sikhsa 'O' Anusandhan Deemed to be University, Bhubaneswar, Odisha, India

Manu Vashistha Advanced Instrumentation Research & Facility, Jawaharlal Nehru University, New Delhi, India

Natural Products for Antibacterial Infections in Current Scenario

1

Ipsa Padhy, Biswajit Banerjee, and Tripti Sharma

Abstract

Bacterial infections have always been a rising life threat from global health perspective. The need for novel antibacterial pharmacotherapeutic candidates is also mounting high with the increasing cases of resistance to frontline antibiotics, often leading to disease resurgence. In the last 10 years, extensive research has been diverted towards medicinal plants, viewing them as potential sources for novel antibacterial drug candidates because of their large functional and structural diversity. The bioactive compounds have a broad spectrum of pharmacological activities ranging from antimicrobial, anticancer, anti-inflammatory, antioxidant, antidiabetic, and immunomodulatory. Moreover, the phytoconstituents have a wider therapeutic window and less toxicity profile. More importantly, several phytoconstituents have been reported to disrupt the quorum-sensing and biofilm-forming capacity of bacteria, thereby emerging as novel and potent therapeutic alternatives for combating antibiotic resistance. The current chapter delves into the present developments in medicinal plant research, which are explored as potential sources of novel antibacterial agents. The chapter emphasizes on different natural compounds and their antibacterial mechanisms. It focuses the state-of-art computational and genomic approaches applied in the

I. Padhy · B. Banerjee
Department of Pharmaceutical Chemistry, School of Pharmaceutical Sciences, Siksha 'O'Anusandhan (Deemed to be University), Bhubaneswar, Odisha, India

T. Sharma (✉)
School of Pharmaceutical Sciences and Research, Chhatrapati Shivaji Maharaj University, Panvel, Navi Mumbai, Maharashtra, India

T. Sharma et al. (eds.), *Natural Products for Antibacterial Drug Development: Recent Advancement of Computational Approach*,
https://doi.org/10.1007/978-981-97-9634-2_1

drug discovery pipeline of novel antibacterial candidates based on natural products.

Keywords

Antibiotic resistance · Phytoconstituents · Computational approaches · Genomic approaches · Drug discovery

Abbreviations

AMDD	Antimicrobial drug database
AMR	Antimicrobial resistance
APD	Antimicrobial peptide database
ATP	Adenosine triphosphate
BacMet	Biocide and metal resistance gene database
CADD	Computer-aided drug designing
CAMP R3	Collection of antimicrobial peptides
CARD	Comprehensive antibiotic resistance
CMAUP	Collective Molecular Activities of Useful Plants
DNA	Deoxyribonucleic acid
DRAMP	Data repository of antimicrobial peptides
GRAS	Generally Recognized as Safe
HIT	Herbal Ingredients Targets
IMPPAT	Indian Medicinal Plants, Phytochemistry And Therapeutics
LC-MS	Liquid chromatography mass spectrometry
MDR	Multidrug resistant
MIC	Minimum inhibitory concentration
MRSA	Methicillin-resistant *Staphylococcus aureus*
NGS	Next generation sequencing
NPASS	Natural Product Activity and Species Source
p-ANAPL	Pan-African Natural Product Library
QS	Quorum sensing
RNA	Ribonucleic acid
SWMD	Seaweed Metabolite Database
TCR	Two-component response regulator
YaTCM	Yet another Traditional Chinese Medicine

1.1 Introduction

Bacterial infections, over the course of time, have become a major health concern worldwide (Cassini et al. 2019). Antibiotics have proved to be powerful bioweapons against microbial superbugs, yet the increasing incidences of resistance to frontline antibiotics have become a matter of serious concern in the last decade (Manesh and Varghese 2021; Patel et al. 2021). Pathogenic bacteria develop resistance to front-line broad-spectrum antibiotics by using intracellular efflux pumps, enzymatic hydrolysis, biofilm formation, quorum sensing, and target modifications through mutations. The mechanisms act as microbial survivable support, thereby reducing the effects of most conventional antibiotics and hindering the rate of novel antibiotic drugs (Upadhyay et al. 2023). For instance, azithromycin resistance occurs by bacterial efflux pumps (Wadsworth et al. 2018), while vancomycin resistance in *Staphylococcus aureus* develops via target modification (Fu et al. 2020). The last 10 years have only witnessed the development of very few antimicrobials against multidrug resistance pathogens, namely, eravacycline (Solomkin et al. 2019), omadacycline (Dougherty et al. 2019), and Lefamulin (Zhanel et al. 2021).

The rapid development of antimicrobial resistance, the resurgence of infections and the paucity of antibiotics against pathogenic resistance pose a daunting challenge to the newer antibacterial drug discovery. In the current scenario, natural products having antimicrobial properties (Silva et al. 2016; Rossiter et al. 2017) and diverse chemical space offer an exquisite source for developing new antibacterial drugs (Niu and Li 2019). The chief sources of natural antibacterial compounds are plants, fungi, algae, and other nonpathogenic bacteria. Plants are the major contributors of bioactive antibacterial compounds. Plant-derived antimicrobial compounds have the edge over the currently used frontline antibiotics because of their multifaceted antibacterial mechanism of action that counter the development of antibacterial resistance in pathogenic bacteria (Rossiter et al. 2017; Martelli and Giacomini 2018). The natural products reportedly potentiate the bactericidal activity of antibiotics by several mechanisms, one being the disruption of biofilms that ultimately aid in overcoming antimicrobial resistance (Zacchino et al. 2017). Additionally, the plant-derived bioactive compounds are tagged as "Generally Recognized as Safe" (GRAS), which are more acceptable to consumers (Guil-Guerrero et al. 2016). Subsequently, the search for bioactive natural compounds to develop novel antibacterial agents seems to be an important strategy in combating antimicrobial resistance (Gonçalves et al. 2023).

Conventional drug discovery protocols incur high costs and are laborious. The current era has witnessed extravagant advancements in the field of computational biology. The multiomics technologies have contributed to linking disease-related indications to precise mutations, epigenetic modifications, and environmental effects, thus recognizing probable pharmacological targets. Computational methods are also amalgamated with artificial intelligence and machine learning methods in the road voyage to phytochemical-based drug discovery (Sarker and Nahar 2018; Singh et al. 2019).

In the present chapter, to explore the antibacterial potencies of natural products. Particular emphasize has been given to phytoconstituents along with addressing their antibacterial mechanism of action. Lastly, a comprehensive discussion is provided regarding the state-of-the-art in silico and genomic methods employed in the natural product-driven antibacterial drug discovery.

1.2 Plants as Sources for Antibacterial Agents

Antimicrobial resistance and associated side effects of currently available antibiotics are addressed using plant-based antimicrobials. The bioactive principles such as alkaloids, glycosides, polyphenols, flavonoids, saponins, tannins, and terpenoids in plants account for their antibacterial properties. The herbal constituents might act as immunomodulators, antioxidants, and nutraceuticals in combating infectious antibiotic-resistant pathogens (Gupta and Birdi 2017). Folklore medicine is primarily dependent on plant-based products for the treatment of bacterial infections affecting integumentary, respiratory, gastrointestinal, excretory, and reproductive systems (Swain and Padhy 2015; Cock and Van Vuuren 2020; Anand et al. 2022). Numerous plants have been screened for their potent antibacterial properties against various pathogenic bacteria, some listed in Table 1.1.

1.3 Bioactive Compounds as Potent Antibacterial Agents

A plethora of biologically active compounds derived from natural sources are used as valuable drugs in alternative medicinal systems (AlSheikh et al. 2020; Genilloud 2019). These biologically active compounds are majorly secondary metabolites presenting diverse chemical space such as alkaloids, glycosides, flavonoids, coumarins, terpenoids, saponins, lignans, tannins, and steroids (Sharma et al. 2022; Srivastav et al. 2020). Apart from plants, natural antibiotics (Fig. 1.1) have also been isolated from marine organisms, fungi, and other microbes (Dai et al. 2020). In the last 10 years, several novel bioactive compounds have been isolated and screened for their antibacterial activities, some significant findings of which are displayed in Table 1.2.

1.4 Antibacterial Mechanism of Natural Products

The phytoconstituents exhibit very potent antibacterial activities by copious mechanisms such as disintegrating microbial cell membranes, disturbing cellular metabolism, and destroying biofilms and anti-quorum sensing techniques (Radulovic et al. 2013; Seukep et al. 2020). Additionally, the bioactive compounds counter the development of antimicrobial resistance to antibiotics, thereby having strong lethality to pathogenic bacteria. In the following sections, we discuss the various modes of action exhibited by the natural compounds in countering antibiotic resistance developed by bacteria (Fig. 1.2).

1.4.1 Disruption of Cell Wall and Lysis

The natural compounds (Fig. 1.3) belonging to flavonoids, quinones, coumarins, tannins, and terpenes destroy the bacterial cell walls through hydroxylation

Table 1.1 Medicinal plants and their antibacterial potencies

Scientific name	Common name	Family	Plant part used	Susceptible pathogen	References
Acacia Arabica	Gum Arabic tree	Leguminosae	Fabricated zinc oxide nanoparticles with aqueous leaf extract	Multidrug-resistant *Escherichia coli*, *Staphylococcus aureus*, *Salmonella entrica*	Hayat et al. (2022)
Acacia catechu	Black catechu, khadir	Fabaceae	Ethyl acetate and aqueous bark extracts	*Staphylococcus aureus*	Aryal et al. (2021)
Achyranthes aspera	Chaffs flower	Amaranthaceae	Methanolic extracts of roots and stems	*Streptococcus mutans*	Murugan et al. (2013)
Aegle marmelos	Wood apple	Rutaceae	Zinc ferrite magnetic nanoparticles of aqueous leaf extract *n*-hexane extracts of essential oil from leaves	*Pseudomonas aeruginosa*, *Escherichia coli*, *Staphylococcus aureus*, and *Bacillus subtilis* *Bacillus subtilis*, *Escherichia coli*, *Klebsiella pneumoniae*, *Proteus vulgaris*, *Pseudomonas denitrificans*, *Sarcinalutea*, and *Xanthomonas campestris*	Sriramulu et al. (2018), Jamal et al. (2017)
Agrimonia eupatoria	Common agrimony	Rosaceae	Acetone and aqueous extract of flowering tops	*Proteus mirabilis*, *Pseudomonas aeruginosa*	Muruzović et al. (2016)
Allium cepa	Onion	Amaryllidaceae	Methanolic bulb extracts	MDR *Escherichia coli*, *Klebsiella pneumoniae*, *Mycobacterium smegmatis*	Sharma et al. (2019)
Allium sativum	Garlic	Amaryllidaceae	Nanoparticles of ultrasonicated aqueous bulb extracts	*Staphylococcus aureus sub. aureus*, *Streptococcus mutans*, *Escherichia coli*, and *Poryphyromonas gingivalis*	Gabriel et al. (2022)
Amomum tsaoko	Chinese cauguo	Zingiberaceae	Essential oil	*Escherichia coli*, *staphylococcus aureus*	Guo et al. (2024)
Anethum graveolens	Dill	Apiaceae	Essential oil	*Staphylococcus aureus*, *Bacillus licheniformis*, *Listeria innocua*, *Enterobacter xiangfangensis*, *Escherichia*, *fergusonii*, *Pseudomonas aeruginosa*	Aati et al. (2022)

(continued)

Table 1.1 (continued)

Scientific name	Common name	Family	Plant part used	Susceptible pathogen	References
Azadirachta indica	Neem, margosa	Meliaceae	Methanolic leaf extract	*P. aeruginosa*, *K. pneumoniae*, *Citrobacter* spp., *E. coli*, *E. faecalis*, *S. epidermidis*, and *S. aureus*	Altayb et al. (2022)
Bahunia kockiana	Climbing bauhinia	Fabaceae	Ethyl acetate extract of flowers	Methicillin-resistant *Staphylococcus aureus*	Chew et al. (2018)
Carica pappaya	Papaya	Caricaceae	Methanolic flower extract	*Escherichia coli*, *Bacillus subtilis*	Asghar et al. (2016), Dwivedi et al. (2020)
Capparis spinosa	Flinders rose	Capparaceae	Aqueous leaf extract	*Escherichia coli*, *Salmonella typhimurium*, *Staphylococcus aureus*, *Bacillus cereus*	Benakashani et al. (2016)
Cassia fistula	Golden shower	Fabaceae	Ethanolic bark extract	*Salmonella typhosa*, *Shigella dysenteriae*	Chaerunisaa et al. (2020)
Chamaemelum nobile	Chamomile	Asteraceae	Flower extract	*Pseudomonas aeruginosa*	Kazemian et al. (2015)
Cinnamomum zeylanicum	Cinnamon	Lauraceae	Essential oil	*Escherichia coli*, *Staphylococcus aureus*	Zhang et al. (2016)
Citrus aurantifolia	Lime	Rutaceae	Essential oil from leaves and fruit peels	*S. mutans* and *L. casei*	Lemes et al. (2018)
Cymbopogan citratus	Lemon grass	Poaceae	Essential oil	MRSA, MRSE, *P. aeruginosa*	Subramaniam et al. (2020)
Dalbergia trichocarpa	Red sandlewood	Fabaceae	Bark extracts	*P. aeruginosa*	Rasamiravaka et al. (2015)
Euphorbia hirta	Asthma plant	Euphorbiaceae	Ethanolic and aqueous extracts aerial parts	MRSA	Kumari and Pandey (2017)
Hydrastis canadensis	Goldenseal	Ranunculaceae	Leaf extract	MRSA	Cech et al. (2012)
Mallotus philippinensis	Kumkum tree	Euphorbiaceae	Dried fruit extract	MRSA	Oyedemi et al. (2016)

Nigella sativa	Black cumin	Ranunculaceae	Seed oil	*S. aureus*	Rafati et al. (2014)
Ocimum gratissimum	Clove basil	Lamiaceae	Essential oils	*S. aureus*, *E. coli*, *K. pneumomiae*	Silva et al. (2022)
Piper betle	Betel vine	Piperaceae	Ethanolic leaf extract	*S. aureus*	Lubis and Wahyuni (2020)
Rosa rugosa	Beach rose	Rosaceae	Aqueous extracts of freez dried fruits	*E. coli*, *B. cereus*, *K. pneumomiae*	Cendrowski et al. (2020)
Salvador apersica	Toothbrush tree	Salvadoraceae	Aqueous extracts of Dried stems	*Staphylococcus aureus*, *Lactobacillus acidophilus*, *Streptococcus mutans*, *Pseudomonas aeruginosa*	Abhary and Al-Hazmi (2016)
Salvia officinalis	Sage	Lamiaceae	Diclorometha-ne extracts of aerial parts	*P. gingivalis*	Mendes et al. (2020)
Syzygicum aromaticum	Clove	Myrtaceae	Essential oil	*S. aureus*, *E. coli*	Bai et al. (2023)
Terminalia chebula	Myrobalan	Combretaceae	Ethanolic extracts of fruit, acetone bark extract	*Streptococcus mutans*, *Salmonella typhi*	Nam and Hwang (2021), Venkatesan et al. (2017)
Zingiber officinale	Ginger	Zingiberaceae	Essential oil	*Escherichia coli*, *Salmonella species*, *Enterobacter* spp., *Shigella* spp., and *Citrobacter*	Yadufashije et al. (2020), Wang et al. (2020a, b)

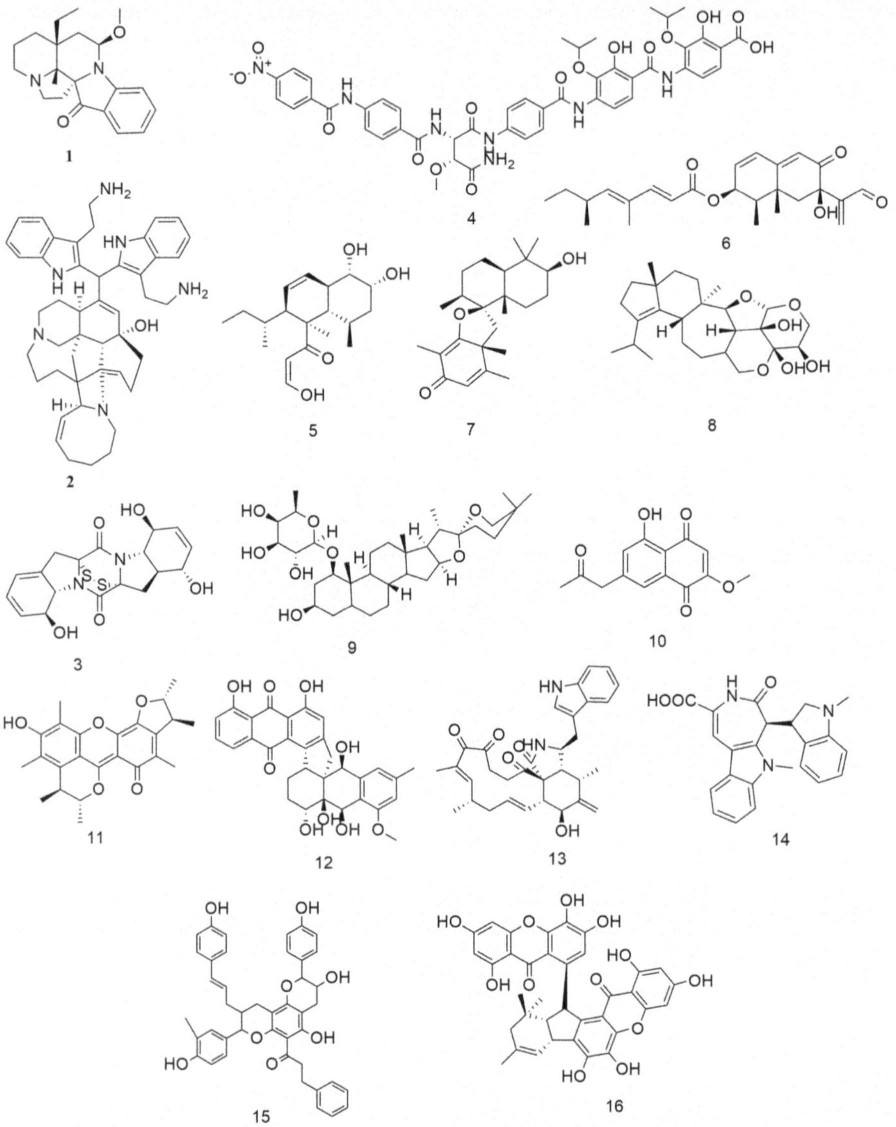

Fig. 1.1 Novel bioactive antibacterial compounds from plant endophytes

reactions, forming irreversible complexes with nucleophilic amino acids, microbial cell surface adhesins, membrane peptides, enzymes, and finally lysis of microbial cells by promoting membrane permeability (Biharee et al. 2020; Campos et al. 2009; Bouarab-Chibane et al. 2019). Thickness reduction in membrane bilayer and enhanced fluidity of bacterial plasma membrane leading to leakage of essential proteins have been induced by flavonoids like quercetin (17), rutin (18), naringenin (19), sophoraflavanone (20), and tiliroside (21) (Tsuchiya and Iinuma 2000; Sanver

Table 1.2 Novel antibacterial bioactive compounds

Bioactive compound	Chemical class	Natural source	Tested pathogen	MIC	References
Melokhanin B (1)	Pentacyclic indole alkaloids	*Melodinus khasianus* leaves (Apocyanaceae)	*Pseudomonas aeruginosa*	2–5 μM	Cheng et al. (2016)
Zamamidine D (2)	Manzamine alkaloid	Okinawan *Amhimedon* (marine sponge)	*Staphylococcus aureus*, *Bacillus subtilis*, *Micrococcus luteus*	8 μM	Kubota et al. (2017)
Brocazine G (3)	Diketopiperazine alkaloid	*Penicillium brocae* (fungi)	*Staphylococcus aureus*	0.25 μM	Meng et al. (2016)
Coralmycin A (4)	Alkaloid	*Corallococcus coralloids* (Myxobacteria)	Methicillin and quinolone-resistant *S. aureus*, *Streptococcus pneumonia*, *Enterococcus faecalis*, *E. coli*, *P. aeruginosa*, *Acinetobacter baumannii*, and *Klebsiella pneumoniae*	0.015–4 μM	Kim et al. (2016)
Eupenicinicol (5)	Alkaloid	*Xanthium sibricum* (Compositae)	*Staphylococcus aureus*	0.1 μM	Li et al. (2017)
Dendryphiellin (6)	Sesquiterpene	*Cochliobolus lunatus* (marine fungi)	*Staphylococcus aureus*	1.5 μM	Fang et al. (2018)
Chermesin A (7)	Spiro-meroterpenoid	*Penicillium chermesinum* (endophytic marine fungi)	*Micrococcus luteus*	8 μM	Liu et al. (2016)
Laxitextrine A (8)	Cyathane diterpenoid	*Laxitextum incrustatum* (tropical fungi)	Methicillin-resistant *Staphylococcus aureus*	7.8 μM	Mudalungu et al. (2016)
Suttonigenin F (9)	Steroid	Fruits of *Cordyline manners-suttoniae* (Asparagaceae)	*Staphylococcus aureus*	5.4 μM	Tran et al. (2019)
2-methoxy-7-acetonyljuglone (10)	Quinone	*Reynoutria japonica* (Polygonaceae)	*Helicobacter pylori*	0.06–0.13 μM	Khalil et al. (2019)
Penicitrinone F (11)	Polyketide	*Penicillium griseofulvum*	*Bacillus subtilis*	6.3 μM	Yang et al. (2018)
Solanrubiellin A (12)	Hydro-anthraquinone	*Solanum lyratum* (*Solanaceae*)	*Staphylococcus aureus*, *Enterococcus faecalis*	1.08–5.44 μM	Chen et al. (2020)

(continued)

Table 1.2 (continued)

Bioactive compound	Chemical class	Natural source	Tested pathogen	MIC	References
ArmochaetoglosinC (13)	Alkaloid	*Chaetomium globosum* (arthropod endophytic fungi)	*Klebsiella pneumoniae*	4 μM	Gao et al. (2019)
Thalicfoetine (14)	Alkaloid	Roots of *Thalictrum foetidum* (Ranunculaceae)	*Escherichia coli*, *Bacillus subtilis*	3–6 μM	Yan et al. (2019)
Balsacone R (15)	Flavonoid	Ethanolic bud extracts of *Populus balsamifera* (Salicaceae)	*Staphylococcus aureus*	6 μM	Simard et al. (2016)
Garmoxanthone (16)	Flavonoid	*Bark of Garcinia mangostana* (Clusiaceae)	Methicillin-resistant *Staphylococcus aureus*	3.9 μM	Wang et al. (2018)

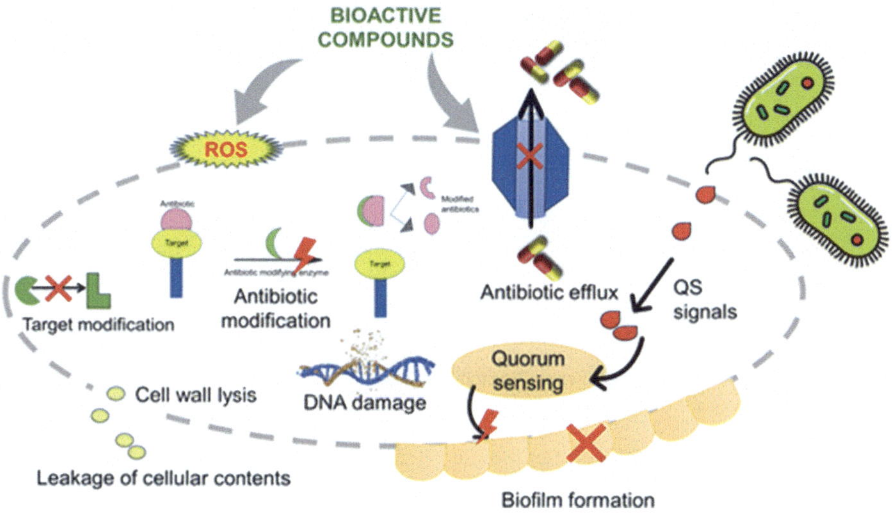

Fig. 1.2 Antibacterial mechanisms of bioactive compounds

et al. 2016). Other flavonoids like acacetin (22), morin (23), and rhamnetin (24) damage the cell wall and disorient the phosphospholipid bilayer (Matijašević et al. 2016). Tannins disengage microbial adhesins, enzymes, and membrane transporter systems (Javed et al. 2020). The terpenoids like thymol (25), eugenol (26), carvacrol (27), and cinnamaldehyde (28) damage the external cell membrane of various gram-negative bacteria, thereby creating pores leading to leakage of essential lipopolysaccharides out of the cytoplasm (Yamaguchi 2022).

1.4.2 Inhibiting Formation of Biofilms

Bacteria developing biofilms are a thousand times more resistant to antibiotics. The plant-derived flavonoids (Fig. 1.3) have been explored as strong disruptors of bacterial biofilms. The flavonoids galangin (29), isovitexin (30), epigallocatechin gallate (31), and 3-O-octanoyl-epicatechin (32) inhibit biofilm formation in *S. aureus* and *S. mutans* by inducing pseudo multicellular aggregation (Stapleton et al. 2004; Cushnie et al. 2007; Hala 2012; Awolola et al. 2014). Enteroaggregative biofilm formation has been inhibited via neutralizing quorum sensing/cellular signalling in *E. coli*, *S. typhi*, and *P. aeruginosa* by flavonoid compounds apigenin, naringenin, and quercetin in concentration dependent manner (IgnÃ and Albarello 2016; Ouyang et al. 2016). *N*-(Heptylsulfanylacetyl)-l-homoserine lactone (33) isolated from garlic extract decreased the expression of virulence factors Lux R and Las R in *P. aeruginosa* thereby inhibiting quorum sensing and biofilm formation (Harjai et al. 2010). Polyphenols identified in *Rosa rugosa* inhibited the formation of biofilm *Chormobacterium violaceum* and *P. aeruginosa* (Zhang et al. 2014). Phloretin reduces biofilm by reducing the expressions of virulence factors; toxin genes,

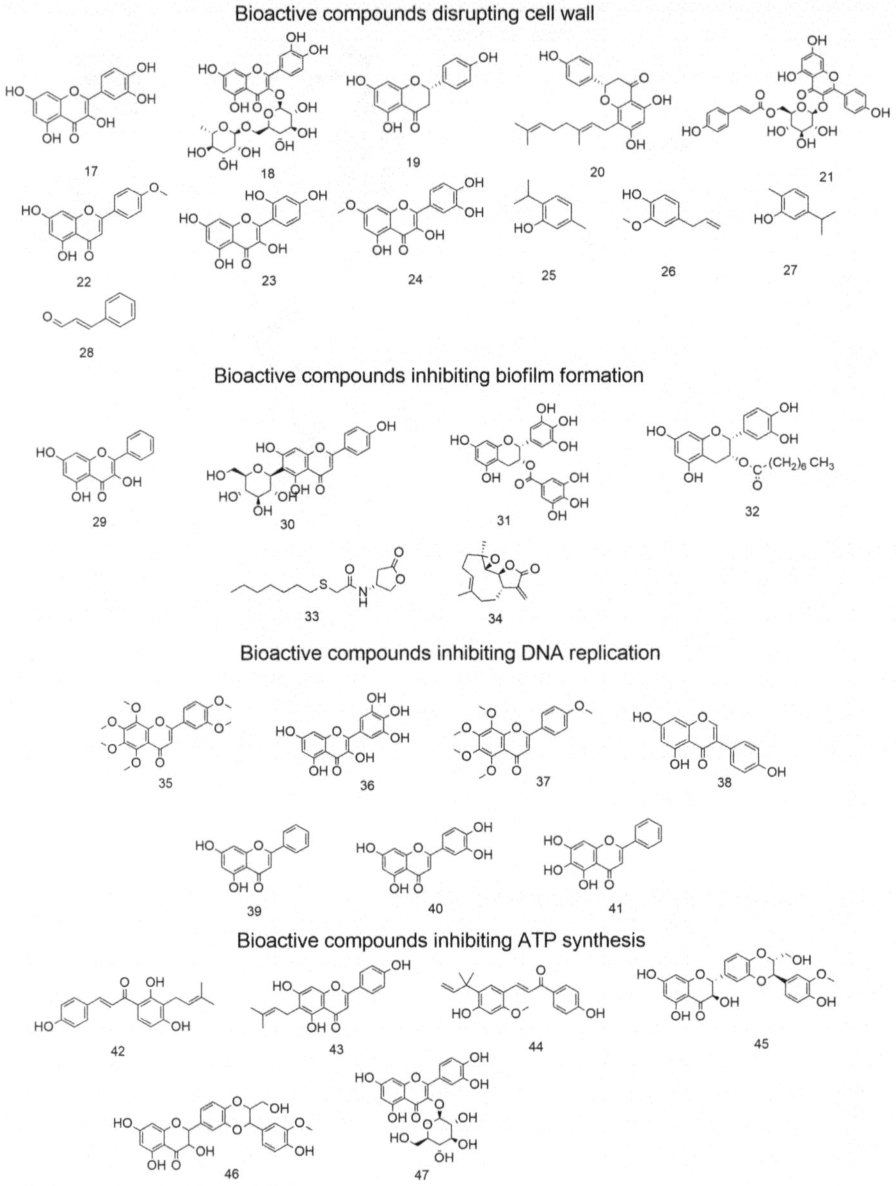

Fig. 1.3 Phytoconstituents with antibacterial potential

autoinducer-2 importer genes, curli genes (csgA and csgB), and prophage genes in *E. coli* and efflux protein genes in *S. aureus* (Lee et al. 2011; Lopes et al. 2017). Parthenolide (34) inhibited QS-related gene expression, including LasI/LasR and RhlI/RhlR, and downregulated extracellular polymeric substances (Mu et al. 2018).

1.4.3 Inhibition of DNA Replication

The isoquinoline alkaloids have been reported for intercalating bacterial DNA and destroy enzymes involved in DNA replication and repair (Zielińska et al. 2019). Many bioactive compounds (Fig. 1.3) have been explored as DNA gyrase inhibitors in the lights of nobiletin (35), myricetin (36), tangeritin (37), genistein (38), chrysin (39), and morin (Yao et al. 2012; Guo 2023; Adesina et al. 2024; Mun et al. 2015). Flavonoids bind to the β-subunit of DNA gyrases of *E. coli*, *M. smegmatis*, *M. tuberculosis*, *B. subtilis*, and *V. harveyi* and simultaneously block the ATP binding pocket thereby hampering DNA replication and repair (Shamsudin et al. 2022). The flavonoids luteolin (40) and morin have been reported to inhibit helicases in *E. coli* (Shadrick et al. 2013). Similarly, baicalein (41) has been identified as an inhibitor of DNA and RNA polymerases (Ozma et al. 2021).

1.4.4 Inhibiting ATP Production

ATP synthesis is crucial for the survival of bacteria. The phytocompounds (Fig. 1.3) also exert their antibacterial action by interrupting the energy production in bacterial cells. Decrease in ATP synthesis by flavonoids isobavachalcone (42) and 6-prenylapigenin (43) leads to membrane depolarization and ultimate lysis of the cell wall in *S. aureus* (Kuete et al. 2011). On a similar note, licochalcones (44) inhibited the electron transport chain, creating an oxygen deficiency and leading to death of bacterial cells (Wu et al. 2019). The FIFO ATP synthase in *E. coli*, *S. aureus*, *S. mutans*, and *P. aeruginosa* is also inhibited by several flavonoids like silibinin (45), silymarin (46), and isoquercetin (47) (Donadio et al. 2021).

1.4.5 Generation of Reactive Free Radicals

Natural products (Fig. 1.4) exhibit antibacterial activity by generating reactive free radicals with bacterial cells (Dryden et al. 2017). Pyochelin (48) isolated from *Burkholderia paludis* enhanced intracellular ROS levels in *E. faecalis* (MIC = 3.13 μM) and *S. aureus* (MIC = 6.26 μM) leading to cell death (Ong et al. 2017). Glabridin (49), a polyphenol isolated from *Glycirhiza glabra* showed antibacterial effects (MIC = 3.12–25 μM) on clinically isolated *S. aureus*. The antibacterial activity of glabridin corresponded to the generation of reactive free radicals (Singh et al. 2015). Naringenin exhibits strong antibacterial action by inducing oxidative damage through the production of reactive oxygen species in *E. coli* (Han and Lee 2022).

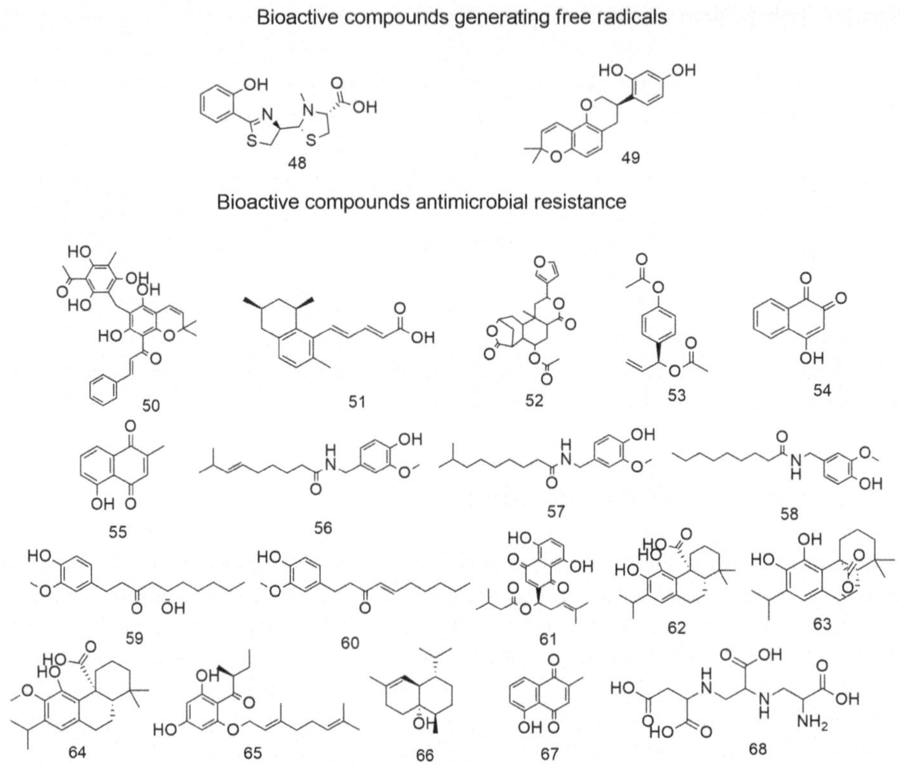

Fig. 1.4 Phytoconstituents as inhibitors of antibiotic resistance

1.4.6 Countering Antimicrobial Resistance

To confront the weighty problem of growing bacterial resistance to frontline clinical antibiotics, antibiotic adjuvants or potentiators in combination with antibiotic therapy has emerged as a resolute strategy (González-Bello 2017). Interestingly, natural compounds (Fig. 1.4) have stolen the limelight as antibiotic adjuvants in countering antimicrobial resistance. The antibiotic resistance can be reversed by implementing methods such as inhibiting plasmid conjugation mechanisms (Lopatkin et al. 2017), plasmid curing (Litake 2022), inhibiting bacterial efflux pumps (Wang et al. 2016), inhibiting antibiotic modifying enzymes (Schaenzer and Wright 2020), and use of membrane permeabilizers (Klobucar and Brown 2022).

Rottlerin (50) isolated from the extracts of *Mallotus philippensis* (Euphorbiaceae) inhibited the conjugal transfer of plasmids, Plasmids pKM101, pUB307, TP114, and R6K in *E. coli* (Patwardhan et al. 2018). Another study mentioned tanzawaic acids A (51) isolated from fungus *Penicillium citrinum* inhibited the conjugal

Table 1.3 Natural compounds targeting resistance plasmids in bacteria

Bioactive compound	Natural source	Bacterial strain	Target plasmids	References
8-Epidiosbulbin (52)	Methanolic extract of *Dioscorea bulbifera*	*E. coli*, *B. subtilis*, *S. typhi*, *P. aeruginosa*	RP4, pUB110, RIP64, and R136	Shriram et al. (2008)
1-Acetoxychavicol acetate (53)	*Alpinia galangal*	*E. coli*, *S. typhi*, *P. aeruginosa*, *Enterococcus*	Plasmid RP4 and pUB110	Latha et al. (2009)
Lawsone (54), Plumbagin (55)	*Plumbago zeylanica*	*E. coli*	Conjugative plasmids pBR322, pUPI281, pRK2013, R136, and pUPI282	Patwardhan et al. (2018)
Capsaicin (56), dihydrocapsaicin (57), nonivamide (58), 6-gingerol (59) and 6-shogaol (60)	*Capsicum annum*	Methicillin-resistant *Staphylococcus aureus*	pUB307, pKM 101 and TP114	Oyedemi et al. (2019)

transfer of plasmids IncW plasmid R388 and the IncFII plasmid R100-1 in *E coli* (Getino et al. 2016). Plasmid curing involves the removal of antibiotic resistance genes from a specific bacterial population. Several plant-derived bioactive compounds (Table 1.3) have been explored possessing plasmid curing abilities.

Numerous plants also exhibit the potential to inhibit efflux pumps. The ethanolic leaf extracts of *Callistemom citrinus* and *Veronia adoensis* effectively inhibited bacterial efflux pumps (Chitemerere and Mukanganyama 2014). Isovaleryl shikonin (61) isolated from a Chinese medicinal herb *Arnebia euchroma* strongly inhibited MRSA efflux pump by suppressing expression of mRNA mrs A. Table 1.4 displays an array of natural compounds and their role in inhibiting antibiotic resistance in multidrug-resistant bacterial strains.

Of note, many plant products also potentiate the antibacterial properties of front-line antibiotics by acting as membrane permeabilizers especially in gram-negative bacteria. Gallic acid and thymol potentiated the antibacterial effects of azithromycin, erythromycin, novobiocin, and nitrofurantoin by disrupting the outer membrane barriers is gram-negative bacteria (Farrag et al. 2019). Essential oils from *Citrus limon* and *Cinnamomum zeylanicum* exhibited membrane permeabilizing effect on *Aceinetobacter* species thereby potentiating antibacterial effects of amikacin and gentamicin (Guerra et al. 2012).

Table 1.4 Natural compounds targeting antibiotic efflux pumps and modification enzymes in bacteria

Bioactive compound	Natural source	AMR inhibition mechanism	Bacterial target	References
Carnosic acid (62) Carnosol (63), 12-methoxy-cis-carnosic acid (64)	*Rosmarinus officinalis L.*	Efflux pump inhibition	Nor(A), Tet(K) and Msr(A) multidrug efflux pumps in *S. aureus*	Oluwatuyi et al. (2004)
Olympicin A (65)	*Hypericum olympicum*	Efflux pump inhibition	Nor(A) multidrug efflux pumps in *S. aureus*	Shiu et al. (2013)
Epi-cubenol (66)	Pilgerodendron uviferum	Efflux pump inhibition	EtBr efflux pumps in *S. aureus*	Espinoza et al. (2019)
Plumbagin (67)	*Plumbago zeylanica*	Efflux pump inhibition	AcrAB-TolC efflux pump in enterococci	Wang et al. (2020a, b)
Aspergillomarasmine A (68)	*Aspergillus versicolor* (mold)	Inhibition of bacterial enzyme modification	NDM-1 and VIM-2 in enterococci, *Pseudomonas* spp.	King et al. (2014)

1.5 State-of-Art Approaches for Natural Product-Based Antibacterial Drug Development

1.5.1 CADD Approaches

The conventional new drug discovery and development process is overcrowded with all-embracing and laborious research (Sun et al. 2022). However, technological advances adjuvated with computer-aided drug designing (CADD) has completely transfigured the drug discovery process in terms of specificity, time, and money (Oselusi et al. 2021). CADD plays a specific role in each drug discovery and development process, starting from target identification and molecular modelling to lead optimization (Katara 2020; Maia et al. 2020). The repertoire of the CADD strategies in accelerating antibacterial drug discovery relies upon an understanding of the pathophysiology underlying infectious diseases, which ultimately facilitates appropriate target and therapeutic lead identification (Sliwoski et al. 2014). The chapter delves into the state-of-art in silico methods accelerating the antibacterial drug discovery from natural sources.

Several databases (Table 1.5) or repositories are available in the public domain containing valuable information regarding pharmacologically important natural compounds (Singh et al. 2019).

Table 1.5 Databases for bioactive compounds

Database	Natural product information	Availability	References
AfroDB	954 natural compounds from traditional African medicinal plants	Open source	Ntie-Kang et al. (2013a, b)
Alkamid	300 plant-derived natural compounds	Open source	Boonen et al. (2012)
BIOFAQUIM	420 bioactive compounds from American medicinal plants and fungi	Open source	Pilón-Jiménez et al. (2019)
BioPhytMol	633 drug-like molecules from Asian medicinal plants	Open source	Sharma et al. (2014)
CamMedNP	>2500 natural compounds from African medicinal plants	Commercial	Ntie-Kang et al. (2013a, b)
Carotenoids database	1117 natural carotenoids from about 680 natural sources	Open source	Yabuzaki (2017)
CEMTDD—Chinese ethnic minority traditional drug database	>4000 compounds from >600 Chinese medicinal herbs	Open source	Huang et al. (2015)
CMAUP	>47,000 natural compounds from >5400 medicinal plants	Open source	Zeng et al. (2019)
ETM-DB	>16,000 natural compounds from about 1000 Ethiopian medicinal plants	Open source	Bultum et al. (2019)
HIT (herbal ingredients targets)	>500 herbal drugs from 1300 Chinese herbs	Open source	Ye et al. (2010)
IMPPAT	9596 phytochemicals from >1700 plant species of Indian medicinal plants	Open source	Karthikeyan et al. (2018)
InflamNat	>600 natural compounds with anti-inflammatory activity	Open source	Zhang et al. (2018)
KNApSaCK	>10,000 plant metabolites	Open source	Nakamura et al. (2013)
MedPserver	>1000 natural products from medicinal plants from northeast India	Open source	Potshangbam et al. (2019)
NANPDB	>4500 natural products from plants, endophytes, fungi, algae, and bacterial sources	Open source	Ntie-Kang et al. (2017)
NeMedplant	About 4000 species of medicinal aromatic plants from northeast India	Open source	Meetei et al. (2012)
NPASS	About 35,000 natural products from 25,000 plant species	Open source	Zeng et al. (2018)
NPEdia	25,000 natural products	Open source	Tomiki et al. (2006)
p-ANAPL (Pan-African natural product library)	>500 natural compounds from African medicinal plants	Open source	Ntie-Kang et al. (2014)
Seaweed metabolite database (SWMD)	1000 compounds produced from marine red algae	Open source	Davis and Vasanthi (2011)

(continued)

Table 1.5 (continued)

Database	Natural product information	Availability	References
StreptomeDB	About 4000 natural compounds produced from *Streptomycetes* (gram-positive actinobacteria)	Open source	Klementz et al. (2016)
TCMDB@Taiwan	20,000 bioactive compounds from Chinese medicinal plants	Open source	Chen (2011)
TMDB	1393 plant metabolites	Open source	Yue et al. (2014)
TriForC	266 plant-based triterpenes	Open source	Miettinen et al. (2018)
YaTCM	47,696 natural compounds from Asian medicinal plants	Open source	Li et al. (2018)

Table 1.6 Databases for antimicrobial compounds

Database	URL	Description
Antimicrobial peptide database (APD)	http://aps.unmc.edu/AP/	Provides information regarding antimicrobial peptides
Antimicrobial drug database (AMDD)	http://www.amddatabase.info	Database for antibacterial and antifungal compounds
Biocide and metal resistance gene database (BacMet)	http://bacmet.biomedicine.gu.se/	Database of antibacterial biocides and metal resistance genes
Collection of antimicrobial peptides ($CAMP_{R3}$)	http://www.camp3.bicnirrh.res.in/	Database provides information regarding different families of antimicrobial peptides along with their structural and functional aspects
Comprehensive antibiotic resistance (CARD)	https://card.mcmaster.ca/	Repository of bacterial-resistant genes/gene products and associated phenotypes
Data repository of antimicrobial peptides (DRAMP)	http://dramp.cpu-bioinfor.org	Contains sequencing, structural and physiochemical information of antimicrobial peptides
MIC database	www.trimslabs.com/mic/index.html	Contains two dimensional structures of synthetic compounds and antibiotics along with MIC values
PhytAMP database	http://phytamp.hammamilab.org/	Presents physicochemical, structural, and antimicrobial data of antibacterial peptides

Several antimicrobial databases containing information regarding genes, microarray data, proteins, key signalling pathways, and genome sequencing are available. Such information plays a significant role in the screening of natural/synthetic compounds in the road voyaged to antimicrobial drug discovery. Table 1.6 enlists numerous biological and phytochemical databases proving beneficial in the virtual screening of potential antibacterial compounds.

The high-throughput computational strategies have been instrumental in exploring novel natural compounds as leads for antibacterial drug discovery (Fig. 1.5).

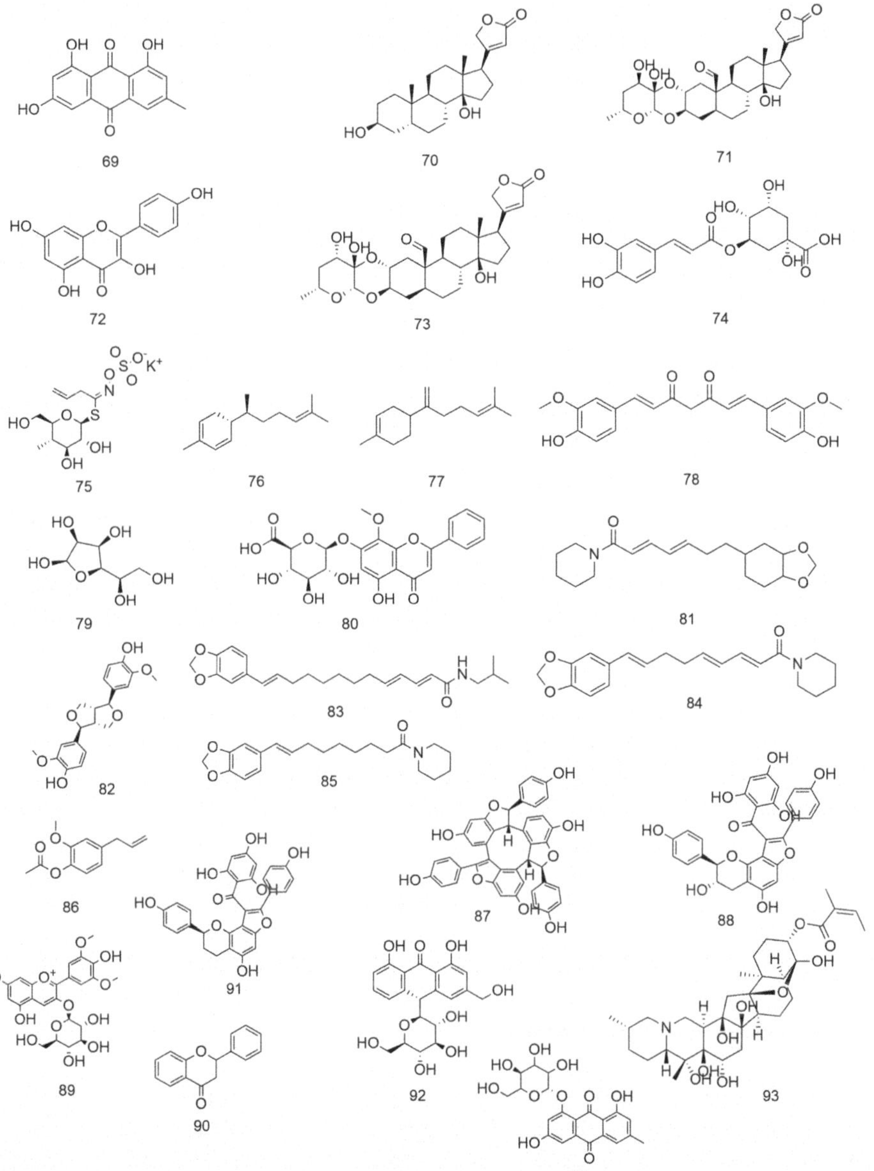

Fig. 1.5 Natural inhibitors of antibiotic resistance

Molecular docking and MD simulation studies revealed that emodin (69), an anthraquinone glycoside from *Rheum emodi*, portrayed strong binding interactions with bacterial (Penicillin-binding protein 3, 3VSL and fungal target (cytochrome P450 14 alpha-sterol demethylase 1EA1) with binding energies −8.2 and −8.0 kcal mol^{-1} respectively (Rolta et al. 2022). The studied proteins are responsible for modifying antibiotics, thereby inducing resistance in bacteria. The in silico study suggested that major phytoconstituents from *Rheum emodi* would effectively counter antibacterial resistance. Additionally, bioactive compounds from *Rheum emodi* produced synergistic antimicrobial action with antibiotics by 4–257 folds (Rolta et al. 2020).

The methanolic extract of *Pergularia tomentosa* contained bioactive components like digitoxigenin (70), calactin (71), kaemferol (72), and calotropagenin (73) as identified in LC-MS investigation. Haddaji and his research team performed molecular docking studies to explore potent inhibitors of tyrosyltRNA synthetase (TyrRS), topoisomerase type IIA, and Candidapepsin-1 (SAP1) enzymes. Inhibition of the above-mentioned proteins negated the expressions of virulence traits in *S. aureus* and *C. albicans*. The docking studies revealed calactin as the strongest inhibitor of the three selected microbial targets, which were also corroborated by the in vitro antimicrobial assays (Haddaji et al. 2021).

In another study, a library of natural compounds were screened against the MRSA protein (6VVA) of methicillin-resistant *S. aureus*. The compounds Luteolin, Kaempferol, Chlorogenic acid (74), Sinigrin (75), Zingiberene (76), 1-Methyl-4-(6-methylhepta-1,5-dien-2-yl)cyclohex-1-ene (77), and Curcumin (78) have found with good binding energies of −8.6 kcal/mol, −8.4 kcal/mol, −8.2 kcal/mol, −7.5 kcal/mol, −7.4 kcal/mol, −7.3 kcal/mol, and −7.2 kcal/mol, respectively. Such computational leads have a better edge in wet laboratories due to their nontoxic nature and high inhibition efficiency against resistant bacteria (Nandhini et al. 2023).

B-d-Mannofuranoside (79) present in methanolic leaf extracts of *Azadirachta indica* was revealed as potent inhibitors of bacterial tyrosyl-tRNA synthetase from molecular docking and MD simulation studies (Altayb et al. 2022). Similarly, chlorogenic acid was explored as a prominent bacterial efflux pump inhibitor of AcrB protein in *E. coli* from in silico screening (Samreen and Ahmad 2023). Another bioactive compound from *Bacopa monnieri* oroxindin (80) was revealed as inhibitor of *Klebsiella pneumoniae* fosfomycin resistance protein (5WEW) and the Zn-dependent receptor-binding domain of *Proteus mirabilis* MR/P fimbrial adhesin MrpH (6Y4F).

1.5.2 Genomic Approaches

Genomics approaches are implemented in computational biology for structurally and functionally analyze the genetic makeup of an individual organism (Mishra and Srivastava 2017). Screening druggable targets for antibacterial drugs based on genomic sequences is challenging since the biological functions of numerous genes remain unexplored. Through comparative genomic study of novel sequences with biochemically characterized protein/enzyme sequences, homologs of known genes have been identified, largely because to the emergence of bioinformatics. Comprehensive genome sequencing provides deeper insight into pathogenicity, virulence, and, more importantly, the mechanism of resistance by bacteria to broad-spectrum antibiotics. A single bacterial genome can be a source of a multitude of antigens that can serve as potential pharmacotherapeutic targets. On such grounds, genomic approaches can excel over classical drug discovery methods in the screening and identification of novel antibacterial drugs against multidrug-resistant pathogenic bacteria. Using whole genome sequencing, the bacterial cell mechanism involved in the therapeutic modification of antibacterial drug targets can also be explored. The genomic methods like reverse vaccinology have been employed in the drug discovery pipeline of novel antibacterial drugs from natural sources and in the development of antibiotic vaccines.

The reverse vaccinology investigations have identified potential therapeutic targets in *Vibrio cholerae* such as the genes ompU and uppP. Bioactive compounds isolated from *Piper betel* have shown potent anticholera activity by inhibiting the identified genes. Piperdardine (81), Pinoresinaol (82), and Guineensine (83) can target both *ompU* and *uppP*, whereas Dehydropipernonaline (84) and Piperrolein B (85) are specific inhibitors of *ompU* and Eugenyl acetate (86) and Chlorogenic acid are specific to *uppP*. Majority of the compounds showcased a dose-dependent and higher antibacterial effect than the standard antibacterial chloramphenicol in nanomolar ranges (Barh et al. 2013).

In another study, chromosomal replication initiator protein DnaA and two-component response regulator (TCR), were revealed as promising therapeutic candidates in *Streptococcus pyogenes*. Among 1000 phytocompounds virtually screened against the two targets, sophorastilbene A (87), daphnodorin B (88), oenin (89), flavumone A (90), and daphnodorin A (91) were found to be strongly bind to DNaA while aloin B (92), chlorogenic acid, triterpenoids, veratrine (93) and 1,6-dihydroxy-3-methyl-8-[(2S,5S)-3,4,5-trihydroxy-6-(hydroxymethyl)oxan-2-yl]oxyanthracene-9,10-dione (94) strongly interacted with TCR protein. The binding affinities of the screened bioactives ranged from about −21.31 to −16.00 kcal/mol for the DNaA protein and −18.02 to −16.84 kcal/mol for the TCR protein. The natural compounds screened in the study serve as novel leads for combating resistant *S. pyogenes* infections (Rehman et al. 2021).

Advancements in genomic methods have unveiled the abundance of biosynthetic gene clusters (BGCs) in numerous microorganisms that encode for specific natural products. The implementation of integrated omics strategies has backed up the discovery of antibiotics from nonpathogenic microbiota (Wright 2017). Several bioactive molecules have been isolated from culturable microbial strains such as insect microbiomes, endophytes, epiphytes, marine microbiota, and human microbiomes (Chevrette et al. 2019; Demers et al. 2018; Xu et al. 2018; Donia et al. 2014). In spite of sincere efforts, the cultivation of such microbial strains under laboratory environment and identification as well as isolation of novel antibiotics has remained as a daunting challenge. In such scenario the culture-based approaches and genome-targeted methods have come to the rescue. Figure 1.6 presents metabolomics-driven approaches for antibiotic drug discovery from microbial resources.

The culture-based methods employ co-cultivation of two different strains of microorganisms or elicitors to trigger antibiotic production by activating silent metabolic pathways in the lights of arcyriaflavin E (95) (from co-cultures of *Rhodococcus* and *Nocardia*) and Keyicin (96) (from co-cultures of *Rhodococcus* and *Micromonospora*) whose structures are displayed in Fig. 1.7 (Wright 2017).

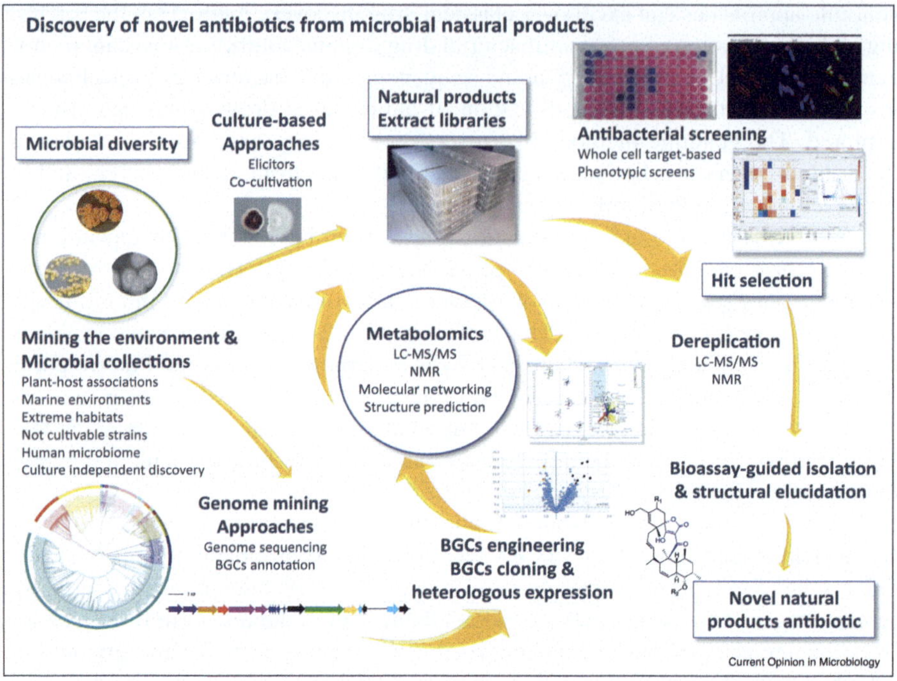

Fig. 1.6 Multidisciplinary approaches in antibiotic discovery from microbial sources. (The figure is Reprinted from Current Opinion in Microbiology, Vol: 51, Author(Olga Genillaud), Title of article: Natural products discovery and potential for new antibiotics, Pages No 81–87, Copyright (2019), with permission from Elsevier)

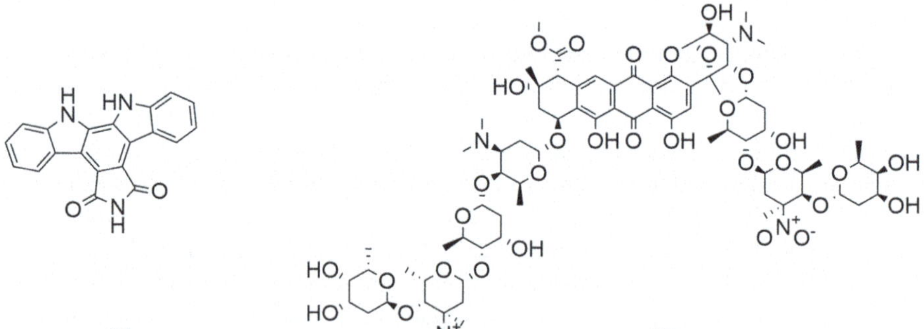

95 96

Fig. 1.7 Structures of antibacterial compounds derived co-cultivated microbial strains

Computational biology platforms like AntiSMASH, RODEO, and eSNAPD have been exclusively used for identification and annotations of biosynthetic gene clusters as well as predicting compound structures from metabolic pathways (Medema and Fischbach 2015).

1.6 Conclusion and Future Prospective

Ubiquitous resistance by pathogenic bacteria to frontline antibacterials has become a global challenge in countering bacterial infections. Such a delinquent issue of antibiotic crisis needs to be addressed through concrete research efforts directed towards novel drug discovery aiming to combat multidrug-resistant bacteria. The natural products have a diverse chemical space, providing a huge platform for screening novel leads for antibacterial drug development. Several potential antibacterial compounds have been identified in plants, endophytes, and marine microorganisms. With multifaceted mechanism of actions, the natural compounds have immense potential to counter antibiotic resistance. As per preclinical reports phytochemicals potentiate the antibacterial activity of conventional antibiotics leading a way forward for developing efficient combination therapies. Despite huge potential, several challenges have been confronted in the exploitation of natural products especially from microbial sources. Chemical complexity, purification and isolation challenges, dereplication, and stability issues have portrayed the natural products less attractive for antibacterial drug discovery. Albeit these challenges the current advancements in fields of synthetic, analytical, and computational chemistry have propelled novel antibiotic drug discovery to new highs. Expanding the scope of antibiotics through natural product research to inhibit biofilms, quorum sensing and bacterial virulence factors offer an exuberant landscape for novel antibiotic discovery.

References

Aati HY, Perveen S, Aati S, Orfali R, Alqahtani JH, Al-Taweel AM, Wanner J, Aati AY (2022) Headspace solid-phase microextraction method for extracting volatile constituents from the different parts of Saudi Anethum graveolens L. and their antimicrobial activity. Heliyon 8(3):e09051

Abhary M, Al-Hazmi AA (2016) Antibacterial activity of Miswak (Salvadora persica L.) extracts on oral hygiene. J Taibah Univ Sci 10(4):513–520

Adesina AF, Adewuyi A, Otuechere CA (2024) Exploratory studies on chrysin via antioxidant, antimicrobial, ADMET, PASS and molecular docking evaluations. Pharmacol Res Mod Chin Med 11:100413

AlSheikh HMA, Sultan I, Kumar V, Rather IA, Al-Sheikh H, Tasleem Jan A, Haq QMR (2020) Plant-based phytochemicals as possible alternative to antibiotics in combating bacterial drug resistance. Antibiotics 9(8):480

Altayb HN, Yassin NF, Hosawi S, Kazmi I (2022) In-vitro and in-silico antibacterial activity of Azadirachta indica (neem), methanolic extract, and identification of Beta. d-Mannofuranoside as a promising antibacterial agent. BMC Plant Biol 22(1):262

Anand U, Tudu CK, Nandy S, Sunita K, Tripathi V, Loake GJ, Dey A, Proćków J (2022) Ethnodermatological use of medicinal plants in India: from ayurvedic formulations to clinical perspectives–a review. J Ethnopharmacol 284:114744

Aryal B, Adhikari B, Aryal N, Bhattarai BR, Khadayat K, Parajuli N (2021) LC-HRMS profiling and antidiabetic, antioxidant, and antibacterial activities of Acacia catechu (Lf) willd. Biomed Res Int 2021:1

Asghar N, Naqvi SAR, Hussain Z, Rasool N, Khan ZA, Shahzad SA, Sherazi TA, Janjua MRSA, Nagra SA, Zia-Ul-Haq M, Jaafar HZ (2016) Compositional difference in antioxidant and antibacterial activity of all parts of the Carica papaya using different solvents. Chem Cent J 10:1–11

Awolola GV, Koorbanally NA, Chenia H, Shode FO, Baijnath H (2014) Antibacterial and anti-biofilm activity of flavonoids and triterpenes isolated from the extracts of Ficus sansibarica Warb. subsp. sansibarica (Moraceae) extracts. Afr J Tradit Complement Altern Med 11(3):124–131

Bai J, Li J, Chen Z, Bai X, Yang Z, Wang Z, Yang Y (2023) Antibacterial activity and mechanism of clove essential oil against foodborne pathogens. LWT 173:114249

Barh D, Barve N, Gupta K, Chandra S, Jain N, Tiwari S, Leon-Sicairos N, Canizalez-Roman A, Rodrigues dos Santos A, Hassan SS, Almeida S (2013) Exoproteome and secretome derived broad spectrum novel drug and vaccine candidates in Vibrio cholerae targeted by Piper betel derived compounds. PLoS One 8(1):e52773

Benakashani F, Allafchian AR, Jalali SAH (2016) Biosynthesis of silver nanoparticles using Capparis spinosa L. leaf extract and their antibacterial activity. Karbala Int J Mod Sci 2(4):251–258

Biharee A, Sharma A, Kumar A, Jaitak V (2020) Antimicrobial flavonoids as a potential substitute for overcoming antimicrobial resistance. Fitoterapia 146:104720

Boonen J, Bronselaer A, Nielandt J, Veryser L, De Tre G, De Spiegeleer B (2012) Alkamid database: chemistry, occurrence and functionality of plant N-alkylamides. J Ethnopharmacol 142(3):563–590

Bouarab-Chibane L, Forquet V, Lantéri P, Clément Y, Léonard-Akkari L, Oulahal N, Degraeve P, Bordes C (2019) Antibacterial properties of polyphenols: characterization and QSAR (quantitative structure–activity relationship) models. Front Microbiol 10:440698

Bultum LE, Woyessa AM, Lee D (2019) ETM-DB: integrated Ethiopian traditional herbal medicine and phytochemicals database. BMC Complement Altern Med 19:1–11

Campos FM, Couto JA, Figueiredo AR, Tóth IV, Rangel AO, Hogg TA (2009) Cell membrane damage induced by phenolic acids on wine lactic acid bacteria. Int J Food Microbiol 135(2):144–151

Cassini A, Högberg LD, Plachouras D, Quattrocchi A, Hoxha A, Simonsen GS, Colomb-Cotinat M, Kretzschmar ME, Devleesschauwer B, Cecchini M, Ouakrim DA (2019) Attributable deaths and disability-adjusted life-years caused by infections with antibiotic-resistant bacteria in the EU and the European Economic Area in 2015: a population-level modelling analysis. Lancet Infect Dis 19(1):56–66

Cech NB, Junio HA, Ackermann LW, Kavanaugh JS, Horswill AR (2012) Quorum quenching and antimicrobial activity of goldenseal (Hydrastis canadensis) against methicillin-resistant Staphylococcus aureus (MRSA). Planta Med 78(14):1556–1561

Cendrowski A, Kraśniewska K, Przybył JL, Zielińska A, Kalisz S (2020) Antibacterial and antioxidant activity of extracts from rose fruits (Rosa rugosa). Molecules 25(6):1365

Chaerunisaa AY, Susilawati Y, Muhaimin M, Milanda T, Hendriani R, Subarnas A (2020) Antibacterial activity and subchronic toxicity of Cassia fistula L. barks in rats. Toxicol Rep 7:649–657

Chen CYC (2011) TCM database@ Taiwan: the world's largest traditional Chinese medicine database for drug screening in silico. PLoS One 6(1):e15939

Chen H, Du K, Sun YJ, Hao ZY, Zhang YL, Bai J et al (2020) Solanrubiellin A, a hydroanthraquinone dimer with antibacterial and cytotoxic activity from Solanum lyratum. Nat Prod Res 34(22):3176–3181

Cheng GG, Li D, Hou B, Li XN, Liu L, Chen YY et al (2016) Melokhanines A–J, bioactive monoterpenoid indole alkaloids with diverse skeletons from Melodinus khasianus. J Nat Prod 79(9):2158–2166

Chevrette MG, Carlson CM, Ortega HE, Thomas C, Ananiev GE, Barns KJ, Book AJ, Cagnazzo J, Carlos C, Flanigan W, Grubbs KJ (2019) The antimicrobial potential of Streptomyces from insect microbiomes. Nat Commun 10(1):516

Chew YL, Mahadi AM, Wong KM, Goh JK (2018) Anti-methicillin-resistance Staphylococcus aureus (MRSA) compounds from Bauhinia kockiana Korth. And their mechanism of antibacterial activity. BMC Complement Altern Med 18:1–9

Chitemerere TA, Mukanganyama S (2014) Evaluation of cell membrane integrity as a potential antimicrobial target for plant products. BMC Complement Altern Med 14:278. https://doi.org/10.1186/1472-6882-14-278

Cock IE, Van Vuuren SF (2020) The traditional use of southern African medicinal plants for the treatment of bacterial respiratory diseases: a review of the ethnobotany and scientific evaluations. J Ethnopharmacol 263:113204

Cushnie TPT, Hamilton VES, Chapman DG, Taylor PW, Lamb AJ (2007) Aggregation of Staphylococcus aureus following treatment with the antibacterial flavonol galangin. J Appl Microbiol 103(5):1562–1567

Dai J, Han R, Xu Y, Li N, Wang J, Dan W (2020) Recent progress of antibacterial natural products: future antibiotics candidates. Bioorg Chem 101:103922

Davis GDJ, Vasanthi AHR (2011) Seaweed metabolite database (SWMD): a database of natural compounds from marine algae. Bioinformation 5(8):361

Demers DH, Knestrick MA, Fleeman R, Tawfik R, Azhari A, Souza A, Vesely B, Netherton M, Gupta R, Colon BL, Rice CA (2018) Exploitation of mangrove endophytic fungi for infectious disease drug discovery. Mar Drugs 16(10):376

Donadio G, Mensitieri F, Santoro V, Parisi V, Bellone ML, De Tommasi N, Izzo V, Dal Piaz F (2021) Interactions with microbial proteins driving the antibacterial activity of flavonoids. Pharmaceutics 13(5):660

Donia MS, Cimermancic P, Schulze CJ, Brown LCW, Martin J, Mitreva M, Clardy J, Linington RG, Fischbach MA (2014) A systematic analysis of biosynthetic gene clusters in the human microbiome reveals a common family of antibiotics. Cell 158(6):1402–1414

Dougherty JA, Sucher AJ, Chahine EB, Shihadeh KC (2019) Omadacycline: a new tetracycline antibiotic. Ann Pharmacother 53(5):486–500

Dryden MS, Cooke J, Salib RJ, Holding RE, Biggs T, Salamat AA, Allan RN, Newby RS, Halstead F, Oppenheim B, Hall T (2017) Reactive oxygen: a novel antimicrobial mechanism for targeting biofilm-associated infection. J Glob Antimicrob Resist 8:186–191

Dwivedi MK, Sonter S, Mishra S, Patel DK, Singh PK (2020) Antioxidant, antibacterial activity, and phytochemical characterization of Carica papaya flowers. Beni-Suef Univ J Basic Appl Sci 9:1–11

Espinoza J, Urzúa A, Sanhueza L, Walter M, Fincheira P, Muñoz P et al (2019) Essential oil, extracts, and Sesquiterpenes obtained from the heartwood of act as potential inhibitors of the NorA multidrug efflux pump. Front Microbiol 10:337. https://doi.org/10.3389/fmicb.2019.00337

Fang W, Wang J, Wang J, Shi L, Li K, Lin X et al (2018) Cytotoxic and antibacterial eremophilane sesquiterpenes from the marine-derived fungus Cochliobolus lunatus SCSIO41401. J Nat Prod 81(6):1405–1410

Farrag HA, Abdallah N, Shehata MMK, Awad EM (2019) Natural outer membrane permeabilizers boost antibiotic action against irradiated resistant bacteria. J Biomed Sci 26:69. https://doi.org/10.1186/s12929-019-0561-6

Fu J, Wang K, Ye C, Chen L, Liang Y, Mao Y, Chen J, Peng R, Chen Y, Shi F, Huang TY (2020) Study on the virulome and resistome of a vancomycin intermediate-resistance Staphylococcus aureus. Microb Pathog 145:104187

Gabriel T, Vestine A, Kim KD, Kwon SJ, Sivanesan I, Chun SC (2022) Antibacterial activity of nanoparticles of garlic (Allium sativum) extract against different bacteria such as Streptococcus mutans and Poryphormonas gingivalis. Appl Sci 12(7):3491

Gao W, He Y, Li F, Chai C, Zhang J, Guo J et al (2019) Antibacterial activity against drug-resistant microbial pathogens of cytochalasan alkaloids from the arthropod-associated fungus Chaetomium globosum TW1-1. Bioorg Chem 83:98–104

Genilloud O (2019) Natural products discovery and potential for new antibiotics. Curr Opin Microbiol 51:81–87

Getino M, Fernández-López R, Palencia-Gándara C, Campos-Gómez J, Sánchez-López JM, Martínez M et al (2016) Tanzawaic acids, a chemically novel set of bacterial conjugation inhibitors. PLoS One 11:e0148098. https://doi.org/10.1371/journal.pone.0148098
Gonçalves AS, Leitão MM, Simões M, Borges A (2023) The action of phytochemicals in biofilm control. Nat Prod Rep 40(3):595–627
González-Bello C (2017) Antibiotic adjuvants – a strategy to unlock bacterial resistance to antibiotics. Bioorganic Med Chem Lett 27:4221–4228. https://doi.org/10.1016/j.bmcl.2017.08.027
Guerra FQS, Mendes JM, Sousa JP, Morais-Braga MFB, Santos BHC et al (2012) Increasing antibiotic activity against a multidrugresistant Acinetobacter spp by essential oils of Citrus limon and Cinnamomum zeylanicum. Nat Prod Res 26:2235–2238. https://doi.org/10.1080/14786419.2011.647019
Guil-Guerrero JL, Ramos L, Moreno C, Zúñiga-Paredes JC, Carlosama-Yepez M, Ruales P (2016) Antimicrobial activity of plant-food by-products: a review focusing on the tropics. Livest Sci 189:32–49
Guo X (2023) Antibacterial and anti-inflammatory effects of genistein in Staphylococcus aureus induced osteomyelitis in rats. J Biochem Mol Toxicol 37(4):e23298
Guo J, Li W, Wan S, Zhou J, Qin Z, Gao H (2024) Antibacterial activity of Amomum tsaoko essential oil and its interaction with Staphylococcus aureus. LWT 191:115700
Gupta PD, Birdi TJ (2017) Development of botanicals to combat antibiotic resistance. J Ayurveda Integr Med 8(4):266–275
Haddaji F, Papetti A, Noumi E, Colombo R, Deshpande S, Aouadi K, Adnan M, Kadri A, Selmi B, Snoussi M (2021) Bioactivities and in silico study of Pergularia tomentosa L. phytochemicals as potent antimicrobial agents targeting type IIA topoisomerase, TyrRS, and Sap1 virulence proteins. Environ Sci Pollut Res 28:25349–25367
Hala EA (2012) Inhibitory effect of grape seed extract (GSE) on cariogenic bacteria. J Med Plant Res 6(34):4883–4891
Han G, Lee DG (2022) Naringin generates three types of reactive oxygen species contributing differently to apoptosis-like death in Escherichia coli. Life Sci 304:120700
Harjai K, Kumar R, Singh S (2010) Garlic blocks quorum sensing and attenuates the virulence of Pseudomonas aeruginosa. FEMS Immunol Med Microbiol 58(2):161–168
Hayat S, Ashraf A, Zubair M, Aslam B, Siddique MH, Khurshid M et al (2022) Biofabrication of ZnO nanoparticles using Acacia arabica leaf extract and their antibiofilm and antioxidant potential against foodborne pathogens. PLoS One 17(1):e0259190
Huang J, Zheng Y, Wu W, Xie T, Yao H, Pang X, Sun F, Ouyang L, Wang J (2015) CEMTDD: the database for elucidating the relationships among herbs, compounds, targets and related diseases for Chinese ethnic minority traditional drugs. Oncotarget 6(19):17675
IgnÃ ACDPR, Albarello N (2016) Cytotoxic, antibacterial and antibiofilm activities of aqueous extracts of leaves and flavonoids occurring in Kalanchoe pinnata (Lam.) Pers. J Med Plant Res 10(41):763–770
Jamal MAHM, Rahman MS, Hossain MB, Sharma SP, Chung HJ, Kim HJ, Hong ST (2017) Antibacterial properties and chemical composition of essential oil from Aegle marmelos (L.) Corr leaves growing in Bangladesh. J Essent Oil Bear Plant 20(1):155–174
Javed B, Nawaz K, Munazir M (2020) Phytochemical analysis and antibacterial activity of tannins extracted from Salix alba L. against different gram-positive and gram-negative bacterial strains. Iran J Sci Technol Trans A: Sci 44(5):1303–1314
Karthikeyan M, Karthikeyan BS, Vivek-Ananth RP, Chand RB, Aparna SR, Pattulingam M, Areejit S (2018) IMPPAT: a curated database of Indian medicinal plants, phytochemistry and therapeutics. Sci Rep 8(1)
Katara P (2020) Computational approaches for drug target identification. Computer-Aided Drug Design:163–185
Kazemian H, Ghafourian S, Heidari H, Amiri P, Yamchi JK, Shavalipour A, Houri H, Maleki A, Sadeghifard N (2015) Antibacterial, anti-swarming and anti-biofilm formation activities of Chamaemelum nobile against Pseudomonas aeruginosa. Rev Soc Bras Med Trop 48:432–436

Khalil AAK, Park WS, Lee J, Kim HJ, Akter KM, Goo YM et al (2019) A new anti-Helicobacter pylori juglone from Reynoutria japonica. Arch Pharm Res 42:505–511
Kim YJ, Kim HJ, Kim GW, Cho K, Takahashi S, Koshino H, Kim WG (2016) Isolation of coralmycins A and B, potent anti-gram negative compounds from the myxobacteria Corallococcus coralloides M23. J Nat Prod 79(9):2223–2228
King AM, Reid-Yu SA, Wang W, King DT, De Pascale G, Strynadka NC et al (2014) Aspergillomarasmine A overcomes metallo-β-lactamase antibiotic resistance. Nature 510:503–506. https://doi.org/10.1038/nature13445
Klementz D, Doering K, Lucas X, Telukunta KK, Erxleben A, Deubel D, Erber A, Santillana I, Thomas OS, Bechthold A, Guenther S (2016) StreptomeDB 2.0 – an extended resource of natural products produced by streptomycetes. Nucleic Acids Res 44(D1):D509–D514
Klobucar K, Brown ED (2022) New potentiators of ineffective antibiotics: targeting the gram-negative outer membrane to overcome intrinsic resistance. Curr Opin Chem Biol 66:102099. https://doi.org/10.1016/j.cbpa.2021.102099
Kubota T, Nakamura K, Kurimoto S, Sakai K, Fromont J, Gonoi T, Kobayashi J (2017) Zamamidine D, a manzamine alkaloid from an Okinawan Amphimedon sp. marine sponge. J Nat Prod 80:1196–1199
Kuete V, Ango PY, Fotso GW, Kapche GD, Dzoyem JP, Wouking AG, Ngadjui BT, Abegaz BM (2011) Antimicrobial activities of the methanol extract and compounds from Artocarpus communis (Moraceae). BMC Complement Altern Med 11:1–5
Kumari I, Pandey RK (2017) Antibacterial activity of Euphorbia hirta L. In: Applications of biotechnology for sustainable development. Springer, Singapore, pp 1–5
Latha C, Shriram VD, Jahagirdar SS, Dhakephalkar PK, Rojatkar SR (2009) Antiplasmid activity of 1'-acetoxychavicol acetate from Alpinia galanga against multi-drug resistant bacteria. J Ethnopharmacol 123:522–525. https://doi.org/10.1016/j.jep.2009.03.028
Lee JH, Regmi SC, Kim JA, Cho MH, Yun H, Lee CS, Lee J (2011) Apple flavonoid phloretin inhibits Escherichia coli O157: H7 biofilm formation and ameliorates colon inflammation in rats. Infect Immun 79(12):4819–4827
Lemes RS, Alves CC, Estevam EB, Santiago MB, Martins CH, Santos TCD, Crotti AE, Miranda ML (2018) Chemical composition and antibacterial activity of essential oils from Citrus aurantifolia leaves and fruit peel against oral pathogenic bacteria. An Acad Bras Cienc 90:1285–1292
Li G, Kusari S, Golz C, Laatsch H, Strohmann C, Spiteller M (2017) Epigenetic modulation of endophytic Eupenicillium sp. LG41 by a histone deacetylase inhibitor for production of decalin-containing compounds. J Nat Prod 80(4):983–988
Li B, Ma C, Zhao X, Hu Z, Du T, Xu X, Wang Z, Lin J (2018) YaTCM: yet another traditional Chinese medicine database for drug discovery. Comput Struct Biotechnol J 16:600–610
Litake GM (2022) Plant-assisted plasmid curing strategies for reversal of antibiotic resistance. In: Antimicrobial resistance: underlying mechanisms and therapeutic approaches, pp 559–575
Liu H, Li XM, Liu Y, Zhang P, Wang JN, Wang BG (2016) Chermesins A–D: Meroterpenoids with a drimane-type spirosesquiterpene skeleton from the marine algal-derived endophytic fungus Penicillium chermesinum EN-480. J Nat Prod 79(4):806–811
Lopatkin AJ, Meredith HR, Srimani JK, Pfeiffer C, Durrett R, You L (2017) Persistence and reversal of plasmid-mediated antibiotic resistance. Nat Commun 8:1689. https://doi.org/10.1038/s41467-017-01532-1
Lopes LAA, dos Santos Rodrigues JB, Magnani M, de Souza EL, de Siqueira-Júnior JP (2017) Inhibitory effects of flavonoids on biofilm formation by Staphylococcus aureus that overexpresses efflux protein genes. Microb Pathog 107:193–197
Lubis RR, Wahyuni DD (2020) Antibacterial activity of betle leaf (Piper betle l.) extract on inhibiting Staphylococcus aureus in conjunctivitis patient. Am J Clin Exp Immunol 9(1):1
Maia EH, Assis LC, De Oliveira TA, Da Silva AM, Taranto AG (2020) Structure-based virtual screening: from classical to artificial intelligence. Front Chem 28(8):343
Manesh A, Varghese GM (2021) Rising antimicrobial resistance: an evolving epidemic in a pandemic. Lancet Microbe 2(9):e419–e420

Martelli G, Giacomini D (2018) Antibacterial and antioxidant activities for natural and synthetic dual-active compounds. Eur J Med Chem 158:91–105

Matijašević D, Pantić M, Rašković B, Pavlović V, Duvnjak D, Sknepnek A, Nikšić M (2016) The antibacterial activity of Coriolus versicolor methanol extract and its effect on ultrastructural changes of Staphylococcus aureus and Salmonella Enteritidis. Front Microbiol 7:1226

Medema MH, Fischbach MA (2015) Computational approaches to natural product discovery. Nat Chem Biol 11(9):639–648

Meetei PA, Singh P, Nongdam P, Prabhu NP, Rathore RS, Vindal V (2012) NeMedPlant: a database of therapeutic applications and chemical constituents of medicinal plants from north-east region of India. Bioinformation 8(4):209

Mendes FSF, Garcia LM, da Silva Moraes T, Casemiro LA, de Alcantara CB, Ambrósio SR, Veneziani RCS, Miranda MLD, Martins CHG (2020) Antibacterial activity of salvia officinalis L. against periodontopathogens: an in vitro study. Anaerobe 63:102194

Meng LH, Wang CY, Mándi A, Li XM, Hu XY, Kassack MU et al (2016) Three diketopiperazine alkaloids with spirocyclic skeletons and one bisthiodiketopiperazine derivative from the mangrove-derived endophytic fungus Penicillium brocae MA-231. Org Lett 18(20):5304–5307

Miettinen K, Inigo S, Kreft L, Pollier J, De Bo C, Botzki A, Coppens F, Bak S, Goossens A (2018) The TriForC database: a comprehensive up-to-date resource of plant triterpene biosynthesis. Nucleic Acids Res 46(D1):D586–D594

Mishra M, Srivastava P (2017) Review on computational approaches for identification of new targets and compounds for fighting against Filariasis. Open Bioactive Compounds J 5(1):72

Mu D, Xiang H, Dong H, Wang D, Wang T (2018) Isovitexin, a potential candidate inhibitor of sortase A of Staphylococcus aureus USA300. J Microbiol Biotechnol 28:1426–1432

Mudalungu CM, Richter C, Wittstein K, Abdalla MA, Matasyoh JC, Stadler M, Süssmuth RD (2016) Laxitextines A and B, cyathane xylosides from the tropical fungus Laxitextum incrustatum. J Nat Prod 79(4):894–898

Mun SH, Lee YS, Han SH, Lee SW, Cha SW, Kim SB, Seo YS, Kong R, Kang DH, Shin DW, Kang OH (2015) In vitro potential effect of morin in the combination with β-lactam antibiotics against methicillin-resistant Staphylococcus aureus. Foodborne Pathog Dis 12(6):545–550

Murugan K, Sekar K, Sangeetha S, Ranjitha S, Sohaibani SA (2013) Antibiofilm and quorum sensing inhibitory activity of Achyranthes aspera on cariogenic Streptococcus mutans: an in vitro and in silico study. Pharm Biol 51(6):728–736

Muruzović MŽ, Mladenović KG, Stefanović OD, Vasić SM, Čomić LR (2016) Extracts of Agrimonia eupatoria L. as sources of biologically active compounds and evaluation of their antioxidant, antimicrobial, and antibiofilm activities. J Food Drug Anal 24(3):539–547

Nakamura K, Shimura N, Otabe Y, Hirai-Morita A, Nakamura Y, Ono N, Ul-Amin MA, Kanaya S (2013) KNApSAcK-3D: a three-dimensional structure database of plant metabolites. Plant Cell Physiol 54(2):e4–e4

Nam YJ, Hwang YS (2021) Antibacterial and antioxidant effect of ethanol extracts of Terminalia chebula on Streptococcus mutans. Clin Exp Dent Res 7(6):987–994

Nandhini P, Gupta PK, Mahapatra AK, Das AP, Agarwal SM, Mickymaray S, Alothaim AS, Rajan M (2023) In-Silico molecular screening of natural compounds as a potential therapeutic inhibitor for methicillin-resistant Staphylococcus aureus inhibition. Chem Biol Interact 374:110383

Niu G, Li W (2019) Next-generation drug discovery to combat antimicrobial resistance. Trends Biochem Sci 44(11):961–972

Ntie-Kang F, Mbah JA, Mbaze LMA, Lifongo LL, Scharfe M, Hanna JN, Cho-Ngwa F, Onguéné PA, Owono LCO, Megnassan E, Sippl W (2013a) CamMedNP: building the Cameroonian 3D structural natural products database for virtual screening. BMC Complement Altern Med 13:1–10

Ntie-Kang F, Zofou D, Babiaka SB, Meudom R, Scharfe M, Lifongo LL, Mbah JA, Mbaze LMA, Sippl W, Efange SM (2013b) AfroDb: a select highly potent and diverse natural product library from African medicinal plants. PLoS One 8(10):e78085

Ntie-Kang F, Amoa Onguéné P, Fotso GW, Andrae-Marobela K, Bezabih M, Ndom JC, Ngadjui BT, Ogundaini AO, Abegaz BM, Meva'a LM (2014) Virtualizing the p-ANAPL library: a step towards drug discovery from African medicinal plants. PLoS One 9(3):e90655

Ntie-Kang F, Telukunta KK, Döring K, Simoben CV, Moumbock AF, Malange YI, Njume LE, Yong JN, Sippl W, Günther S (2017) NANPDB: a resource for natural products from northern African sources. J Nat Prod 80(7):2067–2076

Oluwatuyi M, Kaatz GW, Gibbons S (2004) Antibacterial and resistance modifying activity of Rosmarinus officinalis. Phytochemistry 65:3249–3254. https://doi.org/10.1016/j.phytochem.2004.10.009

Ong KS, Cheow YL, Lee SM (2017) The role of reactive oxygen species in the antimicrobial activity of pyochelin. J Adv Res 8(4):393–398

Oselusi SO, Christoffels A, Egieyeh SA (2021) Cheminformatic characterization of natural antimicrobial products for the development of new lead compounds. Molecules 26(13):3970

Ouyang J, Sun F, Feng W, Sun Y, Qiu X, Xiong L, Liu Y, Chen Y (2016) Quercetin is an effective inhibitor of quorum sensing, biofilm formation and virulence factors in Pseudomonas aeruginosa. J Appl Microbiol 120(4):966–974

Oyedemi BO, Shinde V, Shinde K, Kakalou D, Stapleton PD, Gibbons S (2016) Novel R-plasmid conjugal transfer inhibitory and antibacterial activities of phenolic compounds from Mallotus philippensis (Lam.) Mull. Arg. J Glob Antimicrob Resist 5:15–21

Oyedemi BO, Kotsia EM, Stapleton PD, Gibbons S (2019) Capsaicin and gingerol analogues inhibit the growth of efflux-multidrug resistant bacteria and R-plasmids conjugal transfer. J Ethnopharmacol 245:111871. https://doi.org/10.1016/j.jep.2019.111871

Ozma MA, Khodadadi E, Pakdel F, Kamounah FS, Yousefi M, Yousefi B, Asgharzadeh M, Ganbarov K, Kafil HS (2021) Baicalin, a natural antimicrobial and anti-biofilm agent. J Herbal Med 27:100432

Patel H, Wu ZX, Chen Y, Bo L, Chen ZS (2021) Drug resistance: from bacteria to cancer. Mol Biomed 2:1–19

Patwardhan RB, Dhakephalkar PK, Chopade BA, Dhavale DD, Bhonde RR (2018) Purification and characterization of an active principle, Lawsone, responsible for the plasmid curing activity of Plumbago zeylanica root extracts. Front Microbiol 9:2618. https://doi.org/10.3389/fmicb.2018.02618

Pilón-Jiménez BA, Saldívar-González FI, Díaz-Eufracio BI, Medina-Franco JL (2019) BIOFACQUIM: a Mexican compound database of natural products. Biomol Ther 9(1):31

Potshangbam AM, Polavarapu R, Rathore RS, Naresh D, Prabhu NP, Potshangbam N, Kumar P, Vindal V (2019) Med PS erver: a database for identification of therapeutic targets and novel leads pertaining to natural products. Chem Biol Drug Des 93(4):438–446

Radulovic NS, Blagojevic PD, Stojanovic-Radic ZZ, Stojanovic NM (2013) Antimicrobial plant metabolites: structural diversity and mechanism of action. Curr Med Chem 20(7):932–952

Rafati S, Niakan M, Naseri M (2014) Anti-microbial effect of Nigella sativa seed extract against staphylococcal skin infection. Med J Islam Repub Iran 28:42

Rasamiravaka T, Vandeputte OM, Pottier L, Huet J, Rabemanantsoa C, Kiendrebeogo M, Andriantsimahavandy A, Rasamindrakotroka A, Stévigny C, Duez P, El Jaziri M (2015) Pseudomonas aeruginosa biofilm formation and persistence, along with the production of quorum sensing-dependent virulence factors, are disrupted by a triterpenoid coumarate ester isolated from Dalbergia trichocarpa, a tropical legume. PLoS One 10(7):e0132791

Rehman A, Wang X, Ahmad S, Shahid F, Aslam S, Ashfaq UA, Alrumaihi F, Qasim M, Hashem A, Al-Hazzani AA, Abd-Allah EF (2021) In silico core proteomics and molecular docking approaches for the identification of novel inhibitors against streptococcus pyogenes. Int J Environ Res Public Health 18(21):11355

Rolta R, Sourirajan A, Kumar V, Upadhyay NK, Dev K (2020) Bioassay guided fractionation of rhizome extract of Rheum emodi wall as bio-availability enhancer of antibiotics against bacterial and fungal pathogens. J Ethanopharmacol 257:112867. https://doi.org/10.1016/j.jep.2020.112867

Rolta R, Salaria D, Kumar V, Patel CN, Sourirajan A, Baumler DJ, Dev K (2022) Molecular docking studies of phytocompounds of Rheum emodi wall with proteins responsible for antibiotic resistance in bacterial and fungal pathogens: in silico approach to enhance the bio-availability of antibiotics. J Biomol Struct Dyn 40(8):3789–3803

Rossiter SE, Fletcher MH, Wuest WM (2017) Natural products as platforms to overcome antibiotic resistance. Chem Rev 117(19):12415–12474

Samreen QFA, Ahmad I (2023) In silico screening and in vitro validation of phytocompounds as multidrug efflux pump inhibitor against E. coli. J Biomol Struct Dyn 41(6):2189–2201

Sanver D, Murray BS, Sadeghpour A, Rappolt M, Nelson AL (2016) Experimental modeling of flavonoid–biomembrane interactions. Langmuir 32(49):13234–13243

Sarker SD, Nahar L (2018) An introduction to computational phytochemistry. In: Computational phytochemistry. Elsevier, pp 1–41

Schaenzer AJ, Wright GD (2020) Antibiotic resistance by enzymatic modification of antibiotic targets. Trends Mol Med 26(8):768–782

Seukep AJ, Kuete V, Nahar L, Sarker SD, Guo M (2020) Plant-derived secondary metabolites as the main source of efflux pump inhibitors and methods for identification. J Pharma Anal 10(4):277–290

Shadrick WR, Ndjomou J, Kolli R, Mukherjee S, Hanson AM, Frick DN (2013) Discovering new medicines targeting helicases: challenges and recent progress. J Biomol Screen 18(7):761–781

Shamsudin NF, Ahmed QU, Mahmood S, Ali Shah SA, Khatib A, Mukhtar S, Alsharif MA, Parveen H, Zakaria ZA (2022) Antibacterial effects of flavonoids and their structure-activity relationship study: a comparative interpretation. Molecules 27(4):1149

Sharma A, Dutta P, Sharma M, Rajput NK, Dodiya B, Georrge JJ, Kholia T, Bhardwaj A, OSDD Consortium info@ osdd. net (2014) BioPhytMol: a drug discovery community resource on anti-mycobacterial phytomolecules and plant extracts. J Chem 6:1–10

Sharma D, Rani R, Chaturvedi M, Rohilla P, Yadav JP (2019) In silico and in vitro approach of Allium cepa and isolated quercetin against MDR bacterial strains and mycobacterium smegmatis. S Afr J Bot 124:29–35

Sharma A, Sharma S, Kumar A, Kumar V, Sharma AK (2022) Plant secondary metabolites: an introduction of their chemistry and biological significance with physicochemical aspect. In: Plant secondary metabolites: physico-chemical properties and therapeutic applications. Singapore, Springer Nature Singapore, pp 1–45

Shiu WKP, Malkinson JP, Rahman MM, Curry J, Stapleton P, Gunaratnam M et al (2013) A new plant-derived antibacterial is an inhibitor of efflux pumps in Staphylococcus aureus. Int J Antimicrob Agents 42:513–518. https://doi.org/10.1016/j.ijantimicag.2013.08.007

Shriram V, Jahagirdar S, Latha C, Kumar V, Puranik V, Rojatkar S et al (2008) A potential plasmid-curing agent, 8-epidiosbulbin E acetate, from Dioscorea bulbifera L. against multidrug-resistant bacteria. Int J Antimicrob Agents 32:405–410. https://doi.org/10.1016/j.ijantimicag.2008.05.013

Silva LN, Zimmer KR, Macedo AJ, Trentin DS (2016) Plant natural products targeting bacterial virulence factors. Chem Rev 116(16):9162–9236

Silva JC, Pereira RLS, de Freitas TS, Rocha JE, Macedo NS, Nonato CDFA, Linhares ML, Tavares DSA, da Cunha FAB, Coutinho HDM, de Lima SG (2022) Evaluation of antibacterial and toxicological activities of essential oil of Ocimum gratissimum L. and its major constituent eugenol. Food Biosci 50:102128

Simard F, Gauthier C, Legault J, Lavoie S, Mshvildadze V, Pichette A (2016) Structure elucidation of anti-methicillin resistant Staphylococcus aureus (MRSA) flavonoids from balsam poplar buds. Bioorg Med Chem 24(18):4188–4198

Singh V, Pal A, Darokar MP (2015) A polyphenolic flavonoid glabridin: oxidative stress response in multidrug-resistant Staphylococcus aureus. Free Radic Biol Med 87:48–57

Singh A, Zahra S, Kumar S (2019) In-silico tools in phytochemical research. In: Phytochemistry: an in-silico and in-vitro update: advances in phytochemical research. Springer, pp 351–372

Sliwoski G, Kothiwale S, Meiler J, Lowe EW (2014) Computational methods in drug discovery. Pharmacol Rev 66(1):334–395

Solomkin JS, Sway A, Lawrence K, Olesky M, Izmailyan S, Tsai L (2019) Eravacycline: a new treatment option for complicated intra-abdominal infections in the age of multidrug resistance. Future Microbiol 14(15):1293–1308

Sriramulu M, Shukla D, Sumathi S (2018) Aegle marmelos leaves extract mediated synthesis of zinc ferrite: antibacterial activity and drug delivery. Mater Res Express 5(11):115404

Srivastav VK, Egbuna C, Tiwari M (2020) Plant secondary metabolites as lead compounds for the production of potent drugs. In: Phytochemicals as lead compounds for new drug discovery. Elsevier, pp 3–14

Stapleton PD, Shah S, Hamilton-Miller JM, Hara Y, Nagaoka Y, Kumagai A, Uesato S, Taylor PW (2004) Anti-Staphylococcus aureus activity and oxacillin resistance modulating capacity of 3-O-acyl-catechins. Int J Antimicrob Agents 24(4):374–380

Subramaniam G, Yew XY, Sivasamugham LA (2020) Antibacterial activity of Cymbopogon citratus against clinically important bacteria. S Afr J Chem Eng 34:26–30

Sun D, Gao W, Hu H, Zhou S (2022) Why 90% of clinical drug development fails and how to improve it? Acta Pharm Sin B 12(7):3049–3062

Swain SS, Padhy RN (2015) In vitro antibacterial efficacy of plants used by an Indian aboriginal tribe against pathogenic bacteria isolated from clinical samples. J Taibah Univ Med Sci 10(4):379–390

Tomiki T, Saito T, Ueki M, Konno H, Asaoka T, Suzuki R, Uramoto M, Kakeya H, Osada H (2006) RIKEN natural products encyclopedia (RIKEN NPEdia), a chemical database of RIKEN natural products depository (RIKEN NPDepo). J Comput Aid Chem 7:157–162

Tran TD, Olsson MA, Choudhury MA, McMillan DJ, Cullen JK, Parsons PG et al (2019) Antibacterial 5α-spirostane saponins from the fruit of Cordyline manners-suttoniae. J Nat Prod 82(10):2809–2817

Tsuchiya H, Iinuma M (2000) Reduction of membrane fluidity by antibacterial sophoraflavanone G isolated from Sophora exigua. Phytomedicine 7(2):161–165

Upadhyay A, Pal D, Kumar A (2023) Combinatorial enzyme therapy: A promising neoteric approach for bacterial biofilm disruption. Process Biochem 129:56–66. https://doi.org/10.1016/j.procbio.2023.02.022

Venkatesan A, Kathirvel A, Prakash S, Sujatha V (2017) Antioxidant, antibacterial activities and identification of bioactive compounds from Terminalia chebula bark extracts. Free Radicals and Antioxidants 7(1):43–49

Wadsworth CB, Arnold BJ, Sater MRA, Grad YH (2018) Azithromycin resistance through interspecific acquisition of an epistasis-dependent efflux pump component and transcriptional regulator in Neisseria gonorrhoeae. MBio 9(4):10–1128

Wang Y, Venter H, Ma S (2016) Efflux pump inhibitors: a novel approach to combat efflux-mediated drug resistance in bacteria. Curr Drug Targets 17(6):702–719

Wang W, Liao Y, Huang X, Tang C, Cai P (2018) A novel xanthone dimer derivative with antibacterial activity isolated from the bark of Garcinia mangostana. Nat Prod Res 32(15):1769–1774

Wang Y, Alenzy R, Song D, Liu X, Teng Y, Mowla R et al (2020a) Structural optimization of natural product nordihydroguaretic acid to discover novel analogues as AcrB inhibitors. Eur J Med Chem 186:111910. https://doi.org/10.1016/j.ejmech.2019.111910

Wang X, Shen Y, Thakur K, Han J, Zhang JG, Hu F, Wei ZJ (2020b) Antibacterial activity and mechanism of ginger essential oil against Escherichia coli and Staphylococcus aureus. Molecules 25(17):3955

Wright GD (2017) Opportunities for natural products in 21st century antibiotic discovery. Nat Prod Rep 34(7):694–701

Wu SC, Yang ZQ, Liu F, Peng WJ, Qu SQ, Li Q, Song XB, Zhu K, Shen JZ (2019) Antibacterial effect and mode of action of flavonoids from licorice against methicillin-resistant Staphylococcus aureus. Front Microbiol 10:467228

Xu D, Han L, Li C, Cao Q, Zhu D, Barrett NH, Harmody D, Chen J, Zhu H, McCarthy PJ, Sun X (2018) Bioprospecting deep-sea actinobacteria for novel anti-infective natural products. Front Microbiol 9:343953

Yabuzaki J (2017) Carotenoids database: structures, chemical fingerprints and distribution among organisms. Database 2017:bax004

Yadufashije C, Niyonkuru A, Munyeshyaka E, Madjidi S, Mucumbitsi J (2020) Antibacterial activity of ginger extracts on bacteria isolated from digestive tract infection patients attended Muhoza Health Center. Asian J Med Sci 11(2):35–41

Yamaguchi T (2022) Antibacterial effect of the combination of terpenoids. Arch Microbiol 204(8):520

Yan W, Zhao SS, Ye YH, Zhang YY, Zhang Y, Xu JY, Yin SM, Tan RX (2019) Generation of indoles with agrochemical significance through biotransformation by Chaetomium globosum. J Nat Prod 82(8):2132–2137

Yang JK, Zhang B, Gao T, Yang MY, Zhao GZ, Zhu HJ et al (2018) A pair of enantiomeric 5-oxabicyclic [4.3. 0] lactam derivatives and one new polyketide from the marine-derived fungus Penicillium griseofulvum. Nat Prod Res 32(19):2366–2369

Yao X, Zhu X, Pan S, Fang Y, Jiang F, Phillips GO, Xu X (2012) Antimicrobial activity of nobiletin and tangeretin against pseudomonas. Food Chem 132(4):1883–1890

Ye H, Ye L, Kang H, Zhang D, Tao L, Tang K, Liu X, Zhu R, Liu Q, Chen YZ, Li Y (2010) HIT: linking herbal active ingredients to targets. Nucleic Acids Res 39(suppl_1):D1055–D1059

Yue Y, Chu GX, Liu XS, Tang X, Wang W, Liu GJ, Yang T, Ling TJ, Wang XG, Zhang ZZ, Xia T (2014) TMDB: a literature-curated database for small molecular compounds found from tea. BMC Plant Biol 14:1–8

Zacchino SA, Butassi E, Cordisco E, Svetaz LA (2017) Hybrid combinations containing natural products and antimicrobial drugs that interfere with bacterial and fungal biofilms. Phytomedicine 37:14–26

Zeng X, Zhang P, He W, Qin C, Chen S, Tao L, Wang Y, Tan Y, Gao D, Wang B, Chen Z (2018) NPASS: natural product activity and species source database for natural product research, discovery and tool development. Nucleic Acids Res 46(D1):D1217–D1222

Zeng X, Zhang P, Wang Y, Qin C, Chen S, He W, Tao L, Tan Y, Gao D, Wang B, Chen Z (2019) CMAUP: a database of collective molecular activities of useful plants. Nucleic Acids Res 47(D1):D1118–D1127

Zhanel GG, Deng C, Zelenitsky S, Lawrence CK, Adam HJ, Golden A, Berry L, Schweizer F, Zhanel MA, Irfan N, Bay D (2021) Lefamulin: a novel oral and intravenous pleuromutilin for the treatment of community-acquired bacterial pneumonia. Drugs 81:233–256

Zhang J, Rui X, Wang L, Guan Y, Sun X, Dong M (2014) Polyphenolic extract from Rosa rugosa tea inhibits bacterial quorum sensing and biofilm formation. Food Control 42:125–131

Zhang Y, Liu X, Wang Y, Jiang P, Quek S (2016) Antibacterial activity and mechanism of cinnamon essential oil against Escherichia coli and Staphylococcus aureus. Food Control 59:282–289

Zhang R, Lin J, Zou Y, Zhang XJ, Xiao WL (2018) Chemical space and biological target network of anti-inflammatory natural products. J Chem Inf Model 59(1):66–73

Zielińska S, Wójciak-Kosior M, Dziągwa-Becker M, Gleńsk M, Sowa I, Fijałkowski K, Rurańska-Smutnicka D, Matkowski A, Junka A (2019) The activity of isoquinoline alkaloids and extracts from Chelidonium majus against pathogenic bacteria and Candida sp. Toxins 11(7):406

Natural Products in Management of MDR

2

T. Gayatri, Sumedha Saha, Shivangi, Sandhyarani Patra, Ramchander Merugu, Diptikanta Acharya, and Sagarika Satapathy

Abstract

The escalating prevalence of microorganisms that are multidrug-resistant (MDR) causes a substantial risk to public health worldwide, necessitating novel therapeutic strategies. A large population of microorganisms, including bacteria, viruses, fungi, and parasites, have evolved resistance to multiple antimicrobial agents, making conventional therapies ineffectual and increasing the risk of illness, death, and medical expenses. Prominent MDR pathogens including *Staphylococcus aureus* that is resistant to methicillin, *Enterobacteriaceae* that are resistant to carbapenem, *Mycobacterium tuberculosis*, and long-spectrum beta-lactamase (ESBL)-producing *Enterobacteriaceae* belongs to MDR group, poses serious threat to human health. Various approaches including multidrug combinations, nanotechnological approaches, and many others have been recently developed to fight against the global threat of MDR. However, natural products, derived from plants, microorganisms, and marine organisms, have garnered considerable attention due to their diverse chemical structures and potent biological activities, have been essential to the advancement of the medical

T. Gayatri (✉)
Department of Biotechnology, School of Sciences, OP Jindal University, Raigarh, Chhattisgarh, India
e-mail: tgayatri@opju.ac.in

S. Saha
Advanced Laboratory for Plant Genetic Engineering, Advanced Technology Development Centre, IIT Kharagpur, Kharagpur, West Bengal, India

Shivangi · S. Patra · D. Acharya · S. Satapathy
Department of Biotechnology, School of Sciences, GIET University, Gunupur, Odisha, India

R. Merugu
Department of Biochemistry, Mahatma Gandhi University, Nalgonda, Telangana, India

T. Sharma et al. (eds.), *Natural Products for Antibacterial Drug Development: Recent Advancement of Computational Approach*,
https://doi.org/10.1007/978-981-97-9634-2_2

industry. This chapter reveals the potential of various phytochemical in the management of MDR infections, emphasizing their mechanisms of action, efficacy, and potential as alternative therapies.

Keywords

Multiple drug resistance microorganisms · Diseases · Phytochemicals · Disease management

Abbreviations

AITC	Allyl isothiocyanate
BITC	Benzyl isothiocyanate
CHE	Chelerythrine
CMV	Cucumber mosaic virus
CRAB	Carbapenem-resistant *A. baumannii*
EPI	Efflux pump inhibitors
HGT	Horizontal gene transfer
ITCs	Isothiocyanates
JNK	c-Jun N-terminal kinase
MDR	Multidrug resistant
MIC	Minimal inhibitory concentrations
MRSA	Methicillin-resistant *Staphylococcus aureus*
PAHs	Polycyclic aromatic hydrocarbons
PDR	Pan drug resistance
PDSs	Polyphenol-disaccharides
PITC	Phenyl isothiocyanate
QS	Quorum sensing
QSAR	Quantitative structure activity analysis
RAR	Retinoic acid receptors
ROS	Reactive oxygen species
RXR	Retinoid X receptors
TEM	Transmission electron microscopy
TMV	Tobacco mosaic virus
VRE	Vancomycin-resistant Enterococci
XDR	Extensive-drug resistance

2.1 Introduction

Bacterial illness are the major contributors of human illness around the globe. However, with advent and advancements in medicine and public health, remarkedly reduced the mortality rate due to bacterial infections. In this context, a breakthrough was the discovery of antibiotics which began with penicillin and continued to discovery of many drugs with promising antibacterial properties. With the discovery of antibiotics, it was believed that infectious diseases would soon be under control. However, this belief was overruled with the emergence of new bacterial species possessing the capability of resistance against antibiotics. Over time, pathogens changed and developed resistance to previously identified antibiotics.

Acquired bacterial resistance fall into three categories: (1) multidrug resistance (MDR), (2) pan-drug resistance (PDR), and (3) extensive-drug resistance (XDR). Pathogens that are vulnerable to only one or two antibiotics are classified as members of the extensive-drug resistance (XDR) class. One instance of XDR-M tuberculosis that demonstrates resistance to antibiotics of the second order such as amikacin, kanamycin, capreomycin, and rifampicin in addition to fluoroquinolones and isoniazid. Bacterial species classified under pan-drug resistant (PDR) category shows resistance to many commercially available or commonly used antimicrobial medications. Microbes that can withstand a variety of antibiotics, are known as multidrug-resistant microorganisms (MDR). A common list of multidrug-resistant bacteria (MDRs) includes extended spectrum β-lactamases, methicillin-resistant *Staphylococcus aureus* (MRSA), multidrug-resistant *P. aeruginosa*, carbapenem-resistant *A. baumannii* (CRAB), vancomycin-resistant *Enterococci* (VRE), and *Enterobacteriaceae* that produce CRE (Terreni et al. 2021). Likewise many other bacteria exist which is detailed in Table 2.1. The fact that infections due to resistant forms of bacteria are at grave risk to human health with than those with infections from more delicate species, not to mention the growing expense of treating these infections. It is estimated that worldwide there will be 10 million deaths per year due to antimicrobial resistance infectious diseases by 2050, hence will result in burden on healthcare system which is predicted to be $100.2 trillion (Robinson et al. 2016).

Inappropriate prescription and consumption of antibiotics are the cause behind the acquired resistance species: both exceptionally multidrug-resistant (XDR) and multidrug-resistant (MDR) bacteria (Davies and Davies 2010; Banin et al. 2017). In general, bacterial biofilms exhibit lesser susceptibility to antibiotics, higher resistance to exsiccation, and immune system clearance (Høiby et al. 2011). After a new antibiotic medicine is commercially introduced, resistant strains are found within a few years, which escalates the hazard level (Smith et al. 2015; Davies and Davies 2010).

Bacteria develop antibiotic resistance by one of these two mechanisms: horizontal gene transfer (HGT) and spontaneous mutation (Jian et al. 2021). The bacteria can acquire target protection proteins, efflux pumps, degradative enzymes specific to antibiotics, and bypass pathways (such as supply systems for alternate cell wall manufacturing processes) through horizontal gene transfer. Similarly, by altering the antibiotic's target or degree of expression, or by upregulating resistance genes, like those producing efflux pumps, spontaneous mutations might purposefully result in resistance to an antibiotic. Due to evolved resistance, these bacteria rapidly outnumber the sensitive bacteria and thus eventually leave the drug ineffective. Consequently, fewer antimicrobial drugs are left to cope with the infections brought on by these bacteria (Elmaidomy et al. 2022). In order to manage antibiotic-resistant

Table 2.1 Bacteria exhibiting multidrug resistance

S No	Name of the bacterium	Resistance shown to drugs	References
1.	*Escherichia coli*	Multidrug resistance	Lomovskaya and Lewis (1992)
2.	*Streptococcus pneumonia*	Multidrug resistance	Spratt (1994)
3.	*Neisseria meningitidis*	Multidrug resistance	Spratt (1994)
4.	*Bacillus subtilis*	Tetracyclines	Cheng et al. (1994)
5.	*Staphylococcus aureus*	Lipophilic drugs	Paulsen et al. (1996)
6.	*Pseudomonas aeruginosa*	Multidrug resistance	Tseng et al. (1999)
7.	*Vibrio parahaemolyticus*	Multidrug resistance	Morita et al. (2000)
8.	*Enterobacter aerogenes*	Multidrug resistance	Achouak et al. (2001)
9.	*Klebsiella pneumoniae*	Multidrug resistance	Achouak et al. (2001)
10.	*Pseudomonas putida*	Multidrug resistance	Ramos et al. (2002)
11.	*M. tuberculosis*	Multidrug resistance	Domenech et al. (2004)
12.	*Lactococcus lactis*	Daunomycin, tetracyclines, and macrolides	Mazurkiewicz et al. (2004)
13.	*Streptomyces*	Multidrug resistance	D'Costa et al. (2006)
14.	*Staphylococcus aureus*	Multidrug resistance	Deurenberg et al. (2007)
15.	*Klebsiella pneumoniae*	Carbapenem	Cano et al. (2018)
16.	*Enterococcus faecium*	Multidrug resistance	Dos Santos et al. (2021)

bacteria and stop infections, it is imperative to look for novel antimicrobial medicines.

Nature had always been surprisingly presented itself before the mankind with all the possible solutions. From ancient days, people were dependent on plant chemicals also called phytochemicals for various illnesses such as labor discomfort, diarrhea, the common cold, and dental illness (Jain et al. 2019). Both traditional Chinese medicine and Ayurveda, two historic Indian medical systems, heavily emphasize the use of phytochemicals in disease therapy. Many of these phytochemicals are studied to possess antimicrobial activity. A total of 80% people worldwide, according to the World Health Organization, choose plant-based medicines for their medical needs because they are less expensive and less likely to cause negative effects (Manzo et al. 2019). Extensive research is being conducted on plants as potential sources for the manufacture of novel pharmaceutical compounds and the possible means of treating various medical ailments using the crude extracts (Hemavathy et al. 2019).

2.2 Role of Phytochemicals in Controlling MDR

Phytochemicals are well observed to interfere with the membrane structural integrity, enhances the cell permeability, alters bacterial cell wall and membrane integrity, results in the loss of ATP, affects rate of protein synthesis, quorum sensing, drug efflux pumps, ion binding, free radical production, mobile genetic elements, DNA damage, suppression of DNA gyrase, and inhibition of bacterial gene expression (Fig. 2.1) (Bazzaz et al. 2018; Yu et al. 2020a; Suganya et al. 2022).

2.2.1 Exclusion of Resistant Plasmids

Plasmids are dynamic vehicles to transfer resistance genes among bacterial pathogens via horizontal gene transfer. The propagation of resistance genes is aided by the synthesis, movement, and transmission of resistant plasmids, which are crucial processes that result in broad antibiotic resistance (Suganya et al. 2022). Hence, the removal of R-plasmid, or preventing the propagation of resistant plasmids is one of the most effective approaches to lower antibiotic resistance. The synergistic activity of phytochemicals possessing plasmid curing ability with conventional antibiotics like amoxicillin, polymycin, and lincomycin showed strong antimicrobial activity, thus reducing the risk of drug resistance development in microbes (Si et al. 2008; Skalicka-Wozniak et al. 2018).

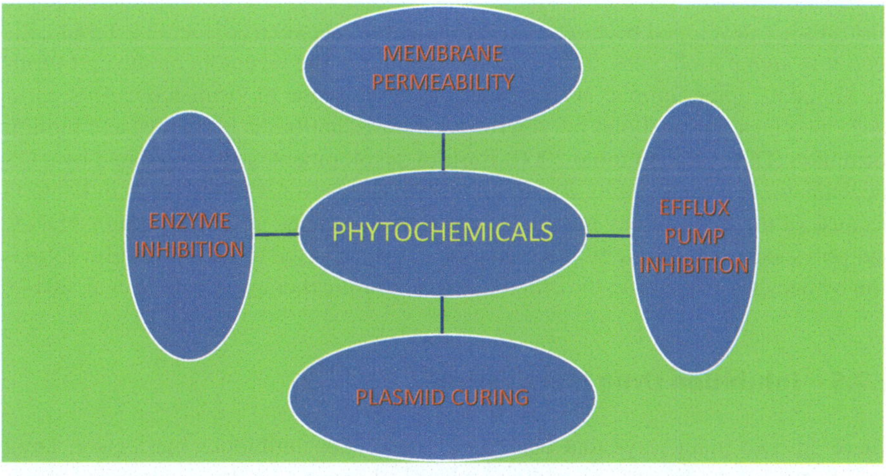

Fig. 2.1 Mode of action of phytochemicals against multidrug-resistant microorganisms

2.2.2 Antibiotic-Resistant Bacteria's Efflux Pump Inhibition

Bacterial efflux pump overexpression can be readily induced by antibiotics, which exacerbates bacterial infection and causes bacteria to release additional antimicrobial medications and considerably lower the amount of medication present in the target area. It has been established that both Gram-positive and Gram-negative bacteria share the same efflux mechanism. Methicillin-resistant *Staphylococcus aureus* (MRSA) expresses more NorA gene in order to improve drug excretion (Markham et al. 1999). It was first found in a clinical context that the NorA efflux pump of the MFS family is a critical mechanism of methicillin- and quinolone-resistant bacteria. The ergoline alkaloid lysergol functions by downregulating and subsequently inhibiting the efflux pump ATPases (Maurya et al. 2013).

2.2.3 Impact on the Cell Membrane's Permeability

Bacterial cell membranes primarily serve as a defense wall against antimicrobial agents. Antibiotic movement into the cell is regulated by the membrane permeability. By changing specific ion channels, it is possible to change the bacterial cell membrane's permeability, which reduces the antibiotic's ability to travel and boosts its potency (Yu et al. 2020b; Suganya et al. 2022). This process permits medications to pass through bacteria's cell walls and fight multidrug-resistant (MDR) infections that have evolved antibiotic resistance.

2.2.4 Inhibition on the Biofilm Formation

Biofilms are developed because of communication between adjacent bacterial cells secreting signal molecules. These molecules bind to the receptors on the cell forming signal receptor complex which induce the expression of virulence factors, virulence enzymes, extracellular polymeric substances, antibiotic pigments, and biofilm formation (Davey and O'toole 2000) Biofilm development is also regulated by a QS mechanism. It is challenging for antimicrobial drugs to get to the resident bacteria because of the thick, sticky biofilm matrix, which is made up of the proteins, DNA, and polysaccharides. Several phytochemicals are reported to inhibit biofilm formation by bacteria, increasing the efficiency of antibiotic therapy (Cushnie et al. 2014).

2.2.5 Inhibition Drug Inactivation

There are two main ways that bacteria might render antibiotics ineffective. Drug modification—the process of adding a chemical group to a drug's chemical structure—or the drug breaking down naturally are the two possible causes of them (for example, a broad class of enzymes b-lactamases hydrolyze tetracycline). *Pseudomonas* develops resistance against β-lactam antibiotics by preventing the amide bond formation by β-lactamases enzymes (Bush and Bradford 2016). Three types of chemical groups are most frequently added to the drug during drug inactivation: adenyl, phosphoryl, and acetyl. The enzymes adenylyltransferase, phosphoryltransferase, and acetyltransferase modifies aminoglycosides of antibiotics and thus interferes with their binding to cell membrane (Bush and Bradford 2016). It has been discovered that there are several transferases that can be employed to combat antibiotic (drug).

2.2.6 Quorum-Sensing Inhibitors (QSIs)

Quorum sensing (QS) in bacteria is a method of intercellular communication that regulates group behaviors based on cell density. Pathogens can cause illness when their transcription patterns change to an invasive phenotype at high concentrations. This includes genes linked to virulence factors and antibiotic resistance (Vivas et al. 2019). Because they frequently control the expression of many virulence genes at once, QS systems are significant targets for anti-pathogenicity. QSI work by obstructing cell-to-cell transmission and, as a result, the progression of illness, enabling the removal of an infection that already exists or the prevention of bacterial colonization by the host immune system. Rather than eliminating the bacteria, this antimicrobial management method focuses on lessening the virulence load (Packiavathy et al. 2014). Over the past two decades, we have seen the characterization of several QSI from plants, animals, and microbes; animal and plant infection models have confirmed their antibacterial activity against QS infections. Hence, QSI could be a useful substitute for treating infections brought on by MDR pathogens. However, further study is still needed before it may be used therapeutically (Packiavathy et al. 2014).

2.3 Products Originating from Plants

Plants produce two kinds of metabolites: primary and secondary. The term "primary metabolites" refers to plant compounds such as those from the glycolysis, tricarboxylic acid cycle, and shikimate pathway that are necessary for the survival, development, and reproduction of plants. On the other hand, plants produce secondary metabolites in response to external stimuli (Pagare et al. 2015; Pott et al. 2019). In order to generate new compounds necessary for plant adaptation to stressful conditions, plants induce metabolic pathways other than their primary metabolic pathways to protect against a range of abiotic (damage, temperature, and moisture) and biotic (bacteria, fungi, insects, and disease) stressors. Intermediates of primary metabolic pathways are the only precursor for many secondary metabolites (Cardoso et al. 2019; Pott et al. 2019). These secondary metabolites have confirmed to show bioactivity against several pathogenic microbes including MDR bacteria (Table 2.2). Numerous solvent-based extracts of medicinal plants including Lemongrass, Neem, Aloe vera, Oregano, Rosemary, Thyme, Tulsi, and Bryophyllum are reported to

Table 2.2 Natural products showing activity against microbes exhibiting multidrug resistance

S No	Name of the compound	Activity against microbes exhibiting multidrug resistance	References
1.	Bergamottin epoxide	MRSA	Abulrob et al. (2004)
2.	Galbanic acid	MDR clinical isolates of *S. aureus*	
3.	Baicalein	*M. smegmatis*, MRSA and *C. albicans*	Chan et al. (2011)
4.	Bergamottin epoxide	MRSA	Roy et al. (2013)
5.	Sophoraflavanone G	MRSA	Mun et al. (2014)
6.	Piperine	MRSA and *S. aureus*	Khameneh et al. (2015)
7.	Kaempferol	MRSA and *C. albicans*	Shao et al. (2016)
8.	Sabinene	Multidrug-resistant strains	Matias et al. (2016)
9.	Catechin	MRSA	Sinsinwar and Vadivel (2020)
10.	Chelerythrine	MRSA and *Escherichia coli*	Wang (2021)
11.	Sanguinarine	Carbapenem-resistant *Serratia marcescens*	Fu et al. (2021)

possess inhibitory effect against ten multidrug-resistant clinical isolates (including Gram-negative and Gram-positive bacteria) (Dahiya and Purkayastha 2012). *S. aureus*, *K. pneumoniae*, *E. coli*, *P. mirabilis*, *B. subtilis*, and *P. aeruginosa* were among the clinical microorganisms that were effectively killed by a variety of solvent extracts of medicinal herbs, such as *Curcuma longa*, *Tinospora cordifolia*, and *Zingiber officinale* (Chakraborty et al. 2014). Hence the plant products have acquired attention of pharmaceutical industry to develop new potential antimicrobial agents that will be effective against MDR bacteria. (Pagare et al. 2015; Khameneh et al. 2019).

Plant secondary metabolites are further classified based on their biosynthetic origin into alkaloids, phenolic compounds, flavenoids, terpenoids, and sulfur-containing compounds (Pagare et al. 2015). These phytochemicals are well known to act as antimicrobial agent, attractants/repellents and prevents from herbivores.

2.3.1 Alkaloids

Alkaloids are organic nitrogen heterocyclic compounds synthesized by a wide range of living beings including microscopic organisms, parasites, plants, and animal. Chemical structure of few of the alkaloids constituted of heterocyclic ring whereas others were simply atypical alkaloid nonheterocyclic. Alkaloids are investigated to be an important secondary metabolite possessing antimicrobial activity for four decades (Schmitz, 1985). Several studies have established that alkaloids possess broad-spectrum antimicrobial activities, and these compounds could be the promising candidate to deal with a multitude of infections (Cushnie et al. 2014; Casciaro et al. 2020).

Alkaloids are unique bioactive compounds because of their capacity to establish hydrogen bonds with proteins, enzymes, and receptors. According to reports available, the antibacterial mechanism of alkaloids revealed their ability to damage the bacterial cell membrane, interfere with DNA synthesis, and prevent the creation of proteins (Kelley et al. 2013; Larghi et al. 2015). These properties make them a promising agent for the development of new antimicrobial drugs. Alkaloid-rich biorational plant extracts with a wide range of families (*Burseraceae*, *Tiliaceae*, *Capparaceae*, *Mimosaceae*, *Amaryllidaceae*, and *Vitaceae*) have shown remarkable antibacterial activity against a variety of microorganisms, including *E. coli*, *P. aeruginosa*, *S. aureus*, *K. pneumonia*, and *S. Typhimurium*, as well as mycobacteria (*M. fortuitum*, *M. smegmatis*, and *M. tuberculosis*) (Thawabteh et al. 2019; Omar et al. 2021). Based on the structural diversity alkaloids are classified into: Quinoline, isoquinoline, quinolizidine, imidazole, indole, pyridine, pyrrolizidine, and piperidine (Ribeiro et al. 2021). Figure 2.2 shows few of the alkaloids extensively studied for their antimicrobial activity.

2.3.1.1 Quinoline

Quinoline alkaloids, categorized as first-generation inhibitors include quinine and quinidine from Cinchona bark. Quinoline alkaloids for instance dictamnine, masculine, and kokusagine isolated from the bark of stem of *Teclea afzelii* has demonstrated strong antibacterial activities (Kuete et al. 2008). Natural and synthetic quinolone alkaloids are also documented to prevent DNA replication by inhibiting type II topoisomerase enzymes (Heeb et al. 2011).

Tomatidine, a steroidal alkaloid found in solanaceous plants such as eggplant (*Solanum melongena*), potato (*Solanum tuberosum*), and tomato (*Solanum lycopersicum*), and has been shown to have significant antibacterial activity against *S. aureus* whether used alone or in conjunction with aminoglycosides (Patel et al. 2021). Its synergistic effect along with potential antibiotic is well documented against bacteria like *P. aeruginosa* and *Enterococcus faecalis* infections (Soltani et al. 2017).

2.3.1.2 Isoquinoline Alkaloids

Isoquinoline alkaloids, also referred as benzylisoquinoline alkaloids form the largest alkaloid group are widely dispersed in the families *Menispermaceae*, *Papaveraceae*, *Berberidaceae*, and *Ranunculaceae* (Wangchuk 2019). Strong antibacterial activity was confirmed by the isolation of new natural products, spathullin A and spathullin B, from the fungus *Penicillium spathulatum*. It was discovered that these were effective against bacteria that were Gram-positive and Gram-negative, comprising *Enterobacter cloacae*, *Klebsiella pneumonia*, *Staphylococcus aureus*, *Escherichia coli*, *Pseudomonas aeruginosa*, and *Acinetobacter baumannii* (Nord et al. 2019). Another effective isoquinoline alkaloid berberine, the primary representative quaternary ammonium salt derived from *Berberis* spp. All investigated strains of methicillin-resistant *Staphylococcus aureus* (MRSA) were susceptible to the antibacterial action of berberine, with 32–128 μg/mL as the MIC, minimal inhibitory concentrations. The antimicrobial activity of berberine is reported to be due to DNA intercalation, targeting gyrase and topoisomerase IV, RNA polymerase,

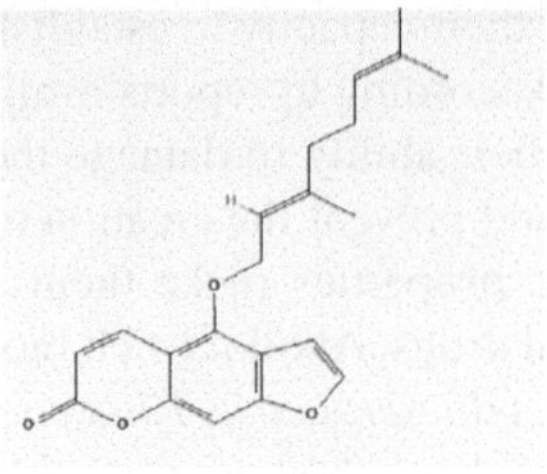

Bergamottin

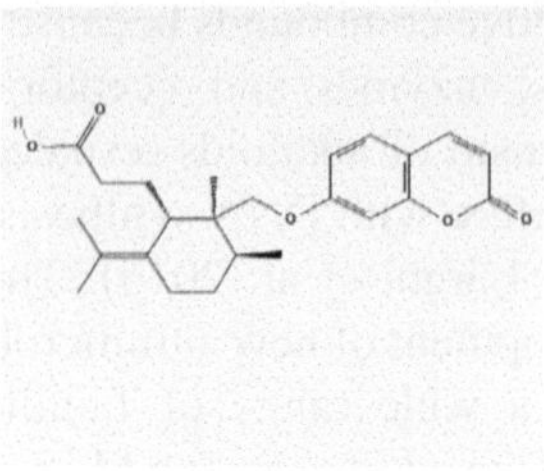

Galbanic acid

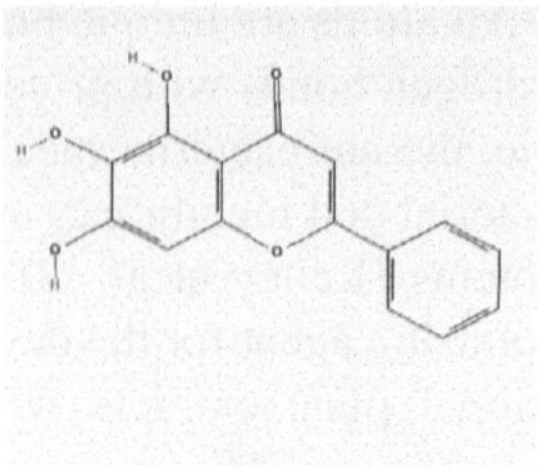

Baicalein

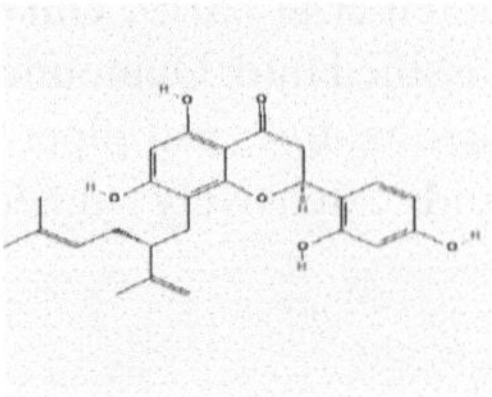

Sophoraflavanone G

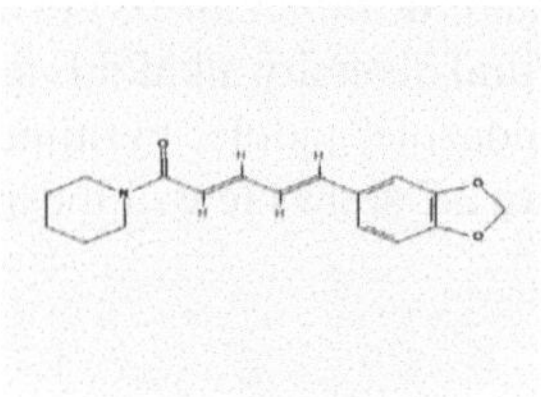

Piperine

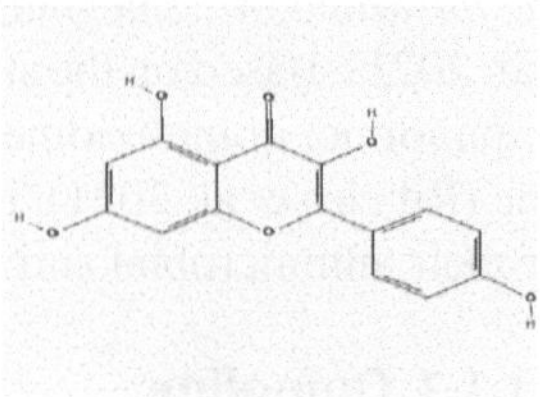

Kaempferol

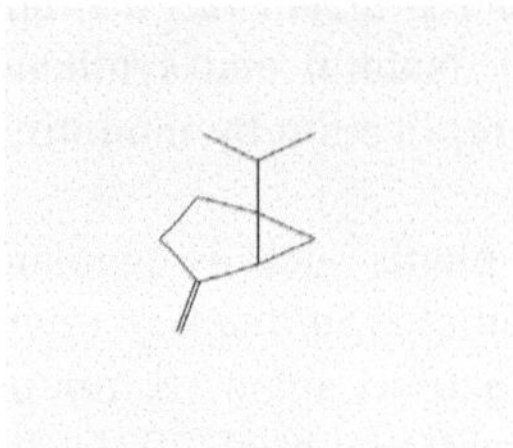

Sabinene

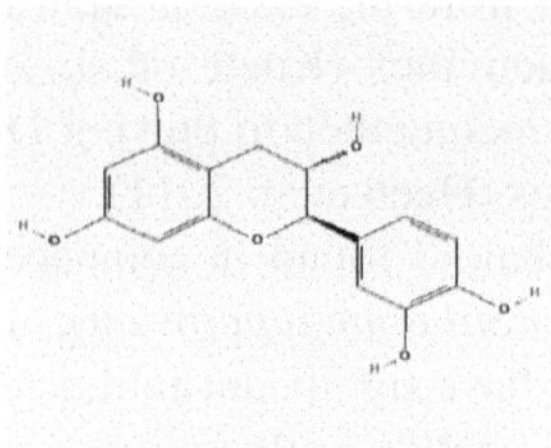

Catechine

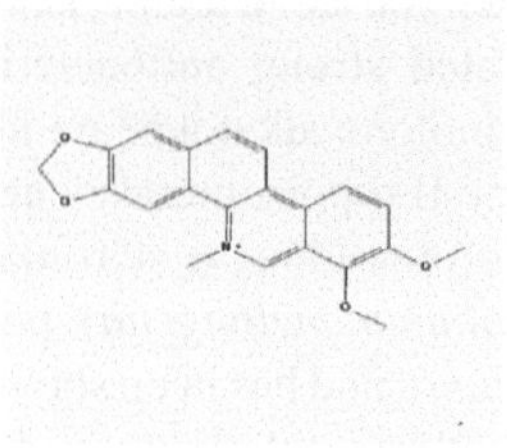

Chelerythrine

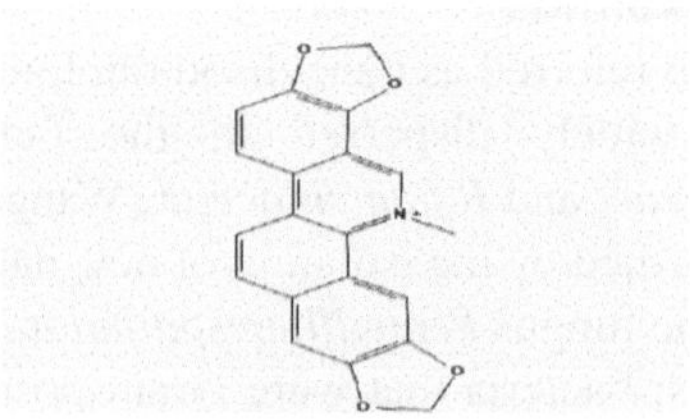

Sanguinarine

Fig. 2.2 Structures of the phytochemicals showing activity against bacteria exhibiting multidrug resistance

and finally, the inhibition of cell division (Khameneh et al. 2019). Among the Isoquinoline alkaloids, obtained from *Chelidonium majus*, was assessed for the antibacterial activity, chelerythrine (CHE) was found to have optimum efficiency against *P. aeruginosa* and sanguinarine against *S. aureus* (Zielińska et al. 2019). The mechanism to act against microbes was similar to that of commercially available antimicrobial drugs (Yan et al. 2021).

2.3.1.3 Indole Alkaloids

Indole alkaloids are the complex structures of tryptophan derivatives with excellent bioactivities. Indole alkaloids have bicyclic structure resulted from the fusion of benzene ring and a five-membered pyrrole ring. Based on the central structure, they are mainly divided into monoterpenoid indole alkaloids, monomeric indole alkaloids, bis-indolic alkaloids, and tryptamine indole alkaloids (Liu et al. 2020). They are further classified based on different groups attached to the ring, position of the groups and unsaturation in the ring into Strychnos alkaloids, yohimbans, heteroyohimbans, Vinca alkaloids, β-carbolines, kratom alkaloids, tryptamines, ergolines, and *Tabernanthe iboga* alkaloids (Casciaro et al. 2020). Strong efflux pump inhibitor (EPI) action is exhibited by the well-known natural substance serpentine, an indole alkaloid that was isolated from the *Rauwolfia serpentina* plant (Abdelfatah and Efferth 2015). Reserpine was used to lower the antibiotic resistance capacities of resistant *Stenotrophomonas maltophilia*, and the overexpression of EP was believed to be the primarily caused resistance to fluoroquinolones (Jia et al. 2015). Astrychnine, an indole alkaloid that was extracted from *Psychotria pilifera* leaves, exhibited bacteriostatic activity in an assay comparable to cefotaxime against *E. coli* ATCC 11775, *Enterococcus faecalis* ATCC 10541, and *S. aureus* ATCC 25922 (Liu et al. 2016). After being administered in combination with novobiocin, carbenicillin, and colistin, harmaline, a β-carboline alkaloid that was isolated from *P. harmala*, exhibited enhanced synergistic EPI action. The authors hypothesized that the secondary amine group in the pyrrole ring along with a methoxy group in the aromatic ring at the C-6 position of indole alkaloids might influence EPI activity.

Monoterpene indole alkaloids isolated from *Voacanga africana*, the voacafricines A and B were tested for antibacterial activity. According to the study, voacafricine B has better antibacterial activity than voacafricine A. Voacafricine B's minimum inhibitory concentrations (MICs) for *S. typhimurium* and *S. aureus* were 0.78 and 3.12 g/mL, respectively. Compared to the well-researched alkaloids fibraurtine and berberine, which were used as the positive control, both alkaloids exhibited higher antibacterial activity. *S. aureus* and *S. typhimurium* had MICs of 3.12 and 6.25 g/mL for the alkaloid berberine, respectively. In addition, fibraurtine's minimum inhibitory concentrations (MICs) for *S. typhimurium* and *S. aureus* were 3.12 and 25 g/mL, respectively (Ding et al. 2018). The potent antibacterial properties of the indole alkaloids may offer a way to manage drug-resistant infections, like MRSA, and may be also used to create therapies that target this kind infections caused by drug resistance (Yan et al. 2021).

2.3.1.4 Piperidine Alkaloid

This class of alkaloid possess saturated piperidine ring is derived from L-lysin. These alkaloids are believed to be potent candidates against multidrug resistance microbes. Being the primary component of both *Piper nigrum* and *Piper longum*, piperine, a piperidine alkaloid, is well known for its ability to inhibit many cytochrome P450-mediated processes as well as human P-glycoprotein (Bhardwaj et al. 2002, Casciaro et al. 2020). In contrast several studies suggested that the efflux inhibitory efficacy against MRSA and *S. aureus* as the main antibacterial mechanism (Khan et al. 2006, Mohtar et al. 2009).

Among the several alkaloids studied for antimicrobial activity by Mohtar et al. (2009), Piperine shown the greatest potential EPI actions, reducing the EtBr MIC against N441 and ATCC 25923 by eightfold and fourfold respectively. The authors hypothesized that the conjugated diene attached to aromatic ring may be responsible for the strong EPI activity. A study conducted on the synergistic antibacterial activity of piperine and gentamicin in combination to combat MRSA and investigate the underlying mechanisms that influences piperine's ability to withstand the bacterial resistance (Khameneh et al. 2015). Because the alkaloid is hydrophobic and the antibiotics are hydrophilic, the blend was delivered in liposomal form. The findings were interpreted to mean that, in comparison to the drug used alone, the MIC value of liposome carrying gentamicin was 32-fold lower, shown more antibacterial activity as well as compared to vancomycin. Thus, it's possible that the piperine can increase gentamicin's antibacterial effectiveness by preventing the antibiotic from effluxing (Khameneh et al. 2015).

2.3.2 Polyphenols

Polyphenols are extensively studied to possess biological activity against a broad group of microbes. Based on number of structural components and phenolic groups, they are divided into lignans, flavonoids, phenolic acids, and stilbenes. These flavonoids and phenolic acids are further subdivided such as: flavones, isoflavones, flavonols, flavanones, chalcones, catechins together form the flavonoids group; phenolic acids include hydroxycinnamic acid and hydroxybenzoic acid (Aslam et al. 2017). Besides possessing antioxidant property, polyphenols are well-studied to have antibacterial, anticancer, anti-inflammatory, and antiallergic properties (Lin et al. 2016). Foods high in flavanols have potent antibacterial qualities that can fight with both Gram-positive and Gram-negative infections, comprising strains that are resistant to treatment (Abbas et al. 2017). Quercetin's synergistic effects with *P. aeruginosa* in relation to gentamycin, amikacin, ceftriaxone, levofloxacin, and tobramycin resulted in $\geq$80% suppression on biofilm development and biofilm cell viability. Subsequently, the bactericidal activity resulted in (68–85%) cell death (Vipin et al. 2020). Tetracyclin, a broad-spectrum drug failed to control multidrug-resistant (MDR) *E. coli*, however in combination with low concentrations of quercetin showed bactericidal activity clinical *E. coli* isolates (Qu et al. 2019). Another polyphenol Kaempferol demonstrated considerable antibacterial activity against various antibiotic-resistant strains such as *P. aeruginosa* and *S. aureus*. Another synergistic study of polyphenols with commercially available antibiotics showed the positive effect of polyphenols. Antibiotics such as amoxicillin, ampicillin, cefixime, cephradine, ciprofloxacin, methicillin, and ceftriaxone when assessed alone were found inactive against clinical isolated MRSA. However, these antibiotics were combined with kaempferol, quercetin, or both, they all demonstrated a valuable anti-MRSA activity (Prakash et al. 2018).

Although the exact process by which polyphenols inhibit bacterial growth is still unknown, several investigations have shown that, within cells, polyphenols and bacterial enzymes generate hydrogen bonds that alter the permeability and structural integrity of bacterial cell walls (Ganesan and Xu 2017).

Chalcones are the plant's precursors to flavonoids and isoflavonoids and are recognized as important secondary metabolites. Research findings indicate that the antibacterial activity of chalcones is greatly enhanced when taken in conjunction with antibiotics. For instance, synergistic effect of a chalcone isolated from *Sophora flavescens* Aiton, a traditional herb root used in Africa and America showed potent antimicrobial activity (Lee et al. 2010). Ethanol extracts of *Uvaria chamae* assessed for antimicrobial activity revealed that the chalcone and dihydrochalcone significantly inhibits the growth of Gram-positive MDR pathogens including VRE, MRSA, and *S. aureus* (Koudokpon et al. 2018).

Flavons are also studied to possess antimicrobial activity against many MDR microbes. Such instances are Licoflavon and artocarpin. Licoflavone, a bioactive flavon compound extracted from the *Retama raetam* flowers, was shown to actively inhibit *E. coli* growth with MIC 7.81 μg/mL (Edziri et al. 2012). Artocarpin is another bioactive flavon isolated from Miq. leaves of *Artocarpus anisophyllus* that exert an antibacterial effect against MDR *P. aeruginosa* and *E. coli* by enzymes inhibition. Synergistic effect of flavons such as Baicalein was found to reinstate the antimicrobial activity of β-lactams and tetracycline against *E. coli* and MRSA using the EPI mechanism (Fujita et al. 2005). The amalgamation of baicalein with vancomycin and methicillin were also assessed to pose bactericidal effect against MRSA and VRE via inhibiting synthesis of bacterial cell wall (Lee et al. 2015).

Resveratrol, the major stilbene polyphenol, is of recent attraction as antimicrobial agent alone or in combination with antibiotics. Resveratrol interferes with microbial growth by inhibiting ATP synthase enzyme and biofilm production, in turn resulting in lesser consumption of energy insufficient to meet the energy requirement for growth. It restricts the toxin production and motility, interferes with quorum sensing as well (Vestergaard and Ingmer 2019). The study suggested that resveratrol stood as potent antimicrobial compound against *P. aeruginosa* and *S. aureus* when compared to norfloxacin taken as reference.

2.3.3 Tannins

Tannin is a polyphenol that dissolves in water and is commonly derived from woody plants. There are two varieties of tannin: nonhydrolyzable (condensed), which is a more prevalent variety that is typically produced from flavonoid dimers like procyanidin, and hydrolyzable tannin, a typical ester of phenolic acid (Serrano et al. 2009). Tannin interferes with a growth of several microbes including bacteria, yeast, and fungi. The antimicrobial function of tannins against several Gram-positive and Gram-negative bacterial is well studied. The tannins are examined to possess bacteriostatic activity (Boakye et al. 2016). Primary strains of MRSA did not exhibit any efflux pump activity in research. But when fusidic acid is used continuously, mutations occur and these mutant organisms become resistant to efflux pumps. It has been demonstrated that administering tannic acid along with tetracycline, erythromycin, and fusidic acid inhibits mutation by giving bacteria a different target and encouraging cell lysis (Myint et al. 2013; Tintino et al. 2017). β-Lactamases inhibited the activity of penicillin and cephalosporin antibiotics by hydrolyzing the β-lactam ring, thus providing resistance against antibiotic to bacteria.

Natural substances known as phenolic acids have an aromatic ring that is connected to several hydroxyl groups and one carboxylic acid functionality. These could be benzoic acid derivatives, such as gallic, phydroxybenzoic, and protocatechuic acids, or derivatives of cinnamic acid, such as ferulic and p-coumaric acids. Gallic acid and ferulic acid reduce the negative surface charge of *E. aeruginosa* membranes, opening pores in the cell membrane and altering their hydrophobicity, which causes the internal contents of the bacteria to seep out. These alterations are permanent (Borges et al. 2013).

2.3.4 Sulfur-Containing Compounds

It has been shown that volatile compounds having an organo-sulfur structure are broadly antibacterial against microorganisms that are Gram-positive and Gram-negative. Although their exact mode of action is unknown, the disulfide bond greatly increases their potency (Kim et al. 2020). These compounds include, among others, ajoene, cysteines, isothiocyanate, and allicin.

Allium sativum L., often known as garlic, is one of the Allium species that is well-known for its antibacterial properties against bacterial resistance (Barbieri et al. 2017). This activity is related to the organo-sulfur elements of garlic, particularly allicin, which is formed in reaction to tissue injury in garlic and gives garlic its distinct scent when crushed (Barbieri et al. 2017). Several trials have demonstrated that allicin inhibits the growth of bacteria that are resistant to drugs (MDR), *P. aeruginosa*, MRSA, *E. coli*, and *S. aureus*. However, it is recommended to use allicin in conjunction with other antibiotics because of its bacteriostatic qualities (Lanzotti et al. 2014; Nakamoto et al. 2020). In this way, allicin suppresses the creation of DNA, proteins, biofilms, and bacterial enzymes like DNA gyrase, RNA polymerase, and alcohol dehydrogenase (Lanzotti et al. 2014; Farag et al. 2017; Reiter et al. 2020).

The Brassicaceae plant family of plants, which includes broccoli, mustard, cauliflower, and cabbage, produces glucosinolate metabolites that the enzyme myrosinase subsequently breaks down into isothiocyanates (ITCs) (Romeo et al. 2018). Antibacterial effects of several ITCs, both natural and synthetic, have been discovered to protect against human illnesses, including those caused by resistant bacteria (Lin et al. 2000). Assessment of the antibacterial activity of a synthetic version of allyl isothiocyanate AITC, tests were conducted using *S. aureus* and an *E. coli* strain. According to a different study on the mechanism of action of AITC, by interacting with thioredoxin's sulfhydryl groups, it may prevent the synthesis of DNA in bacteria, which encourages decreased enzyme activity, and acetate kinase, influences the energy metabolism of bacteria (Luciano and Holley 2009; Evangelista et al. 2021).

It has been demonstrated that phenyl isothiocyanate (PITC) has a moderate impact (MIC 1000 μg/mL) against *S. aureus* and *E. coli* (Abreu et al. 2013). It regulates alterations in the bacterial membrane, which cause the negative charges on the bacterial surface to decrease as well as an increase in surface hydrophilicity and electron donation through interactions with bacterial cells. These changes cause membrane degradation and cellular disruption, which in turn cause cell death. Additionally, the synergistic effects of PITC with ciprofloxacin and erythromycin were investigated. The prior research suggested combining PITC with one of these two antibiotics since it boosts their antibacterial efficacy against resistant *E. coli* and *S. aureus* (Abreu et al. 2013).

Antibacterial efficacy of cruciferous species-derived ITCs was evaluated against 15 clinical isolates of MRSA detected in diabetic foot ulcers (Dias et al. 2014). The structural properties of ITCs primarily dictate their capacity to inhibit bacteria; of them, benzyl isothiocyanate (BITC) exhibits the highest level of activity (bacteriocidal impact of up to 87%) (Dias et al. 2014). The lipophilic and electrophilic characteristics of the BITC structure work as a mechanism to pierce the bacterial outer membrane and impair the bacterium's ability to maintain membrane integrity (Sofrata et al. 2011). Garlic extract has an organo-sulfur characteristic called ajoene. This compound exhibits a broad spectrum of pharmacological properties, including antiviral, antibacterial, anticancer, antioxidant, and antithrombogenic properties (Kaschula et al. 2010). An investigation into the antibacterial effects of ajoene in *M. tuberculosis* showed that the endoplasmic reticulum (ER) experiences an ajoene-induced stress response, which is a component of the host's defense against mycobacterial infection. Despite this, the exact mechanism of ajoene action is still unknown (Choi et al. 2018). Ajoene activates c-Jun N-terminal kinase (JNK), which in turn increases the formation of reactive oxygen species (ROS).

2.3.5 Terpenes

Terpenes are a wide class of intricate chemical compounds that possess outstanding antimicrobial activity (Barbieri et al. 2017). These are believed to interfere with oxygen uptake required for cellular respiration, subsequently affecting oxidative phosphorylation (Zengin and Baysal 2014). Terpenes are chemically hydrocarbons, and the primary building block for their production is a 5-carbon isoprene unit (Mahizan et al. 2019). Terpenes are categorized into four groups based on the number of isoprenes: monoterpenes, diterpenes, triterpenes, and tetraterpenes.

Monoterpenes consist of two isoprene units abundantly found across plants. A study conducted by Griffin et al. (1999) showed monoterpenes such as menthol, thymol, carvacrol, and geraniol were potent inhibitors for many Gram-positive and Gram-negative bacteria. Geraniol is tested to efficiently prevent the growth of the Gram-negative multidrug-resistant (MDR) *Enterobacter aerogenes* by inhibiting the potent efflux pump (Lorenzi et al. 2009). Thymol is the most effective monoterpene against *S. aureus* and *E. coli*, according to numerous studies that demonstrate the antibacterial properties of menthol, thymol, eugenol, and carvacrol by suppressing membrane characteristics, lowering bacterial growth, and blocking the efflux pump (El Atki et al. 2019). Another study with four terpenes p-cymene, carvacrol, thymol, and gamma terpenes using dimyristoylphosphatidylcholine as a mode of delivery, demonstrated the bactericidal effect on *S. aureas* and *E. coli* (Marchese et al. 2017). Furthermore, the study postulated that these monoterpenes may combat MDR infections by rupturing membrane lipid, and that the bacterial membrane's lipid composition would be strongly associated with the extent of this disruption (Chen et al. 2016). These substances interfere with the integrity of the membrane to effectively prevent bacterial growth in vitro, as demonstrated by the results of SEM, flow cytometry, and confocal microscopy.

Furthermore, the formation of bacterial biofilms may be impacted by carvacrol, thymol, and cinnamonaldehyde at concentrations of 0.5 mg/mL or higher. Through their interactions with the cell membrane and inactivation of microbial enzymes, the hydroxyl groups of carvacrol and thymol have been suggested to have a role in interfering with bacterial biofilm formation that leads to permeability of intracellular components (García-Salinas et al. 2018).

This family of terpenes also includes retinoic acid, the active component of retinol (vitamin A). The rich sources are cantaloupe, carrots, sweet potatoes, pumpkin, and squash. These compounds mainly act on the skin, and a number of studies have shown that they are effective in treating infectious skin diseases (Feneran et al. 2011; Chaudhuri and Bojanowski 2014). Vitamin A influences gene expression by attaching to both types of common skin receptors: retinoic acid receptors (RAR) and retinoid X receptors (RXR) (Liggins et al. 2019). The potential bactericidal effect of a novel synthetic type of retinoid CD437 against *E. faecalis* was assessed. It has been demonstrated that CD437 exhibits substantial synergy and anti-biofilm actions in conjunction with gentamycin to combat this infection (Beckenbach et al. 2015).

Diterpens are studied to possess broad bioactivities, showed distinct antimicrobial activity against *E. coli*, *C. albicans*, *P. aeruginosa*, and *S. aureus* (Griffin et al. 1999). Additionally, remarkable synergistic effect of diterpenes with antibiotic against MRSA has also been well documented. Clerodane diterpene, 16α-hydroxycleroda-3, 13 (14)-Z-dien-15, 16-olide (CD), isolated from leaves of *Polyathia longifolia* showed to enhance the effectiveness of oxacillin, tetracycline, daptomycin, and linezolid against clinical isolates of MRSA on synergystic application (Gupta et al. 2016). Two other diterpenes salvipisone and aethiopinone extracted from roots of *Salvia sclarea*, exhibited antibiofilm and antibacterial activities against *Enterococcus faecalis*, *S. aureus*, and *S. epidermidis* (Różalski et al.

2007). In the same year, synergistic activity of salvipisone and aethiopinone were tested in combination with antimicrobial drugs including oxacillin, vancomycin, and linezolid against MRSA and methicillin-resistant *Staphylococcus epidermidis* (MRSE). The results interpreted that they either showed bactericidal or bacteriostatic activity against the both the cultures (Walencka et al. 2007).

Over 20,000 different molecules make up the large and diversified class of phytochemicals known as triterpenes, with wide-ranging pharmacological actions, such as analgesic, cardiotonic, sedative, antibacterial, and anticancer properties (Jesus et al. 2015). Oleanolic acid (OA) is a triterpene synthesized by plant species belonging to Oleaceae family, Olea europaea being the major producer. OA and its derivatives were reported as a potent antimicrobial agent against *S. pneumoniae*, *E. faecalis*, *P. aeruginosa*, *M. tuberculosis*, *Streptococci*, and *E. coli* (Horiuchi et al. 2007; Kim et al. 2012; Ayeleso et al. 2017a, b). The capacity of *Listeria monocytogenes* to form biofilms is inhibited by ursolic acid (UA), another triterpenoid acid that is commonly found in berry peels, especially those of cranberries. However, the exact mechanism behind this effect is yet unknown (Kurek et al. 2014).

In a different investigation, the ursolic acid phenotype 3-O-α-L-arbinopyranoside (URS), which was isolated from *Acanthopanax henryi* (Oliv.) leaves, revealed strong antibiotic efficacy against MRSA by inhibiting the bacterial cell wall and triggering the cell lysis mechanism, as evidenced by the outcomes of the transmission electron microscopy (TEM), time kill assay, and white checkboard dilution tests. When URS and oxacillin worked in partial synergy, the time-kill curve was suppressed below the inhibitory line (Zhou et al. 2017).

Traditional medicinal practices used these essential oils isolated from *Salvia hydrangea* DC. ex Benth. for the treatment of various bacterial and fungal infections. *Salvia hydrangea* DC. ex Benth., belonging to the Lamiaceae family, is an Iranian shrub, that produces essential oils including Spathulenol, β-eudesmol, trans-caryophyllene, 1,8-cineole, caryophyllene oxide, camphor, α-pinene, and β-pinene. Ghavam et al. (2020) investigated the antibacterial effects of these essential oils on *K. pneumoniae* (MIC ~ 62 μg/mL) and *P. aeruginosa* (MIC ~ 16 μg/mL). These findings together demonstrated the use of essential oils in the future creation of novel medicinal agents.

2.3.6 Coumarins

As secondary metabolites, fungus, bacteria, and plants have all been reported to contain coumarins, a class of naturally occurring chemicals. Numerous biological actions are associated with them, such as vasodilation, antioxidant, antiviral, antihypertensive, anticancer, anti-inflammatory, and neuroprotective characteristics (Venugopal et al. 2013). Although coumarin is present throughout the plant, it is particularly abundant in the following parts: roots (*Ferulago campestris* (Besser) Grecescu.), seeds (*Calophyllum inophyllum* L., *Calophyllum cerasiferum* Vesque), fruit (*Aegle marmelos* L. Correa., *Tetrapleura tetraptera*), and leaves (*Murraya paniculata* L.). Six subgroups of coumarins can be distinguished based on their chemical structures: Polycyclic aromatic hydrocarbons (PAHs), dihydrofurano PAHs, simple PAHs, linear and angular pyrano PAHs, phenyl PAHs, and bicoumarins (Venugopal et al. 2013).

In several studies, the sesquiterpene coumarin that was isolated from *Ferula pseudalliacea* Rech.f. roots demonstrated enhanced antibacterial action against Gram-positive infections (Dastan et al. 2016). It has been proposed that these coumarins inhibit the DNA gyrase enzyme (Maxwell 1993). According to a different study, coumarins work by blocking quorum-sensing activity, which reduces bacterial pathogenicity and prevents the formation of biofilms (da SM Forezi et al. 2018). Antibiotic resistance has been shown to be significantly impacted by oxidative stress in prior research (Martelli and Giacomini 2018; Kim et al. 2019). Consequently, it is better to look for substitute drugs that include antimicrobial and antioxidant qualities. Thus, a large number of hydroxy-3-aryl coumarins were chemically synthesized and examined to see their ability to act as antioxidants and hinder microorganisms that were resistant to drugs (Pisano et al. 2019).

Some studies show that when a metal complex formed between the drug molecule and transitional metals such as copper, zinc, nickel, and cobalt and so on, the medication's biological activity is greatly increased (Sahoo and Paidesetty 2017). To clarify what follows: The metal complex carries the drug ligand, and the drug acts more quickly and efficiently in the body due to the metal-ligand connection's facile cleavage (Abou-Dobara et al. 2014; Bhagat et al. 2019). To be clear, the medication operates more quickly and effectively in the body because of the easy cleavage of the metal-ligand connection, which is carried by the metal complex that contains the drug ligand. Recent studies have demonstrated that a new coumarin compound known as 3,3′-(3,4- dichlorobenzylidene)-bis-(4-hydroxycoumarin) inhibits the formation of MRSA biofilms (Qu et al. 2020).

2.4 Case Studies

Attempts in drug discovery are facing challenges due to the rise of multidrug-resistant (MDR) antibiotic-resistant bacteria which in turn have been posing a significant threat to global healthcare. These bacteria have developed new antibiotic-resistance mechanisms, nonetheless, making current medications ineffective. The overuse of antimicrobial drugs is the causal agent of the rising resistant bacterial strains. Although it can be acquired and is passed down within various bacterial species, the ability of pathogens to resist antimicrobial drug therapy is mostly innate. The depletion of powerful antibiotics is further worsening the situation. Even though there is constant research and development going on for new antibiotics, particularly those with novel modes of action, it is unlikely that this would be enough to win the war on its own. Additionally, it's critical to look for other strategies. The advancement of adjuvants development is one such tactic for raising the effectiveness of currently available antibiotics in market. The topic of nano antibiotic methods is being currently explored, emphasizing the significant potential of phenolic compounds, essential oils, alkaloids, and nano antibiotics in the fight against the severe antibiotic resistance (Habib Adam et al. 2023). Phytochemicals, like, phenols, coumarins, terpenes, and alkaloids, have been found promising in suppressing drug-resistant bacteria. However further investigation is required for their commercial success in healthcare (Khare et al. 2021). As an alternative to existing antibiotics in market phytochemicals can be the alternative to it that operates by targeting various key bacterial structures and functions. Appropriately, these strategies include prevention of bacterial biofilms formation, inhibition of efflux pumps, and rupturing of membrane proteins. Phytochemicals have the ability to further regulate the expression of genes linked to virulence factors and resistance mechanisms in bacteria. Numerous in vitro and in vivo studies have demonstrated the effectiveness of phytochemicals in reducing antibiotic resistance; however, before utilization in clinical settings, it is imperative to carefully assess the safety, toxicity, and cost aspect of these phytochemicals. Further research and clinical validation are needed to fully realize the potential of phytochemicals in addressing antibiotic resistance (Suganya et al. 2022). A broad range of activities in herbal medicine have been developed to combat the emerging challenge of developing effective medicines with fewer unfavorable side effects, as a result of the rapid and continuous emergence of drug-resistant pathogens, their unmanageable virulence, and the loss of effectiveness of currently used antimicrobial agents against them over the decade. The "Dogbane" or Apocynaceae family has gained worldwide recognition for producing unique chemicals and some life-saving phytochemicals. Members belonging to this *Apocynaceae* family have also been studied thoroughly for in vitro and in vivo experiments against a multitude of nosocomial pathogenic microorganisms (Anand et al. 2020). Analytical techniques for the separation, characterization, and phytochemical testing of plant constituents that have been found to have ethnopharmacological effects have been developed over years with the aim of treating various diseases. These procedures take into account the quantitative features of plant constituents that attribute health benefits. Polyphenol-disaccharides (PDSs) like polyphenols, alkaloids, and tannins have a significant deal of potential for application as antimicrobials or as modifiers of drug resistance against bacterial infections (AlSheikh et al. 2020). The latest research is conducted on the synergistic interactions between plant extracts and extracted pure chemicals and already-available chemotherapeutics, antibacterials, and antifungal medications. A number of phytochemicals have been shown to function as resistance-modifying agents when combined with currently prescribed medications, and appropriate combinations may restore the effectiveness of crucial, life-saving antimicrobial treatments. One such example is target-site/receptor alteration that reduces the antibacterial activity of medications by modifying the target-site through mutation. Another approach is applying resistance-modifying medications coupled to the application of active medication is one of the most effective ways to combat multidrug resistance (MDR). This method has led to the clinical approval of a number of much helpful medications, like, beta-lactamase inhibitors (clavulanic acid, sulbactam, and tazobactam) when combined with amoxicillin. In order to stop the establishment of bacterial resistance, various combinations of natural chemicals may facilitate the interaction of an antimicrobial drug with its target inside the pathogen. By using lower amounts of both drugs, the method can further decrease toxicity (Ayaz et al. 2019).

Moving from lab research to animal studies and, eventually, clinical trials in human is a protocol for enhancing novel phytocompounds. Numerous medicinal plants with a multitude of phytochemicals demonstrated antibacterial action, which can substitute traditional medical care. Introducing a brand-new, innovative medication or antibiotic to market is accompanied with a significant high cost and takes time. As a result, the significance of drug isolation from natural sources has increased in addition to the identification of various chemical compounds with qualities for resistance (Mandal et al. 2014). To ensure drugs to be safe, effective, and formulated with consistency, multiple preclinical and clinical analytical criteria for phytochemical substances are necessitated. Components taken out of herbal remedies that were preclinically examined for a range of infectious disorders before being granted a license based on the results of those studies. It could consist of co-administration of one chemical or two or more bioactive ingredients. A minute percentage of the various bioactive substances discovered in recent decades have been investigated further in clinical trials. Furthermore, a high concentration of most phytocompounds is needed when taken as monotherapy as opposed to that in case of antibiotics. In order to counter these issues, scientists concentrated on combining fewer synthetic antibiotics with more phytochemicals to block the action of resistance against myriads of microorganisms (Touani et al. 2014; Santiago et al. 2015). As part of preclinical investigations, in silico technique was utilized to choose natural phytocompounds based on their bioactive ingredients, in order to circumvent the time-consuming nature of these active phytocompounds from the market. According to Ahamad et al. (2017), quantitative structure activity analysis (QSAR) aids in the prediction of a specific target in a compound. This works in addition to the interpretation of molecular structural characterizations such as protein-ligand binding interactions. The interaction between ligand and target is also complicated in studies on pharmacophore models, which mimic the three-dimensional groupings of particles with different physicochemical properties. Molecular docking, a popular in silico method, suggests a structure-activity link for phytocompounds to uncover their mode of action and comprehend where a ligand is positioned within a pocket for protein-binding. As per previous research, since 2000, over 16,000 antimicrobial trials have been filed with ClinicalTrials.gov. Antimicrobial mediators were evaluated and examined in about one out of every ten reported scientific investigation studies. The majority being almost 26% of which enrolled children in addition to adults, were randomized interventional trials including pharmaceuticals and biologicals. Just 12% of all interventional medication trials that were completed had their diagonal relationships explained by the findings of the inquiry.

2.5 Conclusions and Future Perspectives

Large amounts of bioactive compounds that are easily accessible, cheaply priced, easy to separate, and almost entirely safe for human consumption can be found in nature. For thousands of years, a variety of infectious diseases have been treated with substances derived from plants. For instance, since the Middle Ages, honey has been applied topically to treat wounds and prevent the spread of infection. For the same purpose, Native People utilized echinacea extract. When traditional antibiotics failed to control bacteria, scientists looked to nature for solutions. The therapeutic value of compounds derived from plants has been established via numerous experiments.

Some of these compounds may be more effective against bacteria when combined with currently existing antibiotics, while others work well on their own. Thus, a potential strategy to address resistance would be to combine conventional antibiotics with phytochemicals in a combination therapy. Several methods have been put out to clarify the activity of substances produced from plants, comprising DNA, protein synthesis, disruption of bacterial membranes, suppression of the efflux pump, and reduction of biofilm development. Exactly what mechanism behind phytochemical action must be determined in order to develop a new antibacterial medication. However, because various factors, including tissue penetration and bioavailability, may affect the effectiveness of phytochemicals, the transition from in vitro research to in vivo investigations and clinical trials remains a controversial topic. Therefore, a deeper comprehension of the natural chemical's pharmacokinetic and pharmacodynamic properties is required.

References

Abbas M, Saeed F, Anjum FM, Afzaal M, Tufail T, Bashir MS, Ishtiaq A, Hussain S, Suleria HA (2017) Natural polyphenols: an overview. Int J Food Prop 20(8):1689–1699. https://doi.org/10.1080/10942912.2016.1220393

Abdelfatah SA, Efferth T (2015) Cytotoxicity of the indole alkaloid reserpine from *Rauwolfia serpentina* against drug-resistant tumor cells. Phytomedicine 22(2):308–318. https://doi.org/10.1016/j.phymed.2015.01.002

Abou-Dobara MI, El-Sonbati AZ, Diab MA, El-Bindary AA, Morgan SM (2014) Thermal properties, antimicrobial activity of azo complexes and ultrastructure study of some affected bacteria. J Microbial Biochem Technol S 3:2. https://doi.org/10.4172/1948-5948.S3-006

Abreu AC, Borges A, Simões LC, Saavedra MJ, Simões M (2013) Antibacterial activity of phenyl isothiocyanate on *Escherichia coli* and *Staphylococcus aureus*. Med Chem 9(5):756–761

Abulrob AN, Suller MT, Gumbleton M, Simons C, Russell AD (2004) Identification and biological evaluation of grapefruit oil components as potential novel efflux pump modulators in methicillin-resistant *Staphylococcus aureus* bacterial strains. Phytochemistry 65(22):3021–3027. https://doi.org/10.1016/j.phytochem.2004.08.044

Achouak W, Heulin T, Pagès JM (2001) Multiple facets of bacterial porins. FEMS Microbiol Lett 199(1):1–7. https://doi.org/10.1111/j.1574-6968.2001.tb10642.x

Ahamad S, Rahman S, Khan FI, Dwivedi N, Ali S, Kim J, Imtaiyaz HM (2017) QSAR based therapeutic management of *M. tuberculosis*. Arch Pharm Res 40:676–694. https://doi.org/10.1007/s12272-017-0914-1

AlSheikh HM, Sultan I, Kumar V, Rather IA, Al-Sheikh H, Tasleem Jan A, Haq QM (2020) Plant-based phytochemicals as possible alternative to antibiotics in combating bacterial drug resistance. Antibiotics 9(8):480. https://doi.org/10.3390/antibiotics9080480

Anand U, Nandy S, Mundhra A, Das N, Pandey DK, Dey A (2020) A review on antimicrobial botanicals, phytochemicals and natural resistance modifying agents from Apocynaceae family: possible therapeutic approaches against multidrug resistance in pathogenic microorganisms. Drug Resist Updat 51:100695. https://doi.org/10.1016/j.drup.2020.100695

Aslam MS, Ahmad MS, Ahmad MA, Akhlaq M (2017) An updated review on phytochemical and pharmacological propeties of *Piper sarmentosum*. Curr Trends Biotechnol Pharm 11(4):345–356

Ayaz M, Ullah F, Sadiq A, Ullah F, Ovais M, Ahmed J, Devkota HP (2019) Synergistic interactions of phytochemicals with antimicrobial agents: potential strategy to counteract drug resistance. Chem Biol Interact 308:294–303. https://doi.org/10.1016/j.cbi.2019.05.050

Ayeleso TB, Matumba MG, Mukwevho E (2017a) Oleanolic acid and its derivatives: biological activities and therapeutic potential in chronic diseases. Molecules 22(11):1915. https://doi.org/10.1016/B978-0-12-809450-1.00009-0

Ayeleso TB, Matumba MG, Mukwevho E (2017b) Oleanolic acid and its derivatives: biological activities and therapeutic potential in chronic diseases. Molecules 22(11):1915. https://doi.org/10.3390/molecules22111915

Banin E, Hughes D, Kuipers OP (2017) Bacterial pathogens, antibiotics and antibiotic resistance. FEMS Microbiol Rev 41(3):450–452. https://doi.org/10.1093/femsre/fux016

Barbieri R, Coppo E, Marchese A, Daglia M, Sobarzo-Sánchez E, Nabavi SF, Nabavi SM (2017) Phytochemicals for human disease: an update on plant-derived compounds antibacterial activity. Microbiol Res 196:44–68. https://doi.org/10.1016/j.micres.2016.12.003

Bazzaz BSF, Khameneh B, Ostad MRZ, Hosseinzadeh H (2018) *In vitro* evaluation of antibacterial activity of Verbascoside, *Lemon verbena* extract and caffeine in combination with gentamicin against drug-resistant *Staphylococcus aureus* and *Escherichia coli* clinical isolates. Avicenna J Phytomed 8:246–253. https://doi.org/10.22038/AJP.2018.15338.1607

Beckenbach L, Baron JM, Merk HF, Löffler H, Amann PM (2015) Retinoid treatment of skin diseases. Eur J Dermatol 25:384–391. https://doi.org/10.1684/ejd.2015.2544

Bhagat K, Bhagat J, Gupta MK, Singh JV, Gulati HK, Singh A, Kaur K, Kaur G, Sharma S, Rana A, Singh H (2019) Design, synthesis, antimicrobial evaluation, and molecular modeling studies of novel indolinedione–coumarin molecular hybrids. ACS Omega 4(5):8720–8730. https://doi.org/10.1021/acsomega.8b02481

Bhardwaj RK, Glaeser H, Becquemont L, Klotz U, Gupta SK, Fromm MF (2002) Piperine, a major constituent of black pepper, inhibits human P-glycoprotein and CYP3A4. J Pharmacol Exp Ther 302(2):645–650. https://doi.org/10.1124/jpet.102.034728

Boakye YD, Agyare C, Hensel A (2016) Anti-infective properties and time-kill kinetics of *Phyllanthus muellerianus* and its major constituent, geraniin. Med Chem Curr Res 6(2):95–104. https://doi.org/10.4172/2161-0444.1000332

Borges A, Ferreira C, Saavedra MJ, Simões M (2013) Antibacterial activity and mode of action of ferulic and gallic acids against pathogenic bacteria. Microb Drug Resist 19(4):256–265. https://doi.org/10.1089/mdr.2012.0244

Bush K, Bradford PA (2016) β-Lactams and β-lactamase inhibitors: an overview. Cold Spring Harb Perspect Med 6(8):a025247. https://doi.org/10.1101/cshperspect.a025247

Cano A, Gutiérrez-Gutiérrez B, Machuca I, Gracia-Ahufinger I, Pérez-Nadales E, Causse M, Castón JJ, Guzman-Puche J, Torre-Giménez J, Kindelán L, Martínez-Martinez L (2018) Risks of infection and mortality among patients colonized with *Klebsiella pneumoniae* carbapenemase–producing K. pneumoniae: validation of scores and proposal for management. Clin Infect Dis 66(8):1204–1210. https://doi.org/10.1093/cid/cix991

Cardoso JC, Oliveira ME, Cardoso FD (2019) Advances and challenges on the in vitro production of secondary metabolites from medicinal plants. Hortic Bras 37(2):124–132. https://doi.org/10.1590/S0102-053620190201

Casciaro B, Mangiardi L, Cappiello F, Romeo I, Loffredo MR, Iazzetti A, Calcaterra A, Goggiamani A, Ghirga F, Mangoni ML, Botta B (2020) Naturally-occurring alkaloids of plant origin as potential antimicrobials against antibiotic-resistant infections. Molecules 25(16):3619. https://doi.org/10.3390/molecules25163619

Chakraborty B, Nath A, Saikia H, Sengupta M (2014) Bactericidal activity of selected medicinal plants against multidrug resistant bacterial strains from clinical isolates. Asian Pac J Trop Med 7:S435–S441. https://doi.org/10.1016/S1995-7645(14)60271-6

Chan BC, Ip M, Lau CB, Lui SL, Jolivalt C, Ganem-Elbaz C, Litaudon M, Reiner NE, Gong H, See RH, Fung KP (2011) Synergistic effects of baicalein with ciprofloxacin against NorA over-expressed methicillin-resistant Staphylococcus aureus (MRSA) and inhibition of MRSA pyruvate kinase. J Ethnopharmacol 137(1):767–773. https://doi.org/10.1016/j.jep.2011.06.039

Chaudhuri RK, Bojanowski K (2014) Bakuchiol: a retinol-like functional compound revealed by gene expression profiling and clinically proven to have anti-aging effects. Int J Cosmet Sci 36(3):221–230. https://doi.org/10.1111/ics.12117

Chen J, Jiang QD, Chai YP, Zhang H, Peng P, Yang XX (2016) Natural terpenes as penetration enhancers for transdermal drug delivery. Molecules 21(12):1709. https://doi.org/10.3390/molecules21121709

Cheng J, Guffanti AA, Krulwich TA (1994) The chromosomal tetracycline resistance locus of *Bacillus subtilis* encodes a Na^+/H^+ antiporter that is physiologically important at elevated pH. J Biol Chem 269(44):27365–27371

Choi JA, Cho SN, Lim YJ, Lee J, Go D, Kim SH, Song CH (2018) Enhancement of the anti-mycobacterial activity of macrophages by ajoene. Innate Immun 24(1):79–88. https://doi.org/10.1177/1753425917747975

Cushnie TT, Cushnie B, Lamb AJ (2014) Alkaloids: an overview of their antibacterial, antibiotic-enhancing and antivirulence activities. Int J Antimicrob Agents 44(5):377–386. https://doi.org/10.1016/j.ijantimicag.2014.06.001

da SM Forezi L, Froes TQ, Cardoso MF, de Oliveira Maciel CA, Nicastro GG, Baldini RL, Costa D, Ferreira VF, Castilho MS, de C da Silva F (2018) Synthesis and biological evaluation of coumarins derivatives as potential inhibitors of the production of *Pseudomonas aeruginosa* virulence factor pyocyanin. Curr Top Med Chem 18(2):149–156. https://doi.org/10.2174/1568026618666180329122704

Dahiya P, Purkayastha S (2012) Phytochemical screening and antimicrobial activity of some medicinal plants against multi-drug resistant bacteria from clinical isolates. Indian J Pharm Sci 74(5):443. https://doi.org/10.4103/0250-474X.108420

Dastan D, Salehi P, Aliahmadi A, Gohari AR, Maroofi H, Ardalan A (2016) New coumarin derivatives from *Ferula pseudalliacea* with antibacterial activity. Nat Prod Res 30(24):2747–2753. https://doi.org/10.1080/14786419.2016.1149705

Davey ME, O'toole GA (2000) Microbial biofilms: from ecology to molecular genetics. Microbiol Mol Biol Rev 64(4):847–867. https://doi.org/10.1128/mmbr.64.4.847-867.2000

Davies J, Davies D (2010) Origins and evolution of antibiotic resistance. Microbiol Mol Biol Rev 74(3):417–433. https://doi.org/10.1128/mmbr.00016-10

D'Costa VM, McGrann KM, Hughes DW, Wright GD (2006) Sampling the antibiotic resistome. Science 311(5759):374–377. https://doi.org/10.1126/science.1120800

Deurenberg RH, Vink C, Kalenic S, Friedrich AW, Bruggeman CA, Stobberingh EE (2007) The molecular evolution of methicillin-resistant *Staphylococcus aureus*. Clin Microbiol Infect 13(3):222–235. https://doi.org/10.1111/j.1469-0691.2006.01573.x

Dias C, Aires A, Saavedra MJ (2014) Antimicrobial activity of isothiocyanates from cruciferous plants against methicillin-resistant *Staphylococcus aureus* (MRSA). Int J Mol Sci 15(11):19552–19561. https://doi.org/10.3390/ijms151119552

Ding CF, Ma HX, Yang J, Qin XJ, Njateng GS, Yu HF, Wei X, Liu YP, Huang WY, Yang ZF, Wang XH (2018) Antibacterial indole alkaloids with complex heterocycles from *Voacanga africana*. Org Lett 20(9):2702–2706. https://doi.org/10.1021/acs.orglett.8b00913

Domenech P, Reed MB, Dowd CS, Manca C, Kaplan G, Barry CE (2004) The role of MmpL8 in sulfatide biogenesis and virulence of *Mycobacterium tuberculosis*. J Biol Chem 279(20):21257–21265. https://doi.org/10.1074/jbc.M400324200

Dos Santos LD, Furlan JP, Gallo IF, Ramos MS, Savazzi EA, Stehling EG (2021) Occurrence of multidrug-resistant *Enterococcus faecium* isolated from environmental samples. Lett Appl Microbiol 73(2):237–246

Edziri H, Mastouri M, Mahjoub MA, Mighri Z, Mahjoub A, Verschaeve L (2012) Antibacterial, antifungal and cytotoxic activities of two flavonoids from *Retama raetam* flowers. Molecules 17(6):7284–7293. https://doi.org/10.3390/molecules17067284

El Atki Y, Aouam I, El Kamari F, Taroq A, Gourch A, Lyoussi B, Abdellaoui A (2019) Antibacterial efficacy of thymol, carvacrol, eugenol and menthol as alternative agents to control the growth of nosocomial infection-bacteria. J Pharm Sci Res 11(2):306–309

Elmaidomy AH, Shady NH, Abdeljawad KM, Elzamkan MB, Helmy HH, Tarshan EA, Adly AN, Hussien YH, Sayed NG, Zayed A, Abdelmohsen UR (2022) Antimicrobial potentials of natural products against multidrug resistance pathogens: a comprehensive review. RSC Adv 12(45):29078–29102. https://doi.org/10.1039/D2RA04884A

Evangelista AG, Bocate KC, Meca G, Luciano FB (2021) Combination of allyl isothiocyanate and cinnamaldehyde against the growth of mycotoxigenic fungi and aflatoxin production in corn. J Food Process Preserv 45(9):e15760. https://doi.org/10.1111/jfpp.15760

Farag MA, Ali SE, Hodaya RH, El-Seedi HR, Sultani HN, Laub A, Eissa TF, Abou-Zaid FO, Wessjohann LA (2017) Phytochemical profiles and antimicrobial activities of *Allium cepa* red cv. and *A. sativum* subjected to different drying methods: a comparative MS-based metabolomics. Molecules 22(5):761. https://doi.org/10.3390/molecules22050761

Feneran AN, Kaufman WS, Dabade TS, Feldman SR (2011) Retinoid plus antimicrobial combination treatments for acne. Clin Cosmet Investig Dermatol 1:79–92. https://doi.org/10.2147/CCID.S13873

Fu Y, Liu W, Liu M, Zhang J, Yang M, Wang T, Qian W (2021) *In vitro* anti-biofilm efficacy of sanguinarine against carbapenem-resistant *Serratia marcescens*. Biofouling 37(3):341–351. https://doi.org/10.1080/08927014.2021.191964

Fujita M, Shiota S, Kuroda T, Hatano T, Yoshida T, Mizushima T, Tsuchiya T (2005) Remarkable synergies between baicalein and tetracycline, and baicalein and β-lactams against methicillin-resistant *Staphylococcus aureus*. Microbiol Immunol 49(4):391–396. https://doi.org/10.1111/j.1348-0421.2005.tb03732.x

Ganesan K, Xu B (2017) A critical review on polyphenols and health benefits of black soybeans. Nutrients 9(5):455. https://doi.org/10.3390/nu9050455

García-Salinas S, Elizondo-Castillo H, Arruebo M, Mendoza G, Irusta S (2018) Evaluation of the antimicrobial activity and cytotoxicity of different components of natural origin present in essential oils. Molecules 23(6):1399. https://doi.org/10.3390/molecules23061399

Ghavam M, Manca ML, Manconi M, Bacchetta G (2020) Chemical composition and antimicrobial activity of essential oils obtained from leaves and flowers of *salvia hydrangea* DC ex Benth. Sci Rep 10(1):15647. https://doi.org/10.1038/s41598-020-73193-y

Griffin SG, Wyllie SG, Markham JL, Leach DN (1999) The role of structure and molecular properties of terpenoids in determining their antimicrobial activity. Flavour Fragr J 14(5):322–332

Gupta VK, Tiwari N, Gupta P, Verma S, Pal A, Srivastava SK, Darokar MP (2016) A clerodane diterpene from *Polyalthia longifolia* as a modifying agent of the resistance of methicillin resistant *Staphylococcus aureus*. Phytomedicine 23(6):654–661. https://doi.org/10.1016/j.phymed.2016.03.001

Habib Adam M, Tandon N, Singh I, Tandon R (2023) The phytochemical tactics for battling antibiotic resistance in microbes: secondary metabolites and nano antibiotics methods. Chem Biodivers 20(9):e202300453. https://doi.org/10.1002/cbdv.202300453

Heeb S, Fletcher MP, Chhabra SR, Diggle SP, Williams P, Cámara M (2011) Quinolones: from antibiotics to autoinducers. FEMS Microbiol Rev 35:247–274. https://doi.org/10.1111/j.1574-6976.2010.00247.x

Hemavathy A, Shanthi P, Sowndharya C, Thiripura SS, Priyadharshni K (2019) Extraction and isolation of bioactive compounds from a therapeutic medicinal plant-*Wrightia tinctoria* (Roxb.) R. Br Int J Pharmacogn Phytochem Res 11(3):199–204

Høiby N, Ciofu O, Johansen HK, Song ZJ, Moser C, Jensen PØ, Molin S, Givskov M, Tolker-Nielsen T, Bjarnsholt T (2011) The clinical impact of bacterial biofilms. Int J Oral Sci 3(2):55–65. https://doi.org/10.4248/IJOS11026

Horiuchi K, Shiota S, Hatano T, Yoshida T, Kuroda T, Tsuchiya T (2007) Antimicrobial activity of oleanolic acid from *Salvia officinalis* and related compounds on vancomycin-resistant *enterococci* (VRE). Biol Pharm Bull 30(6):1147–1149. https://doi.org/10.1248/bpb.30.1147

Jain C, Khatana S, Vijayvergia R (2019) Bioactivity of secondary metabolites of various plants: a review. Int J Pharm Sci Res 10(2):494–504. https://doi.org/10.13040/ijpsr.0975-8232.10(2).494-04

Jesus JA, Lago JH, Laurenti MD, Yamamoto ES, Passero LF (2015) Antimicrobial activity of oleanolic and ursolic acids: an update. Evid Based Complement Alternat Med 2015(1):620472. https://doi.org/10.1155/2015/620472

Jia W, Wang J, Xu H, Li G (2015) Resistance of *Stenotrophomonas maltophilia* to fluoroquinolones: prevalence in a university hospital and possible mechanisms. Int J Environ Res Public Health 12(5):5177–5195. https://doi.org/10.3390/ijerph120505177

Jian Z, Zeng L, Xu T, Sun S, Yan S, Yang L, Huang Y, Jia J, Dou T (2021) Antibiotic resistance genes in bacteria: occurrence, spread, and control. J Basic Microbiol 61(12):1049–1070. https://doi.org/10.1002/jobm.202100201

Kaschula CH, Hunter R, Parker MI (2010) Garlic-derived anticancer agents: structure and biological activity of ajoene. Biofactors 36(1):78–85. https://doi.org/10.1002/biof.76

Kelley C, Lu S, Parhi A, Kaul M, Pilch DS, LaVoie EJ (2013) Antimicrobial activity of various 4-and 5-substituted 1-phenylnaphthalenes. Eur J Med Chem 60:395–409. https://doi.org/10.1016/j.ejmech.2012.12.027

Khameneh B, Iranshahy M, Ghandadi M, Ghoochi Atashbeyk D, Fazly Bazzaz BS, Iranshahi M (2015) Investigation of the antibacterial activity and efflux pump inhibitory effect of co-loaded piperine and gentamicin nanoliposomes in methicillin-resistant *Staphylococcus aureus*. Drug Dev Ind Pharm 41(6):989–994. https://doi.org/10.3109/03639045.2014.920025

Khameneh B, Iranshahy M, Soheili V, Fazly Bazzaz BS (2019) Review on plant antimicrobials: a mechanistic viewpoint. Antimicrob Resist Infect Control 8:1–28. https://doi.org/10.1186/s13756-019-0559-6

Khan IA, Mirza ZM, Kumar A, Verma V, Qazi GN (2006) Piperine, a phytochemical potentiator of ciprofloxacin against *Staphylococcus aureus*. Antimicrob Agents Chemother 50(2):810–812. https://doi.org/10.1128/aac.50.2.810-812.2006

Khare T, Anand U, Dey A, Assaraf YG, Chen Z-S, Liu Z, Kumar V (2021) Exploring phytochemicals for combating antibiotic resistance in microbial pathogens. Front Pharmacol 12:720726. https://doi.org/10.3389/fphar.2021.720726

Kim SG, Kim MJ, Jin DC, Park SN, Cho EG, Freire MO, Jang SJ, Park YJ, Kook JK (2012) Antimicrobial effect of ursolic acid and oleanolic acid against methicillin-resistant *Staphylococcus aureus*. Korean J Microbiol 48(3):212–215

Kim SY, Park C, Jang HJ, Kim BO, Bae HW, Chung IY, Kim ES, Cho YH (2019) Antibacterial strategies inspired by the oxidative stress and response networks. J Microbiol 57:203–212. https://doi.org/10.1007/s12275-019-8711-9

Kim G, Gan RY, Zhang D, Farha AK, Habimana O, Mavumengwana V, Li HB, Wang XH, Corke H (2020) Large-scale screening of 239 traditional Chinese medicinal plant extracts for their antibacterial activities against multidrug-resistant *Staphylococcus aureus* and cytotoxic activities. Pathogens 9(3):185. https://doi.org/10.3390/pathogens9030185

Koudokpon H, Armstrong N, Dougnon TV, Fah L, Hounsa E, Bankolé HS, Loko F, Chabrière E, Rolain JM (2018) Antibacterial activity of Chalcone and Dihydrochalcone compounds from Uvaria chamae roots against multidrug-resistant bacteria. Biomed Res Int 2018(1):1453173. https://doi.org/10.1155/2018/1453173

Kuete V, Wansi JD, Mbaveng AT, Sop MK, Tadjong AT, Beng VP, Etoa FX, Wandji J, Meyer JM, Lall N (2008) Antimicrobial activity of the methanolic extract and compounds from *Teclea afzelii* (Rutaceae). S Afr J Bot 74(4):572–576. https://doi.org/10.1016/j.sajb.2008.02.004

Kurek A, Markowska K, Grudniak AM, Janiszowska W, Wolska KI (2014) The effect of oleanolic and ursolic acids on the hemolytic properties and biofilm formation of *Listeria monocytogenes*. Pol J Microbiol 63(1):21–25

Lanzotti V, Scala F, Bonanomi G (2014) Compounds from allium species with cytotoxic and antimicrobial activity. Phytochem Rev 13:769–791. https://doi.org/10.1007/s11101-014-9366-0

Larghi LE et al (2015) Neocryptolepine: a promising indoloisoquinoline alkaloid with interesting biological activity. Evaluation of the drug and its most relevant analogs. Curr Top Med Chem 15(17):1683–1707

Lee GS, Kim ES, Cho SI, Kim JH, Choi G, Ju YS, Park SH, Jeong SI, Kim HJ (2010) Antibacterial and synergistic activity of prenylated chalcone isolated from the roots of *Sophora flavescens*. J Korean Soc Appl Biol Chem 53:290–296. https://doi.org/10.3839/jksabc.2010.045

Lee YS, Jung EK, Cha JD (2015) Synergistic effect between baicalein and antibiotics against clinic methicillin and vancomycin-resistant *Staphylococcus aureus*. Chemo Open Access 4(141):2

Liggins MC, Li F, Zhang LJ, Dokoshi T, Gallo RL (2019) Retinoids enhance the expression of cathelicidin antimicrobial peptide during reactive dermal adipogenesis. J Immunol 203(6):1589–1597. https://doi.org/10.4049/jimmunol.1900520

Lin CM, Preston JF III, Wei CI (2000) Antibacterial mechanism of allyl isothiocyanate. J Food Prot 63(6):727–734. https://doi.org/10.4315/0362-028X-63.6.727

Lin D, Xiao M, Zhao J, Li Z, Xing B, Li X, Kong M, Li L, Zhang Q, Liu Y, Chen H (2016) An overview of plant phenolic compounds and their importance in human nutrition and management of type 2 diabetes. Molecules 21(10):1374. https://doi.org/10.3390/molecules21101374

Liu L, Song CW, Khan A, Li XN, Yang XW, Cheng GG, Liu YP, Luo XD (2016) A potent antibacterial indole alkaloid from *Psychotria pilifera*. J Asian Nat Prod Res 18(8):798–803. https://doi.org/10.1080/10286020.2016.1158710

Liu Y, Cui Y, Lu L, Gong Y, Han W, Piao G (2020) Natural indole-containing alkaloids and their antibacterial activities. Arch Pharm 353(10):2000120. https://doi.org/10.1002/ardp.202000120

Lomovskaya OL, Lewis KI (1992) Emr, an *Escherichia coli* locus for multidrug resistance. Proc Natl Acad Sci 89(19):8938–8942. https://doi.org/10.1073/pnas.89.19.8938

Lorenzi V, Muselli A, Bernardini AF, Berti L, Pages JM, Amaral L, Bolla JM (2009) Geraniol restores antibiotic activities against multidrug-resistant isolates from gram-negative species. Antimicrob Agents Chemother 53(5):2209–2211. https://doi.org/10.1128/aac.00919-08

Luciano FB, Holley RA (2009) Enzymatic inhibition by allyl isothiocyanate and factors affecting its antimicrobial action against *Escherichia coli* O157: H7. Int J Food Microbiol 131(2–3):240–245. https://doi.org/10.1016/j.ijfoodmicro.2009.03.005

Mahizan NA, Yang SK, Moo CL, Song AA, Chong CM, Chong CW, Abushelaibi A, Lim SE, Lai KS (2019) Terpene derivatives as a potential agent against antimicrobial resistance (AMR) pathogens. Molecules 24(14):2631. https://doi.org/10.3390/molecules24142631

Mandal SM, Roy A, Ghosh AK, Hazra TK, Basak A, Franco OL (2014) Challenges and future prospects of antibiotic therapy: from peptides to phages utilization. Front Pharmacol 5:105. https://doi.org/10.3389/fphar.2014.00105

Manzo LM, Moussa I, Ikhiri K, Yu L (2019) Toxicity studies of *Acacia nilotica* (L.): a review of the published scientific literature. J Herbmed Pharmacol 8(3):163–172. https://doi.org/10.15171/jhp.2019.26

Marchese A, Arciola CR, Barbieri R, Silva AS, Nabavi SF, Tsetegho Sokeng AJ, Izadi M, Jafari NJ, Suntar I, Daglia M, Nabavi SM (2017) Update on monoterpenes as antimicrobial agents: a particular focus on p-cymene. Materials 10(8):947. https://doi.org/10.3390/ma10080947

Markham PN, Westhaus E, Klyachko K, Johnson ME, Neyfakh AA (1999) Multiple novel inhibitors of the NorA multidrug transporter of *Staphylococcus aureus*. Antimicrob Agents Chemother 43:2404–2408. https://doi.org/10.1128/aac.43.10.2404

Martelli G, Giacomini D (2018) Antibacterial and antioxidant activities for natural and synthetic dual-active compounds. Eur J Med Chem 158:91–105. https://doi.org/10.1016/j.ejmech.2018.09.009

Matias EF, Alves EF, Silva MK, Carvalho VR, Figueredo FG, Ferreira JV, Coutinho HD, Silva JM, Ribeiro-Filho J, Costa JG (2016) Seasonal variation, chemical composition and biological activity of the essential oil of *Cordia verbenacea* DC (Boraginaceae) and the sabinene. Ind Crop Prod 87:45–53. https://doi.org/10.1016/j.indcrop.2016.04.028

Maurya A, Dwivedi GR, Darokar MP, Srivastava SK (2013) Antibacterial and synergy of clavine alkaloid lysergol and its derivatives against nalidixic acid-resistant *E scherichia coli*. Chem Biol Drug Des 81(4):484–490. https://doi.org/10.1111/cbdd.12103

Maxwell A (1993) The interaction between coumarin drugs and DNA gyrase. Mol Microbiol 9(4):681–686. https://doi.org/10.1111/j.1365-2958.1993.tb01728.x

Mazurkiewicz P, Driessen AJ, Konings WN (2004) Energetics of wild-type and mutant multidrug resistance secondary transporter LmrP of *Lactococcus lactis*. Biochim Biophys Acta Bioenerg 1658(3):252–261. https://doi.org/10.1016/j.bbabio.2004.06.004

Mohtar M, Johari SA, Li AR, Isa MM, Mustafa S, Ali AM, Basri DF (2009) Inhibitory and resistance-modifying potential of plant-based alkaloids against methicillin-resistant *Staphylococcus aureus* (MRSA). Curr Microbiol 59:181–186. https://doi.org/10.1007/s00284-009-9416-9

Morita Y, Kataoka A, Shiota S, Mizushima T, Tsuchiya T (2000) NorM of *Vibrio parahaemolyticus* is an Na^+-driven multidrug efflux pump. J Bacteriol 182(23):6694–6697. https://doi.org/10.1128/jb.182.23.6694-6697.2000

Mun SH, Joung DK, Kim SB, Park SJ, Seo YS, Gong R, Choi JG, Shin DW, Rho JR, Kang OH, Kwon DY (2014) The mechanism of antimicrobial activity of sophoraflavanone B against methicillin-resistant *Staphylococcus aureus*. Foodborne Pathog Dis 11(3):234–239. https://doi.org/10.1089/fpd.2013.1627

Myint KB, Sing LC, Wei Z (2013) Tannic acid as phytochemical potentiator for antibiotic resistance adaptation. APCBEE procedia 7:175–181. https://doi.org/10.1016/j.apcbee.2013.08.030

Nakamoto M, Kunimura K, Suzuki JI, Kodera Y (2020) Antimicrobial properties of hydrophobic compounds in garlic: Allicin, vinyldithiin, ajoene and diallyl polysulfides. Exp Ther Med 19(2):1550–1553. https://doi.org/10.3892/etm.2019.8388

Nord C, Levenfors JJ, Bjerketorp J, Sahlberg C, Guss B, Öberg B, Broberg A (2019) Antibacterial isoquinoline alkaloids from the fungus *Penicillium spathulatum* Em19. Molecules 24(24):4616. https://doi.org/10.3390/molecules24244616

Omar F, Tareq AM, Alqahtani AM, Dhama K, Sayeed MA, Emran TB, Simal-Gandara J (2021) Plant-based indole alkaloids: a comprehensive overview from a pharmacological perspective. Molecules 26(8):2297. https://doi.org/10.3390/molecules26082297

Packiavathy IA, Priya S, Pandian SK, Ravi AV (2014) Inhibition of biofilm development of uropathogens by curcumin–an anti-quorum sensing agent from Curcuma longa. Food Chem 148:453–460. https://doi.org/10.1016/j.foodchem.2012.08.00

Pagare S, Bhatia M, Tripathi N, Pagare S, Bansal YK (2015) Secondary metabolites of plants and their role: overview. Curr Trends Biotechnol Pharma 9(3):293–304

Patel P, Prasad A, Srivastava K, Singh SS, Chakrabarty D, Misra P (2021) Updates on steroidal alkaloids and glycoalkaloids in *Solanum* spp.: biosynthesis, *in vitro* production and pharmacological values. Stud Nat Prod Chem 69:99–127. https://doi.org/10.1016/B978-0-12-819487-4.00012-4

Paulsen IT, Skurray RA, Tam R, Saier MH Jr, Turner RJ, Weiner JH, Goldberg EB, Grinius LL (1996) The SMR family: a novel family of multidrug efflux proteins involved with the efflux of lipophilic drugs. Mol Microbiol 19(6):1167–1175. https://doi.org/10.1111/j.1365-2958.1996.tb02462.x

Pisano MB, Kumar A, Medda R, Gatto G, Pal R, Fais A, Era B, Cosentino S, Uriarte E, Santana L, Pintus F (2019) Antibacterial activity and molecular docking studies of a selected series of hydroxy-3-arylcoumarins. Molecules 24(15):2815. https://doi.org/10.3390/molecules24152815

Pott DM, Osorio S, Vallarino JG (2019) From central to specialized metabolism: an overview of some secondary compounds derived from the primary metabolism for their role in conferring nutritional and organoleptic characteristics to fruit. Front Plant Sci 28(10):835. https://doi.org/10.3389/fpls.2019.00835

Prakash O, Singh R, Singh N, Verma N, Mahapatra DK, Kumar S, Ved A (2018) Exploring the potentials of quercetin and kaempferol combinations along with regular antibiotics for the effective management of methicillin-resistant *Staphylococcus aureus* (MRSA). Res Rev J Microbiol Virol 8(3):6–9

Qu S, Dai C, Shen Z, Tang Q, Wang H, Zhai B, Zhao L, Hao Z (2019) Mechanism of synergy between tetracycline and quercetin against antibiotic resistant *Escherichia coli*. Front Microbiol 10:2536. https://doi.org/10.3389/fmicb.2019.02536

Qu D, Hou Z, Li J, Luo L, Su S, Ye Z, Bai Y, Zhang X, Chen G, Li Z, Wang Y (2020) A new coumarin compound DCH combats methicillin-resistant *Staphylococcus aureus* biofilm by targeting arginine repressor. Sci Adv 6(30):eaay9597. https://doi.org/10.1126/sciadv.aay959

Ramos JL, Duque E, Gallegos MT, Godoy P, Ramos-Gonzalez MI, Rojas A, Terán W, Segura A (2002) Mechanisms of solvent tolerance in gram-negative bacteria. Ann Rev Microbiol 56(1):743–768. https://doi.org/10.1146/annurev.micro.56.012302.161038

Reiter J, Hübbers AM, Albrecht F, Leichert LI, Slusarenko AJ (2020) Allicin, a natural antimicrobial defence substance from garlic, inhibits DNA gyrase activity in bacteria. Int J Med Microbiol 310(1):151359. https://doi.org/10.1016/j.ijmm.2019.151359

Ribeiro J, Silva V, Aires A, Carvalho R, Igrejas G, Poeta P (2021) Antimicrobial activity of phenolic compounds extracted from Platanus hybrida: exploring alternative therapies for a post-antibiotic era. PRO 66(1):18

Robinson TP, Bu DP, Carrique-Mas J, Fèvre EM, Gilbert M, Grace D, Hay SI, Jiwakanon J, Kakkar M, Kariuki S, Laxminarayan R (2016) Antibiotic resistance is the quintessential one health issue. Trans R Soc Trop Med Hyg 110(7):377–380. https://doi.org/10.1093/trstmh/trw048

Romeo L, Iori R, Rollin P, Bramanti P, Mazzon E (2018) Isothiocyanates: an overview of their antimicrobial activity against human infections. Molecules 23(3):624. https://doi.org/10.3390/molecules23030624

Roy SK, Kumari N, Pahwa S, Agrahari UC, Bhutani KK, Jachak SM, Nandanwar H (2013) NorA efflux pump inhibitory activity of coumarins from *Mesua ferrea*. Fitoterapia 90:140–150. https://doi.org/10.1016/j.fitote.2013.07.015

Różalski M, Walencka E, Różalska B, Wysokińska H (2007) Antimicrobial activity of diterpenoids from hairy roots of *Salvia sclarea* L.: Salvipisone as a potential anti-biofilm agent active against antibiotic resistant *Staphylococci*. Phytomedicine 14(1):31–35. https://doi.org/10.1016/j.phymed.2005.10.008

Sahoo J, Paidesetty SK (2017) Antimicrobial activity of novel synthesized coumarin based transitional metal complexes. J Taibah Univ Med Sci 12(2):115–124. https://doi.org/10.1016/j.jtumed.2016.10.004

Santiago C, Pang EL, Lim KH, Loh HS, Ting KN (2015) Inhibition of penicillin-binding protein 2a (PBP2a) in methicillin resistant *Staphylococcus aureus* (MRSA) by combination of ampicillin and a bioactive fraction from Duabanga grandiflora. BMC Complement Altern Med 15:1–7. https://doi.org/10.1186/s12906-015-0699-z

Schmitz R (1985) Friedrich Wilhelm Sertürner and the discovery of morphine. Pharm Hist 27(2):61–74

Serrano J, Puupponen-Pimiä R, Dauer A, Aura AM, Saura-Calixto F (2009) Tannins: current knowledge of food sources, intake, bioavailability and biological effects. Mol Nutr Food Res 53(S2):S310–S329. https://doi.org/10.1002/mnfr.200900039

Shao J, Zhang M, Wang T, Li Y, Wang C (2016) The roles of CDR1, CDR2, and MDR1 in kaempferol-induced suppression with fluconazole-resistant *Candida albicans*. Pharm Biol 54(6):984–992. https://doi.org/10.3109/13880209.2015.1091483

Si H, Hu J, Liu Z, Zeng ZL (2008) Antibacterial effect of oregano essential oil alone and in combination with antibiotics against extended-spectrum β-lactamase-producing

Escherichia coli. FEMS Immunology & Medical Microbiology 53(2):190–194. https://doi.org/10.1111/j.1574-695X.2008.00414.x

Sinsinwar S, Vadivel V (2020) Catechin isolated from cashew nut shell exhibits antibacterial activity against clinical isolates of MRSA through ROS-mediated oxidative stress. Appl Microbiol Biotechnol 104:8279–8297. https://doi.org/10.1007/s00253-020-10853-z

Skalicka-Woźniak K, Walasek M, Aljarba TM, Stapleton P, Gibbons S, Xiao J, Łuszczki JJ (2018) The anticonvulsant and anti-plasmid conjugation potential of *Thymus vulgaris* chemistry: an in vivo murine and in vitro study. Food Chem Toxicol 120:472–478. https://doi.org/10.1016/j.fct.2018.07.045

Smith JR, Roberts KD, Rybak MJ (2015) Dalbavancin: a novel lipoglycopeptide antibiotic with extended activity against gram-positive infections. Infect Dis Ther 4:245–258. https://doi.org/10.1007/s40121-015-0077-7

Sofrata A, Santangelo EM, Azeem M, Borg-Karlson AK, Gustafsson A, Pütsep K (2011) Benzyl isothiocyanate, a major component from the roots of *Salvadora persica* is highly active against gram-negative bacteria. PLoS One 6(8):e23045. https://doi.org/10.1371/journal.pone.0023045

Soltani R, Fazeli H, Najafi RB, Jelokhanian A (2017) Evaluation of the synergistic effect of tomatidine with several antibiotics against standard and clinical isolates of *Staphylococcus aureus*, *Enterococcus faecalis*, *Pseudomonas aeruginosa* and *Escherichia coli*. Iran J Pharm Res 16(1):290

Spratt BG (1994) Resistance to antibiotics mediated by target alterations. Science 264(5157):388–393. https://doi.org/10.1126/science.815362

Suganya T, Packiavathy IA, Aseervatham GS, Carmona A, Rashmi V, Mariappan S, Devi NR, Ananth DA (2022) Tackling multiple-drug-resistant bacteria with conventional and complex phytochemicals. Front Cell Infect Microbiol 12:883839

Terreni M, Taccani M, Pregnolato M (2021) New antibiotics for multidrug-resistant bacterial strains: latest research developments and future perspectives. Molecules 26:2671. https://doi.org/10.3390/molecules26092671

Thawabteh A, Juma S, Bader M, Karaman D, Scrano L, Bufo SA, Karaman R (2019) The biological activity of natural alkaloids against herbivores, cancerous cells and pathogens. Toxins. 11(11):656. https://doi.org/10.3390/toxins11110656

Tintino SR, Morais-Tintino CD, Campina FF, Costa MD, Menezes IR, de Matos YM, Calixto-Júnior JT, Pereira PS, Siqueira-Junior JP, Leal-Balbino TC, Coutinho HD (2017) Tannic acid affects the phenotype of *Staphylococcus aureus* resistant to tetracycline and erythromycin by inhibition of efflux pumps. Bioorg Chem 74:197–200. https://doi.org/10.1016/j.bioorg.2017.08.004

Touani FK, Seukep AJ, Djeussi DE, Fankam AG, Noumedem JA, Kuete V (2014) Antibiotic-potentiation activities of four Cameroonian dietary plants against multidrug-resistant gram-negative bacteria expressing efflux pumps. BMC Complement Altern Med 14:1–8. https://doi.org/10.1186/1472-6882-14-258

Tseng TT, Gratwick KS, Kollman J, Park D, Nies DH, Goffeau A, Saier MH Jr (1999) The RND permease superfamily: an ancient, ubiquitous and diverse family that includes human disease and development proteins. J Mol Microbiol Biotechnol 1(1):107–125

Venugopala KN, Rashmi V, Odhav B (2013) Review on natural coumarin lead compounds for their pharmacological activity. Biomed Res Int 2013(1):963248. https://doi.org/10.1155/2013/963248

Vestergaard M, Ingmer H (2019) Antibacterial and antifungal properties of resveratrol. Int J Antimicrob Agents 53(6):716–723. https://doi.org/10.1016/j.ijantimicag.2019.02.015

Vipin C, Saptami K, Fida F, Mujeeburahiman M, Rao SS, Athmika AAB, Rekha PD (2020) Potential synergistic activity of quercetin with antibiotics against multidrug-resistant clinical strains of *Pseudomonas aeruginosa*. PLoS One 15(11):e0241304. https://doi.org/10.1371/journal.pone.0241304

Vivas R, Barbosa AA, Dolabela SS, Jain S (2019) Multidrug-resistant bacteria and alternative methods to control them: an overview. Microb Drug Resist 25(6):890–908. https://doi.org/10.1089/mdr.2018.031

Walencka E, Rozalska S, Wysokinska H, Rozalski M, Kuzma L, Rozalska B (2007) Salvipisone and aethiopinone from *Salvia sclarea* hairy roots modulate staphylococcal antibiotic resistance and express anti-biofilm activity. Planta Med 73(06):545–551. https://doi.org/10.1055/s-2007-967179

Wang M, Ma B, Ni Y, Xue X, Li M, Meng J, Luo X, Fang C, Hou Z (2021) Restoration of the antibiotic susceptibility of methicillin-resistant *Staphylococcus aureus* and extended-spectrum β-lactamases *Escherichia coli* through combination with Chelerythrine. Microb Drug Resist 27(3):337–341. https://doi.org/10.1089/mdr.2020.0044

Wangchuk P (2019) Plant alkaloids: classification, isolation, and drug development. In: Medicinal plants. CRC Press, pp 131–138

Yan Y, Li X, Zhang C, Lv L, Gao B, Li M (2021) Research progress on antibacterial activities and mechanisms of natural alkaloids: a review. Antibiotics 10(3):318. https://doi.org/10.3390/antibiotics10030318

Yu Z, Liu X, Chen H, Zhu L (2020a) Naringenin-loaded dipalmitoylphosphatidylcholine phytosome dry powders for inhaled treatment of acute lung injury. J Aerosol Med Pulm Drug Deliv 33(4):194–204. https://doi.org/10.1089/jamp.2019.156

Yu Z, Tang J, Khare T, Kumar V (2020b) The alarming antimicrobial resistance in ESKAPEE pathogens: can essential oils come to the rescue? Fitoterapia 140:104433. https://doi.org/10.1016/j.fitote.2019.104433

Zengin H, Baysal AH (2014) Antibacterial and antioxidant activity of essential oil terpenes against pathogenic and spoilage-forming bacteria and cell structure-activity relationships evaluated by SEM microscopy. Molecules 19(11):17773–17798. https://doi.org/10.3390/molecules191117773

Zhou T, Li Z, Kang O-H et al (2017) Antimicrobial activity and synergism of ursolic acid 3-O-α-L-arabinopyranoside with oxacillin against *methicillin-resistant Staphylococcus aureus*. Int J Mol Med 40(4):1285–1293. https://doi.org/10.3892/ijmm

Zielińska S, Wójciak-Kosior M, Dziągwa-Becker M, Gleńsk M, Sowa I, Fijałkowski K, Rurańska-Smutnicka D, Matkowski A, Junka A (2019) The activity of isoquinoline alkaloids and extracts from *Chelidonium majus* against pathogenic bacteria and *Candida* sp. Toxins 11(7):406. https://doi.org/10.3390/toxins11070406

Natural Product-Based Treatment of Urinary Tract Infections

3

Chita Ranjan Sahoo, Debdutta Bhattacharya, and Sanghamitra Pati

Abstract

Antibiotic use has resulted in the emergence of multidrug-resistant, rendering even the most effective medications ineffective. Urinary tract infections (UTIs) are the second most prevalent infectious ailment, impacting about 150 million individuals globally annually. These represent a substantial clinical and economic burden linked to a decline in patient quality of life. UTIs are a public health risk in both community and hospital environments. The development of innovative treatments to treat drug-resistant infections is urgently needed. Alternative therapies include using antibiotics alone or in conjunction with phytocompound, or lead phytocompound, which is the call of the day. The search for novel phytocompounds as potential antibacterial candidates will be inspired by this updated compilation.

Keywords

Natural products · Phytocompounds · UTI · *E. coli*

Abbreviations

AMR	Antimicrobial resistance
ARSE	Amoxicillin-resistant *Staphylococcus epidermidis*
EP	Efflux pumps

C. R. Sahoo (✉) · D. Bhattacharya · S. Pati
Department of Health Research, Ministry of Health and Family Welfare, Government of India, ICMR-Regional Medical Research Centre, Bhubaneswar, Odisha, India
e-mail: drchita.icmr@gmail.com

T. Sharma et al. (eds.), *Natural Products for Antibacterial Drug Development: Recent Advancement of Computational Approach*,
https://doi.org/10.1007/978-981-97-9634-2_3

EPIs	Extracellular polysaccharide inhibitors
ESKAPE	pathogens—*Enterococcus faecium*, *Staphylococcus aureus*, *Klebsiella pneumoniae*, *Acinetobacter baumannii*, *Pseudomonas aeruginosa* and *Enterobacter* spp.
MDR	Multidrug-resistant
MRSA	Methicillin-resistant *Staphylococcus aureus*
TMP-SMX	Trimethoprim-sulfamethoxazole
UTIs	Urinary tract infections

3.1 Introduction

Currently, the most significant obstacles to public health are microbiological diseases that have developed resistance to antibiotic medications. Microbial infections cause millions of deaths annually on a global scale. A total of 9.2 million deaths have been attributed to infections, accounting for approximately 17% of all reported deaths. The emergence of resistance has rendered the current antibacterial medications less efficacious or completely ineffectual. Antimicrobial resistance (AMR) refers to the ability of microorganisms, including bacteria, viruses, fungi, and parasites, to withstand the impact of medications that were previously capable of effectively combating them. This happens when these microorganisms mutate or evolve mechanisms that make them resistant to drugs. It is a growing problem that makes it difficult to treat infections, increases healthcare costs, and poses a significant threat to public health. AMR can occur naturally or through the misuse or overuse of antimicrobial medicines. To combat AMR, it is essential to promote the responsible use of antibiotics and other antimicrobial medications, enhance infection prevention and control practices, and invest in research to develop new antimicrobial drugs (WHO 2014; Baym et al. 2016). Antibiotics are typically used to treat bacterial infectious infections, but their overuse and misuse have contributed to an alarming rise in antibiotic resistance globally. Therefore, there are increasingly fewer therapeutic choices available due to a dearth of novel antibiotic research (Buchmann et al. 2022). The spread of microorganisms that are multidrug-resistant (MDR) nowadays is a significant global public health issue that motivates the quest for fresh approaches to deal with this complex phenomenon.

Antimicrobial resistance (AMR) is a major concern to global health. *Escherichia coli* is the most frequent pathogen responsible for UTI in adults. The majority of UTI patients receive empirical treatment with common antibiotics. Since these organisms are resistant to all penicillins, cephalosporins, and aztreonam, treating ESBL-UTI is challenging. Additionally, it has been found that these organisms exhibit high levels of resistance to fluoroquinolones and trimethoprim-sulfamethoxazole (TMP-SMX) (Naushad et al. 2022).

Urinary tract infections (UTIs) are among the foremost common sorts of irresistible illness, with around 150–250 million cases all-inclusive per year. 4–6 Around 40–50% of ladies and 5% of men will create a UTI at slightest once during their lifetime. Within the USA alone, the ascribed fetched of UTI related therapeutic costs has been assessed to extend from US$1.6 billion to $3.5 billion every year in case broader societal costs are too considered (Zowawi et al. 2015). Antimicrobial utilisation is the single most vital modifiable chance factor for antimicrobial resistance. The gram-negative bacillus is responsible for the majority of UTIs namely *Escherichia coli* along with some other notorious pathogens *Enterococci*, *Klebsiella*, and *Proteus mirabilis*. One of the most common reasons for using antibiotics in ambulatory care is UTI. Alternative medications for treating uncomplicated UTIs include fluoroquinolones like levofloxacin and ciprofloxacin (Sanchez et al. 2016).

Natural products refer to compounds or substances derived from natural sources such as plants, animals, fungi, or microorganisms (Bhatia et al. 2021). These products can have various applications, including medicinal, cosmetic, and dietary uses. Natural products often contain bioactive compounds that have beneficial effects on human health. Indeed, a large variety of chemicals known as secondary compounds are known to be produced by plants. For their fundamental medical requirements, 80% of the population in poor countries relies on readily available traditional medicines. Metabolites or phytochemicals including quinones, tannins, terpenoids, alkaloids, flavonoids, and polyphenols that help them in their self-defence and interaction with other species in their environment as well as have the ability to prevent disease (Chadha et al. 2022). Since the beginning of time, medicinal plants, herbs, and spices have been utilised to heal infectious diseases and preserve food. Alkaloids, steroids, tannins, glycosides, volatile oils, fixed oils, resins, phenols, terpenoids, and flavonoids are just a few examples of bioactive substances found in plant extracts that have been shown to be effective in diseases.

The phenolic phytochemicals from plants play a key role as antimicrobial agents. Antimicrobial agents decay the protein components of the cell wall, disrupting the work of enzymes and DNA and RNA replication.

Phytocompounds provide a natural alternative to synthetic compounds and pharmaceuticals. Because they are readily available, largely non-toxic, biocompatible, easily degradable, and environmentally friendly, phytocompounds are preferred. For the purpose of finding new drugs, natural substances have been widely researched. Natural substances from pre-biotic, microbial, plant, and animal sources have always fascinated mankind.

It's important to note that while phytocompounds hold great potential for health benefits, individual responses can vary, and more research is needed to fully understand their mechanisms and effects.

3.2 Role of Phytochemicals in UTI

Phytochemicals, also known as phytonutrients, are natural compounds found in plants. These compounds are not essential for our basic nutritional needs but have been found to have beneficial effects on human health. Phytochemicals act as antioxidants and anti-inflammatory agents and may also have anticancer properties.

The rise of antimicrobial resistance has prompted the research of plant products for the discovery of novel antibiotics. The effectiveness of phytochemicals is mostly due to the chemical variety and structure. Since the 1990s, using antimicrobials excessively and irresponsibly has resulted in the worldwide dissemination of bacterial resistance mechanisms, making phytochemicals increasingly fascinating.

Plant compounds are classified into two types namely phytoanticipins and phytoalexins (Singh et al. 2018). Phytoanticipins and phytoalexins are both types of secondary metabolites produced by plants in response to various biotic and abiotic stresses. Phytoanticipins are constitutive chemical compounds present in plants even before any stress or pathogen attack. They act as preformed defence compounds and play a role in deterring and inhibiting potential pathogens. Phytoanticipins can be found in various plant tissues, such as leaves, stems, and roots. Examples of phytoanticipins include alkaloids, terpenoids, and phenolic compounds. On the other hand, phytoalexins are induced chemical compounds that are rapidly synthesised and accumulated in plants in response to stress or pathogen attack. They act as defence compounds to inhibit the growth and spread of pathogens. Phytoalexins are typically produced at the site of infection or injury. Examples of phytoalexins include flavonoids, isoflavonoids, lignans, and stilbenes. Both phytoanticipins and phytoalexins contribute to the defence mechanisms of plants against pathogens and other stresses (Chowdhary and Tank 2023).

They have antimicrobial and antifungal properties, and their production can be influenced by factors such as plant genetics, environmental conditions, and the specific pathogen or stress encountered. In various plant organs, the same molecule can act as a phytoalexin or a phytoanticipin of same plant. Terprenoids, quinones, and tannins are examples of phytoanticipins.

Despite having antibacterial capabilities, several phytochemical classes are not considered as drugs by the medical community. Indeed, some phytochemicals have shown narrow spectrums of antimicrobial actions (Matos et al. 2015) necessary concentrations are too high to have a clinically relevant effect. Variability in the extraction techniques and antibacterial tests utilised is the main issue for the identification of antimicrobial compounds from plants. Furthermore, antimicrobial properties of phytochemicals can also have antiviral (Ghildiyal et al. 2020), antiparasitic (Batiha et al. 2023; Khameneh et al. 2021), and antifungal (Zaidi and Dahiya 2015) activities as well as cytotoxic activity against tumour cells (Al Bratty et al. 2020).

Terpenoids, phenolics, alkaloids, components of essential oils, lectins polypeptides, and polyacetylenes are only a few of the classes of secondary metabolites that plants provide.

The phytochemical products are mostly secondary metabolites of plants that deposit in certain parts of the plant.

One of the most basic classes of bioactive phytochemicals, phenols, and polyphenols are composed of a single substituted phenolic ring (Ogidi et al. 2021). Due to the site(s) and amount of hydroxyl groups on the phenolic ring, this group is likely to be hazardous to microorganisms (Gulcin 2020). The toxicity is increased by increase of the hydroxylation. Quinones are made up of an aromatic ring with two replacements of ketones (Ogidi et al. 2021). These compounds cause the inactivation or loss of function of proteins by interacting with nucleophilic amino acids in proteins (Schefer et al. 2021). A class of compounds known as flavones, flavonoids, and flavonols have phenolic structures and one carbonyl group. Because they can reduce the extracellular polysaccharide activity and acting as extracellular polysaccharide inhibitors (EPIs) (Khameneh et al. 2021). Tannins are another class that can be split into two categories: condensed and hydrolysable. The first group is built around gallic acid, whereas the second is made up of a variety of condensed tannins made from flavonoid monomers (Shnawa et al. 2020). The antimicrobial mode of action is associated with their capacity to render microbial adhesin, enzymes, and cell envelope transport proteins inactive. Terpenoids and essential oils are chemical substances with an isoprene structure. When they contain additional elements, such as oxygen, they are referred to as terpenoids and appear as diterpenes, triterpenes, and tetraterpenes (C20, C30, and C40). These substances have antibacterial, antifungal, antiviral, and antiprotozoal properties (Raut and Karuppayil 2014). Figure 3.1 depicts the leading phytocompounds derived and their inhibitory antibacterial actions. Table 3.1 lists phytocompound inhibitory actions for ESKAPE (*Enterococcus faecium*, *Staphylococcus aureus*, *Klebsiella pneumoniae*, *Acinetobacter baumannii*, *Pseudomonas aeruginosa*, and *Enterobacter* spp.) pathogens.

3.3 Mechanisms and Action in UTI

In addition to being acquired by chromosomal gene mutations and horizontal gene transfer, antibiotic resistance can also be intrinsic (Bello-López et al. 2019). Bacterial intrinsic resistance to a specific antibiotic refers to a species' capacity to withstand the effects of that antimicrobial drug due to intrinsic structural and features that serve a purpose (Abebe et al. 2016). Through spontaneous mutation or the appropriation of genes from other bacteria, microorganisms can also develop or acquire resistance to antibiotics. The reception of free DNA produced by bacterial lysis can be used in transformation, which involves the transfer of plasmids or transposon, as well as transduction (mediated by bacteriophages), conjugation, and direct cell-to-cell contact.

Susceptible bacteria can develop resistance to antibiotics by a variety of intricate methods, such as blocking access to the target by decreasing permeability, increasing the expression of efflux pumps (EPs), degrading or altering the antibiotics, and changing the molecular target (Fig. 3.2).

3.3.1 Prevention of Access to Target by Reducing Permeability

The cell envelop that surrounds microorganisms acts as a selective permeability barrier, successfully shielding them from drug compounds in the extracellular environment while yet supplying enough nutrients to the cell (Thorn et al. 2021). The majority of antibiotics must be able to penetrate the bacterial membrane and target intracellular activities (Kamaruzzaman et al. 2017). Reducing the quantity of the

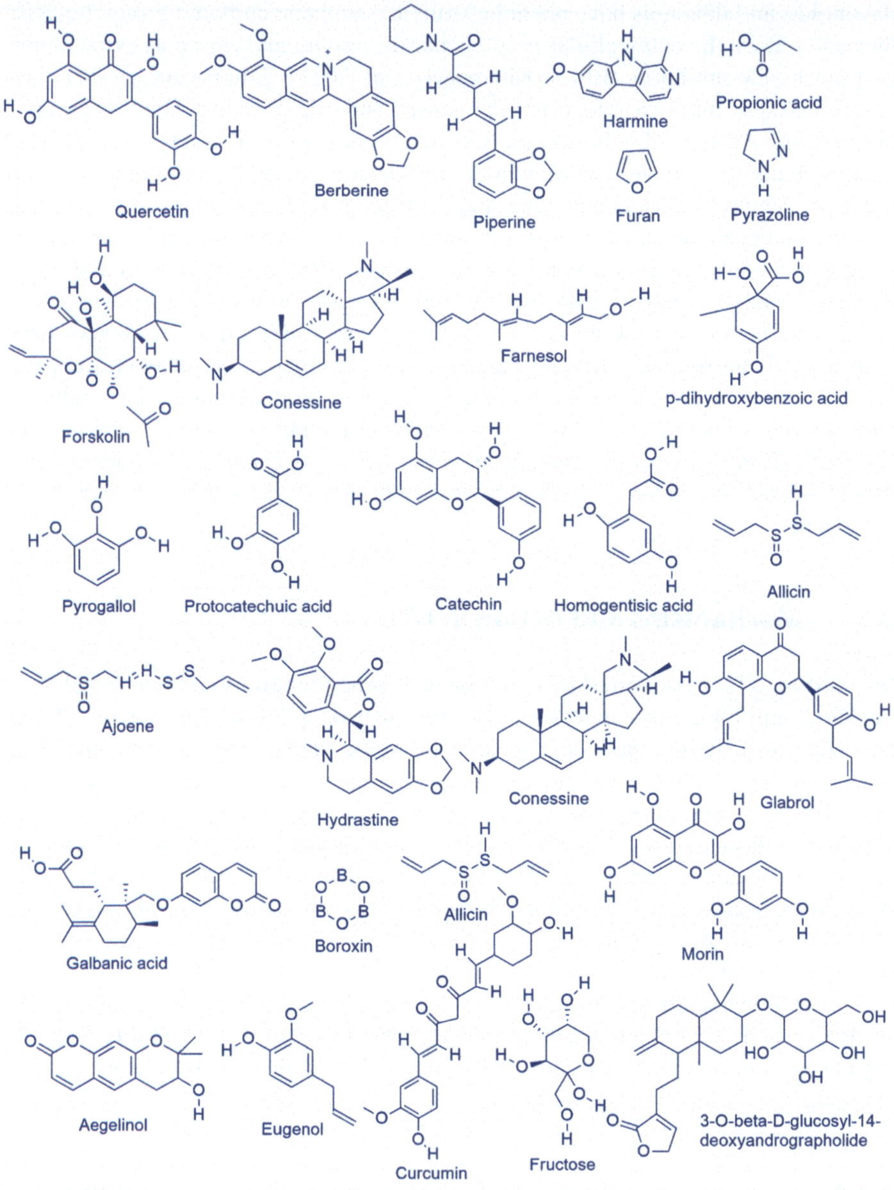

Fig. 3.1 Leading phytocompounds derived and their inhibitory antibacterial action

antibiotic inside the cell as a result of cell surface modification restricts the drug's interaction (for example, lipid A modification) or lowers the number of entry channels (for example, porins) (North and Brown 2021). The outer membrane of Gram-negative bacteria acts as a functional and physical barrier. Such an outer membrane consists of an outer layer made of lipid A and an interior layer made of phospholipids.

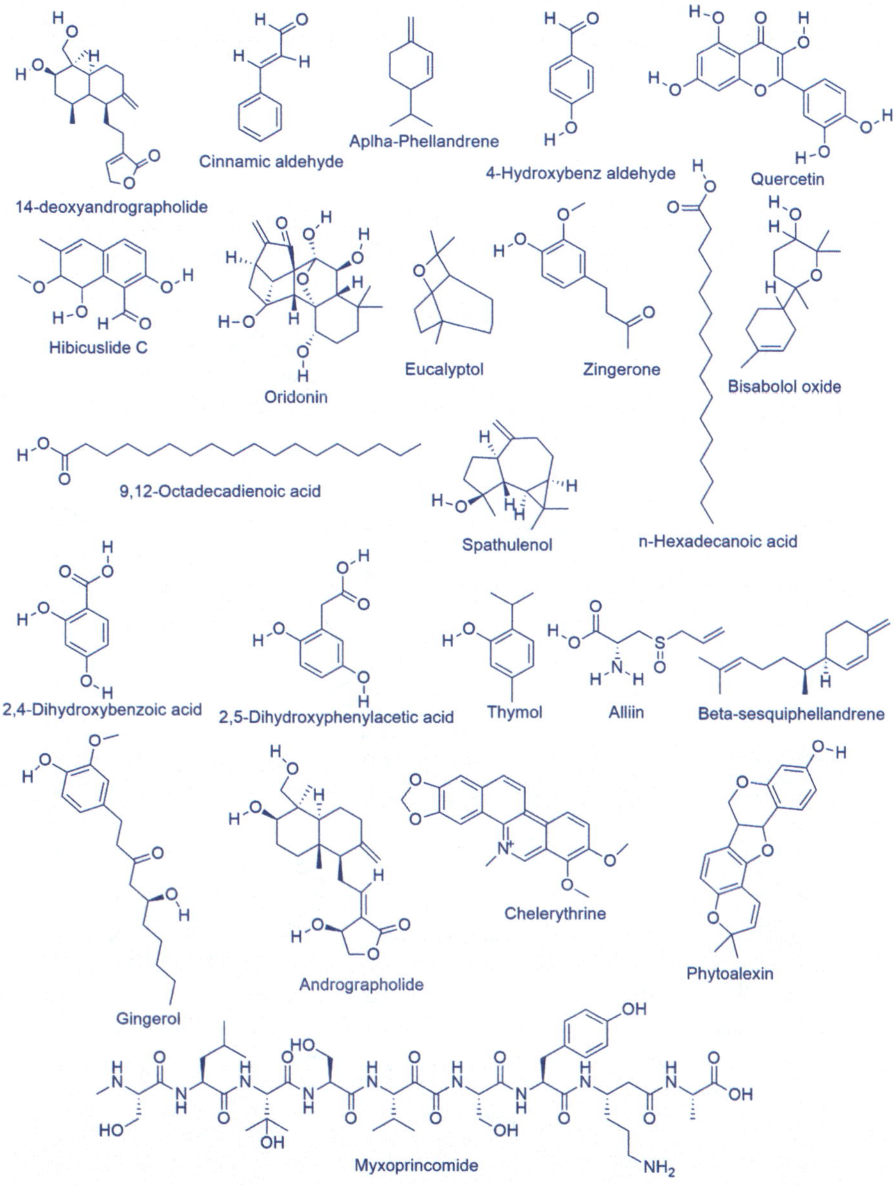

Fig. 3.1 (continued)

The permeability to several drugs is decreased by this combination. The outer-membrane proteins OmpF and OmpC of *Escherichia coli* are an example of this; they are porins found in the Enterobacteriaceae family of bacteria and are hypothesised to serve as non-specific channels (Blair et al. 2015).

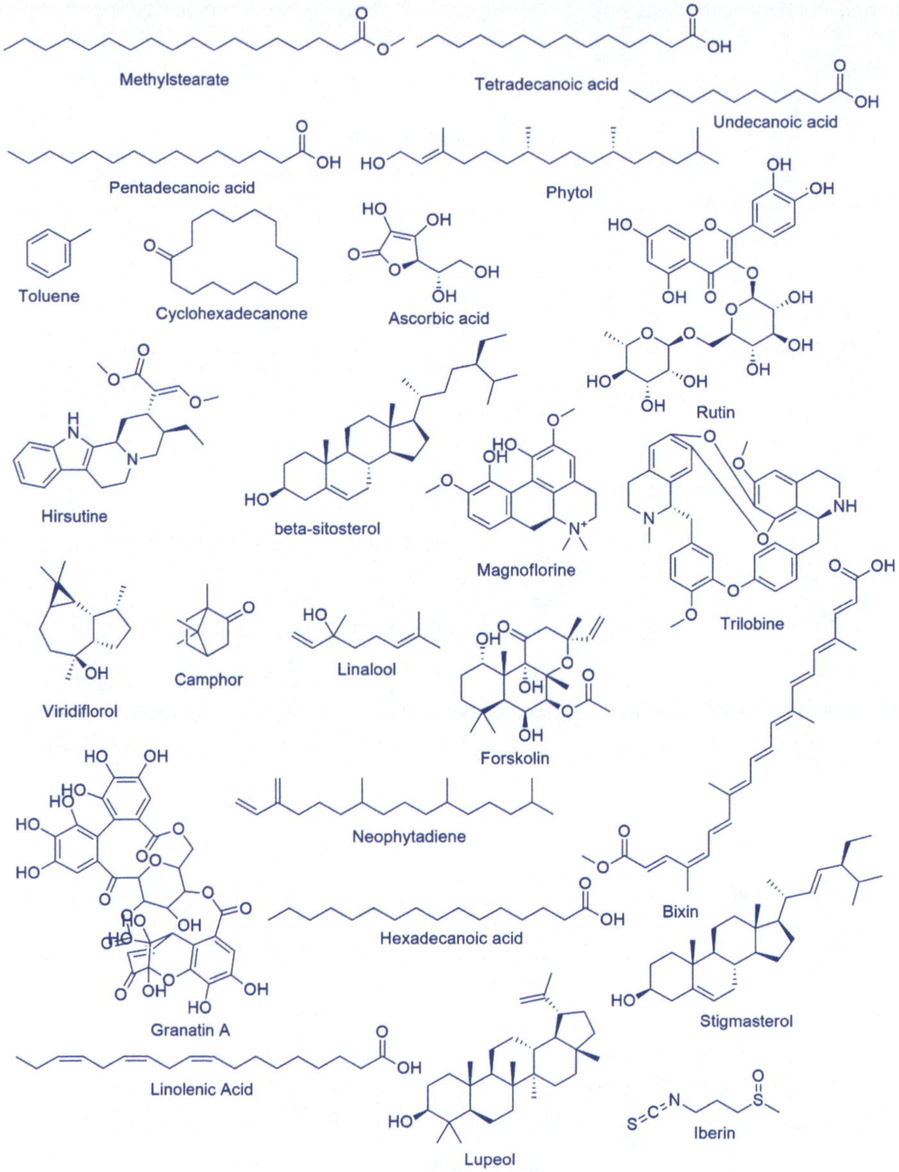

Fig. 3.1 (continued)

3.3.2 Increased Efflux Pumps

Numerous energy-dependent processes in bacteria have been found to lower the concentration of harmful chemicals inside of the cells. EPs are transport proteins that are found in eukaryotic and Gram-positive bacteria and are localised in the

Table 3.1 Phytocompound inhibitory actions for ESKAPE pathogens

Sl No.	Name of plant	Phytocompound	Inhibitory action	References
1.	*Bauhinia tomentosa*	Quercetin	*P. aeruginosa* and *E. coli*	Sharmila et al. (2018)
2.	*Berberis vulgarios*	Berberine	*E. coli* *Candida albicans*	Khameneh et al. (2019)
3.	*Piper nigrum*	Piperine	*Staphylococcus aureus*	Khameneh et al. (2019)
4.	*Peganum harmala*	Harmine	*S. aureus*	Ali et al. (2011)
5.	*Coleus forskohlii*	Furan, propionic acid, Forskoli	*S. aureus*	Chakraborty et al. (2022)
7.	*Matricaria chamomilla*	Conessine	*Pseudomonas aeruginosa*	Khameneh et al. (2019)
8.	*Syzygium aromaticum*	Farnesol	*S. aureus*	Khameneh et al. (2019)
9.	*Coleus forskohlii*	Pyrazoline	*Pseudomonas aeruginosa*, *Staphylococcus aureus*, *Klebsiella pneumoniae*	Chakraborty et al. (2022)
10.	*Boletus edulis*	p-dihydroxybenzoic acid, Pyrogallol, Protocatechuic acid, Catechin, Homogentisic acid	*P. aeruginosa*	Garcia et al. (2022)
11.	*Allium sativum*	Allicin, Ajoene	*P. aeruginosa*	Khameneh et al. (2019)
12.	*Hydrastis canadensis*	Berberine, Hydrastine	*Candida albicans*, *Escherichia coli*	Khameneh et al. (2019)
13.	*Holarrhena antidysenterica*	Conessine	*Pseudomonas aeruginosa*	Khameneh et al. (2019)
14.	*Glycyrrhiza glabra*	Glabrol	*S. aureus*	Khameneh et al. (2019)
15.	*Thymus vulgaris*	Galbanic acid	*S. aureus*	Khameneh et al. (2019)
16.	*Calotropis procera*	Boroxin	*S. aureus* and *Klebsiella pneumoniae*	Saddiq et al. (2022)
17.	*Allium cepa*	Allicin	*P. aeruginosa*	Khameneh et al. (2019)
18.	*Morus alba L.*	Morin	*S. aureus*	Khameneh et al. (2019)
19.	*Dipteryx odorata*	Aegelinol	*Enterobacter aerogenes*, *S. aureus*	Khameneh et al. (2019)

(continued)

Table 3.1 (continued)

Sl No.	Name of plant	Phytocompound	Inhibitory action	References
20.	*Syzygium aromaticum*	Eugenol	*P. aeruginosa*	Khameneh et al. (2019)
21.	*Curcuma longa*	Curcumin, Fructose	*S. aureus*	Khameneh et al. (2019)
22.	*Andrographis paniculata*	3-O-β-D-glucosyl-14-deoxyandrographolide, 14-deoxyandrographolide	*Staphylococcus aureus*, *Pseudomonas aeruginosa*	Tandon et al. (2015)
23.	*Cinnamomum zeylanicum*	Cinnamic aldehyde	*P. aeruginosa*	Rath et al. (2014)
24.	*Curcuma longa*	α-Phellandrene, 4-Hydroxybenzaldehyde	*P. aeruginosa*	Rath et al. (2014)
25.	*Allium cepa*	Quercetin	*S. aureus*	Khameneh et al. (2019)
26.	*Abutilon theophrasti*	Hibicuslide C	*P. aeruginosa*	Álvarez-Martínez et al. (2021)
27.	*Rabdosia rubescens*	Oridonin	MRSA	Álvarez-Martínez et al. (2021)
28.	*Plectranthus amboinicus*	Eucalyptol	*P. aeruginosa*	Sawant et al. (2023)
29.	*Zingiber officinale*	Zingirone	*P. aeruginosa*	Akbar & Akbar (2020)
30.	*Matricaria chamomilla*	Bisabolol oxide, 9,12-Octadecadienoic acid, Spathulenol, *n*-Hexadecanoic acid	*E. coli*	Jafarzadeh et al. (2020)
31.	*Neoboletus luridiformis*	2,4-Dihydroxybenzoic acid, 2,5-dihydroxyphenlacetic acid	*P. aeruginosa*	Garcia et al. (2022)
32.	*Plectranthus amboinicus*	Thymol	*E. coli, E. aerogenes, S. aureus, P. aeruginosa*	Sawant et al. (2023)
33.	*Allium sativum*	Alliin	*Escherichia coli*, *Pseudomons aeruginosa*, and *Klebsiella pneumoniae*	Weber et al. (1992)
34.	*Zingiber officinale Zingiberaceae*	β-Sesquiphellandrene, Gingerol	*Staphylococcus aureus*	Mao et al. (2019)
35.	*Andrographis paniculate*	Andrographolide	*Staphylococcus aureus*, *Escherichia coli*	Álvarez-Martínez et al. (2021)

(continued)

Table 3.1 (continued)

Sl No.	Name of plant	Phytocompound	Inhibitory action	References
36.	*Toddalia asiatica L.*	Chelerythrine	*Staphylococcus aureus*	Álvarez-Martínez et al. (2021)
37.	*Myxococcus xanthus*	Myxoprincomide	*Pseudomonas aeruginosa*	Wu C et al. (2016)
38.	*Mangifera indica*	Phytoalexins	*Staphylococcus aureus*, *Escherichia coli*	Alqahtani YS et al. (2022)
39.	*Macaranga tanarius (L.)*	Propolin D	*Staphylococcus aureus*	Álvarez-Martínez et al. (2021)
40.	*Dioscorea villosa*	Methylstearate, Tetradecanoic acid, Pentadecanoic acid, Phytol, Undecanoic acid, Toluene, Cyclohexadecanone	*S. aureus*, K. *pneumoniae*	Maroyi (2022)
41.	*Solanum lycopersicum*	Ascorbic acid	*Pseudomonas aeruginosa*	Abdallah et al. (2020)
42.	*Bauhinia tomentosa*	Rutin	*Staphylococcus aureus*	Sharmila et al. (2018)
43.	*Cocculus hirsutus*	*Hirsutine,* beta-sitosterol, *Magnoflorine*, *Cohirsinine*, *Trilobine*	*Staphlococcus aureus*, *Pseudomonas aeruginosa*	Shah et al. (2021)
44.	*Cistus monspeliensis*	Viridiflorol	*S. aureus*	Zalegh et al. (2022)
45.	*Cistus salviifolius*	Camphor	*S. aureus*	Zalegh et al. (2022)
46.	*Plectranthus amboinicus*	Linalool	*S. aureus* and *P. aeruginosa*	Sawant et al. (2023)
47.	*Coleus forskohlii*	Forskolin	*Escherichia coli*, *S. aureus*	Chakraborty et al. (2022)
48.	*Bixa Orellana*	Bixin	*Staphylococcus aureus*	Oliveira et al. (2022)
49.	*Punica granatum*	Granatin A	*K. pneumoniae*	Magangana et al. (2022)
50.	*Calotropis procera*	Neophytadiene, Hexadecanoic acid, Stigmasterol, Linolenic acid, Lupeol	*S. aureus, K. pneumoniae*	Saddiq et al. (2022)

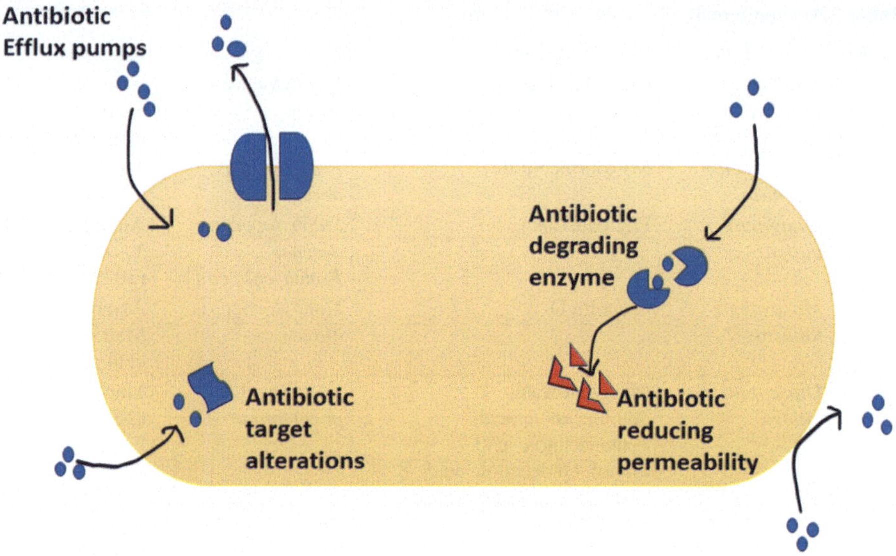

Fig. 3.2 Mechanisms involved in bacterial resistance to antibiotics

cytoplasmatic membrane. These pump systems eliminate harmful substances from the bacterial cell without modifying or degrading the molecules in the process (Zhao et al. 2020). Such latter pumps linked to the emergence of MDR bacteria (Lorusso et al. 2022) can either transport an extensive number of structurally diverse molecules or are specialised for one substrate. The identification and characterisation of EPs remain one of the main issues at the moment. In an operon regulated by a regulatory gene, EP genes are located (Duan et al. 2019).

3.4 Mode of Action of Phytochemicals

In preventing the development of resistance in bacterial pathogens and battling bacterial infections, phytochemicals have favourable results. The antimicrobial properties of phytochemicals relate to chemically interfering with the function or synthesis of essential components and evading the antibacterial resistance process (Khameneh et al. 2021) (Fig. 3.3).

In the case of flavonoids, they inhibit the synthesis of nucleic acids of Gram-negative and Gram-positive bacteria. To investigate the synergy between a number of different flavonoids, such as quercetin or kaempferol, and amoxicillin to see if they may defeat amoxicillin-resistant Staphylococcus epidermis (ARSE). The prevention of peptidoglycan synthesis in the bacterial cell membrane, prevention of beta-lactamase activity, increasing cell membrane permeability, increasing protein amides I and II, and decreasing fatty acids in the bacterial cells, the synergy between quercetin and amoxicillin proved to be effective. The authors emphasised the necessity to test for safety and effectiveness on both human and animal subjects

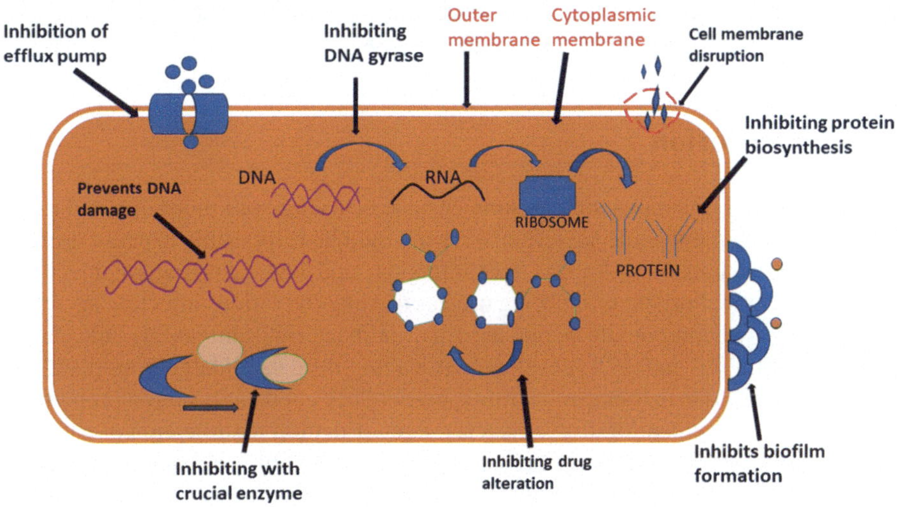

Fig. 3.3 Antibacterial mechanism of action of plant derivative compounds

(Khameneh et al. 2021). Coumarins' antibacterial properties are primarily caused by their inhibition of bacterial DNA gyrase, which prevents supercoiling (Garg et al. 2020).

3.5 In Silico Approaches

Step 1: Selection of UTI-causing bacterial protein from protein data bank. However, several types of bacterial proteins are available in the data bank. Here, *E. coli* DNA gyrase has been chosen PDBID: 1KZN. After retrieving this protein and saved as .pdf format, further removal of hetero atoms, water molecules, and previously presented ligands. Then, the location of active binding pockets in this cleaned protein for docking studies.

Step 2: Individual phytocompounds have been drawn by various chemical drawing software with the input of proper stereochemistry and validation of structural error. The prepared structure saved as .cdx to be converted with .mol for protein-ligand docking studies.

Step 3: Molecular docking—a computational technique called 'molecular docking' predicts the three-dimensional fit of two or more molecules, such as a ligand and a protein. It is a molecular modelling strategy used in drug development to find plausible protein target binding conformations for drug candidates. There are several free access webservers: SwissDock, DockingServer, I-Click Docking, Pardock, PatchDock, Dock Blaster, PLATINUM, IScreen, and others. Moreover, several software are also available for molecular docking studies, namely, Autodock, Autodock Vina, Gold, Glide, ICM, FlexX, FRED, CDocker, and MOE Suit. After screening the lead compound, each derivative further in vitro synthesis and characterisation for confirmatory study with various analytical instruments. Then, the developed product could be evaluated for in vitro antimicrobial activities.

3.6 Conclusion

Today, bacterial infections are common episodes causing both morbidity and mortality. Common pathogenic bacterium *E. coli* would be responsible to cause urinary tract infection, upper respiratory tract infection, and lower respiratory infection, nowadays. Natural products are being used as treating for various infecting agent(s). Moreover, phytocompounds being used as, antibacterial, antifungal, anticancer, antiviral, and a few more. To development leading drug candidates from plant are curio candidates due to therapeutic utility, particularly good antibacterial properties. Moreover, improvement of drug efficacies urges structural modification of parental nucleus with integument of medicinal chemistry and computational chemistry is call of the day. This leading candidate could be a promising antibacterial agent for UTI pathogen in future.

References

Abebe E, Tegegne B, Tibebu S (2016) A review on molecular mechanisms of bacterial resistance to antibiotics. Eur J Appl Sci 8(5):301–310

Akbar, S., & Akbar, S. (2020). Zingiber officinale Rosc. (Zingiberaceae) (Syns.: Amomum zingiber L.; A. Angustifolium Salisb.). Handbook of 200 medicinal plants: a comprehensive review of their traditional medical uses and scientific justifications, pp 1957-1997, Springer

Al Bratty M, Makeen HA, Alhazmi HA, Syame SM, Abdalla AN, Homeida HE et al (2020) Phytochemical, cytotoxic, and antimicrobial evaluation of the fruits of miswak plant, Salvadora persica L. J Chem 2020:1–11

Ali NH, Faizi S, Kazmi SU (2011) Antibacterial activity in spices and local medicinal plants against clinical isolates of Karachi, Pakistan. Pharm Biol 49(8):833–839

Álvarez-Martínez FJ, Barrajón-Catalán E, Herranz-López M, Micol V (2021) Antibacterial plant compounds, extracts and essential oils: an updated review on their effects and putative mechanisms of action. Phytomedicine 90:153626

Alqahtani YS, Bahafi A, Mirajkar KK, Basavaraju RR, Mitra S, More SS et al (2022) In vitro antibacterial activity of green synthesized silver nanoparticles using Mangifera indica aqueous leaf extract against multidrug-resistant pathogens. Antibiotics 11(11):1503

Aydi-Ben-Abdallah R, Jabnoun-Khiareddine H, Daami-Remadi M (2020) Fusarium wilt biocontrol and tomato growth stimulation, using endophytic bacteria naturally associated with Solanum sodomaeum and S. bonariense plants. Egypt J Biol Pest Control 30:1–13

Batiha GES, Wasef L, Teibo JO, Shaheen HM, Zakariya AM, Akinfe OA et al (2023) Commiphora myrrh: a phytochemical and pharmacological update. Naunyn Schmiedeberg's Arch Pharmacol 396(3):405–420

Baym M, Stone LK, Kishony R (2016) Multidrug evolutionary strategies to reverse antibiotic resistance. Science 351(6268):aad3292

Bello-López JM, Cabrero-Martínez OA, Ibáñez-Cervantes G, Hernández-Cortez C, Pelcastre-Rodríguez LI, Gonzalez-Avila LU, Castro-Escarpulli G (2019) Horizontal gene transfer and its association with antibiotic resistance in the genus Aeromonas spp. Microorganisms 7(9):363

Bhatia P, Sharma A, George AJ, Anvitha D, Kumar P, Dwivedi VP, Chandra NS (2021) Antibacterial activity of medicinal plants against ESKAPE: an update. Heliyon 7(2):e06310
Blair JM, Webber MA, Baylay AJ, Ogbolu DO, Piddock LJ (2015) Molecular mechanisms of antibiotic resistance. Nat Rev Microbiol 13(1):42–51
Buchmann D, Schultze N, Borchardt J, Böttcher I, Schaufler K, Guenther S (2022) Synergistic antimicrobial activities of epigallocatechin gallate, myricetin, daidzein, gallic acid, epicatechin, 3-hydroxy-6-methoxyflavone and genistein combined with antibiotics against ESKAPE pathogens. J Appl Microbiol 132(2):949–963
Chadha J, Harjai K, Chhibber S (2022) Repurposing phytochemicals as anti-virulent agents to attenuate quorum sensing-regulated virulence factors and biofilm formation in Pseudomonas aeruginosa. Microb Biotechnol 15(6):1695–1718
Chakraborty A, Haque SM, Dey D, Mukherjee S, Ghosh B (2022) Detection of UTI pathogen-killing properties of Coleus forskohlii from tissue cultured in vitro and ex vitro plants. Proc Natl Acad Sci India Sect B Biol Sci 92(1):157–169
Chowdhary VA, Tank JG (2023) Biomolecules regulating defense mechanism in plants. Proc Natl Acad Sci India Sect B: Biol Sci 93(1):17–25
Duan W, Li X, Ge Y, Yu Z, Li P, Li J et al (2019) Mycobacterium tuberculosis Rv1473 is a novel macrolides ABC efflux pump regulated by WhiB7. Future Microbiol 14(1):47–59
Garcia J, Rodrigues F, Castro F, Aires A, Marques G, Saavedra MJ (2022) Antimicrobial, antibiofilm, and antioxidant properties of boletus edulis and Neoboletus luridiformis against multidrug-resistant ESKAPE pathogens. Front Nutr 8:773346
Garg SS, Gupta J, Sharma S, Sahu D (2020) An insight into the therapeutic applications of coumarin compounds and their mechanisms of action. Eur J Pharm Sci 152:105424
Ghildiyal R, Prakash V, Chaudhary VK, Gupta V, Gabrani R (2020) Phytochemicals as antiviral agents: recent updates. In: Plant-derived bioactives: production, properties and therapeutic applications, pp 279–295
Gulcin İ (2020) Antioxidants and antioxidant methods: an updated overview. Arch Toxicol 94(3):651–715
Jafarzadeh MM, Moghaddam MJM, Bakhshi D (2020) Antimicrobial activity of three plant species against multi-drug resistant E. coli causing urinary tract infection. J Herbal Med 22:100352
Kamaruzzaman NF, Kendall S, Good L (2017) Targeting the hard to reach: challenges and novel strategies in the treatment of intracellular bacterial infections. Br J Pharmacol 174(14):2225–2236
Khameneh B, Iranshahy M, Soheili V, Fazly Bazzaz BS (2019) Review on plant antimicrobials: a mechanistic viewpoint. Antimicrob Resist Infect Control 8(1):1–28
Khameneh B, Eskin NM, Iranshahy M, Fazly Bazzaz BS (2021) Phytochemicals: a promising weapon in the arsenal against antibiotic-resistant bacteria. Antibiotics 10(9):1044
Lorusso AB, Carrara JA, Barroso CDN, Tuon FF, Faoro H (2022) Role of efflux pumps on antimicrobial resistance in Pseudomonas aeruginosa. Int J Mol Sci 23(24):15779
Magangana TP, Makunga NP, Fawole OA, Stander MA, Opara UL (2022) Antioxidant, antimicrobial, and metabolomic characterization of blanched pomegranate peel extracts: effect of cultivar. Molecules 27(9):2979
Mao QQ, Xu XY, Cao SY, Gan RY, Corke H, Beta T, Li HB (2019) Bioactive compounds and bioactivities of ginger (Zingiber officinale roscoe). Food Secur 8(6):185
Maroyi A (2022) Medicinal uses, phytochemistry and pharmacological properties of Dioscorea dregeana (Kunth) T. Durand & Schinz. Med Plants Int J Phytomed Related Industr 14(1):57–63
Matos MJ, Santana L, Uriarte E, Abreu OA, Molina E, Yordi EG (2015) Coumarins – an important class of phytochemicals. In: Phytochemicals-isolation, characterisation and role in human health, vol 25, pp 533–538
Naushad VA, Purayil NK, Wilson GJ, Chandra P, Joseph P, Khalil Z et al (2022) Epidemiology of urinary tract infection in adults caused by extended-spectrum beta-lactamase (ESBL)-producing Enterobacteriaceae–a case–control study from Qatar. IJID Regions 3:278–286
North OI, Brown ED (2021) Phage–antibiotic combinations: a promising approach to constrain resistance evolution in bacteria. Ann N Y Acad Sci 1496(1):23–34

Ogidi OI, Oguoma LMO, Adigwe PC, Anthony BB (2021) Phytochemical properties and in-vitro antimicrobial potency of wild edible mushrooms (Pleurotus ostreatus) obtained from Yenagoa, Nigeri. J Phytopharmacol 10(3):180–184

Oliveira SDSDC, Araújo RDC, da Silva GA, Leitão JH, da Silva Sousa SAB, Fonseca LP et al (2022) Bixa orellana L. from northern Brazil: morphological analysis, phenolic content, antioxidant and antibacterial activities. Brazilian. J Bot 45(3):883–896

Raut JS, Karuppayil SM (2014) A status review on the medicinal properties of essential oils. Ind Crop Prod 62:250–264

Rath S, Padhy RN (2014) Monitoring in vitro antibacterial efficacy of 26 Indian spices against multidrug resistant urinary tract infecting bacteria. Integr Med Res 3(3):133–141

Saddiq AA, Tag HM, Doleib NM, Salman AS, Hagagy N (2022) Antimicrobial, antigenotoxicity, and characterization of calotropis procera and its rhizosphere-inhabiting actinobacteria: in vitro and in vivo studies. Molecules 27(10):3123

Sanchez GV, Babiker A, Master RN, Luu T, Mathur A, Bordon J (2016) Antibiotic resistance among urinary isolates from female outpatients in the United States in 2003 and 2012. Antimicrob Agents Chemother 60(5):2680–2683

Sawant S, Baldwin TC, Khan H, Rahman A (2023) Evaluation of the effect of leaf development in Plectranthus amboinicus L. on antimicrobial activity and virulence factors of Pseudomonas aeruginosa PAO1 and Staphylococcus aureus NCTC8325. Curr Microbiol 80(1):24

Schefer S, Oest M, Rohn S (2021) Interactions between phenolic acids, proteins, and carbohydrates—influence on dough and bread properties. Food Secur 10(11):2798

Shah R, Parmar K, Vaghela H (2021) Analysing antibacterial efficacy of biosynthesized palladium nanoparticles using aqueous leaf extract of Cocculus hirsutus as the reducing agent. Biosc Biotech Res Commun 14. https://doi.org/10.21786/bbrc/14.2.63

Sharmila G, Pradeep RS, Sandiya K, Santhiya S, Muthukumaran C, Jeyanthi J et al (2018) Biogenic synthesis of CuO nanoparticles using Bauhinia tomentosa leaves extract: characterization and its antibacterial application. J Mol Struct 1165:288–292

Shnawa HA, Khalaf MN, Taobi AAH, Panampilly B, Thomas S (2020) General and chemical perspectives and studies on tannins as natural phenolic compounds for some ecoefficient applications. In: Natural products chemistry: biomedical and pharmaceutical phytochemistry, p 115

Singh B, Singh JP, Kaur A, Singh N (2018) Phenolic compounds as beneficial phytochemicals in pomegranate (Punica granatum L.) peel: a review. Food Chem 261:75–86

Tandon C, Mathur P, Sen M (2015) Andrographis paniculata Nees (Kalmegh): a review on its antibacterial activities and phytocompounds. European J Med Plants 8(1):1–10

Thorn CR, Thomas N, Boyd BJ, Prestidge CA (2021) Nano-fats for bugs: the benefits of lipid nanoparticles for antimicrobial therapy. Drug Deliv Transl Res 11:1598–1624

WHO (2014) Antimicrobial resistance: global report on surveillance. WHO, Geneva

Weber ND, Andersen DO, North JA, Murray BK, Lawson LD, Hughes BG (1992) In vitro virucidal effects of Allium sativum (garlic) extract and compounds. Planta Med 58(05):417–423

Wu C, Choi YH, van Wezel GP (2016) Metabolic profiling as a tool for prioritizing antimicrobial compounds. J Ind Microbiol Biotechnol 43(2–3):299–312

Zaidi S, Dahiya P (2015) In vitro antimicrobial activity, phytochemical analysis and total phenolic content of essential oil from Mentha spicata and Mentha piperita. Int Food Res J 22(6):2440

Zalegh I, Bourhia M, Zerouali K, Katfy K, Nayme K, Khallouki F et al (2022) Molecular characterization of gene-mediated resistance and susceptibility of ESKAPE clinical isolates to Cistus monspeliensis L. and Cistus salviifolius L. extracts. Evid Based Complement Alternat Med 2022(1):7467279

Zhao X, Yu Z, Ding T (2020) Quorum-sensing regulation of antimicrobial resistance in bacteria. Microorganisms 8(3):425

Zowawi HM, Harris PN, Roberts MJ, Tambyah PA, Schembri MA, Pezzani MD et al (2015) The emerging threat of multidrug-resistant gram-negative bacteria in urology. Nat Rev Urol 12(10):570–584

Natural Product-Based Treatment for Tuberculosis

4

Ghanishtha Prusty, Somya Ranjan Panda, Rashmi Ranjan Behera, Subha Laxmi Bhushan, Anjali Rath, and Paresh Kumar Prusty

Abstract

Tuberculosis (TB), caused by the bacterial organism *Mycobacterium tuberculosis*, has remained a persistent global health challenge, profoundly impacted millions of lives and posed a significant threat to public health worldwide. Despite notable strides in medical research, TB continues to be a formidable issue, particularly in underdeveloped nations. The existing arsenal of tuberculosis treatment options is notably limited, and the emergence of multi- and extensively drug-resistant TB strains further complicates the situation. The escalating costs of new pharmaceuticals and the adverse effects associated with current prescription treatments have prompted a compelling need to explore alternative approaches. In response, attention has increasingly turned towards investigating natural substances as supplementary treatment options. Derived from diverse biological sources, these natural products, including plant extracts, microbial metabolites, and compounds from marine origins, are gaining popularity as potential complements or substitutes for established anti-mycobacterial interventions. This exploration represents a promising avenue for addressing the complexities of TB treatment while considering cost-effectiveness and mitigating adverse drug reactions.

G. Prusty (✉) · S. R. Panda · R. R. Behera · S. L. Bhushan · A. Rath
Department of Life Science, School of Sciences, Gandhi Institute of Engineering and Technology, Gunupur, Odisha, India
e-mail: prustyghanishta@giet.edu

P. K. Prusty
Department of Microbiology, Aravali Veterinary College, Sikar, Rajasthan, India

T. Sharma et al. (eds.), *Natural Products for Antibacterial Drug Development: Recent Advancement of Computational Approach*,
https://doi.org/10.1007/978-981-97-9634-2_4

Keywords

Tuberculosis · *Mycobacterium tuberculosis* · Multi drug-resistant · Natural compounds

Abbreviations

CADD	Computer-aided drug design
IFN- γ	Interferon-gamma
MTB	Mycobacterium tuberculosis
NBXH	Niu Bie Xiao He
RNTCP	Revised National TB Control Programme
TB	Tuberculosis
TH 1 cells	T helper cells
TNF	Tumor necrosis factor

4.1 Introduction

Tuberculosis, an age-old affliction in humans, is induced by the bacterium *Mycobacterium tuberculosis*, which primarily affects the lungs and causes pulmonary diseases (Terracciano et al. 2020). Tuberculosis has historically held a notorious reputation, being one of the most dreaded diseases, often referred to with ominous titles like the "Captain of All Men of Death" and the "Great White Plague" (Zumla and Grange 1999). Tuberculosis is a multisystemic disease affecting a wide range of systems in the body, displaying a wide variety of manifestations. In addition to its usual occurrence in the respiratory system, tuberculosis can commonly impact various organs such as gastrointestinal, lymphoreticular, musculoskeletal, reproductive, skin, and liver (Zumla and Grange 1999; Mbuh et al. 2019). Presently, tuberculosis stands as a formidable health crisis affecting one-third of the global population, posing not only a risk of mortality but also generating socio-economic repercussions. Recent trends indicate an association between tuberculosis and endocrine as well as metabolic disorders (Racca 2019). The increasing prevalence of drug resistance and the transmission rate of the disease have propelled advanced studies on the disease (Mathiasen et al. 2020).

In tropical and subtropical regions, managing tuberculosis poses significant challenges. Efforts to reduce the cost of tuberculosis treatment face hindrances due to the emergence of drug-resistant tuberculosis. The development of drug resistance in tuberculosis is attributed to mutations in the MTB strain (Mathiasen et al. 2020).

Tuberculosis was considered as a curse in ancient time, better known in Greco roman world, where it is known as "phthisis" "which means consumption". At that time, it was an incurable disease, specifically if it happens in early age, it is mostly fatal. Ancient time scientist suggested for the diet menu based on milk, better nutrition, and purified air most probably in north-African climate. Aretaeus of Cappadocia, a Greek physician, recommended that the patients should travel to high altitudes, travel through the sea, take a good diet and drink plenty of milk (Kashyap et al. 2016).

The traditional tuberculosis treatment primarily relies on the modern antibiotics such as isoniazid, rifampicin, pyrazinamide, and ethambutol. Although, *Mycobacterium tuberculosis* has become resistant to most of these conventional drugs. Furthermore, underdeveloped countries assessing and affording modern medications pose challenges for a significant portion of the population. This underscores the necessity for alternative treatments. This chapter aims to highlight the herbal remedies employed in current and historical traditional medical practices for tuberculosis treatment (Elkington et al. 2009).

4.2 Role of Traditional Medicine for Tuberculosis

Tuberculosis is a potentially contagious disease that can be transmitted to another person while coughing or sneezing. This disease shows various symptoms such as severe cough, chillness, fever, fatigue, sweating, loss of appetite, weight loss, etc. Failure to treat tuberculosis may lead to critical, life-threatening conditions. The disease can be categorized into two situations: active and latent. In the first condition, the disease is communicable, while in the latter, i.e. in the latent condition, the bacterium remains inactive in the host body without showing any symptoms. There are various treatments available, but traditional medications play a key role in combating the disease. In India, the community management of tuberculosis falls under the oversight of a centrally sponsored programme called the Revised National TB Control Programme (RNTCP). The primary objectives of this programme include achieving an 85% cure rate and a 70% case detection rate. Currently, Ayurvedic drugs are not included in RNTCP (Samal 2016). However, many private practitioners do use various types of Ayurvedic medicines.

4.2.1 Ayurvedic Treatments

Ayurveda is one of the ancient practices used in India since more than 3000 years ago, it uses various plant-based medicines and techniques to treat a disease. In Ayurveda *tuberculosis*is known as "Rajyakshma" which means this disease was firstly affected Raja Indra. The Charak Samhita also identifies tuberculosis diseased condition as "shosh".

अयथाबलमारम्भमवेगसंधारणंक्षयम ।यक्ष्मण: कारणंवद्यिाचतुर्थमवषिमाशनम। (च. चटि/१३)
The shloka states cause for Rajyakshma (Tuberculosis).

The verse emphasizes the four causes of Rajyakshma (Tuberculosis):

Overexertion beyond one's physical capacity.
Suppressing natural urges like urine and faeces.
Diminished production of bodily tissues such as rasa, rakta, and mamsa.
Unhealthy eating habits (Choudhary et al. 2021).

For the treatment of tuberculosis there are various methods used such as follows (Nedungadi et al. 2023).

(a) *Snehana*—Massage herbs infused oil on various body parts that helps to liquefy the mucous.
(b) *Swedana*—Inducing heat that promote sweating that helps to relieve congestion, i.e. by steam, pouring hot water to palms and feet, gargling with hot water, etc.
(c) *Vamana*—This method includes the intake of a combination of herbs to induce vomiting that helps to release the mucous with it.
(d) *Virechana*—In this method, specific herbs are administrated orally to relieve mucous and bowels.

4.2.1.1 Ayurvedic Herbs for TB

There are various herbs which are used as medicines for patients affected by tuberculosis and some of them are as follows:

(a) *Vidarikand*—This herb acts directly on the digestive and reproductive systems. As in tuberculosis weight loss is one of the major issue it helps to tackle it.
(b) *Bramhi*—It is used to rejuvenate brain cells and nerves, also cleanse, and improve immunity, as well as help to manage cough, fever, and bronchitis, which are the common symptoms of tuberculosis.
(c) *Rasonam*—It is commonly known as garlic. It helps reduce cough, infection in the lungs and bronchus, and rejuvenates blood and lymph.
(d) *Yashtimadhu*—It improves body healing and reduces bleeding, cough, throat irritation, and bronchitis.
(e) *Ashwagandha*—It boosts immune system, helps to deal with cough and cold.
(f) *Guduchi*—It clears toxins from the body, reduce inflammation, and reduce fever. The above-discussed herbs can treat the disease from its roots. Praanrakshakchurna, kas-harchurna, and tulsi capsules are also available for the proper ayurvedic treatment of tuberculosis.

4.2.1.2 Other Prominent Ayurvedic Medicines for TB

Praanrakshak Churna is a herbal powder comprising ingredients such as Shirish (*Albezzia lebbock*), Vaasa (*Adhatoda vasica*), Dalchini (*Cinnamomum zeylanica*), Kantkari (*Solanum xanthocarpum*), etc., designed to provide effective results in tuberculosis. Aptly named "Praanrakshak", it promotes overall health by addressing various signs and symptoms associated with the disease. The recommended dosage is one to two teaspoonfuls, thrice daily, taken with lukewarm water half an hour after meals.

Kaas-Har Churna serves as a potent remedy for cough and cold. Its ingredients include Pippali (*Piper longum*), Ela (*Eletteria cardamomum*), Dalchini (*Cinnamomum zeylanica*), among others. Balancing all three doshas (Vata, Pitta, and Kapha) in the body acts as a vasodilator, enhancing blood circulation in the lungs. This churn aids in clearing the respiratory tract, alleviating cough, and fortifying the immune system. Regular use effectively manages the symptoms of tuberculosis. The recommended dosage is one to two teaspoonfuls, twice daily, with lukewarm water half an hour after meals.

Tulsi capsules contribute to immune system enhancement in tuberculosis patients, containing standardized Tulsi (*Ocimum sanctum*) extract. These capsules facilitate the balance of doshas, body tissues (dhatus), and body excreta (malas), thereby contributing to the effective treatment of tuberculosis. The suggested dosage is one to two capsules, twice daily, taken with plain water half an hour after meals (Choudhary et al. 2021). This ancient medical science can offer efficient solutions to the problems caused due to deadly bacterium. Ayurvedic can produce a potential medication for the management of TB (Samal 2016).

4.2.2 Traditional Chinese Treatments

The Chinese medicinal treatment for tuberculosis is named Niu Bie Xiao He (NBXH), which focuses on boosting immunity and reducing the treatment period compared to Western treatment methods (Liang et al. 2021).

According to Chinese scientists, "Robert Abramovitch" states that *Mycobacterium tuberculosis* is used to treat patients. It is one of the Chinese herbs. The following are the herbs used in the medicine:

(a) Bulbus Fritillariae cirrhosae has properties that are anti-inflammatory, anti-asthmatic, and antioxidant (Quan et al. 2022).
(b) Rhizoma Bletillae—It treats pulmonary and gastric haemorrhage, cold, and cough (Jiang et al. 2021).
(c) Herba Houttuyniae—It is helpful for the reduction of inflammation and microbial infections (Yang et al. 2022).
(d) Radix Platycodonis treats sore throat, cough with mucous and phlegm, and other respiratory disorders (Ishimaru et al. 2022).
(e) Fructus Arctii—It helps reduce cough and inflammation and detoxifies the body (Li et al. 2022).
(f) Glutinous rice—It increases bone density, reduces inflammation, and increases metabolism.

The NBXH can be extracted with the help of water thrice, and all the filtrates should be mixed and dried in a vacuum medium. This medicine can be consumed as a form of decoction.

4.3 Role of Natural Products

4.3.1 Plant-Based Compounds

Various medicinal plants have been used to treat tuberculosis for thousands of years. Parts of plants, including leaves, roots, bark, stems, flowers, fruits, and latex have been used for traditional treatment against tuberculosis by the tribal community of various countries. Most herbal products and plants were used in this era for commercial treatment, and people were also cured (Sharifi-Rad et al. 2020). However, most studies and research also failed to provide evidence towards the traditional belief.

Traditionally medicinal plants and the parts of plants were used for the treatment of tuberculosis, such as leaves, fruit, and dry seed of *Carica papaya*, roots of *Zanthoxylum case*, the seed of combretum here sense and whole plant of *Myrothamnus flasellifolius* (Rai 2016). Leaves of *Lippiaja vanica* were infused with hot water, and the patient inhaled it as steam. Crushed leaves of the citrus lemon plant were burned and then inhaled by the patient. Such herbal products are used to treat tuberculosis by administering for 1–2 months in the patients (Rai 2016). Medicinal plants have a phytochemical product which has therapeutic properties for tuberculosis. Plant metabolites include compounds like flavonoids, carotenoids, indoles, isothiocyanates, monoterpenes, folic acid, etc., which have antibiotic properties against tuberculosis (Getachew et al. 2022) (Fig. 4.1).

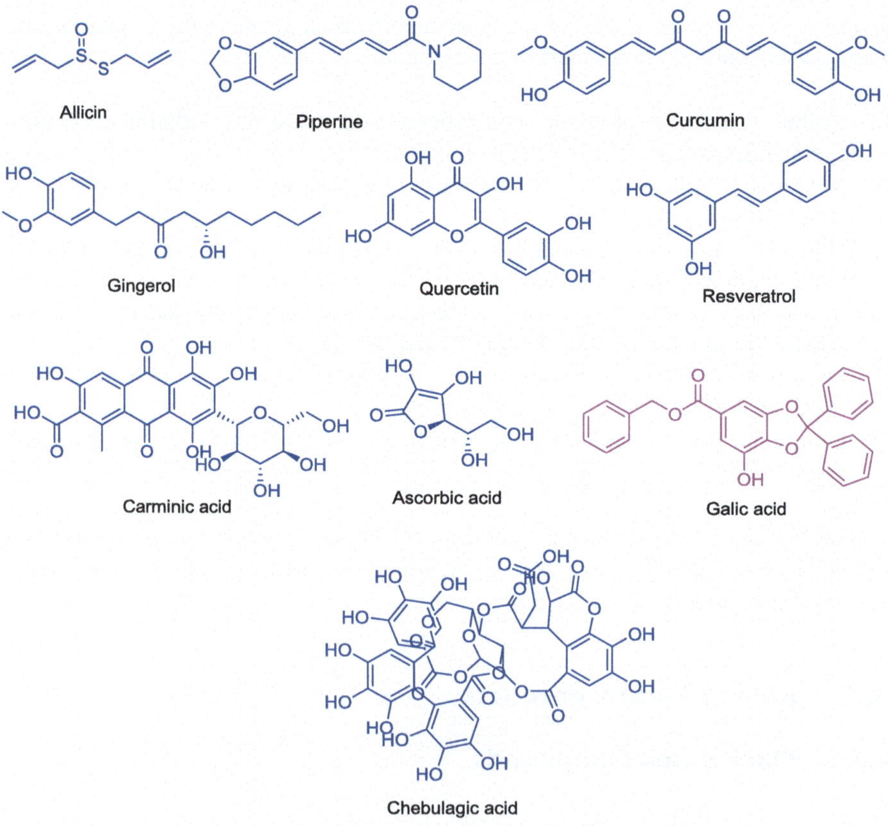

Fig. 4.1 Some phytocompounds and their inhibitory properties for anti-TB

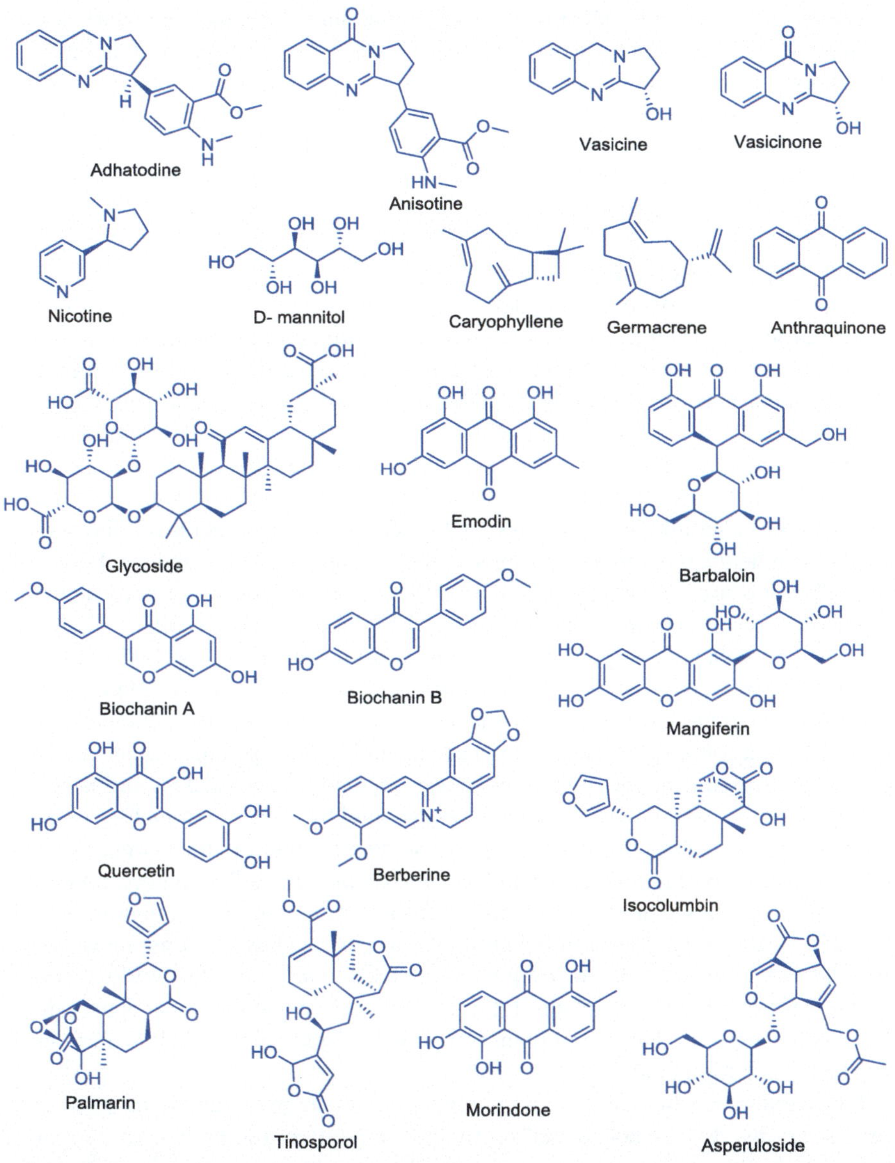

Fig. 4.1 (continued)

These synthetic antibiotic administrations to the patient's bodies have shown various side effects and toxic properties in patients. Mostly, TH1 cells will be suppressed in a patient, the anti-inflammatory cytokines level will be reduced, and the CD4+ TH cells will be reduced, causing immunodeficiency in the patient's body. Hence, these long-term drugs have several side effects, so the researchers were exploring the specific molecule that can act as an immune modulator by regulating the response of TH1 cells. Most plant-based compounds have the characteristics of an immune modulator. The alcoholic extract of Miana leaves immunomodulates by upregulating the -T-lymphocyte proinflammatory response and increasing TNF-α and IFN-γ cytokine levels (Pancu et al. 2021). Allicin from garlic, piperine, an extract of chancapitra, curcumin from turmeric, gingerol from ginger and Rubiaceae species with antioxidant and anti-inflammatory properties must be considered immunomodulators (Elkington et al. 2009).

4.3.1.1 Phytocompounds

The natural polyphenol compound derives from the turmeric plant, a member of the ginger family. Curcumin, a yellowed colour, is a primary compound used in food preparation as traditional medicine. Curcumin has been used for various other diseases for thousands of years as herbal medicine, not just for tuberculosis. It can be used for diabetic ulcers and rheumatism (Abdoul-latif et al. 2023). Phloretin is a flavonoid compound found in fruits, vegetables, legumes, nuts, and plant-derived food (coffee and wine) (Dashti et al. 2014). It has several anti-inflammatory effects (production of chemokines, cytokines, and other factors). Quercetin is a flavonoid-based phytochemical compound in apples, berries, capers, grapes, and tea. In one of the studies, it is used to target glutamine synthetase, an enzyme in nitrogen metabolism and the result is found to be inhibition of *M. tuberculosis* (Akbik et al. 2014). Moreover, polyphenol compounds are found in various plants such as barley, plums, and strawberries (Jeon et al. 2017); tannic acid isolated from the plant extract (*Globularia alypum*) has antioxidant and anti-tuberculosis properties (Németh et al. 2018). Resveratrol is a stilbene polyphenol seen in food like peanuts, wine, cranberries, etc., extracted from the rhubarb plant was obtained and studied for antibacterial activity, and the result concluded that the compound has anti-inflammatory properties (Chung et al. 1998). Similarly, propolis isolated from bee extract contains resin and wax used as herbal medicine and has antioxidant, anti-inflammatory, and antimicrobial properties that can be identified from various research activities (Khlifi et al. 2011). Recently, conventional antibiotic treatments have become more resistant, and natural product-based compounds are used to minimize tuberculosis activity and are more effective than anti-TB drugs. Prevention and treatment approaches, along with natural product-based plant compounds, could possibly help to minimize drug resistance (Sun et al. 2012).

TB treatment might take quite a long time, but it can be extremely hard for several age groups. If it is not treated on time, it may affect the body's vital organs. Hence, for healthy treatment as recommended for antibiotic courses (Teixeira et al. 2019), although it can be treated with the help of natural herbal remedies and contribute towards faster recovery by natural product-based treatments, including active ingredients from plant products like leaves, roots, flowers, etc. (Table 4.1).

Several classes of phytochemical compounds have been reported to have anti-TB activity (Koche et al. 2016).

Table 4.1 Plant-based extract and their compounds for treatment of tuberculosis

Common name	Botanical name and family	Chemical constituent	References
Adulsa (Vasaka)	*Justicavasica*, Acanthaceae	Adhatodine, anisotine, Vasicine, vasicinone	Lin et al. (2009)
Ashwagandha	*Indian ginseng*, Solanaceae	Alkaloid, Saponins, Steroidal lactones.	Tuberculosis Drug Screening Program (2001)
Brahmi (*Indravalli*)	*Bacopa monnieri*, Plantaginacea	Brahmine, Herpestine, Nicotine, D-mannitol, Monnierin	Bishnoi (2016)
Tulsi (Holy basil)	*Ocimumtenuiflorum*, Lamiaceae	Caryophyllene Germacrene Isothymusine	Hasan et al. (2023)
Alovera (Ghirtkumarrika)	*Aloe barbadensis*, Liliaceae	Anthraquinone Glycoside Emodin, Barbaloin Vitamins, enzyme	Kurane et al. (2014)
Garlic (Lahsun)	*Allium sativum*, Amaryllidacea	Aliin, sallylcysteine Vinyldithiines Allylpropyl disulphide	Lin et al. (2009)
Vidirakand (Bhukushmandi)	*Pueraria Tuberosa*, Fabaceae	Biochanin A Biochanin B Mangiferin Quercetin Gentistein	Maji et al. (2014)
Guduchi (Giloy)	*Tinospora Cordifolia*, Menispermaceae	Berberine, Isocolumbin, Palmarin, Tinosporol	Gupta and Kulkarni (2018)
Ashoka (Indian mulberry)	*Morinda Citrifolia*, Rubiaceae	Morindone Flavonoid Asperuloside Carminic acid	Sam-Ang et al. (2023)
Amla (Amalaka)	*Phyllanthus Emblica*, Phyllanthacea	Ascorbic acid Gallic acid Chebulagic acid Quercetin	Srivastava et al. (2015)

(i) Flavonoids are the largest group of plant phenols, which are ubiquitous in nature and occur as aglycones, glucosides, and methylated derivatives. These include vegetables, fruits, and beverages like tea, coffee, and more.
(ii) Tannins are difficult to define in chemical form, as they are heterogeneous substances with high molecular weight and polyphenolic. Which can be classified into four categories: gallo tannins, ellagitannins, complex tannins, and condensed tannins. These are generally found in grapes, blueberries, tea, chocolate, and legumes.
(iii) Alkaloids are heterocyclic nitrogen atoms containing natural products, and their classes are divided based on their ring structure.
(iv) Terpenoids, a derivative form of isoprene with multiple cyclic structures, have antibacterial properties.
(v) Saponin is made up of glycosylated steroids, triterpenoids, and steroid alkaloids that have antimicrobial and antifungal activity.

Table 4.2 shows the various plants and their plant parts utilized for inhibiting TB.

These plants and their secondary metabolites in Table 4.2 show effectiveness against bacteria. The call of the day is the need for or the demand for more alternative ways for drug-resistant TB strains (Mueller-Wieland 1961). Further, a detailed study is required to eliminate these infectious diseases by developing multidisciplinary correlation to include phyto drugs as adjuvants, which may alleviate the problem of drug resistance and improve the recent therapy. Thus, providing an avenue for developing sustainable and low-cost treatments (Samal 2016).

4.3.2 Commercial Drugs for TB

The first antibiotic that was found to be more effective towards tuberculosis was discovered by Selman Waksman in 1943. Streptomycin developed from soil fungus can fight against the tuberculosis bacteria but has some side effects on patients.

Table 4.2 Various plants and their plant parts are utilized for inhibiting TB

Plant	Part utilized	Extract utilized	References
Aframomum melegueta	Seed	Methanolic seed extract	Chomini et al. (2020)
Carica papaya	Leaves	Ethanolic leave extract	Pandit et al. (2013)
Citrus lemon	Fruit	Ethanolic fruit extract	Jahanpour et al. (2015)
Chironia baccipfera	Roots	Ethanolic root extract	Maroyi (2019)
Combretum hereroenseschinz	Bark	Methanolic bark extract	Roy et al. (2014)
Lippia javanica	Leaves	Ethanolic leaves extract	Mujovo (2010)
Salix mucronata	Seeds	Methanolic seeds extract	Semenya and Maroyi (2013)

Later, in 1994, para-aminosalicylic acid was discovered against TB treatment. There are anti-TB drugs and their mode of action against *M. tuberculosis* (Table 4.3) (Fig. 4.2).

M. tuberculosis strain may be resistant to numerous drugs. The most potent first-line medications are rifampin and isoniazid. The second-line drug regimen includes any fluroquinolone. Fig. 4.3 shows the inhibitory pathways of various drugs in the bacterium.

Table 4.3 Anti-TB drugs and their mechanism of action

Drugs	Effect on bacterial cell	Mechanism of action	References
Isoniazid	Bactericidal	Inhibition of cell wall, Other multiple effect on DNA, lipid, carbohydrate	Bhanu (2023)
Rifampin	Bactericidal	Inhibition of RNA synthesis	Velásquez et al. (2018)
Pyrazinamide	Bactericidal	Disruption of membrane Transport and energy depletion	Njire et al. (2016)
Ethambutol	Bactericidal	Inhibition of cell wall	Zhu et al. (2018)
Streptomycin	Bactericidal	Protein synthesis inhibition	Rocha et al. (2021)
Kanamycin	Bactericidal	Protein synthesis inhibition	Mantefardo and Sisay (2021)
Cycloserine	Bactericidal	Inhibit peptidoglycan synthesis	Mulubwa and Mugabo (2019a, b)
4-Aminosalisyclic acid	Bactericidal	Inhibition of iron and folic acid metabolism	Tanneau et al. (2022)
Quinolones	Bactericidal	DNA synthesis inhibition	Imran (2022)
Ethionamide	Bactericidal	Mycolic acid synthesis inhibition	Zhu et al. (2018)
Acebrophylline	Bactericidal	Inhibits cell wall synthesis	Rocha et al. (2021)
Bedaquiline	Bactericidal	Inhibits the ATP synthesis	Mantefardo and Sisay (2021)
Capreomycin	Bactericidal	Inhibits protein synthesis (translation)	Mulubwa and Mugabo (2019a, b)
Delamanid	Bactericidal	Inhibits the synthesis of mycolic acid	Tanneau et al. (2022)
Dexamethasone oral	Bactericidal	Chronic pericardial damage	Imran (2022)
Prothionamide	Bactericidal	Inhibits ATP synthase	Imran (2022)
Rifapentine	Bactericidal	Inhibits DNA-dependent RNA polymerase	Alfarisi et al. (2017)
Terizidone	Bactericidal	Inhibits bacterial cell wall synthesis	Mulubwa and Mugabo (2019a, b)
Tenocordin	Bactericidal	Inhibits the activity of cell wall synthesis of gram-positive bacteria	Grzelak et al. (2019)

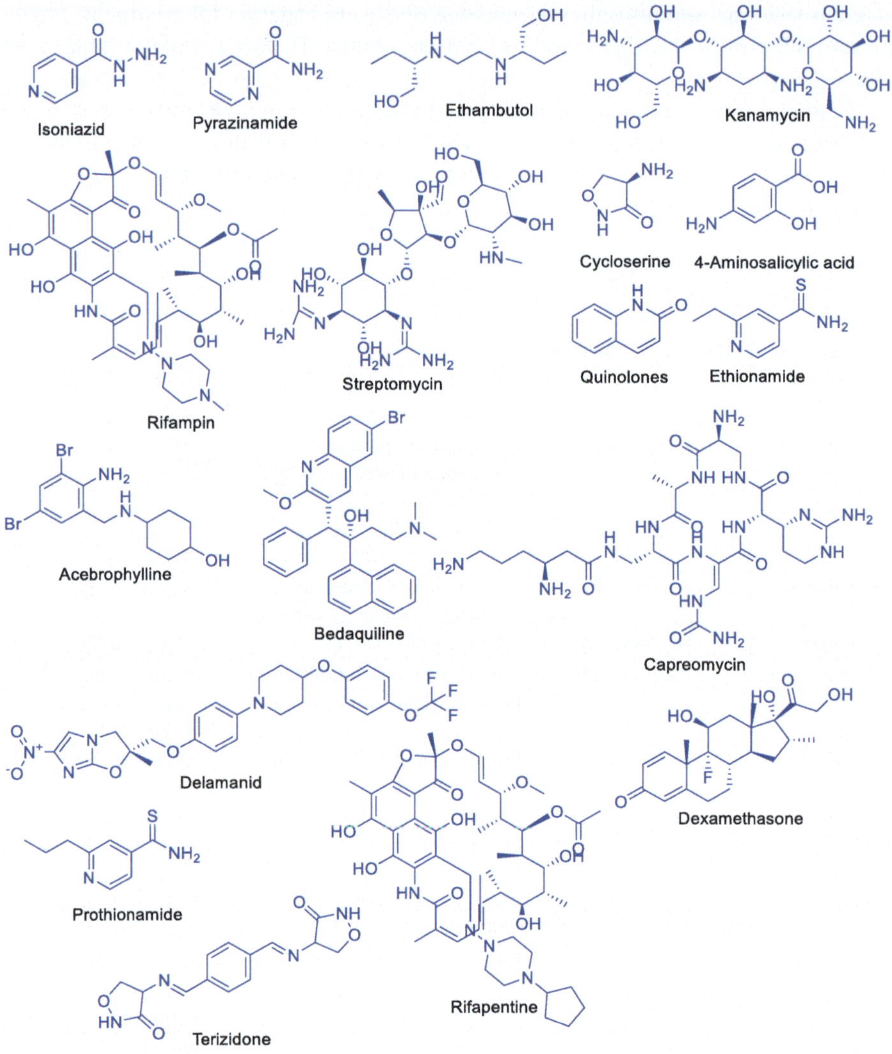

Fig. 4.2 Some anti-TB commercial drug

4.4 Computational Approaches and TB

The combat against TB is being revolutionized by the powerful tools of computational biology. These approaches unlock the secrets hidden within the vast amount of genomic data available for *M. tuberculosis*, the bacterium responsible for TB (Ejalonibu et al. 2021). The first step in drug discovery is identifying and researching the proteins or enzymes required for the pathogen's growth and development. Once researchers have precise information on the target and lead molecule, they use computational tools to screen these proteins against various chemicals or

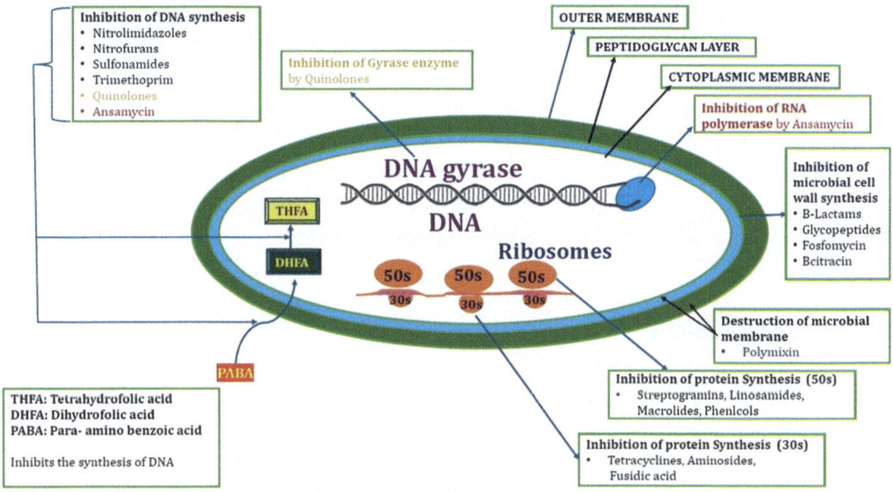

Fig. 4.3 Inhibitory mechanism of drugs on *M. tuberculosis*

Table 4.4 Comparison of the traditional method of drug development with CADD (Swanepoel and Loots 2014)

Traditional method	CADD
It includes more trial-and-error process	It is more logical
It requires blind screening	It is specific and mostly target-based
It is more expensive	It is cost-effective
It consumes more time	It reduces the time for development
It involves sequential steps	It has several sequential, parallel as well as straightforward steps

compounds in the libraries for potency and inhibitory impact, ultimately leading to drug candidate discovery. Pharmaceutical companies, organizations, and research labs may find this process useful in preventing them from pursuing the false hints. Computer-aided drug design (CADD) applications are being used to search for new therapeutics against TB in contrast to the time-consuming and costly traditional drug discovery approach. This is due to a new understanding of the quantitative relationship between structure and biological activity. Table 4.4 shows the advantages of CADD over traditional methodology (Ejalonibu et al. 2021; Sharma and Singh 2021).

By meticulously analysing genetic sequences, researchers can pinpoint vulnerabilities within the bacteria, making them prime targets for new drug development (Fontana et al. 2012). Furthermore, computational techniques allow researchers to unravel the intricate mechanisms by which *M. tuberculosis* develops resistance to existing antibiotics, paving the way for creating more effective therapies. This in silico analysis extends beyond resistance, enabling researchers to predict the emergence of new TB strains and anticipate potential challenges. Several MTB proteins are freely available in the protein data bank, viz., dihydropteroate synthase, to explore and develop therapeutic targets.

Additionally, computational modelling allows for the stimulation of drug interaction at the molecular level, guiding the design of more targeted and potent drug candidates (Lin et al., 2020). This multifaceted approach streamlines the drug discovery process, accelerating the development of novel treatments and ultimately contributing to eradicating TB as a global health threat.

4.5 Conclusion

Despite ongoing efforts to discover novel therapeutic strategies for TB, natural product-based approaches show promise as potential adjuvants to conventional treatment regimens. This consideration may include careful preclinical and clinical observation, which is a major step to ensure these products' efficiency and optimal quantity dosing and safety. These may include an eagle-eye observation of the product's potential and highlight its bioavailability and standardization. Another thing that comes into consideration is describing the mechanism action of these naturally gifted products, which exerts anti-tuberculosis effects, which is a must for optimizing use and guiding its development. Finally, exploration of drug-resistant TB strains and emphasizing the development of drug-resistant TB strains. Finally, ensuring sustainable sourcing and consistent natural product quality is critical for their safe and effective use. Establishing an excellent drug against TB. In conclusion, while natural products offer a valuable avenue for developing novel TB treatment strategies, a comprehensive scientific approach encompassing rigorous evaluation, mechanistic understanding, resistance management, and sustainable sourcing is necessary to fully unlock their potential and ensure their safe and effective integration into TB treatment regimens.

References

Abdoul-Latif FM, Okieh AA, Ali AM, Djama AA (2023) Tuberculosis prevalence among livestock in Djibouti: a regional investigation and herbal treatment arecommendations. J Anal Sci Appl Biotechnol 5(2):5–2

Akbik D, Ghadiri M, Chrzanowski W, Rohanizadeh R (2014) Curcumin as a wound healing agent. Life Sci 116(1):1–7

Alfarisi O, Alghamdi WA, Al-Shaer MH, Dooley KE, Peloquin CA (2017) Rifampin vs. rifapentine: what is the preferred rifamycin for tuberculosis? Expert Rev Clin Pharmacol 10(10):1027–1036

Bhanu MLS (2023) Anti-tuberculosis drugs and mechanisms of action: review. IJ Infectious Disea 4(2):1–7

Bishnoi S (2016) Herbs as functional foods. In: Mudgil D, Barak S (eds) Functional foods: sources and health benefits, pp 141–172

Chomini MS, Peter MK, Ameh M, Chomini AE, Bassey EA, Ayodele AO (2020) Phytochemical screening and antibacterial activities of Aframomum melegueta (K. Schum) seed extracts on salmonella typhi and Klebsiella pneumoniae. J Appl Sci Environ Manag 24(8):1419–1424

Choudhary JM, Parab T, Chorage A, Kesharvani K, Rakhe N, Shaikh KA (2021) Ayurvedic remedies of tuberculosis. World J Adv Res Rev 11(3):280–290
Chung KT, Wong TY, Wei CI, Huang YW, Lin Y (1998) Tannins and human health: a review. Crit Rev Food Sci Nutr 38(6):421–464
Dashti Y, Grkovic T, Quinn RJ (2014) Predicting natural product value, an exploration of anti-TB drug space. Nat Prod Rep 31(8):990–998
Ejalonibu MA, Ogundare SA, Elrashedy AA, Ejalonibu MA, Lawal MM, Mhlongo NN, Kumalo HM (2021) Drug discovery for mycobacterium tuberculosis using structure-based computer-aided drug design approach. Int J Mol Sci 22(24):13259
Elkington BG, Southavong B, Sydara K, Souliya O, Vanthanouvong M, Nettavong K et al (2009) Biological evaluation of plants of Laos used in the treatment of tuberculosis in Lao traditional medicine. Pharm Biol 47(1):26–33
Fontana JM, Alexander E, Salvatore M (2012) Translational research in infectious disease: current paradigms and challenges ahead. Transl Res 159(6):430–453
Getachew S, Medhin G, Asres A, Abebe G, Ameni G (2022) Traditional medicinal plants used in the treatment of tuberculosis in Ethiopia: a systematic review. Heliyon 8(5):e09478
Grzelak EM, Choules MP, Gao W, Cai G, Wan B, Wang Y et al (2019) Strategies in anti-mycobacterium tuberculosis drug discovery based on phenotypic screening. J Antibiot 72(10):719–728
Gupta PK, Kulkarni S (2018) Polysaccharide rich extract (PRE) from Tinospora cordifolia inhibits the intracellular survival of drug resistant strains of mycobacterium tuberculosis in macrophages by nitric oxide induction. Tuberculosis 113:81–90
Hasan MR, Alotaibi BS, Althafar ZM, Mujamammi AH, Jameela J (2023) An update on the therapeutic anticancer potential of Ocimum sanctum L.: "elixir of life". Molecules 28(3):1193
Imran M (2022) Ethionamide and Prothionamide based Coumarinyl-Thiazole derivatives: synthesis, Antitubercular activity, toxicity investigations and molecular docking studies. Pharm Chem J 56(9):1215–1225
Ishimaru N, Suzuki S, Shimokawa T, Iijima K, Kanzawa Y, Nakajima T, Kinami S (2022) Kikyo-to for acute upper respiratory tract infection-associated sore throat pain: a multicenter randomized controlled trial. J Integr Complement Med 28(9):768–774
Jahanpour S, Ghazisaidi K, Davoodi H, Mazandarani M, Samet M, Jahanpour N, Ghaemi EA (2015) Antimicrobial effects of folk medicinal plants from the north of Iran against mycobacterium tuberculosis. Arch Pediatr Infect Dis 3(1 TB)
Jeon D, Jeong MC, Jnawali HN, Kwak C, Ryoo S, Jung ID, Kim Y (2017) Phloretin exerts anti-tuberculosis activity and suppresses lung inflammation. Molecules 22(1):183
Jiang S, Wang M, Jiang L, Xie Q, Yuan H, Yang Y et al (2021) The medicinal uses of the genus Bletilla in traditional Chinese medicine: a phytochemical and pharmacological review. J Ethnopharmacol 280:114263
Kashyap RS, Nayak AR, Husain AA, Shekhawat SD, Satav AR, Jain RK et al (2016) Impact of socioeconomic status and living condition on latent tuberculosis diagnosis among the tribal population of Melghat: a cohort study. Lung India 33(4):372–380
Khlifi D, Hamdi M, Hayouni AE, Cazaux S, Souchard JP, Couderc F, Bouajila J (2011) Global chemical composition and antioxidant and anti-tuberculosis activities of various extracts of Globularia alypum L. (Globulariaceae) leaves. Molecules 16(12):10592–10603
Koche DEEPAK, Shirsat RUPALI, Kawale MAHESH (2016) An overview of major classes of phytochemicals: their types and role in disease prevention. Hislopia J 9(1/2):1–11
Kurane SB, Singh VB, Waghmare SD (2014) Rehabilitation of pulmonary tuberculosis patient with Kharjuradi Ghrita and pranayama: a study. Int J Res Health Sci 2(3):755–760
Li Y, Wang Q, Wei HC, Liang YY, Niu FJ, Li KW et al (2022) Fructus arctii: an overview on its traditional uses, pharmacology and phytochemistry. J Pharm Pharmacol 74(3):321–336
Liang Y, Gong W, Wang X, Zhang J, Ling Y, Song J et al (2021) Chinese traditional medicine NiuBeiXiaoHe (NBXH) extracts have the function of antituberculosis and immune recovery in BALB/c mice. J Immunol Res 2021(1):6234560
Lin G, Li D, De Carvalho LPS, Deng H, Tao H, Vogt G et al (2009) Inhibitors selective for mycobacterial versus human proteasomes. Nature 461(7264):621–626

Lin X, Li X, Lin X (2020) A review on applications of computational methods in drug screening and design. Molecules 25(6):1375
Maji AK, Pandit S, Banerji P, Banerjee D (2014) Pueraria tuberosa: a review on its phytochemical and therapeutic potential. Nat Prod Res 28(23):2111–2127
Mantefardo B, Sisay G (2021) Case report: kanamycin ototoxicity and MDR-TB treatment regimen. Int Med Case Rep J Volume 14:815–817
Maroyi A (2019) Ethnomedicinal uses, phytochemistry and pharmacological properties of Chironia baccifera. J Pharm Sci Res 11(11):3670–3674
Mathiasen VD, Andersen PH, Johansen IS, Lillebaek T, Wejse C (2020) Clinical features of tuberculous lymphadenitis in a low-incidence country. Int J Infect Dis 98:366–371
Mbuh TP, Ane-Anyangwe I, Adeline W, Thumamo Pokam BD, Meriki HD, Mbacham WF (2019) Bacteriologically confirmed extra pulmonary tuberculosis and treatment outcome of patients consulted and treated under program conditions in the littoral region of Cameroon. BMC Pulm Med 19:1–7
Mueller-Wieland K (1961) Tuberculosis and nutrition. Dtsch Med J 12:145–149
Mujovo SF (2010) Antimicrobial activity of compounds isolated from Lippia javanica (Burm. f.) Spreng and Hoslundia opposita against Mycobacterium tuberculosis and HIV-1 reverse transcriptase. Doctoral dissertation, University of Pretoria
Mulubwa M, Mugabo P (2019a) Amount of cycloserine emanating from terizidone metabolism and relationship with hepatic function in patients with drug-resistant tuberculosis. Drugs R&D 19:289–296
Mulubwa M, Mugabo P (2019b) Steady-state population pharmacokinetics of terizidone and its metabolite cycloserine in patients with drug-resistant tuberculosis. Br J Clin Pharmacol 85(9):1946–1956
Nedungadi P, Salethoor SN, Puthiyedath R, Nair VK, Kessler C, Raman R (2023) Ayurveda research: emerging trends and mapping to sustainable development goals. J Ayurveda Integr Med 14(6):100809
Németh E, Nagy Z, Pécsváradi A (2018) Chloroplast glutamine synthetase, the key regulator of nitrogen metabolism in wheat, performs its role by fine regulation of enzyme activity via negative cooperativity of its subunits. Front Plant Sci 9:191
Njire M, Tan Y, Mugweru J, Wang C, Guo J, Yew W et al (2016) Pyrazinamide resistance in mycobacterium tuberculosis: review and update. Adv Med Sci 61(1):63–71
Pancu DF, Scurtu A, Macasoi IG, Marti D, Mioc M, Soica C et al (2021) Antibiotics: conventional therapy and natural compounds with antibacterial activity – a pharmaco-toxicological screening. Antibiotics 10(4):401
Pandit A, Sachdeva T, Bafna P (2013) Ameliorative effect of leaves of Carica papaya in ethanol and antitubercular drug induced hepatotoxicity. Br J Pharm Res 3(4):648–661
Quan Y, Li L, Yin Z, Chen S, Yi J, Lang J et al (2022) Bulbus Fritillariae Cirrhosae as a respiratory medicine: is there a potential drug in the treatment of COVID-19? Front Pharmacol 12:784335
Racca T (2019) The role of gut microbiota in health and disease: how intestinal Dysbiosis contributes to pathogenesis in undernutrition and how it can be treated. In: Infection biology in a global perspective. Uppsala University & International Center Diarrhoeal Research, Bangladesh Infection Biology's global Health Journal, p 2
Rai R (2016) Herbal remedies in cure of tuberculosis prevalent among ethnic communities in Central India. Trop Plant Res 3:344–353
Rocha DM, Viveiros M, Saraiva M, Osório NS (2021) The neglected contribution of streptomycin to the tuberculosis drug resistance problem. Genes 12(12):2003
Roy S, Gorai D, Acharya R, Roy R (2014) Combretum (combretaceae): biological activity and phytochemistry. Am J Pharm Res 4(11):5266–5299
Samal J (2016) Ayurvedic management of pulmonary tuberculosis: a systematic review. J Intercult Ethnopharmacol 5(1):86
Sam-Ang P, Phanumartwiwath A, Liana D, Sureram S, Hongmanee P, Kittakoop P (2023) UHPLC-QQQ-MS and RP-HPLC detection of bioactive alizarin and scopoletin metabolites

from Morinda citrifolia root extracts and their antitubercular, antibacterial, and antioxidant activities. ACS Omega 8(32):29615–29624
Semenya SS, Maroyi A (2013) Medicinal plants used for the treatment of tuberculosis by Bapedi traditional healers in three districts of the Limpopo Province, South Africa. Afr J Tradit Complement Altern Med 10(2):316–323
Sharifi-Rad J, Salehi B, Stojanović-Radić ZZ, Fokou PVT, Sharifi-Rad M, Mahady GB et al (2020) Medicinal plants used in the treatment of tuberculosis-ethnobotanical and ethnopharmacological approaches. Biotechnol Adv 44:107629
Sharma M, Singh P (2021) Use of artificial intelligence in research and clinical decision making for combating mycobacterial diseases. In: Artificial intelligence and machine learning in healthcare, pp 183–215
Srivastava VK, Gupta UD, Tripathi P (2015) Potentiality of anti-mycobacterium (anti-ubercular) activity of ocimum tenuiflorum, justicia adhatoda, aloe vera, phyllanthus emblica, pipper longum, glycyrrhiza uralensis, allium sativum, zingiber officinale and curcuma longa against m tuberculosis. Ind Res J Gen Biotechnol 7(03):350–354
Sun D, Hurdle JG, Lee RE, Lee RE, Cushman M, Pezzuto JM (2012) Evaluation of flavonoid and resveratrol chemical libraries reveals abyssinone II as a promising antibacterial lead. ChemMedChem 7(9):1541
Swanepoel CC, Loots DT (2014) The use of functional genomics in conjunction with metabolomics for mycobacterium tuberculosis research. Dis Markers 2014(1):124218
Tanneau L, Karlsson MO, Rosenkranz SL, Cramer YS, Shenje J, Upton CM et al (2022) Assessing prolongation of the corrected QT interval with bedaquiline and delamanid coadministration to predict the cardiac safety of simplified dosing regimens. Clin Pharmacol Ther 112(4):873–881
Teixeira ÉW, Message D, Meira RM (2019) Methacrylate: an alternative fixing agent for identifying the botanical origin of propolis. Appl Plant Sci 7(12):e11309
Terracciano E, Amadori F, Zaratti L, Franco E (2020) Tuberculosis: an ever present disease but difficult to prevent. Igiene e Sanita Pubblica 76(1):59–66
Tuberculosis Drug Screening Program (2001) Search for new drugs for treatment of tuberculosis. Antimicrob Agents Chemother 45(7):1943–1946
Velásquez GE, Brooks MB, Coit JM, Pertinez H, Vargas Vásquez D, Sanchez Garavito E et al (2018) Efficacy and safety of high-dose rifampin in pulmonary tuberculosis. A randomized controlled trial. Am J Respir Crit Care Med 198(5):657–666
Yang Y, Lai Q, Wang C, Zhou G (2022) Protective effects of Herba Houttuyniae aqueous extract against OVA-induced airway Hyperresponsiveness and inflammation in asthmatic mice. Evid Based Complement Alternat Med 2022(1):7609785
Zhu C, Liu Y, Hu L, Yang M, He ZG (2018) Molecular mechanism of the synergistic activity of ethambutol and isoniazid against mycobacterium tuberculosis. J Biol Chem 293(43):16741–16750
Zumla A, Grange JM (1999) The ‘global emergency’ of tuberculosis. J R Coll Phys Edinb 29(2):104–115

Natural Product-Based Treatment for Leprosy

5

Ipsa Padhy, Ganesh Patro, and Madhulita Panda

Abstract

Hansen's disease or leprosy continues to be a serious health problem globally with high prevalence rate. Epidemiologically, the disease is endemic to tropical and sub-tropical Afro-Asian countries. The emergence of dapsone and rifampicin resistance poses a big challenge to the healthcare system for alleviating this neglected tropical disease. Moreover, leprosy complications, including arthritis, neuropathy and immunological reactions, challenge the multidrug therapeutic approach. In such a scenario, the magical herbs stand high as effective alternatives in the due course of leprosy mitigation. Ethnobotanical surveys provide substantial information regarding several plants or plant parts used in folklore in the treatment of leprosy, especially in developing nations. Alkaloids, saponin glycosides, terpenoids, flavonoids and polyphenols have tremendous potential in combating mycobacterial infections on a higher note than their synthetic counterparts. The current chapter reconnoitres a plethora of medicinal flora and their bioactive constituents addressing anti-leprosy potential with an emphasis on state-of-the-art integrative bioinformatic and computational methods utilized for natural product-based anti-leprosy drug discovery.

I. Padhy
Department of Pharmaceutical Chemistry, School of Pharmaceutical Sciences, Siksha 'O'Anusandhan (Deemed to be University), Bhubaneswar, Odisha, India

G. Patro (✉) · M. Panda
College of Pharmaceutical Sciences, Berhampur, Odisha, India
e-mail: ganeshpatro80@gmail.com

T. Sharma et al. (eds.), *Natural Products for Antibacterial Drug Development: Recent Advancement of Computational Approach*,
https://doi.org/10.1007/978-981-97-9634-2_5

Keywords

Leprosy · Multidrug therapy · Drug resistance · Ethnobotanical · Bioactive · Computational

Abbreviations

DDS	Diaminodiphenyl sulfone
DHPS	Deoxyhypusine synthase
ENR	Enoyl acyl carrier protein reductase
FBA	Flux balance analysis
FDA	Food and drug association
MDT	Multidrug therapy
RMSD	Root mean square deviation
RMSF	Root-mean-square fluctuation
TBDB	T-box riboswitch annotation database

5.1 Introduction

Leprosy, or Hansen's disease, has aggrieved human society since time immemorial (White and Franco-Paredes 2015). It is a chronic infectious ailment triggered by the bacterium *Mycobacterium leprae*. The disease predominantly affects the skin, peripheral nerves and mucous membrane lining of upper respiratory tract in humans. Leprosy stances as a major health concern worldwide with about 2 lakh fresh cases diagnosed annually (Schoenmakers et al. 2020). Socioeconomic factors, such as limited access to resources and nutritious food, play a significant role in the incidence of leprosy (Hannan et al. 2022). The prevalence of leprosy varies regionally, with India, Brazil and Indonesia accounting for more than 81% of new cases reported worldwide (Vengalil et al. 2020). The influence of leprosy outspreads beyond the physical indices, affecting the mental health and life quality of individuals. Humiliation and acumen accompanying the disease continues to be substantial barriers to initial finding and treatment, potentially leading to under-reporting of cases (Xiong et al. 2019; Marahatta et al. 2018). Additionally, the presence of grade 2 disability in leprosy patients remains a concern, highlighting the need for continued efforts in disability prevention and management (Gonçalves et al. 2019; Gunawan et al. 2021).

The conventional treatments of leprosy are associated with several challenges that hinder effective management of the disease. One of the primary challenges is the difficulty in diagnosing leprosy, as it can imitate various other diseases, leading to delays in initiating treatment (Samsudin and Karyanti 2021). Additionally, leprosy reactions, particularly leprosy-related arthritis, often do not respond to conventional therapy, contributing to the complexity of managing the disease (do Espírito Santo et al. 2020). The use of multiple medications, including analgesics, anti-inflammatories and antibiotics, in the course of leprosy treatment presents a significant challenge, especially in primary healthcare facilities (Cerqueira et al. 2020). Occurrence of adverse events, such as haemolytic anaemia, jaundice and methemoglobinaemia, is associated with the use of dapsone (Maymone et al. 2020). Additionally, the management of leprosy-related complications, such as neuropathy and immunologically mediated reactions, poses a significant challenge (Khadilkar et al. 2021). These reactions can occur before, during, or after completion of multidrug therapy (MDT) and can result in severe peripheral nerve injury (Aboagye et al. 2018). The duration of MDT itself is also a challenge, as it can last for 12–24 months (Alemu Belachew and Naafs 2019). Furthermore, the financial burden associated with advanced treatment modalities, such as hyperbaric oxygen therapy and reconstructive surgery, can be a challenge for leprosy-affected patients. Overall, these challenges highlight the need for improved treatment options, prompt diagnosis, and effective management of complications in leprosy patients.

The use of natural products for leprosy treatment has been a budding area of research. Traditional medicines are perfect alternatives for therapeutic management of infectious diseases owing to their safety, ready availability and minor or no undesirable adverse effects. The bioactive constituents stand high as promising leads for the development of novel drug candidates. The exploration of bioactive compounds as promising therapeutic alternatives necessitates careful assessment of their ethnomedicinal, chemico-biological, pharmacological and toxicity profiles (Barbosa-Filho et al. 2007; Swain et al. 2022).

In the current chapter we emphasize on plethora of medicinal plants with anti-leprosy potential. The chapter also highlights upon mechanism of action of the bioactive compounds against *M. leprae.* Focus has also been given to state-of-art computational methods applied for exploring potent bioactive compounds posing as leads for anti-leprosy drug development.

5.2 Ethnopharmacological Uses of Plants with Anti-Leprosy Potential

Phytomedicines have gained a huge impetus in the clinical management of several acute and chronic ailments with a safer therapeutic index. Traditionally, plants have been used as an important source of medicines in almost all traditional systems of healthcare. Natural product-based drugs are vital components in the folklore healing remedies. Reportedly 2.5–7.5 million species of higher plants currently existing are largest sources of bioactive medicinal compounds.

Numerous infectious diseases such as leprosy, venereal infections, febrile conditions, gastrointestinal and respiratory infections are therapeutically managed by natural product remedies in traditional folklore (Sundar et al. 2018). Ghosh and co-researchers provided comprehensive data regarding 40 families of plants possessing the potential to counter leprosy, which is used in traditional Indian folklore (Ghosh et al. 2017). Table 5.1 enlists some plants and their ethnomedicinal uses in the treatment of leprosy. Plants have been exclusively used to alleviate leprosy symptoms like skin lesions, thick dry skin, swelling, nodulations and ulcerations.

Table 5.1 Plants used in traditional medicine for leprosy treatment

Plant name	Family	Common name	Plant part used	References
Achyranthes aspera	Amaranthaceae	Prickly chaff flower	Dried roots and leaf powder	Akbar (2020)
Amaranthus spinosus	Amaranthaceae	Spiny amaranth	Roots	Kumar et al. (2014)
Adhatoda vasica	Acanthaceae	Malabar nut, vasaka	Leaf	Ghosh et al. (2017)
Andrographis paniculata	Acanthaceae	Kalmegh	Leaves	Akbar (2011)
Aegle marmelos	Rutaceae	Bael	Leaves, seeds	Sundar et al. (2018)
Alangium lamarckii	Cornaceae	Ankol	Root, bark, seeds	Shravya et al. (2017)
Ageratum conyzoides	Asteraceae	Neelam	Stem and leaves	Yadav et al. (2012)
Albizia lebbeck	Fabaceae	Shiris	Bark	Mokat and Kharat (2023)
Anacardium occidentale	Anacardiaceae	Cashew tree	Leaves, roots, barks	de Sousa et al. (2016)
Abutilon indicum	Malvaceae	Indian mallow	Leaves	Saini et al. (2015)
Azadirachta indica	Meliaceae	Neem	Leaves and seed oil	Alzohairy (2016)
Butea monosperma	Fabaceae	Palash	Flowers and bark	Hegde et al. (2014)
Bahunia variegata	Fabaceae	Kachnar	Roots, stem-bark	Sharma and Kumar (2012)
Cassia fistula	Fabaceae	Golden shower	Leaves	Sahu et al. (2014)
Calotropis procera	Apocyanaceae	Small crown flower	Latex and leaves	Khairnar et al. (2012)
Calotropis gigantea	Apocyanaceae	Giant milkweed	Latex, bark, leaves, roots	Sivapalan et al. (2023)
Cassia tora	Caesalpiniaceae	Sickle senns	Leaves, seeds	Sarwa et al. (2014)
Centella asiatica	Apiaceae	Indian pennywort, gotu kola	Entire herb	Süntar et al. (2012)
Celastrus paniculata	Celastraceae	Oriental bittersweet	Seed	Ghosh et al. (2017)
Curcuma longa	Zingiberaceae	Turmeric	Rhizhomes	Sabale et al. (2013)
Commelina bengalensis	Commelinaceae	Wandering jew	Leaves	Sharma and Kumar (2012)
Clitoria ternatea	Fabaceae	Butterfly plea	Roots	Lijon et al. (2017)

(continued)

Table 5.1 (continued)

Plant name	Family	Common name	Plant part used	References
Eclipta alba	Asteraceae	Bhringraj	Entire plant	Wagh and Jain (2020)
Ficus racemosa	Moraceae	Fig tree	Fruits	Deep et al. (2013)
Ficus bengalensis	Moraceae	Banyan tree	Bark	Logesh et al. (2023)
Gloriosa superba	Colchicaceae	Flame lily, tiger claw	Leaf	Tirkey et al. (2023)
Hydnocarpus wightiana	Flacourtiaceae	Jangli almond (Indian common name)	Seed oil, chaulmoogra oil	Sahoo et al. (2014)
Humulus lupulus	Cannabinaceae	Common hop	Seed oil	Carbone and Gervasi (2022)
Hemidesmus indicus	Apocyanaceae	Indian sarsaparilla	Roots	Nandy et al. (2020)
Ipomoea eriocarpa	Convolvulaceae	Tiny morning glory	Entire plant	Ganesan et al. (2022)
Kigelia africana	Bignoniaceae	Sausage tree	Fruit, stem bark	Van Wyk (2011), Kumawat et al. (2015)
Lawsonia intermis	Lythraceae	Henna	Leaves	Semwal et al. (2014)
Mimosa pudica	Fabaceae	Touch-me-not	Root	Adurosakin et al. (2023)
Momordica charantia	Cucurbitaceae	Bitter gourd	Fruit	Deka and Deka (2007)
Nerium indicum	Apocyanaceae	Indian oleander	Roots	Malik et al. (2019)
Withania sommnifera	Solanaceae	Indian ginseng, ashwagandha, winter cherry	Roots	Panchal and Singh (2015)

5.3 Bioactive Phytoconstituents with Anti-Leprosy Potency

Chaulmoogra oil is the most used in traditional medicine for the treatment of leprosy. The essential components of chaulmoogra oil are hydnocarpus acid, chaulmoogric acid and gorlic acid, while small amounts of oleic, linoleic and palmitic acid are also present (Sahoo et al. 2014). It is apprehended that both hydnocrpic and chaulmoogric acids stimulate lipase production in host thereby disrupting microbial cell wall lipids. Reportedly hydnocarpic acid also inhibits mycobacterial multiplication (Port et al. 2023). Asiaticoside isolated from *Centella asiatica* showed antimycobacterial activity in vitro and in vivo, *probably by enhancing the* proliferation of dermal fibroblasts and increasing collagen content and vascularization of lesions (Lee et al. 2012). Boswellic acid (triterpene from *Boswellia serrata*), dalbergin (flavonoid from *Dalbergia latifolia*) and curcumin (from *Curcuma longa*) presented anti-leprotic potential in vivo (Barbosa-Filho et al. 2007). Saponin glycosides (sarsaparilloside B, sarsaparilloside C, sarsaparilloside, $\Delta^{20(22)}$-sarsaparilloside and parillin) isolated from Smilax roots have shown promising lethal effects against *Mycobacterium leprae* in vitro (Challinor et al. 2012). Fusidic acid (tetracyclic triterpenoid isolated from microbe, *Fusidium coccineum*) was found active against the leprosy causative bacterium *M. leprae* in macrophage cultures (IC_{50} = 1.25–2.5 μg/mL) (Gurel et al. 2015; Barbosa-Filho et al. 2007).

There are a plethora of natural compounds possessing antimycobacterial potency. The broad-spectrum bactericidal activities of the phytochemicals apprehend their extensive use in the traditional healthcare in the alleviation of the neglected tropical diseases such as leprosy and Buruli's ulcers. Polyphenolics, triterpenoids, alkaloids and especially monoterpenes possess strong bactericidal action against Mycobacterium species by inhibiting the bacteria's growth and replication, and sometimes, causing cell death (Okunade et al. 2004). The bioactive compounds (Fig. 5.1) disrupt bacterial cell walls and membranes, reduce enzyme activity and interfere with essential metabolic processes, thereby decreasing mycobacterial cell viability (Han et al. 2022; Ansari et al. 2023). Besides the bactericidal, anti-inflammatory and wound healing properties of the bioactive compounds have been attributed to their anti-leprosy activities.

5.4 Natural Product-Based Anti-Leprosy Drug Development: State-of-Art Approaches

Reductive evolution, genome downsizing and gene decay have eliminated several catabolic pathways, thereby leaving a few functional genes in *M. leprae* (Acebrón-García-de-Eulate et al. 2021). Presently, the genome of *M. leprae* comprises 1614 genes, out of which only 1010 genes code for functional proteins while the rest, 592 genes, code for hypothetical proteins studied through homology modelling. Culturing the leprosy bacilli in vitro is challenging (Scollard et al. 2006). In this scenario, computational approaches integrated with genomic strategies play a crucial role in discovering novel drugs against MDT resistant *M. leprae* (Fig. 5.2) (Anusuya and Natarajan 2013). Assorted facts from homology knockout trials, flux balance and chokepoint analysis experiments could help identify essential genes critical for inducing infection and ensure no off-target activity on the host (Acebrón-García-de-Eulate et al. 2021).

Numerous repositories have been utilized for integrating genomic and structural data linked to drug resistance in *Mycobacteria* (Table 5.2) in the due process of leprosy drug discovery.

These vital resources contribute towards better understanding of molecular mechanisms involved in drug resistance and improvement in the selection of potential drug targets in the antimycobacterial drug development pipeline.

Structure-based drug design has played a crucial role in the in designing novel anti-leprosy lead compounds (Fig. 5.3).

Kaur and his research team virtually screened the compound against the LipU protein of *M. tuberculosis* and *M. leprae* (Kaur et al. 2019). The lipolytic protein is

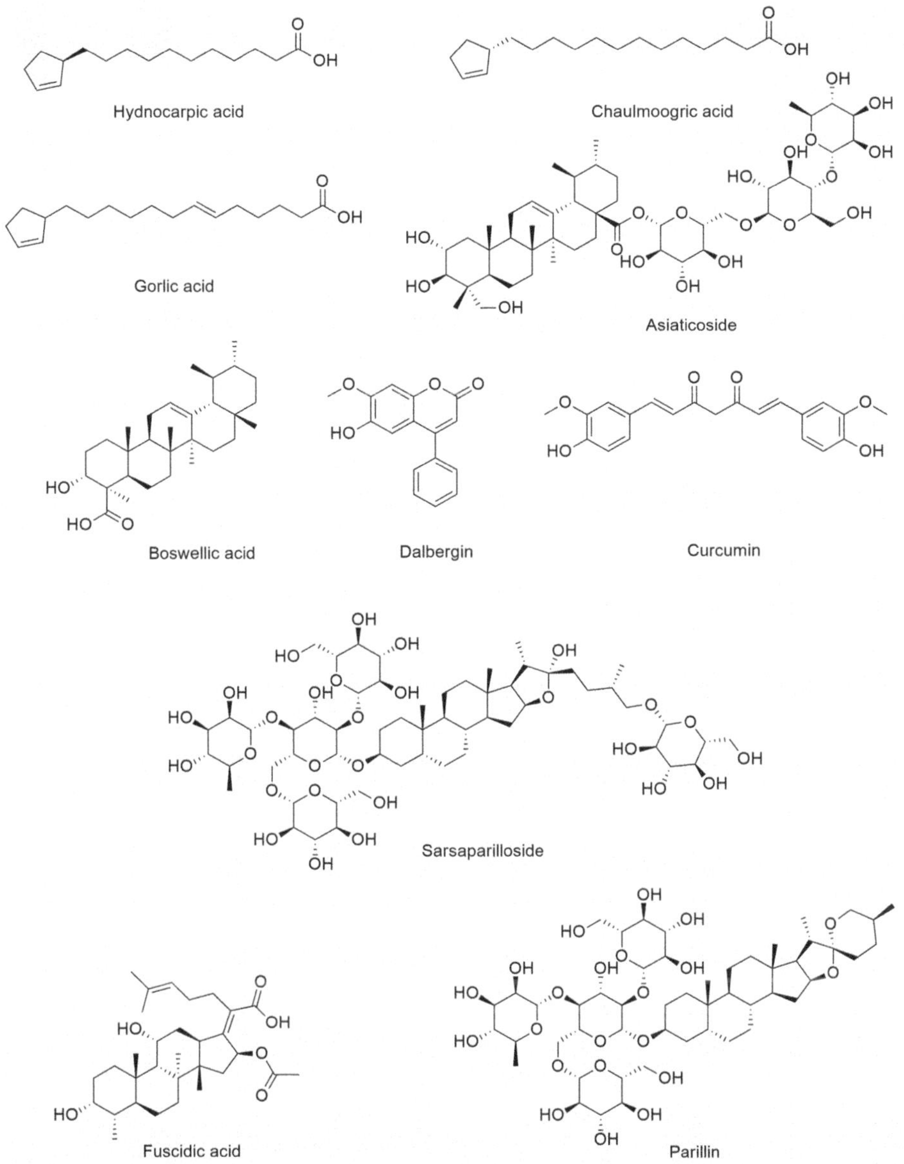

Fig. 5.1 Bioactive compounds with anti-leprosy potential

specific to mycobacteria and is an essential enzyme regulating lipid catabolism crucial for survival of bacilli. In the study some FDA approved drugs (for other ailments) were virtually screened against modelled LipU proteins of *M. tuberculosis* (MTBLipU) and its ortholog *M. leprae* (MLPLipU). Interestingly, the bioactive compounds showed higher binding affinities with the binding pockets of MTBLipU

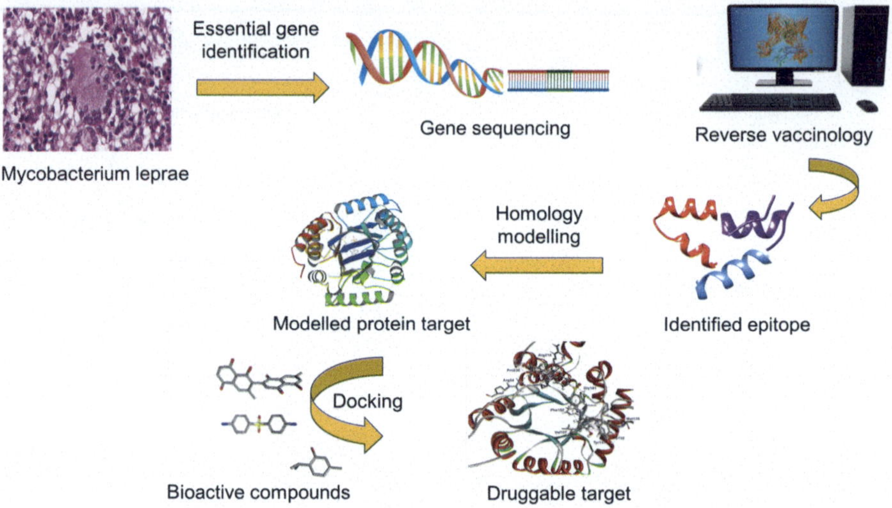

Fig. 5.2 Integrated genomic and computational strategies for anti-leprosy drug development

Table 5.2 Databases for genomic analysis of *Mycobacteria*

Database	Description	Information	References
TDR targets	Curated database of pathogens causing tropical diseases and relevant molecular targets for drug development.	Genomic data of pathogens and data regarding synthetic as well as natural compounds as anti-infection agents.	Urán Landaburu et al. (2020)
MycoResistance	Curated web resource of drug resistance molecules in mycobacteria.	Drug resistance molecules in six different mycobacterial species with annotations.	Dai et al. (2019)
TBDB	Genomic data repository.	Gene expression analysis data of *mycobacterium tuberculosis* and other strains.	Reddy et al. (2009)
PolyTB	Webserver to detect genomic diversity polymorphism in mycobacterial strains.	Consists genomic datasets, phylogenetic lineages.	Coll et al. (2014)
Modbase	Database of modelled proteins.	Consists of 432 modelled proteins of *M. leprae.*	Pieper et al. (2014)
CHOPIN	Webserver for structural and functional proteomics.	Contain homology models of mycobacterial proteins and structural analysis of drug resistance mutations.	Ochoa-Montano et al. (2015)

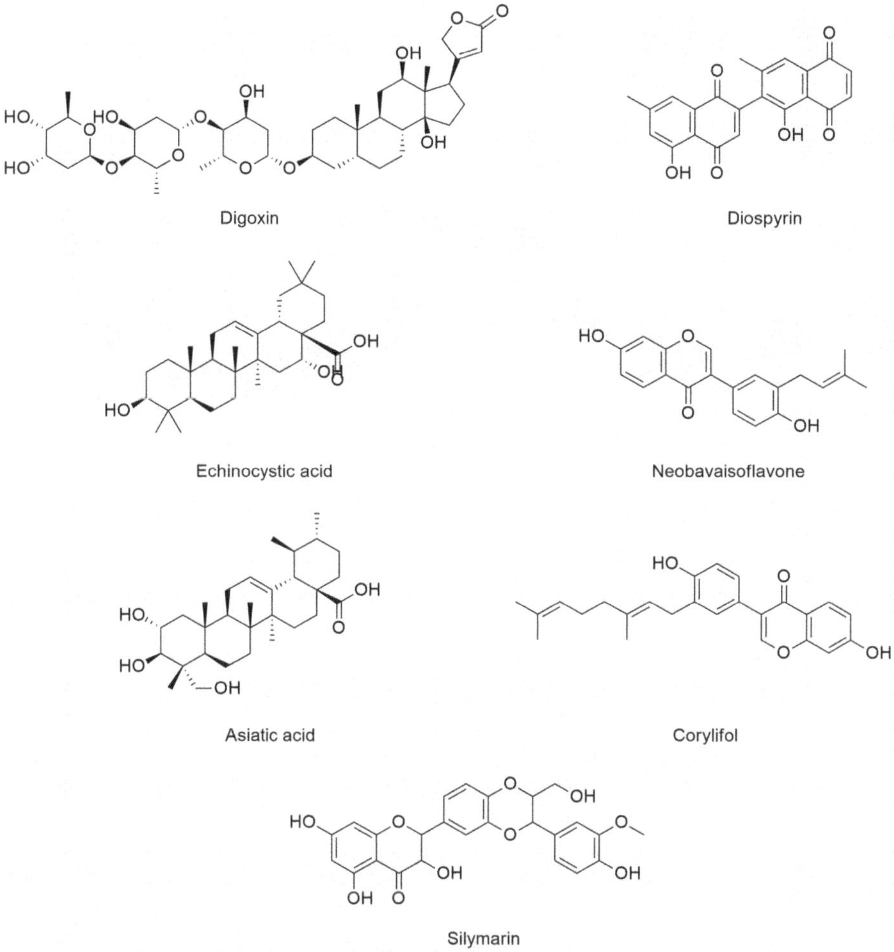

Fig. 5.3 Bioactive compounds as potential leads for anti-leprosy drug development

and MLPLipU, especially digoxin for MLPLipU (binding score − 9.2 kcl/mol). Bigger binding cavity of MLB LipU easily permitted large molecules such as Digoxin (Fig. 5.3) to strongly interact. Stable RMSD, RMSF and Rg values were obtained throughout the molecular dynamic simulation period.

In another study, Jaiswal and co-researchers employed a combined reverse vaccinology and substractive genomic approach for detecting 12 supposed antigenic proteins as vaccine targets and three common drug targets against *Mycobacterium leprae* and *Mycobacterium lepromatosis.* The selected proteins were absolutely specific to the pathogen and non-homologous with the host genome. Consequently, 28 natural compounds were screened against three selected drug targets: Two-component system regulator protein, Phosphoglycerate mutase family protein and Thiamine biosynthesis protein. Among the 28 bioactive compounds screened

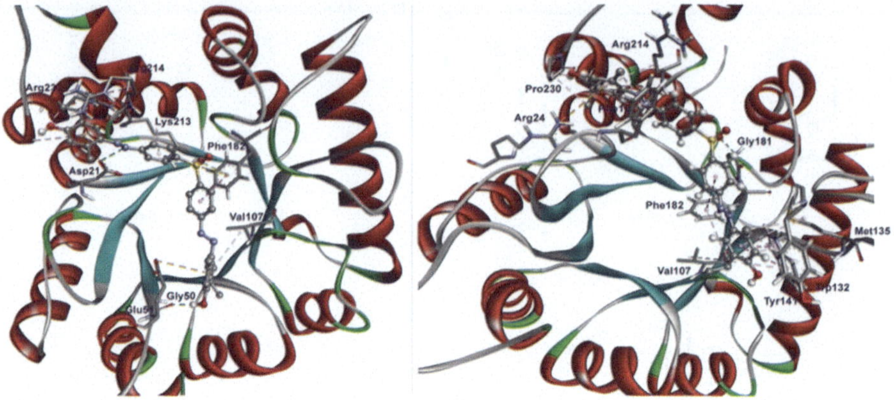

Fig. 5.4 Molecular interactions of DPC4 with mutant dihydropteroate synthase. (Figure adapted from Swain et al. 2020)

against the three putative targets, diospyrin (Fig. 5.4) as the most promising compound with the best interactions with identified drug targets (Jaiswal et al. 2021).

Point mutation in the dihydropteroate synthase, a putative enzyme responsible for folic acid synthase in mycobacteria, is responsible for emergence of dapsone resistance (Chaitanya et al. 2015). Reportedly, the chemical alternation of dapsone produced potent inhibitors for both wild-type and mutant forms of dihydropteroate synthase (Swain et al. 2018). Continuing with, Swain and team synthesized five dapsone-phytochemical conjugates by azo-coupling of 4-hydroxy coumarin, eugenol, salicylic acid, thymol and vanillin with dapsone to yield compounds DPC1–DPC5 respectively. The three-dimensional structures of designed conjugates were docked into the binding pockets of the modelled proteins (both wild and mutants) to apprehend the nature of interactions with predicted binding pocket residues. The molecular dynamic simulation analysis revealed DPC4-protein complex being most stable. Moreover, number of viable bacilli cells in DDS-resistant infected mice footpads significantly reduced. Additionally, the in vitro host toxicity study also confirmed that the DCP4 up to 5000 mg/L level was safety for oral administration (Swain et al. 2020). Conjugating phytochemicals with small-molecule drugs could pave the way for novel anti-leprosy pharmacological agents.

In another study about 25 phytochemicals were screened against modelled dihydropteroate synthase of *Mycobacterium leprae* to establish their anti-leprosy potential (Swain et al. 2022). Selective phytoconstituents were docked with wild and mutant types of the dihydropteroate synthetases (DHPS) of *Mycobacterium leprae*. The docking scores indicated asiaticoside, an active saponin (−11.3, −11.2 and −11.2 kcal/mol), was proved to be the lead phytochemical against both wild and mutant types DHPS. The binding interactions of phytoconstituents (Fig. 5.3) including echinocystic acid (−9.6, −9.5 and −9.5 kcal/mol), neobavaisoflavone (−9.2, −9.0 and −9.0 kcal/mol), boswellic acid (−8.90, −8.90 and −8.90 kcal/mol), asiatic acid (−8.9, −8.8 and −8.9 kcal/mol) and corylifol A (−8.8, 8.0, and −8.0) also backed up their anti-leprosy potency.

Blocking biosynthesis of long chain fatty acids via inhibiting enoyl-acyl carrier protein reductase (ENR), has emerged as a promising strategy in developing novel *M. leprae* inhibitors (Acebrón-García-de-Eulate et al. 2021). ENR (EC. 1.3.1.9) is a homodimeric protein comprised of 268 amino acid residues and an active binding site in each monomer. The enzyme has a deep hydrophobic pocket in which the active site present. The bioactive compounds (Fig. 5.3) exhibited good feasibility as *M. leprae* enoyl-acyl carrier protein reductase inhibitors (Al-Madhagi 2022). Molecular docking results revealed that all of the tested compounds showed better binding energy than the reference drug dapsone. The best of which was silymarin.

5.5 Future Perspectives and Conclusion

Leprosy is a complex disease caused by *Mycobacterium leprae* and in some cases, *Mycobacterium lepromatosis*. The disease presents in various types, each with distinct clinical and genetic characteristics. Understanding the causative agents, characteristics, and types of leprosy is essential for developing effective diagnostic, treatment and prevention strategies. Existing pharmaceutical interventions for leprosy include the use of active pharmaceutical ingredients such as dapsone, single-dose rifampicin-post exposure prophylaxis, and multidrug therapy. However, addressing social stigma, knowledge deficits, and the management of leprosy-related complications are equally important in the overall control and treatment of leprosy.

The current treatment of leprosy primarily relies on multidrug therapy, but the use of natural products has been a subject of interest due to their potential therapeutic benefits. Traditional medicine, particularly the use of natural products, has been a focus of research for leprosy treatment. For instance, *Centella asiatica*, a plant traditionally used in Nepal, has been reported to have various medicinal properties, including the treatment of leprosy (Biswas et al. 2021; Diniz et al. 2023). Additionally, other natural products such as *Punica granatum* L. (pomegranate) and *Calotropis gigantea* have been historically used for the treatment of leprosy in traditional medicine (Farghaly et al. 2020; Chitme et al. 2005). These examples highlight the potential of natural products in the management of leprosy. Furthermore, the use of natural products in traditional medicine, such as *Hydnocarpus anthelminthicus* and *Euphorbia nivulia* Buch-Ham, has been documented for the treatment of leprosy (Leelawat and Leelawat 2018; Lodhi et al. 2006). These traditional remedies have been used for various ailments, including leprosy, demonstrating the historical significance of natural products in managing this disease. Moreover, the potential of natural products in the treatment of leprosy is further supported by the increasing focus of natural product chemists on exploring the biological activities of natural compounds, such as resveratrol oligomers (Shen et al. 2017). This emphasis on natural product research underscores the potential for discovering novel compounds with therapeutic benefits for leprosy treatment. While the current standard of care for leprosy primarily involves drug therapy, the historical and contemporary use of natural products in traditional medicine for leprosy treatment warrants further exploration. The potential of natural products in managing leprosy presents an opportunity for further research and development of complementary or alternative treatment options.

The application of advanced computational strategies collaboratively with current state-of-art genomic approaches have greatly elucidated the structural impacts of point mutations in the drug target proteins and their influence on function of the drug targets in mycobacteria is vital to understanding molecular mechanisms of drug resistance. A multitude of natural compounds have been successfully screened against antimycobacterial targets which regulate crucial metabolic pathways essential for microbial survival, virulence and developing resistance to frontline antibiotics. Evidently, the natural compounds present themselves as excellent alternatives for the pharmacotherapy of leprosy.

References

Aboagye SY, Kpeli G, Tuffour J, Yeboah-Manu D (2018) Challenges associated with the treatment of Buruli ulcer. J Leukoc Biol 105(2):233–242. https://doi.org/10.1002/jlb.mr0318-128

Acebrón-García-de-Eulate M, Blundell TL, Vedithi SC (2021) Strategies for drug target identification in mycobacterium leprae. Drug Discov Today 26(7):1569–1573

Adurosakin OE, Iweala EJ, Otike JO, Dike ED, Uche ME, Owanta JI et al (2023) Ethnomedicinal uses, phytochemistry, pharmacological activities and toxicological effects of Mimosa pudica-a review. Pharmacol Res Mod Chin Med 7:100241

Akbar S (2011) Andrographis paniculata: a review of pharmacological activities and clinical effects. Altern Med Rev 16(1):66–77

Akbar S (2020) Achyranthes aspera L. (Amaranthaceae). In: Handbook of 200 medicinal plants. Springer, Cham. https://doi.org/10.1007/978-3-030-16807-0_8

Alemu Belachew W, Naafs B (2019) Position statement: LEPROSY: diagnosis, treatment and follow-up. J Eur Acad Dermatol Venereol 33(7):1205–1213. https://doi.org/10.1111/jdv.15569

Alzohairy MA (2016) Therapeutics role of Azadirachta indica (Neem) and their active constituents in diseases prevention and treatment. Evid Based Complement Alternat Med 2016

Al-Madhagi H (2022) In silico evaluation of natural bioactive compounds as Mycobacterium leprae enoyl acyl carrier protein reductase inhibitors. JNCB 3(1):1–10. https://doi.org/10.48185/jcnb.v3i1.512

Ansari MA, Shoaib S, Alomary MN, Ather H, Ansari SMA, Hani U et al (2023) Deciphering the emerging role of phytocompounds: implications in the management of drug-resistant tuberculosis and ATDs-induced hepatic damage. J Infect Public Health 16:1443

Anusuya S, Natarajan J (2013) The eradication of leprosy: molecular modeling techniques for novel drug discovery. Expert Opin Drug Discov 8(10):1239–1251

Barbosa-Filho JM, Nascimento Júnior FAD, Tomaz ACDA, Athayde-Filho PFD, Silva MSD, Cunha EV et al (2007) Natural products with antileprotic activity. Rev Bras 17:141–148

Biswas D, Mandal S, Chatterjee Saha S, Tudu CK, Nandy S, Batiha GES et al (2021) Ethnobotany, phytochemistry, pharmacology, and toxicity of Centella asiatica (L.) urban: a comprehensive review. Phytother Res 35(12):6624–6654

Carbone K, Gervasi F (2022) An updated review of the genus Humulus: a valuable source of bioactive compounds for health and disease prevention. Plan Theory 11(24):3434

Cerqueira SR, Santos LS, Morelo EF, de Santos Júnior A, Sousa CA, Gonçalves RT, Hans Neto G, da Marques D, Sampaio RN, Kurizky PS, Gomes CM (2020) The interference of polypharmacy and the importance of clinical pharmacy advice in the treatment of leprosy: a case-control study. Rev Soc Bras Med Trop 53:e20200114. https://doi.org/10.1590/0037-8682-0114-2020

Chaitanya S, Das M, Bhat P, Ebenezer M (2015) Computational modelling of dapsone interaction with dihydropteroate synthase in mycobacterium leprae; insights into molecular basis of dapsone resistance in leprosy. J Cell Biochem 116(10):2293–2303

Challinor VL, Parsons PG, Chap S, White EF, Blanchfield JT, Lehmann RP, De Voss JJ (2012) Steroidal saponins from the roots of smilax sp.: structure and bioactivity. Steroids 77(5):504–511

Chitme HR, Chandra R, Kaushik S (2005) Evaluation of antipyretic activity of Calotropis gigantea (Asclepiadaceae) in experimental animals. Phytother Res 19(5):454–456

Coll F, Preston M, Guerra-Assunção JA, Hill-Cawthorn G, Harris D, Perdigão J et al (2014) PolyTB: a genomic variation map for mycobacterium tuberculosis. Tuberculosis 94:346–354. https://doi.org/10.1016/j.tube.2014.02.005

Dai E, Zhang H, Zhou X, Song Q, Li D, Luo L et al (2019) MycoResistance: a curated resource of drug resistance molecules in mycobacteria. Database 2019:baz074

de Sousa Leite A, Islam MT, Júnior ALG, e Sousa JMDC, de Alencar MVOB, Paz MFCJ, Rolim HML, de Medeiros MDGF, de Carvalho Melo-Cavalcante AA, Lopes JAD (2016) Pharmacological properties of cashew (Anacardium occidentale). Afr J Biotechnol 15(35):1855–1863

Deep P, Singh KA, Ansari MT, Raghav P (2013) Pharmacological potentials of Ficus racemosa – a review. Int J Pharm Sci Rev Res 22(1):29–34

Deka SJ, Deka SP (2007) Survey of medicinal plants used against leprosy disease by the tribal (Lalung) people of Myong area of Morigaon District, Assam, India

Diniz L, Calado L, Duarte A, Sousa D (2023) Centella asiatica and its metabolite asiatic acid: wound healing effects and therapeutic potential. Meta 13(2):276. https://doi.org/10.3390/metabo13020276

do Espírito Santo RB, Serafim RA, Bitran JB, Collin SM, Deps P (2020) Case report: leprosy osteo-articular alterations mimicking rheumatoid arthritis. Am J Trop Med Hyg 102(6):1316–1318. https://doi.org/10.4269/ajtmh.19-0723

Farghaly F, Radi A, Al-Kahtany F (2020) Impacts of zinc oxide nano and bulk particles on redox-enzymes of the punica granatum callus. Sci Rep 10(1):19722. https://doi.org/10.1038/s41598-020-76664-4

Ganesan T, Subban M, Christopher Leslee DB, Kuppannan SB, Seedevi P (2022) Structural characterization of n-hexadecanoic acid from the leaves of Ipomoea eriocarpa and its antioxidant and antibacterial activities. Biomass Convers Biorefinery:1–12

Ghosh G, Narayan B, Bengal W (2017) Traditional use of plants against leprosy in India: a review of the recent literature. J Innov Pharm Biol Sci 4(4):55–64

Gonçalves FG, de Belone A, Rosa PS, Laporta GZ (2019) Underlying mechanisms of leprosy recurrence in the western amazon: a retrospective cohort study. BMC Infect Dis 19(1):1–10. https://doi.org/10.1186/s12879-019-4100-6

Gunawan H, Kamilia A, Menaldi SL, Marissa M, Prakoeswa CR, Alinda MD, Widasmara D, Basuki S, Anum Q, Ariani T, Mamuaja EH, Rusyati LM (2021) Characteristics of grade 2 disability in Indonesian children with leprosy: a five-year multicenter retrospective study. Clin Cosmet Investig Dermatol 14:1149–1153. https://doi.org/10.2147/ccid.s325858

Gurel MS, Naycı S, Turgut AV, Bozkurt ER (2015) Comparison of the effects of topical fusidic acid and rifamycin on wound healing in rats. Int Wound J 12(1):106–110

Han J, Liu X, Zhang L, Quinn RJ, Feng Y (2022) Anti-mycobacterial natural products and mechanisms of action. Nat Prod Rep 39(1):77–89

Hannan M, Notobroto HB, Hidayat S, Mumpuningtias ED, Hasina SN (2022) Behavior of patients with leprosy in disability prevention through patients' self care: a literature review. Int J Health Sci 6(S7):3504–3527. https://doi.org/10.53730/ijhs.v6ns7.12622

Hegde SV, Hegde GR, Mannur S, Poti SS (2014) Pharmacognostical studies on Butea monosperma (lam.) Taub (Faboideae) flower. Int J Pharm Phytopharmacol Res 4(1):34–36

Jaiswal AK, Tiwari S, Jamal SB, Oliveira LDC, Sales-Campos H, Andrade-Silva LE et al (2021) Reverse vaccinology and subtractive genomics approaches for identifying common therapeu-

tics *against mycobacterium leprae* and *mycobacterium lepromatosis*. J Venom Anim Toxins Incl Trop Dis 27:e20200027
Kaur G, Pandey B, Kumar A, Garewal N, Grover A, Kaur J (2019) Drug targeted virtual screening and molecular dynamics of LipU protein of mycobacterium tuberculosis and mycobacterium leprae. J Biomol Struct Dyn 37(5):1254–1269
Khadilkar SV, Patil SB, Shetty VP (2021) Neuropathies of leprosy. J Neurol Sci 420:117288. https://doi.org/10.1016/j.jns.2020.117288
Khairnar AK, Bhamare SR, Bhamare HP (2012) Calotropis procera: an ethnopharmacological update. Adv Res Pharm Biol 2(2):142–156
Kumar RP, Jindal S, Gupta N, Rana R (2014) An inside review of Amaranthus spinosus Linn: a potential medicinal plant of India. Int J Res Pharm Chem 4(3):643–653
Kumawat RB, Mali PC, Sharma RA (2015) Ethanomedicine and pharmacological activities of five traditionally used Indian medicinal plants: a review. Adv Pharmacol Toxicol 16(3):45
Lee JH, Kim HL, Lee MH, You KE, Kwon BJ, Seo HJ, Park JC (2012) Asiaticoside enhances normal human skin cell migration, attachment and growth in vitro wound healing model. Phytomedicine 19(13):1223–1227
Leelawat S, Leelawat K (2018) Cytokine secretion of peripheral blood mononuclear cells by hydnocarpus anthelminthicus seeds. J Trop Med 2018:1–8. https://doi.org/10.1155/2018/6854835
Lijon MB, Meghla NS, Jahedi E, Rahman MA, Hossain I (2017) Phytochemistry and pharmacological activities of Clitoria ternatea. Int J Nat Soc Sci 4(1):1–10
Lodhi M, Hussain J, Abbasi M, Jassbi A, Choudhary M, Ahmad V (2006) A newbacillus pasteuriiurease inhibitor fromeuphorbia decipiens. J Enzyme Inhib Med Chem 21(5):531–535. https://doi.org/10.1080/14756360600586072
Logesh R, Sathasivampillai SV, Varatharasan S, Rajan S, Das N, Pandey J, Devkota HP (2023) Ficus benghalensis L. (Moraceae): a review on ethnomedicinal uses, phytochemistry and pharmacological activities. Curr Res Biotechnol 6:100134
Malik K, Ahmad M, Zafar M, Ullah R, Mahmood HM, Parveen B et al (2019) An ethnobotanical study of medicinal plants used to treat skin diseases in northern Pakistan. BMC Complement Altern Med 19:1–38
Marahatta SB, Amatya R, Adhikari S, Giri D, Lama S, Kaehler N, Rijal KR, Marahatta S, Adhikari B (2018) Perceived stigma of leprosy among community members and health care providers in Lalitpur district of Nepal: A qualitative study. PLoS One 13(12):e0209676
Maymone MBC, Venkatesh S, Laughter M, Abdat R, Hugh J, Dacso MM, Rao PN, Stryjewska BM, Dunnick CA, Dellavalle RP (2020) Leprosy: treatment and management of complications. J Am Acad Dermatol 83(1):17–30. https://doi.org/10.1016/j.jaad.2019.10.138
Mokat DN, Kharat TD (2023) Albizia lebbeck (L.) Benth.: Shirish tree with its Phytochemistry, ethnobotany and pharmacology. In: Bioactives and pharmacology of legumes. Apple Academic Press, pp 339–352
Nandy S, Mukherjee A, Pandey DK, Ray P, Dey A (2020) Indian sarsaparilla (Hemidesmus indicus): recent progress in research on ethnobotany, phytochemistry and pharmacology. J Ethnopharmacol 254:112609
Ochoa-Montano B, Mohan N, Blundell TL (2015) CHOPIN: a web resource for the structural and functional proteome of mycobacterium tuberculosis. Database 2015:1–10
Okunade AL, Elvin-Lewis MP, Lewis WH (2004) Natural antimycobacterial metabolites: current status. Phytochemistry 65(8):1017–1032
Panchal P, Singh K (2015) Antimicrobial activity of Withania somnifera and Calotropis procera on pathogenic strains. Int J Curr Pharm Res 7(4):76–78
Pieper U, Webb BM, Dong GQ, Schneidman-Duhovny D, Fan H, Kim SJ et al (2014) ModBase, a database of annotated comparative protein structure models and associated resources. Nucleic Acids Res 42:D336–D346
Port BCB, Argenta DF, Porto DS, Rauber GS, Zuchi IDP, Silva IT, Caon T (2023) Chaulmoogra oil-based nanoemulsions for leprosy treatment: a case study with the dapsone. J Drug Deliv Sci Technol 88:104955

Reddy TB, Riley R, Wymore F, Montgomery P, DeCaprio D, Engels R et al (2009) TB database: an integrated platform for tuberculosis research. Nucleic Acids Res 37:D499–D508
Sabale P, Modi A, Sabale V (2013) Curcuma longa Linn. A phytochemical and phytopharmacological review. Res J Pharmacog Phytochem 5(2):59–68
Sahoo MR, Dhanabal SP, Jadhav AN, Reddy V, Muguli G, Babu UV, Rangesh P (2014) Hydnocarpus: an ethnopharmacological, phytochemical and pharmacological review. J Ethnopharmacol 154(1):17–25
Sahu PK, Masih V, Gupta S, Sen DL, Tiwari A (2014) Ethnomedicinal plants used in the healthcare systems of tribes of Dantewada, Chhattisgarh India. Am J Plant Sci 2014:1632
Saini A, Gahlawat DK, Chauhan C, Gulia SK, Ganie SA, Yadav SS (2015) Ethnomedicinal uses and phytochemistry of Abutilon indicum (Linn.) sweet: an overview. J Pharmacogn Phytochem 3(5):66–72
Samsudin DD, Karyanti MR (2021) Atypical cinical manifestation of leprosy in Indonesian male adolescent: a case report. Paediatr Indones 61(3):175–178. https://doi.org/10.14238/pi61.3.2021.175-8
Sarwa KK, Mithun Rudrapal MR, Manabendra Debnath MD, Atul Kumar AK, Verma VK (2014) Phytochemical and biological potential of Cassia tora Linn. Eur J Med Plant 4(8):946–963
Schoenmakers A, Mieras L, Budiawan T, van Brakel W (2020) The state of affairs in post-exposure leprosy prevention: a descriptive meta-analysis on immuno-and chemo-prophylaxis. Res Rep Trop Med 11:97–117. https://doi.org/10.2147/rrtm.s190300
Scollard DM, Adams LB, Gillis TP, Krahenbuhl JL, Truman RW, Williams D (2006) The continuing challenges of leprosy. Clin Microbiol Rev 19(2):338–381
Semwal RB, Semwal DK, Combrinck S, Cartwright-Jones C, Viljoen A (2014) Lawsonia inermis L. (henna): ethnobotanical, phytochemical and pharmacological aspects. J Ethnopharmacol 155(1):80–103
Sharma S, Kumar AS (2012) Tribal uses of medicinal plants of Rajashthan: Kachnar. Int J Life Sci Pharma Res 2(4):69–76
Shen J, Zhou Q, Li P, Wang Z, Liu S, He C et al (2017) Update on phytochemistry and pharmacology of naturally occurring resveratrol oligomers. Molecules 22(12):2050. https://doi.org/10.3390/molecules22122050
Shravya S, Vinod BN, Sunil C (2017) Pharmacological and phytochemical studies of Alangium salvifolium Wang.–a review. Bull Fac Pharm Cairo Univ 55(2):217–222
Sivapalan S, Dharmalingam S, Venkatesan V, Angappan M, Ashokkumar V (2023) Phytochemical analysis, anti-inflammatory, antioxidant activity of Calotropis gigantea and its therapeutic applications. J Ethnopharmacol 303:115963. https://doi.org/10.1016/j.jep.2022.115963
Sundar RDV, Settu S, Shankar S, Segaran G, Sathiavelu M (2018) Potential medicinal plants to treat leprosy-a review. Res J Pharm Technol 11:813–821
Süntar I, Akkol EK, Nahar L, Sarker SD (2012) Wound healing and antioxidant properties: do they coexist in plants? Free Radicals Antioxidants 2(2):1–7
Swain SS, Paidesetty SK, Dehury B, Sahoo J, Vedithi SC, Mahapatra N, Hussain T, Padhy RN (2018) Molecular docking and simulation study for synthesis of alternative dapsone derivative as a newer antileprosy drug in multidrug therapy. J Cell Biochem 119(12):9838–9852
Swain SS, Paidesetty SK, Dehury B, Das M, Vedithi SC, Padhy RN (2020) Computer-aided synthesis of dapsone-phytochemical conjugates against dapsone-resistant mycobacterium leprae. Sci Rep 10(1):6839
Swain SS, Sahoo G, Mahapatra PK, Panda SK (2022) Disease burden and current therapeutical status of leprosy with special emphasis on phytochemicals. Curr Top Med Chem 22(19):1611–1625. https://doi.org/10.2174/1568026621666210909162435
Tirkey R, Saraf S, Saraf S (2023) Ethnopharmacological insights on therapeutic potential of Gloriosa superba. Curr Tradit Med 9(5):77–89
Urán Landaburu L, Berenstein AJ, Videla S, Maru P, Shanmugam D, Chernomoretz A et al (2020) TDR targets 6: driving drug discovery for human pathogens through intensive chemogenomic data integration. Nucleic Acids Res 48:D992–D1005

Van Wyk BE (2011) The potential of south African plants in the development of new medicinal products. S Afr J Bot 77(4):812–829

Vengalil S, Lavania M, Singh I, Nashi S, Preethish-Kumar V, Polavarapu K, Mahajan NP, Raju S, Pradeep-Chandra-Reddy C, Keerthipriya M, Mahadevan A, Yasha TC, Nandeesh B, Gnanakumar K, Parry GJ, Sengupta U, Nalini A (2020) Appropriately selected nerve in suspected leprous neuropathy yields high positive results for mycobacterium leprae DNA by polymerase chain reaction method. Am J Trop Med Hyg 103(1):209–213. https://doi.org/10.4269/ajtmh.19-0746

Wagh VV, Jain AK (2020) Ethnopharmacological survey of plants used by the Bhil and Bhilala ethnic community in dermatological disorders in Western Madhya Pradesh. Int J Herb Med 19:100234

White C, Franco-Paredes C (2015) Leprosy in the 21st century. Clin Microbiol Rev 28(1):80–94

Xiong M, Wang X, Su T, Yang B, Li M, Zheng D (2019) Relationship between psychological health and quality of life of people affected by leprosy in the community in Guangdong Province, China: a cross-sectional study. BMC Public Health 19(1):1–8. https://doi.org/10.1186/s12889-019-6672-x

Yadav M, Khan KK, Beg MZ (2012) Ethnobotanical plants used for curing skin diseases by tribals of Rewa district (Madhya Pradesh). Indian J Life Sci 2(1):123–127

Natural Product-Based Treatment of Ear Infection

6

Smarita Lenka, Santosh Kumar Swain,
and Debasmita Dubey

Abstract

Otitis media (OM) has a complex anatomy and numerous presentations in children worldwide, especially in developing countries. A rising number of alternative treatments apart from antibiotics can be offered, taking into consideration their toxicity and adverse effects on human health. Several Indian traditional plants are reported with ethnopharmacological properties and have served as a valuable source of secondary metabolites for decades. This chapter focuses on the rate of prevalence of OM in developing countries, risk factors, complications, and the treatments that rely on OM. The adverse effects of antibiotic therapy and possible therapeutics of natural compounds are also discussed. Significantly, the promising novel alternative for antibiotic treatment using Indian medicinal plant parts with computational tools is the major corner of this chapter. Moreover, using computational techniques like artificial intelligence, molecular docking, and molecular dynamic simulation, the noteworthy advancements in biotechnology for comprehending the pharmacological roles of phytocompounds and developing a roadmap for drug design and development are also covered. The overview of this chapter is the computational research development toward drug development with Indian medicinal plants, moving from static methods to a range of advanced machine learning to confront OM.

S. Lenka · D. Dubey (✉)
Department of Medical Research, IMS & SUM Hospital, Siksha 'O' Anusandhan Deemed to be University, Bhubaneswar, Odisha, India

S. K. Swain
Department of Otorhinolaryngology and Head and Neck Surgery, All India Institute of Medical Sciences, Bhubaneswar, Odisha, India

T. Sharma et al. (eds.), *Natural Products for Antibacterial Drug Development: Recent Advancement of Computational Approach*,
https://doi.org/10.1007/978-981-97-9634-2_6

Keywords

Otitis media · Natural products · Computational approach · Drug development

Abbreviations

AOM	Acute otitis media
COM	Chronic otitis media
COMC	Cholesteatoma development
CSOM	Chronic suppurative otitis media
ET	Eustachian tube
OM	Otitis media
OME	Otitis media with effusion

6.1 Introduction

India is renowned for having a large collection of herbal medical techniques and many therapeutic plants named "living traditions." These medicinal plants are used to treat various chronic illnesses with minimal or no harmful side effects. They have contributed significantly to global health systems. Herbal plants can be used as a natural remedy for various illnesses, such as hypertension, tuberculosis, cancer, wound healing, heart disease, diabetes mellitus, asthma, pharyngitis, and others. Plants rich in bioactive phytomedicine components, such as polyphenols, tannins, flavonoids, and alkaloids have been utilized in treatments because of their diverse pharmacological characteristics (Prasathkumar et al. 2021).

The most prevalent condition affecting the middle ear is called otitis media (OM). Its pathophysiology is influenced by a variety of factors, including cleft palate, viral and bacterial infections, allergies, morphological and functional alterations in the nasal tube, and the eustachian tube (ET) (Khasawneh et al. 2021). A large proportion of outpatient otolaryngology practice includes both acute otitis media (AOM) and otitis media with effusion (OME). After upper respiratory tract infections, OM is the second most commonly diagnosed condition in children and the most frequent pediatric presentation to an otolaryngologist (Rosenfeld et al. 2022; Cleere et al. 2023; Meherali et al. 2019). At least 60–80% of children under the age of five experience an infection of the middle ear, once in their lifetime (Kamphorst et al. 2023; Crawford et al. 2017; Pershad et al. 2023). It is the most commonly diagnosed infection, which also continues to be the primary cause of hearing loss in children under the age of five—an age group in which speech, language, and cognitive development are most critical (Haile et al. 2021). It may result in irreversible hearing loss at any age and also predisposes individuals to sensorineural hearing loss and tinnitus in childhood (Aarhus et al. 2020). Children under five still require over 9 million medical visits each year in the USA due to OM. Therefore, even in developed nations, but particularly in poor nations where access to quality, modern healthcare is limited, OM poses a substantial health and financial burden (Luo et al. 2022; Borre et al., 2022; Wasserman et al. 2021).

The annual cost of treating this condition comes to billions of euros for society. The prevalence of AOM is still high, and its expenses are rising despite research on its prevention and treatment. The traditional management of AOM has to be reevaluated in light of the rise of bacteria that are resistant to antibiotics. But for good reason, the general agreement is that the best first treatment for AOM is antibiotics. Three approaches can be used for surgical management: myringotomy, tympanocentesis, and myringotomy with or without the insertion of a breathing tube. Surgery may be necessary, particularly if a child's condition worsens or doesn't improve while on antibiotics (Tähtinen et al. 2017; Schilder et al. 2016; Cetinkaya and Topsakal 2022).

Antibiotics are classified as antimicrobial agents and are used extensively in both human and veterinary medicine. However, antibiotics were reportedly harmful in various aspects. Even when, due to their extensive use, large-scale manufacture, and inadequate treatment procedures, many antibiotics have recently been found in the environment. Antibiotics found in the environment have drawn interest from all around the world because of the potential harm they could do to the environment and human health (Yang et al. 2021; Polianciuc et al. 2020). Despite the use of various antibiotics or new-generation antibiotics plant-based therapeutic approaches will help to eradicate this problem (AlSheikh et al. 2020; Qassadi et al. 2023; Álvarez-Martínez et al. 2021). In this chapter, we have focused on the recent advancements (computational approach) in the findings of antibacterial agents in the treatment of OM with Indian traditional plants.

6.2 Types of OM

Acute otitis media (AOM), otitis media with effusion (OME), and chronic suppurative otitis media (CSOM) are among the illnesses that fall under this spectrum. Acute signs of infection include fever and middle ear discomfort. The most frequent reason to prescribe antibiotics to children is acute otitis media (AOM), which affects 60% of them by the time they are 4 years old. The clinical decision to treat and the diagnostic techniques used to assess AOM are very subjective, which can lead to overdiagnosis, underdiagnosis, and antibiotic misuse (Chonmaitree et al. 2016; Kaur et al. 2017).

6.2.1 OM with Effusion

Pediatric OME is a common condition. The illness is frequently asymptomatic, making it simple to overlook. However, OME can cause hearing loss, which hinders a child's ability to speak and behave. The tympanic membrane is not perforated in this type of chronic otitis media. Localized inflammation results in the metaplasia of the epithelium and the accumulation of fluids in the spaces within the middle ear.

The middle ear effusion is not purulent; rather, it is mucous or seromucous. The difference between this condition and persistent effusion following acute media otitis, which goes away in 90% of patients after 2 months, is that the condition lasts for at least 3 months (Blanc et al. 2018). OME can have an impact on the very young: 60% of instances affect children under the age of two, and 50% affect newborns under the age of one. Particularly in children with craniofacial deformities (notably trisomy 21 and cleft palate), the frequency is particularly high (between 60 and 85%). Complications from persistent OME include tympanic membrane injury (atrophy, retraction pockets, and cholesteatoma) and hearing loss. It may also cause behavioral issues and delay the acquisition of language (Aarhus et al. 2015; Vanneste and Page 2019).

6.2.2 Chronic OM

A frequent condition affecting both adults and children is chronic otitis media (COM). It results in serious morbidity and, thankfully, uncommon but possibly deadly consequences. Chronic otitis media with cholesteatoma development (COMC) and chronic suppurative otitis media (CSOM) are two subtypes of chronic otitis media. A benign keratinizing cyst that grows from the eardrum into the middle ear and mastoid, cholesteatomas cause a characteristic inflammatory response that can result in osteitis and bone erosion. Surgery is required for COMC, whereas medication management of CSOM is occasionally possible. If medical management of CSOM is insufficient, surgery may potentially be undertaken. The efficacy of surgery in treating COM has been demonstrated (Wallis et al. 2015). Recurrent illness, particularly in COMC, is a significant issue in COM, though (Dornhoffer et al. 2013). The condition appears to have an age-related effect. Children are especially susceptible to COMC's destructiveness and likelihood of recurrence. This is thought to be caused by a greater rate of keratinocyte growth in pediatric cholesteatoma. Growth factor levels that are higher in childhood may have an impact on cholesteatoma. The effects of COM on aging in adults are not well understood (Jolink et al. 2023).

There is a dearth of literature, particularly regarding the behavior and treatment outcomes of COM in the elderly. Certain writers assert that older patients are more likely to appear with problems at the time of COM presentation. It's critical to determine whether older patients' treatment outcomes for COM are influenced by their age. Thus, adjustments in preoperative counseling can be made as needed (Jolink et al. 2018).

6.3 Epidemiology and Global Burden of OM

The burden of OM is greater worldwide, but it is challenging to pinpoint the precise number of cases reported annually because of variations in disease frequency among different geographic areas and in coding and recording systems. However, it is believed that "preventable" illnesses like meningitis and chronic otitis media kill between 30 and 90,000 children annually (Monasta et al. 2012; Kurabi et al. 2023). Its incidence is estimated to be 10.85% or 709 million cases, each year worldwide for AOM and 31 million new cases per year for COM, with children under the age of five accounting for 22.6% of these cases. According to estimates, there are 4.8 new instances of suppurative COM for every 1000 persons per year, and there are between 65 and 300 million cases altogether (Monasta et al. 2012; de Araújo et al. 2024). According to contemporary estimates, over 20,000 individuals pass away each year from complications related to OM, with children under the age of five having the highest death rates. Chronic or recurrent forms of the illness can cause significant hearing loss, which has a detrimental impact on academic performance and learning capacity (Monasta et al. 2012). The prevalence rate of OM from developing countries is mentioned in Table 6.1.

6.4 Risk Factors

The most prevalent cause of conductive hearing loss in children is OME. It is defined as a build-up of liquid substance inside the middle ear without signs of acute infection and with intact tympanic membrane. Such problems have been linked to numerous causes, such as allergies, sinusitis, eustachian tube dysfunction, and adenoid hypertrophy (Hemdan et al. 2024). In industrialized nations, OM is the most common cause for children to receive an antibiotic prescription or require surgery. Furthermore, there are no regulations governing the use of antibiotics for OM and respiratory tract infections in the majority of poor nations, and in a few of these nations, parents are free to buy any antibacterial medication from a pharmacy without a prescription from a doctor.

Over the past 4 years, a few research have examined environmental factors that impact OM. The following variables have been reexamined: air pollution, seasonal variation, breastfeeding, passive smoking exposure, creche attendance, siblings, and household pets (Nokso-Koivisto et al. 2023). The detailed risk factors associated with OM are mentioned in Fig. 6.1.

6.5 The Pathogenetic Mechanism of OM

AOM has a quick start, manifests locally and systemically, and is the main reason why children in high-income countries need antibiotic treatment. While CSOM necessitates continuous middle ear inflammation that results in otorrhea that lasts for at least 2 weeks and tympanic membrane perforation, OME reoccurs after the medication of AOM; however, bacteria may still be present after resolutions (Wallis et al. 2015).

It is possible that the youngster already has a non-symptomatic nasopharyngeal bacterial colonization. A common cold in children is caused by a viral infection that inflames the nasopharynx and the ET, increasing bacterial adhesion and

Table 6.1 Prevalence rate of OM from developing countries

Region	Country	Age group	Diagnosis	Responsible pathogens	Prevalence (%)	Sources
Africa	Nigeria	<5Y	AOM, ASOM	*Streptococcus pneumoniae*, *Hemophylus influenzae*, *Staphylococcus aureus*, *Bacillus* spp., *Corynebacterium* spp., *P. mirabilis*, *Pseudomonas aeruginosa*, *Klebsiella oxyotoca*, *Citrobacter* spp.	9.2	DeAntonio et al. (2016), Alabi et al. (2009), Iseh and Adegbite (2004), Amusa et al. (2005)
	Ethiopia	1–18Y	Bacterial otitis media	*S. aureus*, *Proteus* spp.	69.8	Henok et al. (2023), Tadesse et al. (2019), Seid et al. (2013)
South Asia	Bangladesh	<2Y	AOM CSOM	*H. influenzae*, *S. aureus*, *P. aeruginosa*	7.39	Naziat et al. (2018), Roy et al. (2007), Kamal et al. (2004)
	India	ND	CSOM	*S. aureus*, *P. aeruginosa*, *K. pneumoniae*, *E. coli*, *Acinetobacter baumannii*	9.2	Prakash et al. (2013)
	Pakistan	<2Y	AOM	*H. influenzae*, *S. aureus*, *P. aeruginosa*, *S. pneumoniae*, *S. pyogenes*, *S. epidermis*	27	Shakoor et al. (2016), Shah et al. (2023), Riaz et al. (2022)
Eastern Asia	China	All age group	CSOM	*S. aureus*, *P. aeruginosa*, *K. pneumoniae*, *E. coli*, *A. Baumannii*	6.7	Xu et al. (2020a, b, 2021)

(continued)

Table 6.1 (continued)

Region	Country	Age group	Diagnosis	Responsible pathogens	Prevalence (%)	Sources
	Vietnam	<12 M	AOM	*S. pneumoniae*	17.2	Satoh et al. (2021), Toizumi et al. (2022)
Middle East	Saudi Arabia	<6Y	AOM	*S. epidermidis*, *P. aeruginosa*, *S. aureus*, *S. pneumoniae*, *H. Influenzae*, and *M. catarrhalis*	7.5	Addas et al. (2019), Alshammari et al. (2017)
	Oman	<5Y	OME	*H. Influenzae*, *M. catarrhalis*, and *S. pnemoniae*	ND	Al-Sibani et al. (2021), DeAntonio et al. (2016)
	Iran	All age group	CSOM	*Staphylococci* spp., *Klebsiella* spp., *P. aeruginosa*	9.1	Mofatteh et al. (2018)
	Turkey	Pediatric age group	AOM	*H. Influenzae*, *M. catarrhalis*, and *S. pnemoniae*	4.6	Büyükcam et al. (2018)

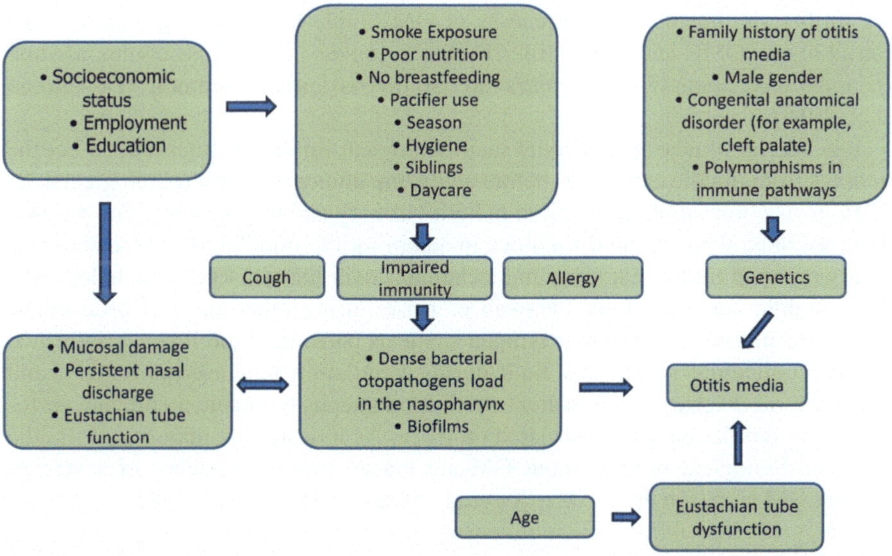

Fig. 6.1 Risk factors associated with OM

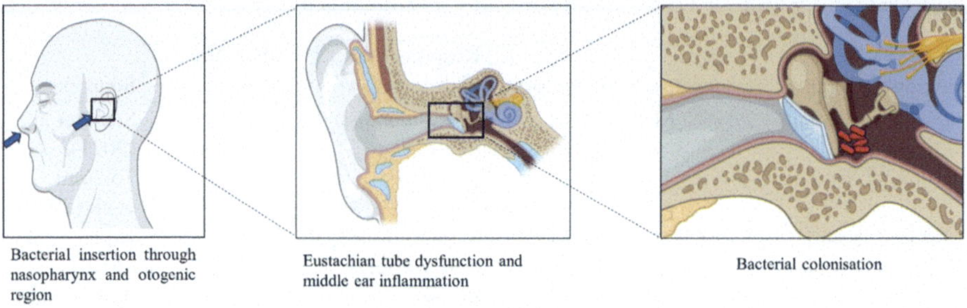

Fig. 6.2 Bacterial-induced pathogenesis of OM

colonization as well as other activating processes. Following eustachian tube failure, which results in negative middle ear pressure, germs and/or viruses from the nasopharynx can enter the middle ear and cause inflammation or infection (Fig. 6.2) (Schilder et al. 2016).

6.5.1 Microbiology

Generally, viruses or bacteria can cause OM. Non-typeable *Haemophilus influenzae* (NTHi) and *Streptococcus pneumoniae* are the two most common bacterial pathogens. *Moraxella catarrhalis*, *Streptococcus pyogenes*, and *Staphylococcus aureus* are further significant bacterial causes of OM (Table 6.1). When it comes to the causal agent, OME and AOM differ in that *S. pneumoniae* is less common while *H. influenzae* and *M. catarrhalis* are somewhat more common (DeAntonio et al. 2016).

It is crucial to take into account methodological limitations that may affect the findings from middle ear tissues before analyzing clinical OM microbiological data. These restrictions apply to research in both virology and bacteriology. The primary thing to think about is the difficulties in sampling the middle ear. The most commonly reported middle ear sampling techniques over the last 4 years included sampling middle ear epithelium by swab or rinse during other surgical procedures, collecting middle ear fluid after tympanocentesis performed forclinically indicated reasons, collecting middle ear fluid during ventilation tubeinsertionsurgery, and swabbing ear discharge from patients with spontaneous tympanic membrane perforations or ventilation tube-associated otorrhea. As a result, the majority of middle ear microbiological reports about OM are biased toward instances of severe or chronic OM (Jörissen et al. 2021; Xu et al. 2020a, b; Taylor et al. 2022).

6.5.2 Diagnosis

A significant constraint in etiology investigation is swabs of the nasopharyngeal region as a proxy indicator of potential pathogens within the middle ear. This is not always correct, as evidenced by earlier research, and this is a restriction to take into account when evaluating NPS data to identify OM-causing substances. Nasopharyngeal sampling is preferred in certain investigations, nevertheless, because it is a far less intrusive technique than tympanocentesis, which is unfeasible in many places and nations and unsuitable for longitudinal analysis. Even though sampling during ventilation tube insertion is more indicative of the infectious reservoir, it is also limited because it is typically not done during acute episodes of illness (Tamir et al. 2023).

The most frequent cause of childhood hearing impairment and disability is OME. Inaccurate diagnosis and the lack of obvious infectious signs, such as fever or earache, might cause delayed diagnosis and inadequate supervision. As a result, there may be a delay in speech and language development as well as potential behavioral and academic issues. Grais et al. looked at several machine learning models to categorize the wide band absorbance immittance data as coming from a normal or OME patient's ear to address the issue of low diagnosis accuracy in patients with OME (Grais et al. 2021, 2024). However, other bacterial infections can be identified by the use of molecular techniques, which have a higher sensitivity than bacterial cultures. For example, PCR or 16S rRNA gene sequencing has more frequently identified *Alloiococcus* otitis and *Turicella* otitidis from samples of OM patients (Xu et al. 2019; Sillanpää et al. 2017).

6.6 Treatment

OM can be treated either with antibiotics or surgical management. However, no such study reports have an association between a complete cure and the treatment procedures (Pershad et al. 2023). In contrast to COMC, which requires surgery, CSOM can occasionally be treated medically. Surgery is an additional option for treating CSOM when medication management is insufficient. For the treatment of COM, surgery has shown to be successful. Recurrent illness, however, plays a significant role in COM, particularly in COMC cases. It appears that the condition has an age-related effect (Wallis et al. 2015).

6.6.1 Complications

Rare complications from AOM can happen even though the majority of patients are successfully treated with antimicrobial medication or watchful supervision. Subperiosteal abscess, venous sinus thrombosis, labyrinthitis, meningitis, facial nerve paralysis, sensorineural hearing loss, mastoiditis, epidural abscess, and cerebral abscess are a few examples of these (Leiva et al. 2021; Leung and Wong 2017). When children with AOM exhibit unusual symptoms and/or examination results, such as postauricular swelling/pain, stiff neck, altered mental state, vertigo, facial paralysis, or such consequences should be taken into account. Timely medical and surgical intervention for these AOM problems may significantly reduce morbidity and mortality.

6.6.2 Antibiotics Over OM

Antibiotic use is probably not going to help many AOMs. Furthermore, over 90% of *M. catarrhalis* isolates and 30–50% of *H. influenzae* isolates produce beta-lactamase, making them resistant to amoxicillin, which is now the first-line antibiotic recommended by the American Academy of Paediatrics for AOM (Kaur et al. 2017). New combinations between one β-lactam and one β-lactamase inhibitor (meropenem and vaborbactam), Tetracycline derivatives (eravacycline), siderophore cephalosporins (cefiderocol), fourth-generation fluoroquinolones (delafloxacin), new aminoglycosides (plazomicin) are the major recent groups of antibiotics used against bacterial infection (Terreni et al. 2021).

6.6.3 Adverse Effects of Antibiotics

Despite their widespread usage in the treatment of illnesses, antibiotics can have potentially fatal adverse effects such as anaphylactic shock and cardiac death. Apart from drug-drug interactions, the most common toxicity pathway through which antibiotics produce side effects is the hypersensitive reaction. Furthermore, host-related factors like age, sex, comorbidity, genetic variability, and drug toxicity may intensify antibiotic-associated toxicity; as a result, toxicity characteristics will differ between hepatotoxicity, neurotoxicity, nephrotoxicity, and antibiotic resistance (Lalanne et al. 2023). Effective strategies for managing antibiotic-associated toxicity can be developed by understanding the various types of toxicity and the mechanisms underlying adverse effects (Ali et al. 2023).

6.6.4 Drug Resistance

One of the most significant medical advances of the twentieth century has been the creation of antibiotic therapy; yet, the tremendous progress made by the availability of antibiotic therapy is at risk due to the emergence of resistance in healthcare facilities and the general population. When compared to identical infections produced by susceptible strains, infections caused by resistant bacteria can result in up to two times higher likelihood of unfavorable consequences. The failure or postponement of antibiotic treatment is the primary cause of these unfavorable consequences, which might be either clinical or financial. As illness severity, strain virulence, or host vulnerability rise, the severity of these unfavorable consequences will also rise (Friedman et al. 2016).

Now, the fight against antibiotic-resistant microorganisms is losing. Prokaryotic cells undergo faster adaptation due to increased environmental exposure and a higher mutation fraction in their DNA, which results from their more rapid generational cycle compared to eukaryotic cells. Numerous adaptive features involved in resistance mechanisms for bacteria continuously exposed to environmental stressors (such as antibiotics) are linked to nucleoid-associated proteins, which modify and control DNA and RNA and can affect the expression profile of genes. For bacteria to have any kind of long-term survival benefit, they must inherit these phenotypic variances, which suggests genetic modifications (Salazar et al. 2024; Graf et al. 2023; Cylke et al. 2022).

6.7 Drug Development with Natural Resources

For many years, natural products have been recognized as an essential source of active components for medicinal medicines. Specifically, the use of plant-derived active substances in microbial natural products dates back long before the development of modern medicine. Natural product chemistry, biochemistry, and synthetic activities have long been linked to the effectiveness of natural products. The distinctive qualities of these natural products are thus being utilized to take advantage of the chemical and structural diversity and biodiversity of these types of products about their therapeutic effect, thanks to scientific advancements in modern molecular and cellular biology, analytical chemistry, and pharmacology (Atanasov et al. 2021; Mathur and Hoskins 2017; Xu et al. 2022).

The inadequate pharmacological assessment, preclinical investigation, and clinical analysis of herbal medicines hinder their incorporation into modern medical procedures. Preclinical research primarily conducted in-vitro and in-vivo, examines a range of characteristics, including gene expression studies, metabolomics fingerprints of produced natural medicines, intracellular activity, cell-environment interaction, and cell cytotoxicity. Strong data and preclinical assessments support the safe, long-term use of herbal medicine to cure horrible illnesses. Furthermore, several contemporary methods are being explored for the accurate and efficient commercial manufacturing of bioactive substances (Aware et al. 2022).

6.7.1 Recent Advancements with Indian Medicinal Plants

Plant antimicrobials that can have physiological effects on the body are composed of a variety of secondary metabolites, including fatty acids, alkaloids, steroids, resins, tannins, phenolics, and flavonoids (Othman et al. 2019).

Complementary and alternative medicine has been the treatment of choice for many people with recurrent diseases in recent years; in the USA, it costs approximately \$22 million annually. According to research, in 1997, 42% of Americans used complementary and alternative medicine, with 12% of people utilizing herbal medicine. Furthermore, 17.7% of people reported using herbs, according to a 2007 National Centre for Complementary and Alternative Medicine Survey (Gruber et al. 2014; Son et al. 2017). In addition, more kids who get antibiotics than those who get a placebo report experiencing adverse symptoms like rash, vomiting, and diarrhea (Venekamp et al. 2015). Additionally, the usage of antibiotics has the negative effect of making resistant bacterial pathogens more prevalent. Antihistamines, steroids, and decongestants are also ineffective in treating AOM (Son et al. 2017).

There are several randomized control trials have evaluated the effects of herbal medicines over OM (Son et al. 2016, 2017). Unfortunately, there have not been reported clinical evaluations of all plants and plant parts with proper clinical trials over OM. However, both national and international studies reported various activities of plant-based drugs (Atanasov et al. 2021; Nasim et al. 2022). To our knowledge, the therapeutic efficacies of various ethnical plants are still not conclusive yet, due to the lack of controlled scientific investigation. Therefore, there should be made careful investigations using recent computational techniques to enhance the therapeutic properties of Indian traditional plants.

Taking into consideration recent advancements in antibacterial treatment, in-vivo, in-vitro, and *in-silico*; all the approaches carry equal responsibility toward therapeutics. Concerning the well-being of the scientific world using the proper direction to develop a drug should be followed by ethical terms and conditions. Among all discussions of drug development, there should be another thing about the high-yielding and genotypic development of the plant breed; which is another concern to prevent the plant species. So, the requirement for effectively managing and utilizing germplasm is the variability of morphological and qualitative features, such as antioxidants, among plant germplasm, which is widely employed in breeding programs to generate cultivars that are favored by consumers. The genetic variety found in the germplasm population is essential for the success of excellent breeding and the selection of promising germplasm. Variation within the population aids in the identification of acceptable germplasm for essential features as well as the conservation and classification of genetic variation in plant germplasm. Therefore, in crop improvement, the level of genetic variety in the base population affects the production of high-yielding genotypes (Teshome et al. 2020; Bouargalne et al. 2022).

6.7.2 Plant Bioactive Compounds Against OM

OM stands for intracranial problems, primarily intracranial abscess, and meningitis, which can cause brain damage or even death. Infratemporal complications, such as mastoiditis, are represented by OM. It is now necessary to research safer phytomedicines and biologically active chemicals with acceptable therapeutic indices to produce new pharmaceuticals due to the rise in the emergence of antibiotic-resistant bacteria. Plants have long been used for medicinal purposes, which has prompted researchers to study their bioactive chemicals and find numerous therapeutic benefits (Obuotor et al. 2023).

The numerous biological effects of plant bioactive compounds, including their antimicrobial, antioxidant, anticarcinogenic, antiallergenic, and anti-inflammatory properties, make them vital to human health. These properties can also help prevent and treat a range of noncommunicable diseases, including autoimmune, inflammatory, cardiovascular, cancer, metabolic, and neurodegenerative disorders. Scientific research is highly popular when it comes to identifying these components and proving their positive health impacts. A crucial component of many drug development initiatives has been the screening of natural sources for novel physiologically active metabolites (Dincheva et al. 2023).

There are numerous studies suggesting natural therapeutics using plant extracts against OM (Wu et al. 2023; Tresch et al. 2019; Obuotor et al. 2023). Although one chapter by Gürünlü 2022, regarding "β-glucans for treatment of ear infection: acute OM" was in the book entitled, "*Medical Applications of Beta-glucan*," there were not any elaborative discussions (Gürünlü 2022). However, there is no such significant research that focuses on the particular bioactive compounds of plants against OM. Still, there could be designed some studies for the isolation and standardization of bioactive pure compounds from plants to treat ear infections.

6.8 Possible Computational Approaches to Drug Discovery

Natural products now play a bigger role in drug discovery thanks to advances in genomics, proteomics, transcriptomics, and metabolomics. Drug action mechanisms are interpreted, new medications and their targets are found, and developed drugs and their therapeutic benefits are tracked down through the use of metabolic research. In the twenty-first century, computational research focuses more on medicinal plants to create novel drugs by screening bioactive compounds. *In-silico* tools that are important in medicinal plant research include computational techniques like Artificial Intelligence (AI), Molecular Docking, and Molecular Dynamic (MD) Simulation. To clarify the interactions and biologically beneficial characteristics of prospective phytochemicals, the molecular docking technique takes advantage of their mode of entry into the target active site. The dynamic behavior of biomolecules at the atomic level is illuminated by MD simulation, which provides high-quality biomolecule representation (Fig. 6.5).

One of the most crucial bioinformatics tools for determining the binding score and binding pose is docking (Fig. 6.3). In the modern day, docking has become increasingly important, being used for tasks like phytocompound discovery, drug candidate lead generation, and more. The optimal scoring functions are produced by the precise binding of both compounds, which aids in sorting the best candidate complex and, as a result, reduces the amount of effort needed for experimental work. For example, *C. grandis*, an Indian traditional plant, used in various regions of India for culinary purposes also reported various phytocompounds with several beneficiaries. A study by Momin and Yeligar (2022), analyzed the effective phytocompounds of *C. grandis*. By using column chromatography, compound (9Z, 12Z)-octadeca-9, 12-dienoic acid was separated from *Coccinia grandis* (L.) extract. A good docking score of -112.37 kcal mol^{-1} was obtained against the human pancreatic alpha-amylase enzyme (PDB: 3OLE). This score indicates a high affinity of the molecule to the receptor via hydrogen bonds, hydrophobic interactions, and

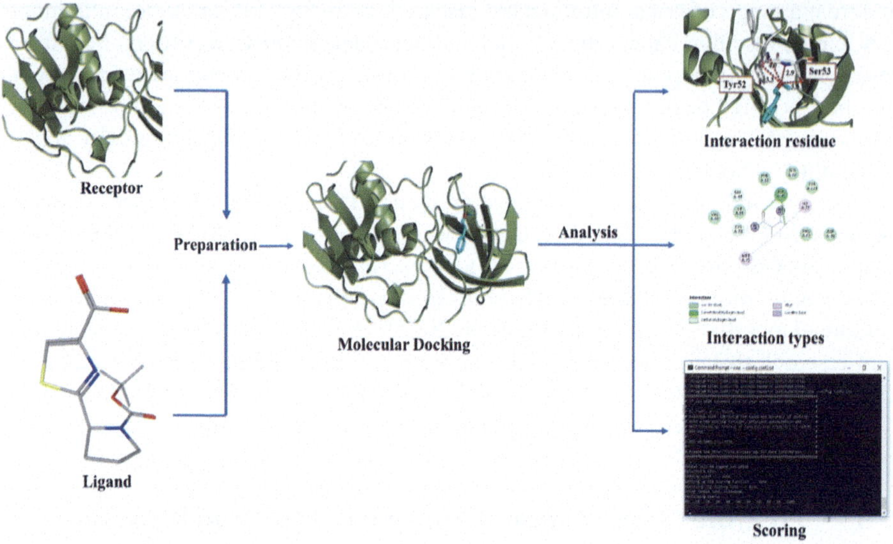

Fig. 6.3 Schematic representation of molecular docking steps

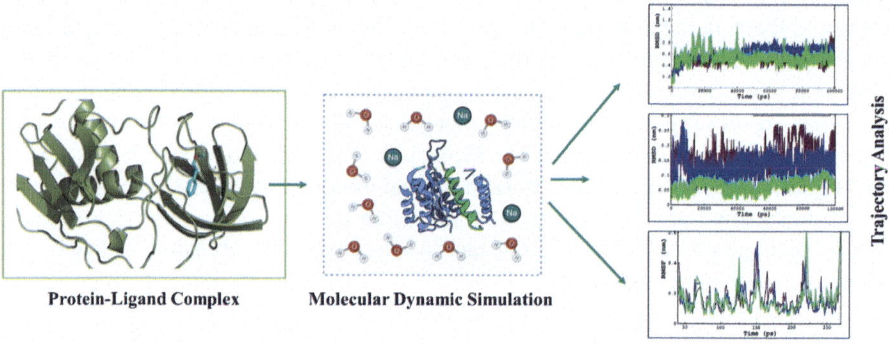

Fig. 6.4 Schematic presentation of molecular dynamic simulation

electrostatic interactions. The compound's hydroxyl group formed a hydrogen connection with glutamine 8, while its carbonyl group and C5–C9 chain demonstrated steric interactions with aspartate 402 and glutamine 8, respectively. Thus, using different solvents with parts of *C. grandis* and other ethno-plants should be considered for some new phytocompounds following the same process.

MD simulations are another method that is frequently utilized in drug discovery and design (Borah et al. 2019). It offers enough dynamic information about proteins and sheds light on the energy roles played by interactions between proteins and their ligands (Fig. 6.4). These data aid in a clearer understanding of the dynamic interaction between ligand and protein as well as in the detailed investigation of a particular binding site to increase docking accuracy. Computational techniques have

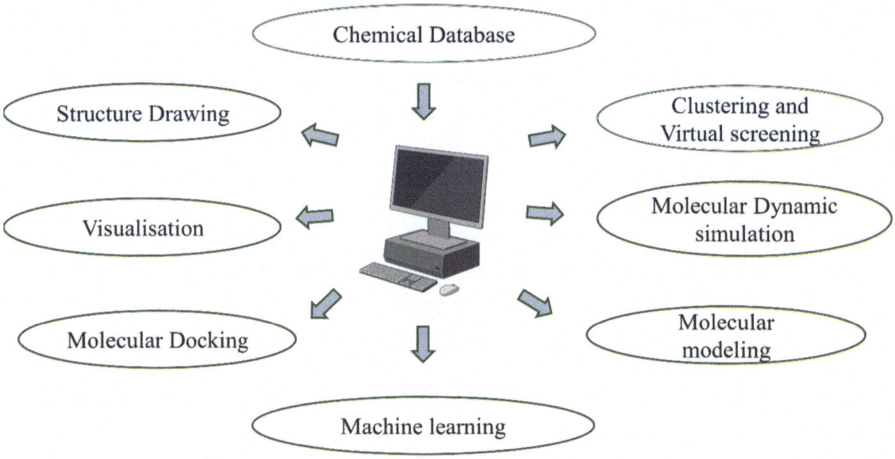

Fig. 6.5 Graphical representation of computational approaches to drug development

proven to be useful and rewarding in the anticancer medications development of derived from naturally occurring plants, and plant-based chemicals for lung and prostate cancer (Cui et al. 2020). In addition to determining the particle trajectory and dynamic stability, MD simulation also determines the radius of gyration, RMSD, and hydrogen bonds between the ligand and targets to improve the *in-silico* screening process and validate the medicinal ligand and target stability in a physiologically relevant setting before confirming it as a therapeutic purpose. Thus, in drug designing, simulation study of docked phytocompounds can play an incredible role in the journey of therapeutics.

There is already discussion about the preservation of plant species in the above section in this chapter. AI additionally predicts the imminent extinction of therapeutic plant species, thereby aiding in their preservation (Wearn et al. 2019). Identification of medicinal plant leaves is also assisted by image processing. AI also predicts interactions between herbal targets. In medicinal plants with therapeutic significance, ML technology enhances secondary metabolites for their widespread industrial application. A medicinal plant's class is determined by an artificial neural network using characteristics like leaf size, color, and measurements. In tissue culture, in-vitro plant growth is dependent on plant nutrition crucially. A comprehensive report on the relationship between phytochemistry and its activity was proposed by machine learning techniques such as artificial neural networks and support vector machines (Singh and Bharadvaja 2021; Aitwadkar et al. 2018). A study conducted by García-Pérez et al. (2020), reported that machine learning has enhanced secondary metabolite production in *Bryophyllum* in plant tissue culture. Therefore, it will be relaxing to know that the essential phytocompounds can be preserved and also enhance the therapeutic value of the secondary metabolites (Fig. 6.5).

6.9 Conclusion

Therapeutic approaches with Indian medicinal plants over several health complications have already been reported earlier. The raw form of all parts of this plant serves both orally and even tropically. Apart from various medicinal implements, the major advantage of the Indian plants is almost all plants that can be used in the therapeutic approach are edible. Although various plants have ethnomedicinal claims, there should be a proper in-vivo toxicity analysis for the commercialization of our traditional plant parts. So, that can be used as an antibacterial agent against OM with no toxicity issue. So, when it comes to being used for medicinal purposes against OM, there is probably not a second think of any life risk as most OM patients are from the pediatric group.

More databases and pharmacological information will be produced by computational work on plant components, which will aid in the development of a product that is appropriate for treating all phases of OM in all age groups. The growing number of biomolecule databases continues to be a problem for computational research since they provide divergent types of information. This in turn encourages software to be updated and changed to identify various plant databases and get an optimal result. Thus, greater investigation, software upgrades, and understanding of pharmacological significance turn into more drug possibilities for subsequent studies on drug design and discovery. These computational methods allow for the connection of medicinally significant plants with various ailments. The pharmacological significance of medicinal plants will be strengthened by performing additional studies using computational techniques, which will also aid in the production of fast-acting, inexpensive, and herbal drugs.

References

Aarhus L, Tambs K, Kvestad E, Engdahl B (2015) Childhood otitis media: a cohort study with 30-year follow-up of hearing (the HUNT study). Ear Hear 36(3):302. https://doi.org/10.1097/AUD.0000000000000118

Aarhus L, Homøe P, Engdahl B (2020) Otitis media in childhood and disease in adulthood: a 40-year follow-up study. Ear Hear 41(1):67–71. https://doi.org/10.1097/AUD.0000000000000729

Addas F, Algethami M, Mahmalji N, Zakai S, Alkhatib T (2019) Bacterial etiology and antimicrobial sensitivity patterns of ear infections at king Abdulaziz university hospital, Jeddah, Saudi Arabia. J Nat Sci Med 2(3):147–152. https://doi.org/10.4103/JNSM.JNSM_42_18

Aitwadkar PP, Deshpande SC, Savant AV (2018) Identification of Indian medicinal plant by using artificial neural network. Int Res J Eng Technol 5(4):1669–1671

Alabi BS, Abdulkarim AA, Fatai O, Abdulmajeed SO (2009) Prevalence of acute otitis media among children with pyrexia in a Nigerian hospital. Auris Nasus Larynx 36(5):532–535. https://doi.org/10.1016/j.anl.2008.12.007

Ali SS, Ali A, Elsamahy T, Okasha KM, Sun J (2023) Toxicity consideration of antibiotics. In: Antibiotics-therapeutic spectrum and limitations. Academic Press, pp 297–328. https://doi.org/10.1016/B978-0-323-95388-7.00008-5

Alshammari MS, Alshaya HK, Alkhalaf AA, Al Marshadi JA, Alshdokhi AM, Ahmed HG (2017) The common microbial causes of otitis media with especial focus in Saudi Arabia. Res Otolaryngo l6(5):67–72. https://doi.org/10.5923/j.otolaryn.20170605.03

AlSheikh HM, Sultan I, Kumar V, Rather IA, Al-Sheikh H, Tasleem Jan A, Haq QM (2020) Plant-based phytochemicals as possible alternative to antibiotics in combating bacterial drug resistance. Antibiotics 9(8):480. https://doi.org/10.3390/antibiotics9080480

Al-Sibani A, Al-Lawati A, Al-Siyabi N, Kumar R, Al-Rashdi A, Al-Jardani A (2021) The prevalence of bacterial etiologies causing otitis media with effusion among omani children. ENT Open A Open J2(1):31–34. https://doi.org/10.33169/ent.ENATOAOJ-2-108

Álvarez-Martínez FJ, Barrajón-Catalán E, Herranz-López M, Micol V (2021) Antibacterial plant compounds, extracts and essential oils: an updated review on their effects and putative mechanisms of action. Phytomedicine 90:153626. https://doi.org/10.1016/j.phymed.2021.153626

Amusa YB, Ogunniyi TA, Onayade OO, Okeowo PA (2005) Acute otitis media, malaria and pyrexia in the under five age group. West African J medicine 24(3):239–241. https://doi.org/10.4314/wajm.v24i3.28205

Atanasov AG, Zotchev SB, Dirsch VM, Supuran CT (2021) Natural products in drug discovery: advances and opportunities. Natreviews Drug Discov 20(3):200–216. https://doi.org/10.1038/s41573-020-00114-z

Aware CB, Patil DN, Suryawanshi SS, Mali PR, Rane MR, Gurav RG, Jadhav JP (2022) Natural bioactive products as promising therapeutics: a review of natural product-based drug development. S Afr J Bot 151:512–528. https://doi.org/10.1016/j.sajb.2022.05.028

Blanc F, Ayache D, Calmels MN, Deguine O, François M, Leboulanger N, Lescanne E, Marianowski R, Nevoux J, Nicollas R, Tringali S (2018) Management of otitis media with effusion in children. Société française d'ORL et de chirurgiecervico-faciale clinical practice guidelines. Eur Ann Otorhinolaryngol Head Neck dis 135(4):269–273. https://doi.org/10.1016/j.anorl.2018.04.008

Borah K, Sharma S, Silla Y (2019) Structural bioinformatics-based identification of putative plant based lead compounds for Alzheimer disease therapy. Comput Biol Chem 1(78):359–366. https://doi.org/10.1016/j.compbiolchem.2018.12.012

Borre ED, Ayer A, Der C, Ibekwe T, Emmett SD, Dixit S, Shahid M, Olusanya B, Garg S, Johri M, Saunders JE (2022) Validation of the decision model of the burden of hearing loss across the lifespan (DeciBHAL) in Chile, India, and Nigeria. Eclinicalmedicine 50:101502. https://doi.org/10.1016/j.eclinm.2022.101502

Bouargalne Y, Ben Mrid R, Bouchmaa N, Zouaoui Z, Benmrid B, Kchikich A, El Omari R, Kabach I, Mohamed N (2022) Genetic diversity for agromorphological traits, phytochemical profile, and antioxidant activity in Moroccan sorghum ecotypes. Sci Reports 12(1):5895. https://doi.org/10.1038/s41598-022-09810-9

Büyükcam A, Kara A, Bedir T, Gülhan B, Özdemir H, Sütçü M, Düzgöl M, Arslan A, Tekin T, Çelebi S, Kukul MG (2018) Pediatricians′ attitudes in management of acute otitis media and ear pain in Turkey. Int J Pediatr Otorhinolaryngol 107:14–20. https://doi.org/10.1016/j.ijporl.2018.01.011

Cetinkaya EA, Topsakal V (2022) Acute otitis media. In: Pediatric ENT infections, pp 381–392. https://doi.org/10.1007/978-3-030-80691-0_33

Chonmaitree T, Trujillo R, Jennings K, Alvarez-Fernandez P, Patel JA, Loeffelholz MJ, Nokso-Koivisto J, Matalon R, Pyles RB, Miller AL, McCormick DP (2016) Acute otitis media and other complications of viral respiratory infection. Pediatrics 137(4). https://doi.org/10.1542/peds.2015-3555

Cleere EF, Crotty TJ, Lang J, Young O, Keogh IJ (2023) Clinical decision making in paediatric otitis media: a pilot quality improvement study. Int J Pediatr Otorhinolaryngol 164:111395. https://doi.org/10.1016/j.ijporl.2022.111395

Crawford B, Hashim SS, Prepageran N, See GB, Meier G, Wada K, Coon C, Delgleize E, DeRosa M (2017) Impact of pediatric acute otitis media on child and parental quality of life and associated productivity loss in Malaysia: a prospective observational study. Drugs-Real World Outcomes 4:21–31. https://doi.org/10.1007/s40801-016-0099-9

Cui W, Aouidate A, Wang S, Yu Q, Li Y, Yuan S (2020) Discovering anti-cancer drugs via computational methods. Front Pharmacol 11:733. https://doi.org/10.3389/fphar.2020.00733

Cylke C, Si F, Banerjee S (2022) Effects of antibiotics on bacterial cell morphology and their physiological origins. Biochem Soc Trans 50(5):1269–1279. https://doi.org/10.1042/BST20210894

de Araújo AL, Cordeiro FP, da Costa MR, de Oliveira PN (2024) Audiometric evaluation in different clinical presentations of otitis media. Braz J Otorhinolaryngol 90(1):101359. https://doi.org/10.1016/j.bjorl.2023.101359

DeAntonio R, Yarzabal JP, Cruz JP, Schmidt JE, Kleijnen J (2016) Epidemiology of otitis media in children from developing countries: a systematic review. Int J Pediatr Otorhinolaryngol 85:65–74. https://doi.org/10.1016/j.ijporl.2016.03.032

Dincheva I, Badjakov I, Galunska B (2023) New insights into the research of bioactive compounds from plant origins with nutraceutical and pharmaceutical potential. Plan Theory 12(2):258. https://doi.org/10.3390/plants12020258

Dornhoffer JL, Friedman AB, Gluth MB (2013) Management of acquired cholesteatoma in the pediatric population. Curr Opin Otolaryngol Head Neck Surg 21(5):440–445. https://doi.org/10.1097/MOO.0b013e32836464bd

Friedman ND, Temkin E, Carmeli Y (2016) The negative impact of antibiotic resistance. Clin Microbiol Infect 22(5):416–422. https://doi.org/10.1016/j.cmi.2015.12.002

García-Pérez P, Lozano-Milo E, Landin M, Gallego PP (2020) Machine Learning unmasked nutritional imbalances on the medicinal plant Bryophyllum sp. cultured in vitro. Frontiers in Plant Science 11:576177

Graf D, Laistner L, Klingel V, Radde NE, Weirich S, Jeltsch A (2023) Reversible switching and stability of the epigenetic memory system in bacteria. The FEBS J290(8):2115–2126. https://doi.org/10.1111/febs.16690

Grais EM, Wang X, Wang J, Zhao F, Jiang W, Cai Y, Zhang L, Lin Q, Yang H (2021) Analysing wideband absorbance immittance in normal and ears with otitis media with effusion using machine learning. Sci Reports 11(1):10643. https://doi.org/10.1038/s41598-021-89588-4

Grais EM, Nie L, Zou B, Wang X, Rahim T, Sun J, Li S, Wang J, Jiang W, Cai Y, Yang H (2024) An advanced machine learning approach for high accuracy automated diagnosis of otitis media with effusion in different age groups using 3D wideband acoustic immittance. Biomed Signal Proces Cont 87:105525. https://doi.org/10.1016/j.bspc.2023.105525

Gruber M, Ben-Arye E, Kerem N, Cohen-Kerem R (2014) Use of complementary alternative medicine in pediatric otolaryngology patients: a survey. Int J Pediatr Otorhinolaryngol 78(2):248–252. https://doi.org/10.1016/j.ijporl.2013.11.016

Gürünlü B (2022) Medical applications of Beta-Glucan. Bentham Science Publisher. https://doi.org/10.2174/97898150392381220101

Haile LM, Kamenov K, Briant PS, Orji AU, Steinmetz JD, Abdoli A, Abdollahi M, Abu-Gharbieh E, Afshin A, Ahmed H, Rashid TA (2021) Hearing loss prevalence and years lived with disability, 1990–2019: findings from the global burden of disease study 2019. Lancet 397(10278):996–1009. https://doi.org/10.1016/S0140-6736(21)00516-X

Hemdan A, Megahed M, Elokda M, Moneir W (2024) Predictors of effusion viscosity in otitis media with effusion: neutrophil lymphocyte ratio versus mean platelet volume. Acta Otorrinolaringolo (English Edition) 75:205. https://doi.org/10.1016/j.otoeng.2023.11.001

Henok EA, Manilal A, Oumer Y, Keyta G, Birru M, Desta Araya B, Aklilu A, Alahmadi RM, Raman G, Idhayadhulla A (2023) Factors associated with otitis media among pediatrics in two government hospitals in Arba Minch, Southern Ethiopia. Infect Drug Res Volume 16:6405–6426. https://doi.org/10.2147/IDR.S424927

Iseh KR, Adegbite T (2004) Pattern and bacteriology of acute suppurative otitis media in Sokoto, Nigeria. Ann Afr Med 3(4):164–166

Jolink C, Zwemstra MR, de Wolf MJ, Ebbens FA, van Spronsen E (2018) Success rate of tympanic membrane closure in the elderly compared to younger adults. Otol Neurotol 39(1):e34–e38. https://doi.org/10.1097/MAO.0000000000001649

Jolink C, Huijsman A, Dreschler WA, de Wolf MJ, Ebbens FA, van Spronsen E (2023) Surgery for chronic otitis media in the elderly. Am J Otolaryngol 44(2):103698. https://doi.org/10.1016/j.amjoto.2022.103698

Jörissen J, van den Broek MF, De Boeck I, Van Beeck W, Wittouck S, Boudewyns A, Van de Heyning P, Topsakal V, Van Rompaey V, Wouters I, Van Heirstraeten L (2021) Case-control microbiome study of chronic otitis media with effusion in children points at streptococcus salivarius as a pathobiont-inhibiting species. MSystems 6(2):10–128. https://doi.org/10.1128/mSystems.00056-21

Kamal N, Joarder AH, Chowdhury AA, Khan AW (2004) Prevalence of chronic suppurative otitis media among the children living in two selected slums of Dhaka City. Bangladesh Med Res Counc Bull 30(3):95–104

Kamphorst K, Oosterloo BC, van't Riet E, Reichwein LC, Vlieger AM, van Elburg RM (2023) The association between exposure to antibiotics in the first week of life and later otitis media: the INCA study. Int J Pediatr Otorhinol 164:111415. https://doi.org/10.1016/j.ijporl.2022.111415

Kaur R, Morris M, Pichichero ME (2017) Epidemiology of acute otitis media in the postpneumococcal conjugate vaccine era. Pediatrics 140(3). https://doi.org/10.1542/peds.2017-0181

Khasawneh L, Khassawneh AH, Kheirallah KA, Neri G, Pignatelli GF, Al-Balas HI, Martinotti S, Al-Mistarehi AH (2021) Otitis media with effusion: the role of helicobacter pylori in its pathogenesis. Ann Med Surg 62:278–282. https://doi.org/10.1016/j.amsu.2021.01.056

Kurabi A, Dewan K, Kershner JE, Leichtle A, Li JD, Santa Maria PL, Preciado D (2023) PANEL 3: otitis media animal models, cell culture, tissue regeneration & pathophysiology. Int J Pediatr Otorhinol 176. https://doi.org/10.1016/j.ijporl.2023.111814

Lalanne S, Bouzillé G, Tron C, Revest M, Polard E, Bellissant E, Verdier MC, Lemaitre F (2023) Amoxicillin-induced neurotoxicity: contribution of a healthcare data warehouse to the determination of a toxic concentration threshold. Antibiotics 12(4):680. https://doi.org/10.3390/antibiotics12040680

Leiva LM, Delgado H, Holguín LV, Rojas C (2021) Cranial complications of otitis media with paralysis of the contralateral sixth cranial pair in pediatrics. Biomedica 41(2):218–224. https://doi.org/10.7705/biomedica.5763

Leung AK, Wong AH (2017) Acute otitis media in children. Recent Patents Inflamm Allergy Drug Discov 11(1):32–40. https://doi.org/10.2174/1874609810666170712145332

Luo Y, He P, Wen X, Gong R, Hu X, Zheng X (2022) Otitis media and its association with hearing loss in Chinese adults: a population-based study of 4 provinces in China. Front Pub Health 10:1150. https://doi.org/10.3389/fpubh.2022.852556

Mathur S, Hoskins C (2017) Drug development: lessons from nature. Biomed Dermatol 6(6):612–614. https://doi.org/10.3892/br.2017.909

Meherali S, Campbell A, Hartling L, Scott S (2019) Understanding parents' experiences and information needs on pediatric acute otitis media: a qualitative study. J Patient Exp 6(1):53–61. https://doi.org/10.1177/2374373518771362

Mofatteh MR, Moghaddam FS, Yousefi M, Namaei MH (2018) A study of bacterial pathogens and antibiotic susceptibility patterns in chronic suppurative otitis media. J Laryngol Otolog 132(1):41–45. https://doi.org/10.1017/S0022215117002249

Momin YH, Yeligar VC (2022) Molecular docking study of compound isolated from Coccinia Grandis (L.) Voigt. Indian Drugs 59(11):15. https://doi.org/10.53879/id.59.11.13160

Monasta L, Ronfani L, Marchetti F, Montico M, Vecchi Brumatti L, Bavcar A, Grasso D, Barbiero C, Tamburlini G (2012) Burden of disease caused by otitis media: systematic review and global estimates. PLoS One 7(4):e36226. https://doi.org/10.1371/journal.pone.0036226

Nasim N, Sandeep IS, Mohanty S (2022) Plant-derived natural products for drug discovery: current approaches and prospects. Nucleus 65(3):399–411. https://doi.org/10.1007/s13237-022-00405-3

Naziat H, Saha S, Islam M, Saha S, Uddin MJ, Hussain M, Luby SP, Darmstadt GL, Whitney CG, Gessner BD, Saha SK (2018) Epidemiology of otitis media with otorrhea among Bangladeshi children: baseline study for future assessment of pneumococcal conjugate vaccine impact. Pediatr Infect Dis J 37(7):715–721. https://doi.org/10.1097/INF.0000000000002077

Nokso-Koivisto J, Ehrlich GD, Enoksson F, Komatsu K, Mason K, Melhus Å, Patel J, Vijayasekaran S, Ryan A (2023) Otitis media: interactions between host and environment, immune and inflammatory responses. Int J Pediatr Otorhinolaryngol 27:111798. https://doi.org/10.1016/j.ijporl.2023.111798

Obuotor TM, Adeyanju FO, Kolawole AO, Idowu GP, Oluwamuyiwa FA, Afolabi FO (2023) In vitro and *in-silico* antibacterial activities of Syzgium aromaticum essential oil against bacteria associated with otitis media in children. Afr J Infect Dis 17(1):27–44. https://doi.org/10.21010/Ajidv17i1.3

Othman L, Sleiman A, Abdel-Massih RM (2019) Antimicrobial activity of polyphenols and alkaloids in middle eastern plants. Front Microbio 110:911

Pershad AR, Knox EC, Shah RK, Zalzal HG (2023) Disparities in the prevalence and management of otitis media: a systematic review. Int J Pediatr Otorhinolaryngol 13:111786. https://doi.org/10.1016/j.ijporl.2023.111786

Polianciuc SI, Gurzău AE, Kiss B, Ştefan MG, Loghin F (2020) Antibiotics in the environment: causes and consequences. Med Pharm Rep 93(3):231. https://doi.org/10.15386/mpr-1742

Prakash R, Juyal D, Negi V, Pal S, Adekhandi S, Sharma M, Sharma N (2013) Microbiology of chronic suppurative otitis media in a tertiary care setup of Uttarakhand state, India. North Am J Med Sci 5(4):282. https://doi.org/10.4103/1947-2714.110436

Prasathkumar M, Anisha S, Dhrisya C, Becky R, Sadhasivam S (2021) Therapeutic and pharmacological efficacy of selective Indian medicinal plants–a review. Phytomed Plus 1(2):100029

Qassadi FI, Zhu Z, Monaghan TM (2023) Plant-derived products with therapeutic potential against gastrointestinal bacteria. Pathogens 12(2):333. https://doi.org/10.3390/pathogens12020333

Riaz N, Ajmal M, Khan MS (2022) Frequency of otitis media with effusion among children aged 1-5 years presenting to immunization center of tertiary care hospitals, Rawalpindi. World J Otorhinolaryngol Head Neck Surg 8(04):315–320. https://doi.org/10.1016/j.wjorl.2021.03.002

Rosenfeld RM, Tunkel DE, Schwartz SR, Anne S, Bishop CE, Chelius DC, Hackell J, Hunter LL, Keppel KL, Kim AH, Kim TW (2022) Clinical practice guideline: tympanostomy tubes in children (update). Otolaryngol Head Neck Surg 166:S1–S55. https://doi.org/10.1177/01945998211065662

Roy E, Hasan KZ, Haque F, Siddique AK, Sack RB (2007) Acute otitis media during the first two years of life in a rural community in Bangladesh: a prospective cohort study. J Health Popul Nutr 25(4):414

Salazar JA, Díaz JR, Ferrara JG (2024) Who is winning the war: science or the adaptive molecular mechanisms of bacteria, evolving to survive antibiotic therapy? Gene Rep 34:101882. https://doi.org/10.1128/microbiolspec.VMBF-0016-2015

Satoh C, Toizumi M, Nguyen HA, Hara M, Bui MX, Iwasaki C, Takegata M, Kitamura N, Suzuki M, Hashizume M, Dang DA (2021) Prevalence and characteristics of children with otitis media with effusion in Vietnam. Vaccine 39(19):2613–2619. https://doi.org/10.1016/j.vaccine.2021.03.094

Schilder AG, Chonmaitree T, Cripps AW, Rosenfeld RM, Casselbrant ML, Haggard MP, Venekamp RP (2016) Otitis media. Nat Rev Dis Prim 2(1):1–8. https://doi.org/10.1038/nrdp.2016.63

Seid A, Deribe F, Ali K, Kibru G (2013) Bacterial otitis media in all age group of patients seen at Dessie referral hospital, north East Ethiopia. Egyptian J Ear Nose Throat Allied Sci 14(2):73–78. https://doi.org/10.1016/j.ejenta.2013.02.005

Shah MZ, Mustafa G, Iqbal M, Qasim M, Abbas K, Umair M, Baig HM (2023) Prevalence of gram positive bacteria in the affected individuals of otitis media with effusion from the indigenous population of southern Punjab, Pakistan: first report. Braz J Biol 84:e267874. https://doi.org/10.1590/1519-6984.267874

Shakoor S, Malik FR, Khan E (2016) Bacterial aetiology of otitis media in children in Pakistan aged 0–59 months; laboratory surveillance data from 2004 to 2013: comparison between before and after the introduction of Hib vaccination. Paediatr Int Child Health 36(1):34–38. https://doi.org/10.1179/2046905514Y.0000000170

Sillanpää S, Kramna L, Oikarinen S, Sipilä M, Rautiainen M, Aittoniemi J, Laranne J, Hyöty H, Cinek O (2017) Next-generation sequencing combined with specific PCR assays to determine the bacterial 16S rRNA gene profiles of middle ear fluid collected from children with acute otitis media. Msphere 2(2):10–128. https://doi.org/10.1128/mSphere.00006-17

Singh H, Bharadvaja N (2021) Treasuring the computational approach in medicinal plant research. Prog Biophys Mol Biol 164:19–32. https://doi.org/10.1016/j.pbiomolbio.2021.05.004

Son MJ, Choi S, Kim YE, Kim YH (2016) Herbal medicines for the treatment of otitis media with effusion: a systematic review of randomised controlled trials. BMJ Open 6(11):e011250. https://doi.org/10.1136/bmjopen-2016-011250

Son MJ, Kim YE, Song YI, Kim YH (2017) Herbal medicines for treating acute otitis media: a systematic review of randomised controlled trials. Complement Ther Med 35:133–139. https://doi.org/10.1016/j.ctim.2017.11.001

Tadesse B, Shimelis T, Worku M (2019) Bacterial profile and antibacterial susceptibility of otitis media among pediatric patients in Hawassa, Southern Ethiopia: cross-sectional study. BMC Pediatr 19:1–8. https://doi.org/10.1186/s12887-019-1781-3

Tähtinen PA, Laine MK, Ruohola A (2017) Prognostic factors for treatment failure in acute otitis media. Pediatrics 140(3):e20170072. https://doi.org/10.1542/peds.2017-0072

Tamir SO, Bialasiewicz S, Brennan-Jones CG, Der C, Kariv L, Macharia I, Marsh RL, Seguya A, Thornton R (2023) ISOM 2023 research Panel 4-diagnostics and microbiology of otitis media. Int J Pediatr Otorhinolaryngol 174:111741. https://doi.org/10.1016/j.ijporl.2023.111741

Taylor SL, Papanicolas LE, Richards A, Ababor F, Kang WX, Choo JM, Woods C, Wesselingh SL, Ooi EH, MacFarlane P, Rogers GB (2022) Ear microbiota and middle ear disease: a longitudinal pilot study of aboriginal children in a remote south Australian setting. BMC Microbiol 22:1–3. https://doi.org/10.1186/s12866-022-02436-x

Terreni M, Taccani M, Pregnolato M (2021) New antibiotics for multidrug-resistant bacterial strains: latest research developments and future perspectives. Molecules 26(9):2671. https://doi.org/10.3390/molecules26092671

Teshome Z, Terfa MT, Tesfaye B, Shiferaw E, Olango TM (2020) Genetic diversity in anchote (Coccinia abyssinica (Lam.) Cogn) using microsatellite markers. Current. Plant Biol 24:100167. https://doi.org/10.1016/j.cpb.2020.100167

Toizumi M, Satoh C, Quilty BJ, Nguyen HA, Madaniyazi L, Le LT, Ng CF, Hara M, Iwasaki C, Takegata M, Kitamura N (2022) Effect of pneumococcal conjugate vaccine on prevalence of otitis media with effusion among children in Vietnam. Vaccine 40(36):5366–5375. https://doi.org/10.1016/j.vaccine.2022.07.047

Tresch M, Mevissen M, Ayrle H, Melzig M, Roosje P, Walkenhorst M (2019) Medicinal plants as therapeutic options for topical treatment in canine dermatology? A systematic review. BMC Vet Res 15:1–9. https://doi.org/10.1186/s12917-019-1854-4

Vanneste P, Page C (2019) Otitis media with effusion in children: pathophysiology, diagnosis, and treatment. A review. J Otol 14(2):33–39. https://doi.org/10.1016/j.joto.2019.01.005

Venekamp RP, Sanders SL, Glasziou PP, Del Mar CB, Rovers MM (2015) Antibiotics for acute otitis media in children. Cochrane Database Syst Rev 6:CD000219. https://doi.org/10.1002/14651858.CD000219.pub4

Wallis S, Atkinson H, Coatesworth AP (2015) Chronic otitis media. Postgrad Med 127(4):391–395. https://doi.org/10.1080/00325481.2015.1027133

Wasserman M, Chapman R, Lapidot R, Sutton K, Dillon-Murphy D, Patel S, Chilson E, Snow V, Farkouh R, Pelton S (2021) Twenty-year public health impact of 7-and 13-valent pneumococcal conjugate vaccines in US children. Emerg Infect Dis 27(6):1627. https://doi.org/10.3201/eid2706.204238

Wearn OR, Freeman R, Jacoby DM (2019) Responsible AI for conservation. Nat Mach Intell 1(2):72–73. https://doi.org/10.1038/s42256-019-0022-7

Wu Y, Zhang J, Liu Q, Miao Z, Chai R, Chen W (2023) Development of Chinese herbal medicine for sensorineural hearing loss. Acta Pharm Sin B 14:455. https://doi.org/10.1016/j.apsb.2023.11.001

Xu Q, Gill S, Xu L, Gonzalez E, Pichichero ME (2019) Comparative analysis of microbiome in nasopharynx and middle ear in young children with acute otitis media. Front Genet 10:1176. https://doi.org/10.3389/fgene.2019.01176

Xu F, Kong W, Peng J, Gu H, Zheng H (2020a) Analysis of main pathogenic bacteria and drug sensitivity in patients with chronic suppurative otitis media and middle ear cholesteatoma in China. Biotechnol Lett 42:1559–1566. https://doi.org/10.1007/s10529-020-02880-7

Xu J, Dai W, Liang Q, Ren D (2020b) The microbiomes of adenoid and middle ear in children with otitis media with effusion and hypertrophy from a tertiary hospital in China. Int J Pediatr Otorhinol 134:110058. https://doi.org/10.1016/j.ijporl.2020.110058
Xu J, Du Q, Shu Y, Ji J, Dai C (2021) Bacteriological profile of chronic suppurative otitis media and antibiotic susceptibility in a tertiary care hospital in Shanghai, China. Ear Nose Throat J 100(9):NP391-6. https://doi.org/10.1177/0145561320923823
Xu Z, Eichler B, Klausner EA, Duffy-Matzner J, Zheng W (2022) Lead/drug discovery from natural resources. Molecules 27(23):8280. https://doi.org/10.3390/molecules27238280
Yang Q, Gao Y, Ke J, Show PL, Ge Y, Liu Y, Guo R, Chen J (2021) Antibiotics: an overview on the environmental occurrence, toxicity, degradation, and removal methods. Bioengineered 12(1):7376–7416. https://doi.org/10.1080/21655979.2021.1974657

Natural Product-Based Treatment for Skin Infections

7

Mamalisa Sahoo, Gyanranjan Mahalik,
and Monali Priyadarsini Mishra

Abstract

Skin is the largest multifunctional organ of the human body; it consists of three major layers: epidermis, dermis, and hypodermis. Skin acts as the first line of the body's defence mechanism. Skin infection occurs due to exposure to the environment; it profoundly affects human health and places a heavy burden on the healthcare system. Bacteria, viruses, parasites, and fungi can lead to skin infection. Medicinal plants are nature's gift to mankind to live disease-free healthy lives. Nowadays, there is growing interest in using ethnomedicinal plants as they represent a valuable alternative to treat skin infections because of their low cost, high effectiveness, well-tolerated, fewer side effects, and can treat chronic skin infections. The treatment of skin infections using medicinal plants and metabolites is derived from ancient times. Compounds with potential therapeutic advantages against microorganisms causing skin infections can be efficiently identified and optimized by researchers by using computational techniques such as molecular docking, quantitative structure-activity relationship (QSAR), molecular dynamics simulations, virtual screening, pharmacophore modelling, and network pharmacology. These methods help us for better understanding the antimicrobial activity on molecular basis and speed up the process of discovery a new drug and helps in the development of novel plant-based drug.

M. Sahoo · M. P. Mishra (✉)
School of Paramedics and Allied Health Sciences, Centurion University of Technology and Management, Bhubaneswar, Odisha, India
e-mail: monali.mishra88@gmail.com

G. Mahalik
Department of Botany, School of Applied Sciences, Centurion University of Technology and Management, Bhubaneswar, Odisha, India

T. Sharma et al. (eds.), *Natural Products for Antibacterial Drug Development: Recent Advancement of Computational Approach*,
https://doi.org/10.1007/978-981-97-9634-2_7

Keywords

Skin infection · Bacteria · Viruses · Parasites · Fungi · Allopathic medicine · Ethnomedicinal plants · Computational techniques

Abbreviations

LBDD	Ligand-based virtual screening
MD	Molecular dynamics
QSAR	Quantitative structure-activity relationship
SBDD	Structure-based virtual screening
THC	Tetrahydrocannabinol
VS	Virtual screening

7.1 Introduction

Human skin, being the largest multifunctional organ in the body, consists of the epidermis, dermis, and hypodermis layers which have different types of cells, including immune cells, which carry out various types of functions (Tabassum and Hamdani 2014). The skin's outermost layer, the epidermis, which is subdivided into stratum corneum, stratum granulosum, stratum basale, and stratum lucidum, is capable of synthesizing keratin. The stratum corneum is hydrophobic in nature, which acts as a barrier that blocks the passage of water and water-soluble substances. The dermis, which is vital for supporting and protecting other layers, contains fibroblasts, immune cells that facilitate the migration of immune cells, including mast cells, lymphocytes, and macrophages, as well as an abundance of extracellular matrix rich in collagen and elastin. The innermost layer is also known as hypodermis layer; as the subcutaneous tissue, it comprises adipocytes and fibrocytes and is densely packed with blood vessels and nerves. This layer comprises many immune cells, which also produce mediators such as adipokines, growth factors, and cytokines. Fat has a low heat conductivity, so it is essential for stopping heat (Fernandes et al. 2023). Skin acts as the first line of the body's defence mechanism (Majtan et al. 2021; Mohd Zaid et al. 2022). It protects the body from various external factors such as pathogens (bacteria, viruses, fungi, and parasites), physical stresses, and chemicals (Dawid-Pać 2013). It helps in regulating body temperature, preventing water loss, electrolytes, synthesis and storage of essential vitamins (vitamin D), sensation, absorption of oxygen and drugs (Fernandes et al. 2023; Mohd Zaid et al. 2022; Tabassum and Hamdani 2014). Skin infections, being common affect all age groups. Mostly, skin infections are caused by bacteria, viruses, fungi, and parasites. Skin is home to various indigenous microbial flora. Pathogenic bacteria are inhibited from competing by the normal flora (Dryden 2010). The skin has a substantial covering of dry epithelial cells, which inhibit the microorganisms from easily penetrating it (Tabassum and Hamdani 2014). Microbial invasion of the epidermis occurs when there is skin trauma from wounds, burns, ulcers, skin thinning and scratches, allows colonization of bacteria which can weaken the skin's defence, resulting in mild to serious skin infection (Orchard and Van Vuuren 2017). An individual with impaired immune systems is more susceptible to getting skin infections.

Nowadays, there is increasing resistance to antibiotic treatment, creating a challenging situation in the healthcare sector for the treatment of skin infections. Currently, the use of antibiotics has several drawbacks, such as high treatment cost, higher rate of treatment failure, and more side effects. Our ancestors used natural products to treat various kinds of skin infections thousands of years ago. They used special natural products that were grown regionally to treat skin infections naturally. Several parts of the plant, like leaf, stem, root, bark, flowers, and fruits, possess medicinal properties to treat various skin infections. Natural therapy has gained attention in the last two decades as an alternative treatment to allopathic medicine because of the low-cost effect, fewer side effects, higher rate of treatment success, and readily available, and the natural products become a part of the green revolution. Natural products have antibacterial, antifungal, antiparasitic, and antiviral properties.

7.2 Types of Skin Infections

7.2.1 Bacterial Skin Infection

When bacteria penetrate our skin either from an external source or our own skin, it results in bacterial skin infection. *Staphylococci* sp. and *Streptococci* sp. are the most common bacteria in bacterial skin infection (Tabassum and Hamdani 2014). Infections with bacteria can be localized or systemic; localized infections are limited to a specific area, whereas systemic infections show symptoms, such as fever, throughout the body. Certain bacteria can start infection in one area and later spread throughout the body (Chakraborty and Burns 2024).

7.2.1.1 Impetigo

Impetigo is a primary, superficial epidermal infection. Gram-positive bacteria are the most common cause of this highly contagious infection (Nardi and Schaefer 2024; Sunderkötter and Becker 2015). It is extremely contagious and especially common among children. There are two forms:

Non-bullous Impetigo—It usually affects the face and limbs, it is characterized as yellow, crusting lesions which may be itchy and painful (Sukumaran and Senanayake 2016).

Bullous Impetigo—The most frequent causative agent is *Staphylococcus aureus*, which appears as bullae that bursts to produce a dark crust (Sukumaran and Senanayake 2016).

7.2.1.2 Cellulitis and Erysipelas

Staphylococcal aureus and *Streptococci* sp. are the typical causal agents of cellulitis and erysipelas; in some cases, gram-negative bacteria and other *Streptococci* sp. are found to be the causative agents. It usually affects the lower limb and is clinically presented as fever and chill, along with abrupt onset of pain, oedema, erythema, and elevated heat of the affected location. Bullae can also appear as skin lesions (Sapuła et al. 2020).

7.2.1.3 *Staphylococcal* Scaled Skin Syndrome

Staphylococcal scaled skin syndrome is also referred as Ritter disease. It is an uncommon but serious skin infection caused by the *S. aureus* strain by releasing exfoliative toxins. It is characterized by skin depletion, erythematous rashes, and large superficial blisters. Young children have a higher incidence because of their immature renal function and lack of protective antibodies against exfoliative toxins (Brazel et al. 2021; Medugu et al. 2023).

7.2.1.4 Folliculitis

A frequent, most benign skin condition called folliculitis results in hair follicle inflammation and infection, erythematous capsule, and pustule formation on the skin that is covered by skin. *Staphylococcus aureus* is the major causative agent, less frequently by *Staphylococcus epidermidis* and sometimes by some gram-negative bacteria like *Pseudomonas aeruginosa* when exposed to hot tubs and spas. It is mainly divided into superficial and deep folliculitis (Jappa and Kutre 2018; Sukumaran and Senanayake 2016).

Superficial folliculitis

Staphylococcal folliculitis is the most superficial kind of skin infection manifests as tiny, erythematous follicular pustules that do not extend to the surrounding skin. Most commonly, it affects the scalp and extremities (Winters and Mitchell 2024).

Deep folliculitis

Deep follicle infection results in perifolliculitis, which elicits a more inflammatory response than superficial folliculitis. In sycosis barbae a follicular pustule pierced by hair is the main lesion (Jappa and Kutre 2018).

7.2.1.5 Erythrasma

Corynebacterium minutissimum is the primary cause of erythrasma, a persistent superficial skin infection. It is gram-positive bacteria typically present in warm, moist areas of the skin. Erythrasma infection generally affects the axilla, toe web, groin, and inguinal regions. The majority of the lesions are asymptomatic, but some common symptoms are burning and itching. The patches are scaly, dry, and uneven; they start off pink and eventually turn to brown (Forouzan and Cohen 2020).

7.2.1.6 Mycobacterium Ulcerans

Buruli ulcers are caused by *Mycobacterium ulcerans*, most commonly affecting arms or legs. It starts as a skin lesion that necrotizes and develops into painless ulcers (Muleta et al. 2021).

7.2.1.7 Mycobacterium Marinum

The skin and soft tissues are the primary target of *Mycobacterium marinum*, a non-tuberculous mycobacterium that mainly affects the skin and soft tissues. Direct contact with contaminated fish or water from fish tanks or swimming pools is the primary source of this infection, which also causes skin damage (Hendrikx et al. 2022).

7.2.2 Viral Skin Infections

7.2.2.1 Chickenpox

Varicella zoster is the causative agent of chicken pox, a highly contagious airborne disease. It presents as a blister-like rash and fever and is the most prevalent disease in children. Pneumonia, encephalitis, or secondary bacterial infection can aggravate chickenpox, particularly in immunocompromized patients, adults, and adolescents. Varicella zoster-infected pregnant women may experience difficulties and serious consequences for the foetus (Lo Presti et al. 2019).

7.2.2.2 Molluscum Contagiosum

Molluscum contagiosum is otherwise known as water warts. The causative agent is the molluscum contagiosum virus, which belongs to the poxvirus family. It mostly affects children, sexually active adults, and persons with impaired immunity. The lesion appears round, dome-shaped, and has a pink-purple colour. It can spread through direct or indirect contact and by autoinoculation (Badri and Gandhi 2024; Meza-Romero et al. 2019).

7.2.2.3 Shingles

Shingles is otherwise known as herpes zoster. However, reactivation of the latent varicella zoster virus is the aetiology of this disease. There is a correlation between the immunological state and the occurrence of herpes zoster. Shingles are less common among individuals with stronger immunity. The virus produces blisters and develops localized inflammation in the skin. The inflammation of the virus-affected nerves causes zoster pain (Nair and Patel 2024).

7.2.2.4 Warts

The causative agent for warts is human papillomavirus, which commonly causes skin and mucous membrane growth. The warts caused by the human papillomavirus are common, flat, genital, and deep palmoplantar warts. It is easily transmissible through direct or indirect contact. The risk of acquiring warts increases in response to the events that compromise the natural epithelial barrier (Al Aboud and Nigam 2024).

7.2.2.5 ORF

ORF is otherwise known as ecthyma contagiosum is primarily a zoonotic disease that affects human skin. It is caused by parapoxvirus orfviridae. It mostly affects sheep and goats, then from infected animal to the human by direct contact. The majority of ORF infections only affect the skin's epidermal layer. Usually, viral ecthyma appears on the exposed area of the body such as, face, hand, finger, forearm or face. The lesion starts of as little papules and progress to become an ulcerative lesion. Mostly, the lesion starts on finger and hand as a small, hard, red to blue papule at the ORFV entrance, which is typically on a finger or hand. The papule develops into a haemorrhagic pustule or bulla, which may haemorrhage and quickly develops a central crust. The lesion become nodular in later stages, which leaks blood and fluid, usually covers with thin layer of crust, which thickens with time. Along with moderate lymphadenopathy, erythema multiforme or lymphangitis, a low-grade fever lasting for few days could be connected to the skin lesions (Efridi et al. 2014; Kassa 2021).

7.2.2.6 Pityriasis Rosea

Pityriasis rosea is otherwise known as pityriasis circinate, severe form of self-limiting illness. Usually, a herald patch appears and within 2 weeks it develops into scaly oval patches. On the trunk and near extremities, the patches typically spread in the shape of a Christmas tree pattern. Two weeks after the herald patch, the generalized rash appears. At this stage the patient may show clinical manifestations like fever, sore throat, joint pain, swollen lymph nodes, nausea, and overall malaise (Litchman et al. 2024; Villalon-Gomez 2018).

7.2.3 Fungal Skin Infections

7.2.3.1 Ringworm

Ringworm otherwise known as tinea corporis, is a superficial skin infection. Most commonly, dermatophytes namely *Trichophyton*, *Microsporum*, and *Epidermophyton* genera are the most common cause of ring worm infection. Usually, tinea corporis present as a clearly defined, oval or circular, slightly erythematous, scaly patch or plaque with an elevated front edge. Ringworm is caused by an annular lesion, which is first observed as flat, scaly patches, that later spread centrifugally and clears centrally (Leung et al. 2020).

7.2.3.2 Athlete's Foot

Athlete's foot, otherwise called as tinea pedis, arises from the foot infection by dermatophytes. The causative agents include *Trichophyton rubrum*, *Trichophyton mentagrophytes*, *Trichophyton interdigitale*, and *Epidermophyton floccosum*. Usually, the infection develops by direct contact with the organism while walking barefoot.

Itchy scales and erosions between the toes are the first signs of tinea pedis. On the median and lateral part of the foot as well as sole, some patients may have hyperkeratosis with underlying erythema. Some patients may experience painful bullous sores (Nigam et al. 2024).

7.2.3.3 Jock Itch

Jock itch, otherwise called tinea cruris *Trichophyton*, *Epidermophyton*, and *Microsporum* are the common causative agents. The infection involves pubic, genital, perianal, and perineal skin. It most commonly affects the keratinized structure of the body, like the epidermis and hair. Intertriginous areas are favourable habitats for fungi, with sweating, alkaline pH, and maceration being the reasons behind the groin's susceptibility to infection. It is highly contagious and easily spread from person to person. It is characterized by a red, scaly, circular rash with burning and itching (Pippin et al. 2024).

7.2.3.4 Tinea Versicolor

A superficial skin infection called tinea versicolor, otherwise known as pityriasis versicolor, is a superficial skin infection. The causative agent is *Malassezia*, a lipophilic dimorphic fungus. It is characterized by either hypopigmentation or hyperpigmentation of the skin. Most commonly, it affects the abdomen, neck, chest, back, and upper arms (Karray and McKinney 2024).

7.2.3.5 Yeast Infection

Cutaneous candidiasis is the other name for Yeast infections of the skin, the causative agent is candida species, mainly by *Candida albicans.* It is an opportunistic fungus that resides inside human skin, and it develops into a serious health issue in people with HIV/AIDS or in immunocompromized patients. The groin and armpits are the warm, moist, and wrinkled parts of the human body, where infections are most common. This frequently occurs in obese and diabetic individuals. Diaper rashes in babies can be caused by *Candida.* It is characterized by rash, itching, pimple-like bumps, burns and spots that leak a transparent fluid (Lu et al. 2023).

7.2.3.6 Sporotrichosis

Sporotrichosis, otherwise known as "gardener's disease", mostly impacts people who work in rose cultivation. *Sporothrix schenckii*a dimorphic fungus is the causative agent of Sporotrichosis. At temperatures below 37 ° C, *S. schenckii* exists in hyphal form; at temperatures, 37 ° C or above, it exists as a budding yeast. The fungus gets into the body through thorns, pricks, scratches, and other minor wounds, resulting in infection. It is characterized by nodule lesions on the skin, ranging from pink to purple colour. The nodule is usually painless but may be developed into an ulcer (Arenas et al. 2018; Sizar and Talati 2024).

7.2.4 Parasitic Skin Infection

7.2.4.1 Scabies

Human scabies is a parasitic skin infestation the causative agent is *Sarcoptes scabiei*. Most commonly, it affects the wrist, hands, feet, and ankles and is characterized by severe itching, especially at night, and skin eruption with nodules, vesicles, and papules; in children, it presented as large rashes, especially in ankles, soles of the feet, scalp, and palms. Direct skin-to-skin contact with the infected individual can spread it (Chandler and Fuller 2019; Engelman and Steer 2018).

7.2.4.2 Cutaneous Larva Migrans

The term cutaneous larva migrans which was also used to describe the clinical manifestation of creeping eruption is an infectious disease caused by multiple types of hookworm. This is mostly transmitted to humans, with larva entering, through direct contact with animal faeces containing eggs in the soil. The animal vector nematode hookworms *Ancylostoma braziliense*, *Uncinaria stenocephala*, and *Ancylostoma caninum* are among the organisms. *Necator americanus* and *Ancylostoma duodenale* are human hookworms, that can also spread disease (Maxfield and Crane 2024).

7.2.4.3 Onchocerciasis

Onchocerciasis is known as river blindness. Worldwide, after trachoma, it is considered as the infectious blindness's second most frequent cause. It is transmitted by black fly of the genus *Simulium damnosum* repeated bite and is characterized by itchy skin, rashes, nodules under the skin, lesions on the eyes, visual impairment, low vision or complete blindness (Gyasi et al. 2024).

7.2.4.4 Tungiasis

The penetration of female sand fleas is the causative agent of Tungiasis, a parasitic skin infection. People get it from walking barefoot or resting on the soil, where sand flea have completed their host cycle. Most frequently, they burrow into the heels, outside of the foot, in between the toes or into the sole of the feet. It may appear as a white circle with a black dot in the centre that resembles a ring or halo once it has buried itself into the skin (Feldmeier and Keysers 2013).

7.3 Ethnomedicinal Treatment for Skin Infection

In traditional medicine, instead of using isolated compounds, whole plant or mixture of plants are utilized (Mala et al. 2021). Plant-based natural medications are becoming more popular due to their several advantages, such as having fewer side effects, being less expensive, patient tolerance, and being acceptable because of their extensive usage. Herbal medicines offer sensible alternatives to conventional medicine for treating several incurable (Tabassum and Hamdani 2014).

Achyranthes aspera

The common name for *Achyranthes aspera* is Prickly Chaff flower; belongs to the Amaranthaceae family and is used to treat various diseases. Bioactive compound reported in *Achyranthes aspera* is Oleanolic acid which has antiviral activity. Traditionally, the plant is used to treat ringworm, scabies, boils, blisters, eczema, wounds, skin eruption, and any other skin diseases. The leaves of *Achyranthes aspera* were used to extract alkaloid, non-alkaloid, MeOH, and saponin fractions, showed notable suppression of Epstein-Baar virus (Mughal et al. 2013).

Azadirachta indica

Azadirachta indica, known as neem, belongs to Meliaceae family incurable (Tabassum and Hamdani 2014). In India, neem trees are widely used as a traditional remedy for the treating infectious skin. The phytochemicals present in *Azadirachta indica* are polyphenolic flavonoids, quercetin, and sitosterol. *A. indica* has various bioactive compounds such as quercetin, beta-sterol, nimbin, nimbolide, and limonoids. *A. indica* and its ingredients have antibacterial, antiviral, and antifungal activity. Blisters, scabies, chicken pox, and boils are treated externally using leaf extract. Psoriasis and ringworm infection were treated effectively by using oil extracted from *A. indica*'s leaf. For the treatment of eczema, scabies, and other dermatological problems, applying neem and turmeric lotion to the skin at night has proven to be beneficial with no negative side effects (Ahuja et al. 2021).

Cannabis sativus

The common name for *Cannabis sativua* is Charas; it belongs to Cannabinaceae family. Wounds and sores can be treated with the powder of leaves. The most important bioactive compound present in *Cannabis sativus* is delta-9-tetrahydrocannabinol (THC), a psychoactive cannabinoid that includes most common cannabinoids such as cannabidiol, cannabichromene, cannabinol, and cannabigerol. The external use of Ganja is used to treat itchy skin diseases. Use of hemp seed oil can withstand bacterial, viral, and fungal skin infection. Hemp oil is a good treatment for eczema and other skin problems such as psoriasis, dermatitis, seborrheic dermatitis/cradle cap, lichen planus, acne rosacea, and varicose eczema. Scabies is controlled by rubbing crushed leaves on the afflicted areas incurable (Tabassum and Hamdani 2014).

Echinacea angustifolia

The common name for *Echinacea angustifolia* is purple coneflower, which belongs to the Asteraceae family. Different bioactive compounds from *Echinacea angustifolia*, such as flavonoids, alkylamides, and caffeic acid, have been isolated. Active compounds reported from *Echinacea angustifolia* are chicoric acid, Echinacoside, caftaric acid, and caffeic acid. It is applied to treat different types of skin infections, such as ulcers, skin boils, wounds, herps, psoriasis, and burns. There are tablet, tea, and juice forms of Echinacea incurable (Tabassum and Hamdani 2014).

Curcuma longa

The common name for *Curcuma longa* is turmeric; it belongs to the Zingiberaceae family. For ages, the Ayurvedic medicine system has utilized turmeric as therapeutic cure, cosmetic product and for cooking. The bioactive components of turmeric are curcumin, followed by demethoxycurcumin and bisdemethoxycurcumin. Curcumin is a principle bioactive compound of *Curcuma longa* which have antibacterial, antifungal, antiviral, antioxidant and anti- inflammatory properties. It has antibacterial, antifungal, and antiviral effects. It has anti-dermatophytic properties, and skin conditions like, tinea, rash, ringworm, and itching have been treated with this (Jankasem et al. 2013). Applying turmeric during smallpox and chicken pox, helps in speed up the scabbing process (Zanwar et al. 2013). It is used to treat different types of bacterial skin infections such as impetigo and boils.

Embilica officinalis

The common name for *Embilica officinalis* is Indian gooseberry. It is known as "Rasayan" for its ability to produce anti-ageing effects and is recommended by doctors to rejuvenating body, boost immunity, and promote wellness. This fruit is active against bacterial, fungal, and viral infections, in addition to being a strong antioxidant, and rich in vitamin C. The phytochemicals present in *Embilica officinalis* are flavonoids, saponins, tannins, and gallic acid. The bioactive compounds isolated from *Embilica officinalis* are Patuletin glucoside, Patuletin gallocatechin, caftaric acid, quercitrin, and emblicanin A and from ancient times it served as a natural therapy for scabies, warts, skin infections, reducing itching, and providing healing against wound infections (Ahuja et al. 2021).

Ocimum sanctum

Ocimum sanctum, locally known as Tulsi, belongs to the Lamiaceae family. *Ocimum sanctum* has become the preferred medication with a wide range of therapeutic uses and as a cost-effective herbal therapy. The *Ocimum sanctum* contains the following phytochemicals such as phlobatannins, flavonoids, saponins, terpenoids, glycosides, and steroids. The leaf extract of Tulsi contains various bioactive compounds, including eugenol, urosolic acid, carvacrol, and linalool. It has antiviral, antibacterial, and antifungal activity. It served as natural remedy for ringworm infections, blisters, wound infections, and chickenpox. It was also found to be effective against boil when crushed leaves were used along with mustard oil on the affected part of the body (Mahalik 2019).

Sarco asoca

The other name for *Sarco asoca* is Ashoka belongs to the Caesalpiniaceae family. Phytochemicals reported in Sarco also are saponins, tannins, flavonoids, steroids, glycosides, alkaloids, and phenolic compounds. Bioactive compounds present in *Sarco asoca* are Catechin, epicatechin, and epigallocatechin. *Sarco asoca* root, bark, and seed extracts are used as natural remedies for various skin condition such as dermatitis, eczema, pruritis, scabies, herps-kushta/visarga, and tinea pedis. To cure eczema and scabies, dried flowers of *S. asoca* and *L. inermis* leaves are boiled with coconut oil and applied externally twice daily (Nyeem 2017).

Portulaca oleraceae

The common name for *Portulaca oleracea* is Purslane; it is a member of Portulacaceae family. In hot weather, the herb's cooling properties help to relieve skin irritation and rashes by soothing the skin. From portulaca, many compounds like polysaccharides, flavonoids, sterols, fatty acids, proteins, alkaloids, vitamins, and minerals have been isolated. Many bioactive compounds have been isolated from *Portulaca oleraceae* such as kaempferol, apigenin, dopamine, luteolin, genistin, genistein, and quercetin. The aqueous extract has antibacterial and antifungal activity when applied topically to the skin. The juice is effective against eczema, ulcers, skin sores, abscess, and pruritis (Zhou et al. 2015).

Lawsonia inermis

The common name for *Lawsonia inermis* is Heena, which belongs to the Lythraceae family. Heena is a traditional Middle Eastern herb, that applies to the hand and feet. It has antibacterial, antifungal, and antiviral activity. Many phytochemicals like cardio glycosides, carbohydrates, flavonoids, tannins, saponins, phenols, sterols, alkaloids and quinones have been isolated. Various bioactive compounds have been isolated from various parts of *Lawsonia inermis* such as Lawsone, isoplumbagin, lawsochylin, and lawsonaringenin. In the traditional medical system, Impetigo is treated by applying leaf paste twice a day to the affected areas. It also treats Scabies, Erysipelas, Smallpox, and Measles (Akhtar et al. 2021).

Ficus carica

Ficus carica is commonly known as Fig; it belongs to the family Moraceae. It acts as a natural remedy for warts. Many phytochemicals such as phenolic acids (ferulic acid, 3-O- and 5-O- caffeoylquinic acid, psoralen), organic acids (oxalic, malic, citric, and fumaric acids), coumarin, flavones, and flavanols are present. A study conducted on wart patients has shown that this treatment has several benefits such as short treatment duration, no side effects, ease of use, low recurrence rate, and patient compliance and to treat skin ulcers, eczema, skin dryness and it also has anti-helminthic activity (Ahuja et al. 2021).

Honey

Honey is produced by honey bees by using the nectar of flowers. It is primarily made up of glucose and fructose, and it also contains amino acids, organic acids, vitamins, minerals, and enzymes. A wide variety of active compounds are found in honey, such as organic acids, phenolic acid, vitamins, flavonoids, and enzymes, which may improve the healing of wounds and act as a natural remedy for diabetic foot ulcers and boils. Bioactive compounds present in honey are cinnamic acid, syringic acid, caffeic, and coumaric, which have antibacterial activity (Tashkandi 2021).

Canthium parviflorum Lam

Canthium parviflorum Lam. belongs to the Rubiaceae family. It is a widely dispersed, prickly, sub-scandent shrub that grows in forest and arid areas. The fruits and leaves are consumable and also utilized in natural remedies. A wide variety of phytochemicals are found in *Canthium parviflorum* such as flavonoids, alkaloids, saponins, terpenoids, steroids, tannins, and cardiac glycosides. To treat ringworms and scabies, a handful of leaf paste is applied externally twice a day (Pingale et al. 2017).

Lavender oil

Lavender oil is formally called as Labiatea; it belongs to the Lamiaceae family. Because of its essential oil qualities, lavender is one of the most widely grown plants worldwide. Lavender is mostly grown in Europe, Asia, North Africa, and the Middle East. A variety of bioactive compounds have been reported, such as α-pinene, camphene, verbenene, camphor, sabinene, P-cymene, linalool oxide, linalool oxide cis, linalool, α-campholene aldehyde. It has antiviral and antibacterial properties. It is used to treat oral herpes, a viral infection caused by Herpes simplex virus Type 1, and it is also used to treat skin ulcers, wounds, and burns (Wińska et al. 2019).

Thyme vulgaris

The common name for *Thyme vulgaris* is Thyme, which belongs to the Lamiaceae family. Thyme is a native of South Europe and is found all over the world. The plant has been utilized in cooking due to its flavour. Many bioactive compounds such as thymol, terpineol, alpha- Thujene, alpha-Pinene, beta-Pinene, para-Cymene, caryophyllene, para-Cymene have been isolated. Folk medicine uses *Thymus vulgaris* essential oil as an antiseptic, antiviral, and antibacterial agent. It acts as a natural remedy for cellulitis, athlete's foot, and eczema (Tabassum and Hamdani 2014).

Carica papaya L.

The common name for *Carica papaya* L. is papaya; it belongs to Caricaceae family. When it comes to growth stimulation, health benefits, and therapeutic capabilities, *Carica papaya* is regarded as a powerhouse of all the essential elements. It is an important topical fruit with roots, pulp, bark, seeds, and peels that are used to treat a variety of illnesses because it has a lot of vitamins and the proteolytic enzymes papain and chymotrypsin, which have antiviral, antifungal, and antioxidant properties. The leaf of *Carica papaya* L. contains many phytochemicals such as alkaloids, flavonoids, tannins, saponins, starch, and phenol. Several bioactive compounds have been derived from *Carica papaya* such as quercetin, kaempferol, p- coumaric acid, benzyl isothiocyanate, lycopene, and protocatechuic acid. It is used to treat eczema, wounds, mouth ulcers, and warts (Ahuja et al. 2021).

Hemidesmus indicus (L.) R. Br. Var. indicus

The root of *Hemidesmus indicus* is otherwise known as Indian Sarsaparilla or Anatamul; it belongs to Asclepiadaceae. According to Ayurveda Anatamul is very important in Ayurvedic medicine because of its healing and blood purifying properties. Variety of Phytochemicals are present in *Hemidesmus indicus* such as terpenoids, steroids, flavonoids, aldehydes, coumarins, pregnane glycosides, and benzoids. *Hemidesmus indicus* constituents of several bioactive compounds such as 2-Hydroxy-4-methoxybenzaldehyde, 2-Hydroxy-4-methoxybenzoic acid, 4-Hydroxy-3-methoxybenzaldehyde, and isovanillin. With 100 ml of boiled coconut oil, mix a handful of dried Anatamul root and apply externally to the impacted area of the skin twice a day, which is helpful in treating eczema, ringworm infection, and scabies (Anitha et al. 2008).

Aegle marmelos (L.)

The common name for *Aegle marmelos* is Bael; it belongs to the Rutaceae family. In Indian medicine, it has been widely used due to its medicinal properties. Besides its medicinal uses, Bael fruit is rich in nutrients as they contain various minerals (phosphorous, calcium, potassium, iron, magnesium, copper, zinc, and chromium), fat, fibres (cellulose, lignin, pectin, and hemicellulose), vitamins (B1, B2, B3, C), proteins, carbohydrates, fatty acids, and amino acids. This plant is also rich in many phytochemicals such as alkaloids, tannins, flavonoids, saponins, phenols, and terpenoids. The major phytoconstituents present in *Aegle marmelos* are 1-Dodecanol, tetradecanoic acid, 1-Heptadecanol, 1-Butanol,-3-methyl,-acetate. Each part of this plant like leaves, stems, fruit, and bark has medicinal properties. It is used to treat scabies and similar types of skin infections (Agrawal et al. 2022; Rahman et al. 2024).

Jasminum grandiflorum L.

The common name for *Jasminum grandiflorum* L. is Chameli; it belongs to the Oleaceae family. In pharmaceutical industries, different parts of this plant, including stems, flowers, leaves, and bark, have medicinal properties. It is native to India, China, Japan, Italy, Morocco, France, and Egypt. This plant is rich in many phytochemicals such as flavonoids (rutoside), phenol (cresol), glycosides, phenylpropanoid (eugenol), steroid, alkaloids, monoterpenoids (geraniol, iridoids), tannins, and saponins. It is useful for the treatment of mouth ulcers, wound infections, leprosy, boils, and eczema (Mhaskar and Sathe 2019; Sandeep et al. 2009).

Adiantum capillus- vernis L.

The common name for *Adiantum capillus-vernis* L. is Hansraj; it belongs to the Pteridaceae family. This herb is high in moisture content and distributed from cold climates to hot topical areas. This plant is rich in many phytochemicals, such as flavonoids, phenolic acid, phospholipids, chlorophylls, and carotenoids. Bioactive constituents of *Adiantum capillus* are d-Mannose, D-Carvone, Estragole, Tetraacetyl-d-xylonic-nitrile, Tributyl acetyl citrate. It is used for the treatment of wound infections, chicken pox, boils, and eczema (Nilforoushzadeh et al. 2014; Ranjan et al. 2017).

Bombax cieba

It is known as the silk cotton tree, and belongs to the Bombacaceae family. It is widely distributed in Asia, Australia, and Africa. In Ayurvedic medicines, different parts of the plant-like stem, bark, root, leaf, flower, fruit, and gum, have ethnomedicinal properties. Different phytochemicals were detected in *Bombax cieba* such as reducing sugar, phlobatanins, terpenoids, glycosides, saponins, tannins, alkaloids, and phenolic components (Kaemferol, mangiferin, shamimin, and quercetin). Seed paste is used for the treatment of smallpox boils and chicken pox; bark juice act as natural remedy for wound infections, and fruit is used to treat leprosy (Karole et al. 2017; Rani et al. 2016).

Aloe barbadensis Mill

Popularly known as Aloe vera, it belongs to the Liliaceae family. It is distributed widely in Asia, Africa, America, and Europe. For the production of nutraceuticals, skincare, and cosmetic products, aloe vera is widely used. The presence of various phytochemicals such as alkaloids, flavonoids, tannins, saponins, and glycosides has been reported. Bioactive compounds of *Aloe barbadensis* are Hypoxanthine, β-sitosterol, Vanillic, Homovanilllic, p-coumaric, Isoferulic, Ferulic, p-Salicylic, and Phenyl pyruvic. It acts as natural remedies for wound infections, diabetic ulcer foot, and genital herpes (Hekmatpou et al. 2019; Kumar et al. 2019).

Veronica cinerea (L.) Less

It is otherwise known as "little ironweed"; it belongs to the Asteraceae family. This plant has some phytochemicals like flavonoids, sterols, terpenoids, and sesquiterpene lactones. It is useful for the treatment of eczema, herpes, ringworm infections, leprosy, scabies, wounds, and boils (Sonibare et al. 2016).

Wrightia tinctoria (Roxb.) R. Br

Wrightia tinctoria (R0xb.) R. Br. is generally called as "sweet Indrajao"; it belongs to the Apocynaceae family. In India, it is widely distributed. Various parts of this plant like leaves, bark, and seed, have medicinal properties. Leaves of *Wrightia tinctoria* indicated the presence of phenolic acids, flavonoids, glycoflavones, and iso-orientin. Bioactive compounds isolated from various parts of *Wrightia tinctoria* are lupeol, β-amyrin, stigmasterol, campetosterol, indigotin, isatin, Tryptanthrin, and wrightial. It is used for the treatment of scabies and eczema; 100 gm of crushed leaves are boiled with 200 ml of coconut oil and applied externally thrice a day for the treatment. To treat various skin infections, bark paste is used (Hemavathy et al. 2019; Srivastava 2014).

Debregeasia salicifolia

Debregeasia salicifolia otherwise known as White Bully; it belongs to the Urticaeae family. It is widely distributed in south-west India and Laccadive Islands. It is used to treat boils, eczema, and skin rashes (Almubayedh and Ahmad 2019).

Mimosa pudica L.

Mimosa pudica L. generally called as touch-me-not; it belongs to the Mimosaceae family. It is widely distributed in Australia, America, and India. This herb has been used in Ayurveda for ages for the treatment of various diseases. *Mimosa pudica*

L. was found to have some phytochemicals such as sterols, non-protein amino acids, fatty acids, alkaloids, and flavonoids C-glycosides. The major bioactive compounds reported in *Mimosa pudica* are L-mimosine, crocin, ethyl gallate, jasmonic acid, and gallic acid. In India, leaves and roots are used for the treatment of ringworm infections, eczema, leprosy, and smallpox (Majeed et al. 2021).

Kigelia africana (Lam.) Benth

Kigelia africana (Lam.) Benth otherwise known as the cucumber tree; it belongs to the Bignoniaceae family. It is native to Africa but widely found in India, the Philippines, China, Iraq, and Pakistan. Different phytochemicals have been isolated from different parts of the plant such as flavonoids, terpenes, phenylethanoglycosides, and iridoids. *Kigelia africana* is reported to contain a variety of bioactive compounds such as 7-hydroxy eucommiol, elaidic acid, elaidoic acid, trans-phytol, and β-tocopherol. It is used to treat skin ulcers, abscesses, wound infections, eczema, ringworm infections, and boils (Bello et al. 2016).

Boerhaavia diffusa L.

Boerhaavia diffusa L. otherwise known as Punarnava; it belongs to the Nyctaginaceae family. The leaves, roots, and seeds have medicinal properties. This plant has a large number of phytochemicals such as alkaloids, steroids, carbohydrates, proteins, glycoproteins, lipids, and triterpenoids. Bioactive compounds present in *Boerhaavia diffusa* are Phthalic acid, monoamide, pentyl ester, dipyridamole, and α-Tocopherol-β-D-mannoside. It is used to treat blisters, wound infections, skin ulcers, scabies, and ringworm infections (Topno and Sinha 2018).

Clerodendrum infortunatum L.

Clerodendrum infortunatum L. otherwise known as Bhant; it belongs to the Verbanaceae family. It is widely distributed throughout India. *Clerodendrum infortunatum* L. plant has phytochemicals like flavonoids, sterols, saponin, oleic, stearic, lignoceric acid, and triterpenes. A wide variety of bioactive compounds were isolated from *Clerodendrum infortunatum* are luperol, β- sitosterol, Acetoside, fumaric acid, apigenin, and clerodin. It is used to treat scabies, skin ulcers, and other skin infections (Khatun et al. 2024).

Heliotropium indicum L.

Heliotropium indicum L. locally known as Hatisur; it belongs to the Boraginaceae family. It is distributed throughout India, Sri Lanka, Nepal, Thailand, Bangladesh, and in some parts of Asia and Africa. *H. indicum* L. has many phytochemicals like saponins, steroids, glycosides, oils, and tannins. *H. indicum* was reported to contain a wide variety of bioactive compounds such as acetyl indicine, echinitine, Europine-N-oxide, β-sitosterol, and β-linalool. It applied as natural therapy for scabies, eczema, and other skin infections (Sarkar et al. 2021).

***Sida rhombifolia* Linn**

Sida rhombifolia Linn. locally known as Bala or Atibala; it belongs to the Malvaceae family and is found in tropical and warm regions of the world. Different parts of the plant, such as saponins, tannins, flavonoids, amino acids, glycosides, lignans, carbohydrates, sterolic compounds, fatty acids, phenolics, and alkaloids, have been reported. A wide variety of bioactive compounds are present in *S. rhombifolia* such as cryptolepinone, kaempferol, scoporone, scopoletin, and phaeophytin. It has antibacterial, antifungal, and anti-inflammatory properties. Leaves, flower, fruits, stems, and roots have effective medicinal properties, are used for the treatment of chickenpox, wounds, ulcers, and boils (Singh et al. 2018).

***Ricinus communis* L.**

Ricinus communis L. locally known as castor plant (Euphorbiaceae family) largely present worldwide. The phytochemicals like alkaloids, flavonoids, tannins, saponins, polyuronides, and terpenoids help in the healing of candidiasis, eczema, and wound infections. Diverse bioactive compounds have been reported in *Ricinus communis* such as ingenol, triterpenoids, quercetin, gallic acid, α-thujone, β-thujone, and camphore (Ghani et al. 2023; Suurbaar et al. 2017).

***Vitex negundo* L.**

Vitex nengundo locally known as Punjgusht or Sambhalu in India; it belongs to Verbenaceae family and widely distributed in India, China, Indonesia, Japan, Bangladesh, Afghanistan, Nepal, Bhutan, Philippines, Myanmar, Pakistan, Japan, Sri Lanka, Taiwan, Vietnam, Thailand, and Tanzania. The leaf of *Vitex negundo* constituents of casticin, carotene, beta-sitosterol, and globulol. It acts as natural therapy for wound infections and skin ulcers, and for the treatment of eczema, boiled leaf paste is applied externally (Bano et al. 2015). The details of the ethnomedicinal uses of medicinal plants are mentioned in Table 7.1.

The details of the reported phytochemical compounds and the structure of some plants are mentioned in Table 7.2.

7.4 Computational Biology

In order to discover new antimicrobial agent, against pathogens causing skin infection, modern scientific methods have been concentrated more on ethnomedicinal plants. For identifying and screening active compounds from ethnomedicinal plants having antimicrobial activity computational technique offers high-throughput and successful method.

7.4.1 Molecular Docking

Molecular docking is crucial for illustrating how natural substances interact molecularly with various receptors. This method helps in the identification of potential

Table 7.1 Ethnomedicinal uses of plants against skin infections

Plants	Ethnomedicinal uses	References
Croton bonplandianus Baill.	Scabies, wounds, ringworm infections, boils, and chicken pox	Anand et al. (2022), Ghosh et al. (2018)
Acalypha indica L.	Eczema, ringworm infections, skin ulcers, and wounds	Agrawal and Kulkarni (2023)
Santalum album L.	Eczema, psoriasis, wart, and molluscum contagiosum	Moy and Levenson (2017)
Achyranthes aspera	Ringworm, scabies, boils, blisters, eczema, wounds, skin eruption, and other skin diseases	Tabassum and Hamdani (2014), Mughal et al. (2013)
Azadirachta indica	Blisters, scabies, chicken pox, boils, psoriasis, ringworm infection, eczema, scabies, and other dermatological problems	Ahuja et al. (2021)
Xanthium strumarium L.	Eczema, small pox	Mughal et al. (2013)
Piper nigrum L.	Ringworm infections, scabies	Mohanty et al. (2019)
Echinacea angustifolia	Ulcers, skin boils, wounds, herps, psoriasis, and burns	Ahuja et al. (2021)
Annona squamosa L.	Boils, abscess	Mohanty et al. (2019)
Clitoria ternatea L.	Sabies, ringworm infections, and eczema	Zagórska-Dziok et al. (2021)
Curcuma longa	Tinea, rash, ringworm, smallpox and chicken pox, impetigo, boils, and itching	Zanwar et al. (2013)
Embilica officinalis	Scabies, warts, skin infections, reducing itching, and providing healing against wound infections	Tabassum and Hamdani (2014)
Plumbago zeylanica L.	Ulcer, eczema, scabies, ringworm infections	Bloch et al. (2022)
Pongamia pinnata (L.) *Pierre*	Eczema, rashes, scabies, and ringworm infections	Anitha et al. (2008)
Lavender oil	Skin ulcers, wound, oral herpes, and burns	Wińska et al. (2019)
Thyme vulgaris	Cellulitis, athlete's foot, and eczema	Tabassum and Hamdani (2014)
Tridax procumbens (L.)	Boils	Singh et al. (2018)

inhibitors by predicting the orientation and binding affinity phytochemicals within the active site of the microbial enzyme or receptors (Singab et al. 2022). Molecular docking is widely divided into three categories such as flexible docking, semi-flexible docking, and rigid docking. In case of rigid docking molecule structure remains unchanged. It is more applicable for researching macromolecular system such as protein–protein and protein–nucleic acid systems, because the calculation approach is simple and primarily focuses on conformation matching. In case of semi-flexible docking it is preferable for managing the interaction between tiny

Table 7.2 Structure of reported phytocompounds and their medicinal properties

Plant name	Reported phytochemical compound	Structure	Medicinal properties	References
Achyranthes aspera	Oleanolic acid	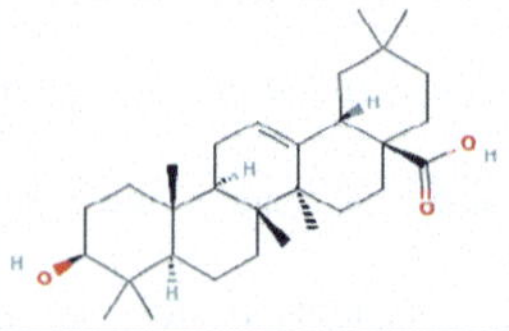	Antiviral activity	Ashraf et al. (2022)
Azadirachta indica	17-hydroxy azadiradione	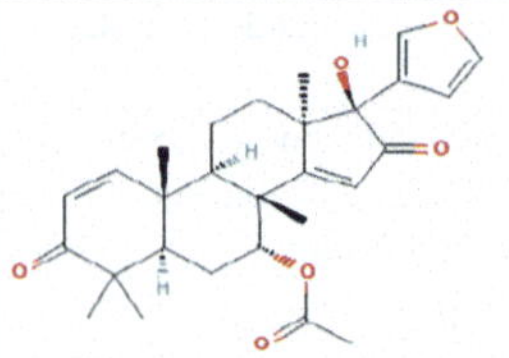	Antibacterial and antifungal activity	Alzohairy (2016)
Cannabis sativus	Cannabidiol	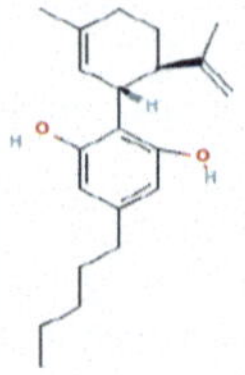	Antibacterial activity	Karas et al. (2020)
Echinacea angustifolia	Echinacoside	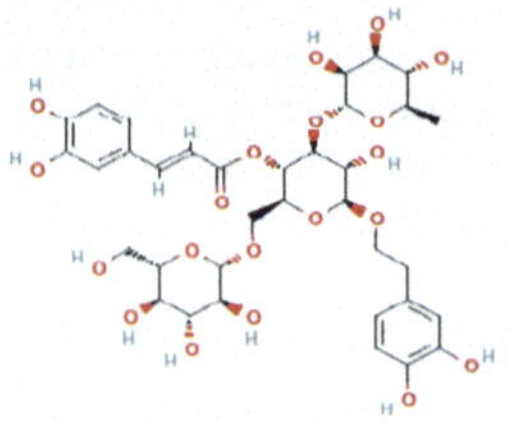	Antimicrobial activity	Burlou-nagy et al. (2023)
Curcuma longa	Curcumin	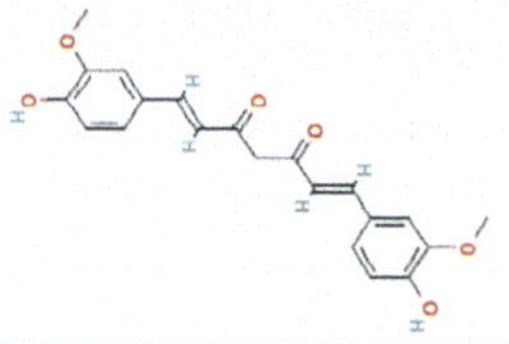	Antibacterial, Antifungal, Antiviral and Antiparasitic activity	Adamczak et al. (2020)
Embilica officinalis	Quercetin	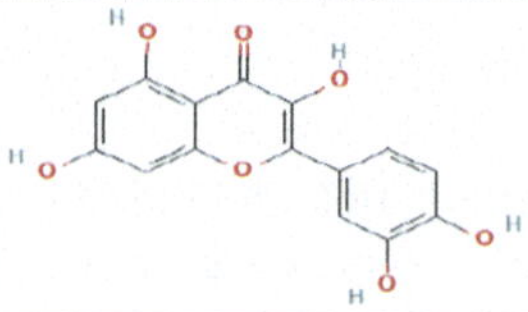	Antibacterial activity	Sing et al. (2022)

(continued)

Table 7.2 (continued)

Plant name	Reported phytochemical compound	Structure	Medicinal properties	References
Ocimum sanctum	Eugenol	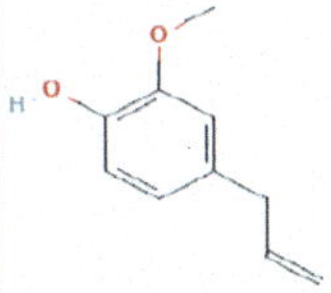	Antimicrobial activity	Raghav and Siani (2018)
Sarco asoca	Catechin	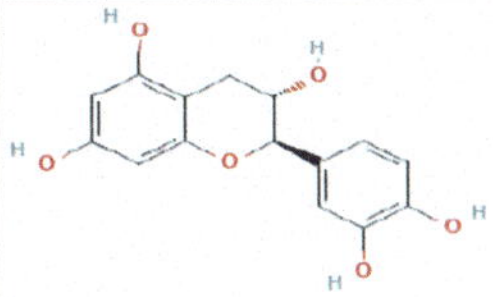	Antimicrobial activity	Shirolkar et al. (2013)
Portulaca oleracea	Kaempferol	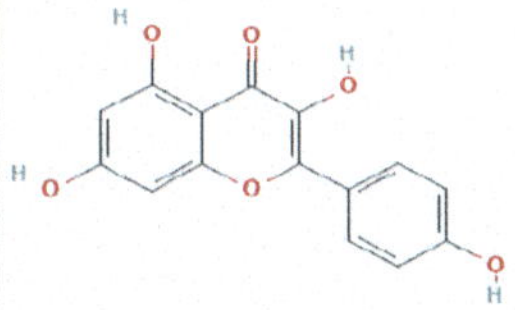	Antibacterial activity	Zhou et al. (2015)
Lawsonia inermis	Lawsone	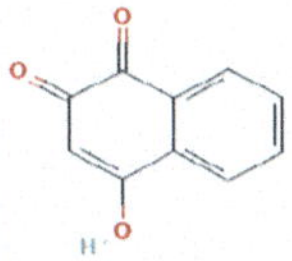	Antibacterial activity	Moutawalli et al. (2023)
Honey	P-coumaric acid	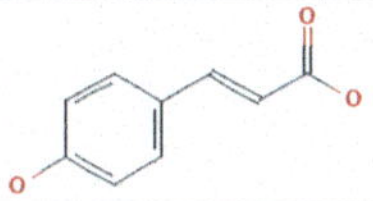	Antibacterial activity	Kunat-Budzynska et al. (2023)
Lavender oil	Linalool	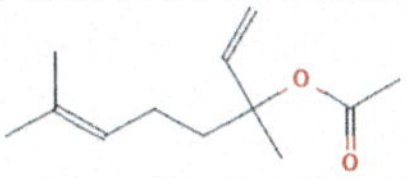	Antibacterial activity	Kwiatkowski et al. (2019)
Thymus vulgaris	Thymol	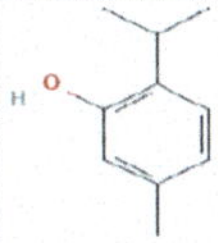	Antibacterial and Antifungal activity	Boruga et al. (2014)
Carica papaya	Quercetin	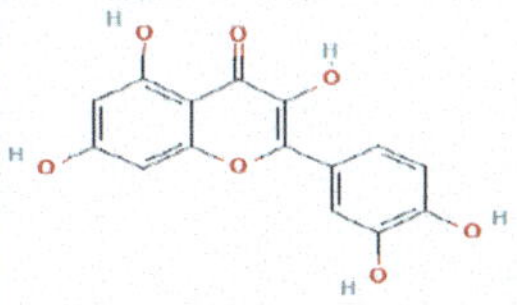	Antibacterial activity	Singh et al. (2020)
Hemidesmus indicus	3-hydroxy-4-methoxy benzaldehyde	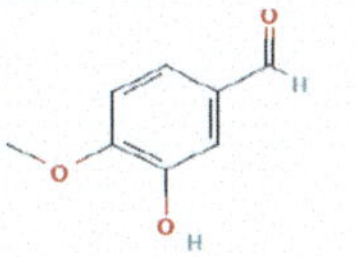	Antifungal activity	Nandy et al. (2020)

(continued)

Table 7.2 (continued)

Plant name	Reported phytochemical compound	Structure	Medicinal properties	References
Aegle marmelos	Quinic acid	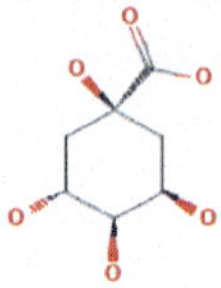	Antimicrobial activity	Mujeeb et al. (2014)
Jasminum grandiflorum	Eugenol	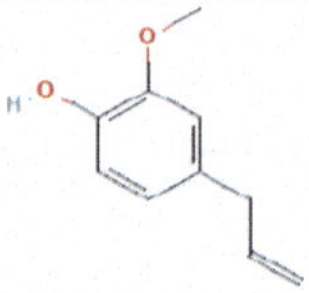	Antibacterial activity	Balkrishnaet al. (2021)
Adiantum capillus-veneris	D-carvone	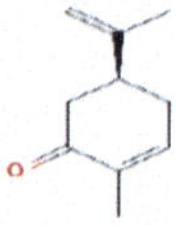	Antimicrobial activity	Hussein et al. (2016)
Bombax cieba	Bombax quinone B	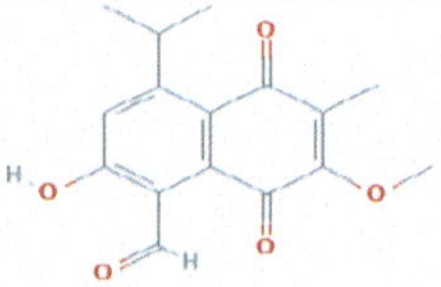	Antibacterial activity	Panwar et al. (2020)
Aloe barbadensis Mill.	P-coumaric acid	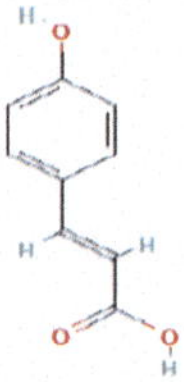	Antibacterial activity	Nejatzadeh-Barandozi (2013)
Mimosa pudica	L-mimosine	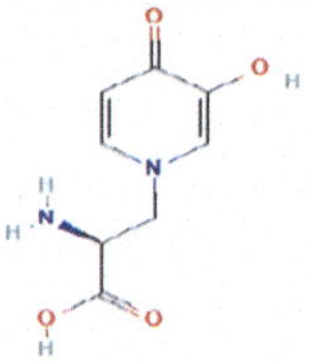	Antibacterial activity	Mandal et al. (2022)
Kigelia africana	2,4-Di-tert-butylphenol	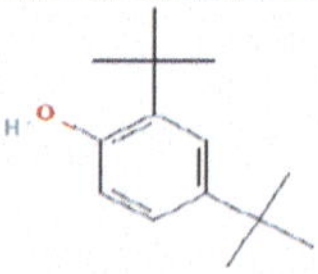	Antibacterial activity	Assanti et al. (2022)
Boerhaavia diffusa	Alpha-tocopherol	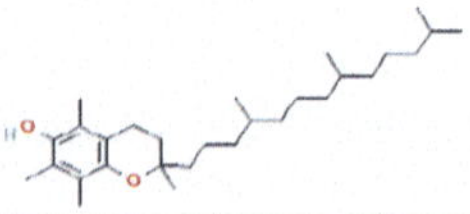	Antimicrobial activity	Kaviya et al. (2022)

(continued)

Table 7.2 (continued)

Plant name	Reported phytochemical compound	Structure	Medicinal properties	References
Clerodendrum infortunatum	Acetoside	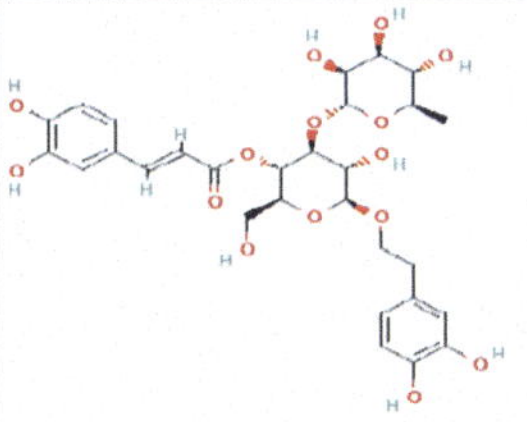	Antibacterial activity	Das et al. (2014)
Sida rhombifolia	Cryptolepinone	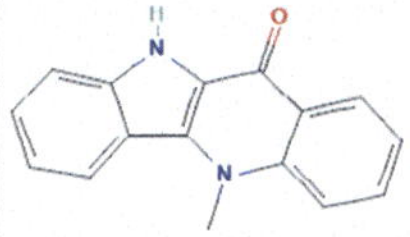	Antibacterial activity	Aminah et al. (2021)
Ricinus communis	Alpha pinene	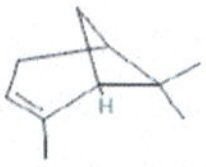	Antibacterial activity	Abdul et al. (2018)
Vitex negundo	4-terpineol	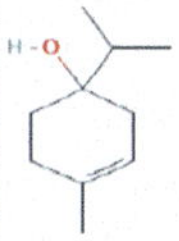	Antibacterial activity	Kapse and Chandekar (2020)
Acalypha indica	Quercetin	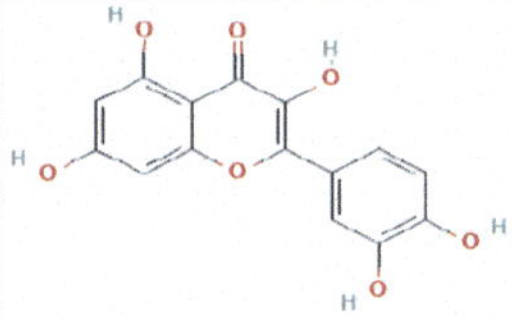	Antibacterial and Antifungal activity	Nag et al. (2018)
Santalum album	Santalbic acid	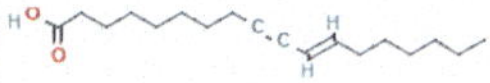	Antibacterial and Antifungal activity	Sharifi-Rad et al. (2023)

molecules and protein because it permits the molecules to change their shape in a certain range. In general, small molecules often have free-changeable structures, for computational efficiency macromolecules are stiff and maintain some rotatable amino acid residues. Flexible docking allows the simulated system conformation to change freely, which increase accuracy but requires more computer resources (Lin et al. 2020).

Docking process is represented in a flow chart as shown in Fig. 7.1.

7.4.2 Quantitative Structure-Activity Relationship (QSAR)

QSAR study is a statistical analysis of the correlation between the chemical structure and biological activities of the compound obtained from plant against microbes. Pharmaceutical industry uses medicinal plants as a component to make new

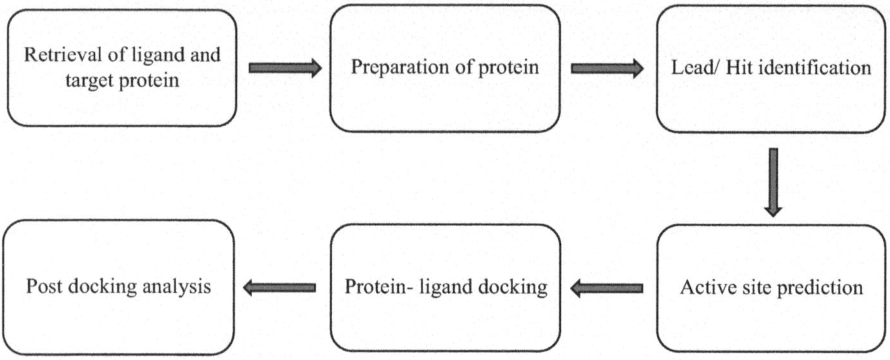

Fig. 7.1 Schematic presentation of the docking process

medications. But efficacy and safety are the two main obstacles that prohibit the use of plants before they are commercialized. Approximately one-third of the product failure are due to preclinical toxicity and adverse drug effect. Computational method for evaluation of toxicity approaches the early drug development by enabling the early identification of the compound. QSAR is a mathematical model that utilized to predict the structural activities of a compound. In system biology QSAR programmes are regarded as effective, useful, and enhanced core prediction. These could provide beneficial approach to expand our understanding about the potentially harmful effect of phytochemicals (Nisekhoto et al. 2023).

7.4.3 Molecular Dynamics

Modern drug discovery process usually begins with the conformation and identification of physiologically relevant target which can be modulated with drug, in order to prevent, treat or lessen the symptoms of a disease. Drug targets are typically proteins like enzymes and receptors but it can also be molecules of DNA or RNA. In ligand design protein conformation is the most significant approximates, because a range of conformational changes can occur in protein due to its dynamic nature. The complementarity between proteins binding site and ligand can be affected by even small conformational changes including movements of the side chain. For the drug design purpose MD simulations can give a valuable insight into the dynamic nature of the target, the range of potential target conformational states for ligand binding is greatly influenced by the flexibility of protein (Salo-Ahen et al. 2021; Badar et al. 2020).

7.4.4 Virtual Screening (VS)

Virtual screening is widely employed in the development of novel pharmaceuticals and it has greatly aided in the production of drugs that are available commercially.

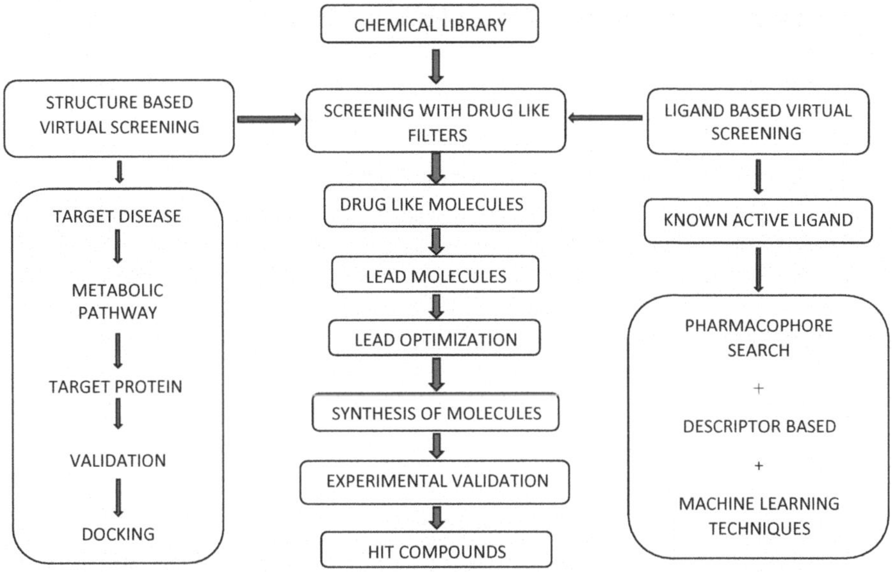

Fig. 7.2 Schematic diagram of VS process for SBDD and LBDD

In structure-based virtual screening (SBDD) it analyses the best relationship between a bacterial target and phytochemicals to form complexes. The phytochemicals with strongest interaction appear first and the other phytochemicals were organized according to how well they bind to the bacterial protein. According to the affinity for the target site phytochemicals are categorized and selected according to that. Consequently, it is possible to identify the phytochemicals that are most likely to interact pharmacologically with the bacterial target. The probability of binding domain describing the ligand-target attraction is checked by using score functions. In drug discovery and development structure-based virtual screening is the frequently used method because of its efficiency in both time and cost (Turabi et al. 2023) in case of ligand-based virtual screening (LBDD) it does not examine repositories of small molecules. Rather, it focuses on the preliminary information of the compound that bound to the target protein. These identified molecules can be used to generate pharmacological frame work, which outlines the minimum structural properties a ligand must has to possess in order to bind to the targeted molecule (Lin et al. 2020). Figure 7.2 shows schematic diagram of VS process for SBDD and LBDD.

7.4.5 Pharmacophore Modelling

Pharmacophore modelling that used in drug design is of two types: structure-based pharmacophore modelling and ligand-based pharmacophore modelling.

7.4.5.1 Structure-Based Pharmacophore Modelling

Target protein like enzyme or receptors the structural details were employed in structure-based pharmacophore modelling. The structure-based approach finds compound that can be used as a drug.

7.4.5.2 Ligand-Based Pharmacophore Modelling

In case of ligand-based pharmacophore modelling only uses physiochemical properties of the known ligand molecule for the development of drug, entails the development of 3D pharmacophore model (Giordano et al. 2022; Muhammed and Aki 2021; Noha and Schuster 2013). Fig. 7.3 depicts the steps involved in pharmacophore modelling workflow.

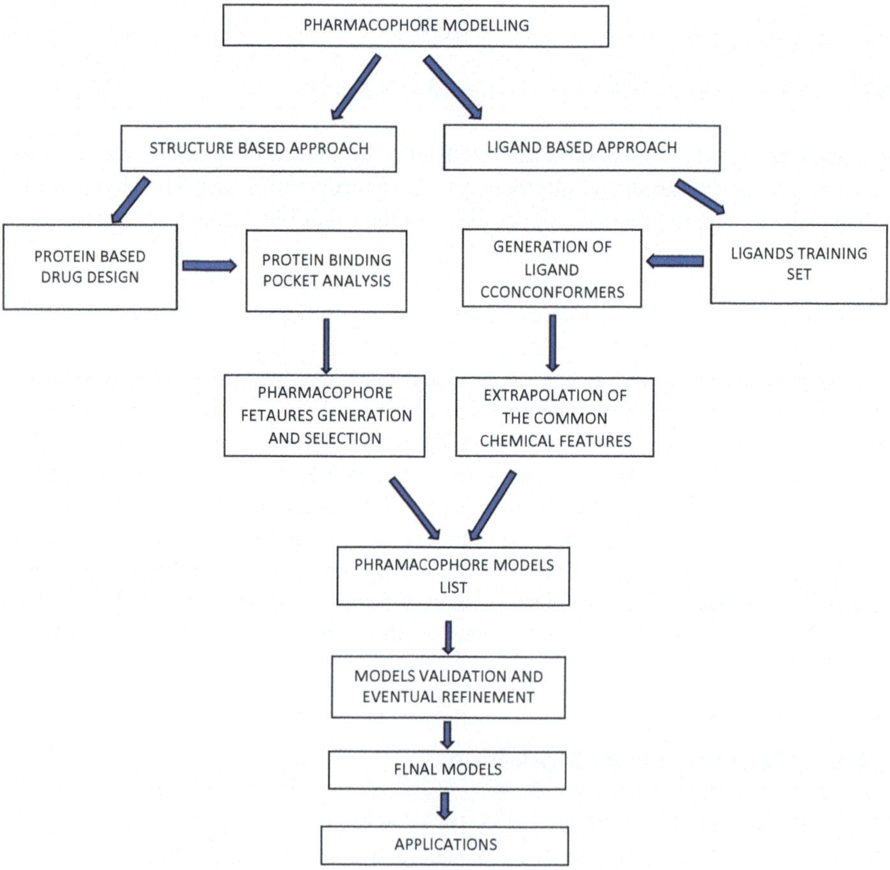

Fig. 7.3 Presenting the steps involved in pharmacophore modelling workflow

7.4.6 Network Pharmacology

A new opportunity to investigate the complicated molecular interactions of herbal remedies and the systemic link between the herbal remedies and complex disorders. Network-based techniques enable to understand the novel treatment that use natural product as the active compound that are responsible for synergetic and cumulative effect of drug. By integrating network pharmacology with traditional medicine will prove a unique and innovative chance to identify biomarkers, active compounds which intricate biological system of the human body (Bonthu et al. 2023; Lotfi Shahreza et al. 2018; Yu et al. 2017).

7.5 Conclusion

Skin infection being the most common disease among human beings, it can be caused by different kinds of bacteria, virus, fungi, and parasites. Nowadays, there is increasing resistance to various kind of antibiotics which is the major rising issues in healthcare sectors, besides this there are some other drawbacks of using allopathic medicine such as high-cost treatment, more side effects and unable to treat various kind of infections due to its increasing resistance. Nowadays, ethnomedicinal plants act as a valuable alternative to treat skin infections due to low-cost effect, can treat chronic skin infections, less side effects, and well tolerated. By using computational techniques like molecular docking, QSAR, molecular dynamics simulations, virtual screening, and pharmacophore modelling researchers are able to identify quickly and optimized compound having therapeutic effect against pathogens causing skin infections. These methods helps in finding a new drug in short duration of time and helps in understanding the molecular activity of antimicrobial actions, which helps in the discovery of novel plant-based treatment.

References

Abdul W, Hajrah N, Sabir J, Al-Garni S, Meshaal JS, Kabil S, Saini K, Bora R (2018) Therapeutic role of Ricinus communis L. and its bioactive compounds in disease prevention and treatment. Asian Pac J Trop Med 11:177

Adamczak A, Ożarowski M, Karpiński TM (2020) Curcumin, a natural antimicrobial agent with strain-specific activity. Pharmaceuticals (Basel) 13(7):153

Agrawal A, Kulkarni GT (2023) Topical application of aerial portion of Acalypha indica Linn ameliorates psoriasis in rodents: evidences from in vivo and in silico studies. J Ethnopharmacol 315:116685

Agrawal S, Kadam A, Sonavale S, Mulik P (2022) Formulation and evaluation of Peel off mask of Bael fruit. Int J All Res Edu Sci:2455–6211

Ahuja A, Gupta J, Gupta R (2021) Miracles of herbal Phytomedicines in treatment of skin disorders: natural healthcare perspective. Infect Disord Drug Targets 21(3):328–338

Akhtar J, Bashir F, Bi S (2021) Scientific basis for the innovative uses of henna (Lawsonia inermis L.) mentioned by Unani scholars in different ailments. J Complement Alternat Med Res 14:1–21. https://doi.org/10.9734/JOCAMR/2021/v14i130234

Al Aboud AM, Nigam PK (2024) In: StatPearls [Internet]. Treasure Island (FL): StatPearls Publishing. Available from: https://www.ncbi.nlm.nih.gov/books/NBK431047/

Almubayedh H, Ahmad R (2019) Ethnopharmacological uses, phytochemistry, biological activities of Debregeasia salicifolia: a review. J Ethnopharmacol 1(231):179–186

Alzohairy MA (2016) Therapeutics role of Azadirachta indica (neem) and their active constituents in diseases prevention and treatment. Evid Based Complement Alternat Med 2016:7382506

Aminah NS, Laili ER, Rafi M et al (2021) Secondary metabolite compounds from Sida genus and their bioactivity. Heliyon 7:e06682

Anand U, Tubu CK, Nandy S et al (2022) Ethnodermatological use of medicinal plants in India: from ayurvedic formulations to clinical perspectives – a review. J Ethnopharmacol 10(284):114744

Anitha B, Mohan VR, Athiperumalsami T, Sutha S (2008) Ethnomedicinal plants used by the Kanikkars of Tirunelveli District, Tamil Nadu, India to treat skin diseases. Ethnobotanical Leaflets 12:171–180

Arenas R, Sánchez-Cardenas CD, Ramirez-Hobak L, Ruíz Arriaga LF, Vega Memije ME (2018) Sporotrichosis: from KOH to molecular biology. J Fungi (Basel) 4(2):62

Ashraf S, Rahman M, Ahmed P et al (2022) Prospective Asian plants with corroborated antiviral potentials: position standing in recent years. Beni-Suef Univ J Basic Appl Sci 11:47

Assanti G, Kaur R, Nizard S, Pollack-Blackwood E, Rafferty B, Priano C, Fernández Romero JA, Koroch AR (2022) Biology, chemistry, and pharmacological activity of *Kigelia africana* (Bignoniaceae) and *Garcinia kola* (Clusiaceae) – a review. J Med Act Plants 11(1):1–21

Badar M, Shamsi S, Ahmed J, Alam A (2020) Molecular dynamics simulations: concept, methods, and applications. In: Transdisciplinarity. Springer International Publishing, Cham, pp 131–151. https://doi.org/10.1007/978-3-030-94651-7_7

Badri T, Gandhi GR (2024) Molluscum Contagiosum. In: StatPearls [Internet]. Treasure Island (FL): StatPearls Publishing; Available from: https://www.ncbi.nlm.nih.gov/books/NBK441898/

Balkrishna A, Rohela A, Kumar A, Kumar A, Arya V, Thakur P, Oleksak P, Krejcar O, Verma R, Kumar D, Kuca K (2021) Mechanistic insight into antimicrobial and antioxidant potential of *Jasminum* species: a herbal approach for disease management. Plants (Basel) 10(6):1089

Bano U, Jabeen A, Ahmed A, Siddiqui MA (2015) Therapeutic uses of Vitex Nigundo. World J Pharm Res 4:12

Bello I, Shehu MW, Musa M, Zaini Asmawi M, Mahmud R (2016) Kigelia africana (Lam.) Benth. (sausage tree): Phytochemistry and pharmacological review of a quintessential African traditional medicinal plant. J Ethnopharmacol 189:253–276

Bloch K, Parihar VS, Kellomäki M, Ghosh S (2022) Natural compounds from *Plumbago zeylanica* as complementary and alternative medicine. In: Chakraborti S (ed) Handbook of oxidative stress in cancer: therapeutic aspects. Springer, Singapore. https://doi.org/10.1007/978-981-16-1247-3_33-1

Bonthu S, Pulichintha S, Raju M, Reddy S (2023) Network pharmacology approach for herbal drugs intended for the therapy of diseases: a comprehensive review. Asian J Biol 19:63

Borugă O, Jianu C, Mişcă C, Goleţ I, Gruia AT, Horhat FG (2014) Thymus vulgaris essential oil: chemical composition and antimicrobial activity. J Med Life 7:56–60

Brazel M, Desai A, Are A, Motaparthi K (2021) Staphylococcal scalded skin syndrome and bullous impetigo. Medicina 57(11):1157

Burlou-Nagy C, Bănică F, Negrean RA, Jurca T, Vicaş LG, Marian E, Bácskay I, Kiss R, Fehér P, Vicaş SI, Miere Groza F, Memete AR, Pallag A (2023) Determination of the bioactive compounds from *Echinacea purpurea* (L.) Moench leaves extracts in correlation with the antimicrobial activity and the in vitro wound healing potential. Molecules 28(15):5711

Chakraborty RK, Burns B (2024) Systemic inflammatory response syndrome. In: StatPearls [Internet]. Treasure Island (FL): StatPearls Publishing; Available from: https://www.ncbi.nlm.nih.gov/books/NBK547669/

Chandler DJ, Fuller LC (2019) A review of scabies: an infestation more than skin deep. Dermatology 235(2):79–90

Das B, Pal D, Haldar A (2014) A review on biological activities and medicinal properties of clerodendrum infortunatum Linn. Int J Pharm Pharm Sci 6:41–43

Dawid-Pać R (2013) Medicinal plants used in treatment of inflammatory skin diseases. Postepy Dermat Alergol 30(3):170–177

Dryden MS (2010) Complicated skin and soft tissue infection. J Antimicrob Chemother 65(3):35–44

Efridi W, Jain H, Sathe NC et al (2014) Orf viral infection. In: StatPearls [Internet]. Treasure Island (FL): StatPearls Publishing. Available from: https://www.ncbi.nlm.nih.gov/books/NBK562191/

Engelman D, Steer AC (2018) Control strategies for scabies. Trop Med Infect Dis 3(3):98

Feldmeier H, Keysers A (2013) Tungiasis – a Janus-faced parasitic skin disease. Travel Med Infect Dis 11(6):357–365

Fernandes A, Rodrigues PM, Pintado M, Tavaria FK (2023) A systemic review of natural products for skin applications: targeting inflammation, wound healing, and photo-aging. Phytomedicine 115:154824

Forouzan P, Cohen PR (2020) Erythrasma revisited: diagnosis, differential diagnoses, and comprehensive review of treatment. Cureus 12(9):e10733

Ghani JN, Zaman A, Mohammed K (2023) Study of effects leaf extract of Ricinus communis in skin disease. Eur Chem Bull 12:7166–7174

Ghosh T, Kumar B, Roy M, Guin P (2018) A review on traditional and pharmacological uses of Croton bonplandianum with special reference to phytochemical aspect. Eur J Med Pla 22:1–10

Giordano D, Biancaniello C, Argenio MA, Facchiano A (2022) Drug design by pharmacophore and virtual screening approach. Pharmaceuticals (Basel) 15(5):646

Gyasi ME, Okonkwo ON, Tripathy K (2024) Onchocerciasis. In: StatPearls [Internet]. Treasure Island (FL): StatPearls Publishing. Available from: https://www.ncbi.nlm.nih.gov/books/NBK559027/

Hekmatpou D, Mehrabi F, Rahzani K, Aminiyan A (2019) The effect of aloe Vera clinical trials on prevention and healing of skin wound: a systematic review. Iran J Med Sci 44(1):1–9

Hemavathy A, Shanthi P, Sowndharya C, Thiripura Sundari S, Priyadharshni K (2019) Extraction and isolation of bioactive compounds from a therapeutic medicinal plant – Wrightia tinctoria (Roxb.) R. Br Int J Pharmacogn Phytochem Res 11(3):199–204

Hendrikx L, van Hees CLM, de Steenwinkel JEM et al (2022) Treatment and outcome of culture-confirmed *Mycobacterium marinum* disease. Open Forum Infect Dis 9(4):ofac077

Hussein HM, Hameed IH, Ibraheem O (2016) Antimicrobial activity and spectral chemical analysis of methanolic leaves extract of adiantum capillus-veneris using GC-MS and FT-IR spectroscopy. Int J Pharm Phyto Res 8:369–385

Jankasem M, Wuthi-Udomlert M, Gritsanapan W (2013) Antidermatophytic properties of Ar-Turmerone, turmeric oil, and Curcuma longa preparations. ISRN Dermatol 250597:1

Jappa LS, Kutre RS (2018) A clinical and bacteriological study of bacterial folliculitis. PJMS 8(2):54–58

Kapse AN, Chandekar CJ (2020) Antimicrobial activity of leaves of Vitex Negundo against pathogenic organisms compared with control drug. Wjpmr 6(10):219–224

Karas JA, Wong LJM, Paulin OKA, Mazeh AC, Hussein MH, Li J, Velkov T (2020) The antimicrobial activity of cannabinoids. Antibiotics (Basel) 9(7):406

Karole S, Gautam G, Gupta S (2017) Pharmacognostic and pharmacological profile of Bombax Ceiba. AJPER 6:3

Karray M, McKinney WP (2024) Tinea versicolor. In: StatPearls [Internet]. Treasure Island (FL): StatPearls Publishing; Available from: https://www.ncbi.nlm.nih.gov/books/NBK482500/

Kassa T (2021) A review on human Orf: a neglected viral zoonosis. Res Rep Trop Med 12:153–172

Kaviya M, Balasubramanian B, Bharathi K, Malaisamy A, Al-Dhabi NA, Mariadhas VA, Anand AV, Liu W (2022) Evaluation of nutritional substances and investigation of antioxidant and antimicrobial potentials of *Boerhavia diffusa* with in Silico molecular docking. Molecules 27(4):1280

Khatun MS, Mia N, Al Bashera M, Murad MA, Zahan R, Parvin S, Akhtar MA (2024) Evaluation of anti-inflammatory potential and GC-MS profiling of leaf extracts from Clerodendrum infortunatum L. J Ethnopharmacol 320:117366

Kumar R, Singh AK, Gupta A, Bishayee A, Pandey AK (2019) Therapeutic potential of Aloe vera – a miracle gift of nature. Phytomedicine 60:152996

Kunat-Budzyńska M, Rysiak A, Wiater A, Grąz M, Andrejko M, Budzyński M, Bryś MS, Sudziński M, Tomczyk M, Gancarz M, Rusinek R, Ptaszyńska AA (2023) Chemical composition and antimicrobial activity of new honey varietals. Int J Environ Res Public Health 20(3):2458

Kwiatkowski P, Łopusiewicz Ł, Kostek M, Drozłowska E, Pruss A, Wojciuk B, Sienkiewicz M, Zielińska-Bliźniewska H, Dołęgowska B (2019) The antibacterial activity of lavender essential oil alone and in combination with Octenidine Dihydrochloride against MRSA strains. Molecules 25(1):95

Leung AKC, Lam JM, Leong KF, Hon KL (2020) Tinea corporis. Drugs Context 9:5–6

Lin X, Li X, Lin X (2020) A review on applications of computational methods in drug screening and design. Molecules 25(6):1375

Litchman G, Nair PA, Le JK (2024) Pityriasis Rosea. In: StatPearls [Internet]. Treasure Island (FL): StatPearls Publishing. Available from: https://www.ncbi.nlm.nih.gov/books/NBK448091/

Lo Presti C, Curti C, Montana M et al (2019) Chickenpox: an update. Med Mal Infect 49(1):1–8

Lotfi Shahreza M, Ghadiri N, Mousavi SR, Varshosaz J, Green JR (2018) A review of network-based approaches to drug repositioning. Brief Bioinform 19:878–892

Lu H, Hong T, Jiang Y, Whiteway M, Zhang S (2023) Candidiasis: from cutaneous to systemic, new perspectives of potential targets and therapeutic strategies. Adv Drug Deliv Rev 199:114960

Mahalik G (2019) Ethnomedicinal plants used by traditional healers for the treatment of skin diseases in Jajpur district of Odisha. India J Bas Appl Res Int 9:738–746

Majeed I, Rizwan K, Ashar A, Rasheed T, Amarowicz R, Kausar H, Zia-Ul-Haq M, Marceanu LG (2021) A comprehensive review of the Ethnotraditional uses and biological and pharmacological potential of the genus *mimosa*. Int J Mol Sci 22(14):7463

Majtan J, Bucekova M, Jesenak M (2021) Natural products and skin diseases. Molecules 26(15):4489

Mala L, Lalouckova K, Skrivanova E (2021) Bacterial skin infections in livestock and plant-based alternatives to their antibiotic treatment. Animals (Basel) 11(8):2473

Mandal AK, Pandey A, Sah RK, Baral A, Sah P (2022) In vitro antioxidant and antimicrobial potency of *Mimosa pudica* of Nepalese Terai region: insight into L-Mimosine as an antibacterial agent. Evid Based Comple Altern Med 2022:6790314

Maxfield L, Crane JS(2024) Cutaneous Larva Migrans. In: StatPearls [Internet]. Treasure Island (FL): StatPearls Publishing; Available from: https://www.ncbi.nlm.nih.gov/books/NBK507706/

Medugu N, Imran J, Mussa-Booth TO (2023) A review of staphylococcal scalded skin syndrome. AJOL 24(3):235–242

Meza-Romero R, Navarrete-Dechent C, Downey C (2019) Molluscum contagiosum: an update and review of new perspectives in etiology, diagnosis, and treatment. Clin Cosmet Investig Dermatol 12:373–381

Mhaskar M, Sathe N (2019) Antimicrobial and wound healing activity of *Jasminum grandiflorum* Linn: a review. WJPR 8:12

Mohanty KKGR, Dash G, Sahoo D (2019) Ethnomedicinal plants used by traditional healers for the treatment of skin diseases in Jajpur district of Odisha, India. IJBASR 9:6

Mohd Zaid NA, Sekar M, Bonam SR, Gan SH (2022) Promising natural products in new drug design, development, and therapy for skin disorders: an overview of scientific evidence and understanding their mechanism of action. Drug Des Devel Ther 16:23–66

Moutawalli A, Benkhouili FZ, Doukkali A et al (2023) The biological and pharmacologic actions of Lawsonia inermis L. Phytomed Plus 3:100468

Moy RL, Levenson C (2017) Sandalwood album oil as a botanical therapeutic in dermatology. J Clin Aesthet Dermatol 10(10):34–39

Mughal SB, Arshad N, Shoaib M et al (2013) Ethnobotanical literature survey of plants used to cure skin diseases. World Appl Sci J 27(4):474–478
Muhammed MT, Aki E (2021) Pharmacophore modeling in drug discovery: methodology and current status. J Turk Chem Soc Sect A: Chem 8:759–772
Mujeeb F, Bajpai P, Pathak N (2014) Phytochemical evaluation, antimicrobial activity, and determination of bioactive components from leaves of Aegle marmelos. Biomed Res Int 497606
Muleta AJ, Lappan R, Stinear TP, Greening C (2021) Understanding the transmission of mycobacterium ulcerans: a step towards controlling Buruli ulcer. PLoS Negl Trop Dis 15(8):e0009678
Nag A, Anoop M, Sharma K, Verma K (2018) Acalypha Indica L. an important medicinal plant with antimicrobial agents: a review. IJRAR 5(4)
Nair PA, Patel BC (2024) Herpes Zoster. Treasure Island (FL): StatPearls Publishing; 2024 Jan–. PMID: 28722854. http://www.ncbi.nlm.nih.gov/books/nbk441824/
Nandy S, Singh J, Pandey DK et al (2020) *Hemidesmus indicus* L. Br.: critical assessment of in vitro biotechnological advancements and perspectives. Appl Microbiol Biotechnol 104:8517–8548
Nardi NM, Schaefer TJ (2024) Impetigo. In: StatPearls [Internet]. Treasure Island (FL): StatPearls Publishing; Available from: https://www.ncbi.nlm.nih.gov/books/NBK430974/
Nejatzadeh-Barandozi F (2013) Antibacterial activities and antioxidant capacity of Aloe vera. Org Med Chem Lett 3(1):5
Nigam PK, Syed HA, Saleh D (2024) Tinea Pedis. In: StatPearls [Internet]. Treasure Island (FL): StatPearls Publishing; Available from: https://www.ncbi.nlm.nih.gov/books/NBK470421/
Nilforoushzadeh MA, Javanmard SH, Ghanadian M, Asghari G, Jaffary F, Yakhdani AF, Dana N, Fatemi SA (2014) The effects of Adiantum capillus-veneris on wound healing: an experimental in vitro evaluation. Int J Prev Med 5(10):1261–1268
Nisekhoto N, Roy D, Chettri A et al (2023) Computational toxicology and food safety assessment of Parkia timoriana phytoconstituents using quantitative structure-activity relationship (QSAR) modeling approaches. Ind J Biochem Biophys 60:896–918
Noha SM, Schuster D (2013) Pharmacophore modeling. In: Lill MA (ed) In Silico drug discovery and design, pp 80–93
Nyeem M (2017) Ashoka (Saraca indica) as women friendly plant: a review. Nat J Adv Resch 3:03–07
Orchard A, van Vuuren S (2017) Commercial essential oils as potential antimicrobials to treat skin diseases. Evid Based Complement Alternat Med 2017:4517971
Panwar A, Singh Y, Naik B (2020) Nutritional and pharmacological health benefits of Bombax ceiba L. The Pharm Innova J 9(6):392–396
Pingale R, Sharma AK, Dash GK (2017) Antidiabetic activity of methanol extract of Canthium parviflorum Lam. Leaves Indian J Pharm Biol Res 5(1):51–54
Pippin MM, Madden ML, Das M (2024) Tinea Cruris. In: StatPearls [Internet]. Treasure Island (FL): StatPearls Publishing; Available from: https://www.ncbi.nlm.nih.gov/books/NBK554602/
Raghav PK, Saini M (2018) Antimicrobial propeties of Tulsi (Ocimum sanctum). IJGHC 7:20–32. https://doi.org/10.24214/IJGHC/HC/7/1/02032
Rahman MT, Halim MA, Mozumder NHMR, Ove TA, Khatun AA (2024) Phytochemicals and antioxidant properties of bael (Aegle marmelosL.) pulp powder and its products. J Agric Food Res 15:100971
Rani S, Raheman K, Sultana A (2016) Ethnomedicinal and pharmacological activities of Mochrus (Bombax ceiba Linn.): an overview. CellMed 6:21–29
Ranjan V, Vats M, Kapoor S (2017) Hansraj (adiantum capillusveneris Linn.): a systematic review on its ethnobotany, phytochemical and pharmacological profile. Int J Ayur Pharma Research 5(6):5–21
Salo-Ahen OMH, Alanko I, Bhadane R et al (2021) Molecular dynamics simulations in drug discovery and pharmaceutical development. PRO 9:71
Sandeep P, Paarakh M, Gavani U (2009) Antibacterial activity of Jasminum grandiflorum Linn leaves. J Pharm Res 2(7):1206–1207

Sapuła M, Krankowska D, Wiercińska-Drapało A (2020) In search of risk factors for recurrent erysipelas and cellulitis of the lower limb: a cross-sectional study of epidemiological characteristics of patients hospitalized due to skin and soft-tissue infections. Interdiscip Perspect Infect Dis 2020:1307232

Sarkar C, Mondal M, Khanom B, Hossain MM, Hossain MS, Sureda A, Islam MT, Martorell M, Kumar M, Sharifi-Rad J, Al-Harrasi A, Al-Rawahi A (2021) *Heliotropium indicum* L.: from farm to a source of bioactive compounds with therapeutic activity. Evid Based Comple Alternat Med:9965481

Sharifi-Rad J, Quispe C, Turgumbayeva A et al (2023) Santalum genus: phytochemical constituents, biological activities and health promoting-effects. Z Naturforsch 78(1–2):9–25

Shirolkar A, Gahlaut A, Chhillar AK, Dabur R (2013) Quantitative analysis of catechins in *Saraca asoca* and correlation with antimicrobial activity. J Pharm Anal 3(6):421–428

Singab ANB, Mostafa NM, Fawzy IM, Bhatia D, Suryawanshi PT, Kabra A (2022) Herbal arsenal against skin ailments: a review supported by in Silico molecular docking studies. Molecules 27(19):6207

Singh A et al (2018) Traditional uses, antimicrobial potential, pharmacological properties and phytochemistry of Sida rhombifolia Linn.: a review. J Phytopharmacol 7(1):88–91

Singh SP, Kumar S, Mathan SV, Tomar MS, Singh RK, Verma PK, Kumar A, Kumar S, Singh RP, Acharya A (2020) Therapeutic application of Carica papaya leaf extract in the management of human diseases. Daru 28(2):735–744

Singh PG, Setty PBS et al (2022) Antimicrobial, antioxidant and anti-inflammatory activities of seeds from *emblica officinalis* (Gaertn.). Bioinformation 18(8):683–691

Sizar O, Talati R (2024) Sporotrichosis. [In: StatPearls [Internet]. Treasure Island (FL): StatPearls Publishing; Available from: https://www.ncbi.nlm.nih.gov/books/NBK532255/

Sonibare MA, Aremu OT, Okorie PN (2016) Antioxidant and antimicrobial activities of solvent fractions of Vernonia cinerea (L.) less leaf extract. Afr Health Sci 16(2):629–639

Srivastava R (2014) A review on phytochemical, pharmacological, and pharmacognostical profile of Wrightia tinctoria: adulterant of kurchi. Pharmacogn Rev 8(15):36–44

Sukumaran V, Senanayake S (2016) Bacterial skin and soft tissue infections. Aust Prescr 39:5

Sunderkötter C, Becker K (2015) Frequent bacterial skin and soft tissue infections: diagnostic signs and treatment. J Dtsch Dermatol Ges 13(6):501–524

Suurbaar J, Mosobil R, Donkor AM (2017) Antibacterial and antifungal activities and phytochemical profile of leaf extract from different extractants of Ricinus communis against selected pathogens. BMC 10:660. https://doi.org/10.1186/s13104-017-3001-2

Tabassum N, Hamdani M (2014) Plant used to treat skin diseases. Pharmacogn Rev 8(15):52–60

Tashkandi H (2021) Honey in wound healing: an updated review. Open Life Sci 16(1):1091–1100

Topno SC, Sinha MR (2018) Study of medicinal plants used to heal skin diseases by tribes of west Singhbhum district of Jharkhand (India). J Pharmacogn Phytochem 7(1):371–376

Turabi KS, Sannakki JA, Aich J, Garse S, Iyer D, Devarajan S (2023) Computational screening of phytochemicals for anti-bacterial drug discovery. In: Phytochemistry, computational tools and databases in drug discovery. Elsevier, pp 213–243

Villalon-Gomez JM (2018) Pityriasis Rosea: diagnosis and treatment. Am Fam Physician 97(1):38–44

Wińska K, Mączka W, Łyczko J, Grabarczyk M, Czubaszek A, Szumny A (2019) Essential oils as antimicrobial agents-myth or real alternative? Molecules 24(11):2130

Winters RD, Mitchell M (2024) Folliculitis. In: StatPearls [Internet]. Treasure Island (FL): StatPearls Publishing; Available from: https://www.ncbi.nlm.nih.gov/books/NBK547754/

Yu G, Zhang Y, Ren W, Dong L, Li J, Geng Y, Zhang Y (2017) Network pharmacologybased identification of key pharmacological pathways of yin–Huang–Qing–Fei capsule acting on chronic bronchitis. Int J Chronic Obstr Pulm Dis 12:85

Zagórska-Dziok M, Ziemlewska A, Bujak T, Nizioł-Łukaszewska Z, Hordyjewicz-Baran Z (2021) Cosmetic and dermatological properties of selected Ayurvedic plant extracts. Molecules 26(3):614. https://doi.org/10.3390/molecules26030614

Zanwar AA, Badole SL, Meena F (2013) *Curcuma longa*: use for skin disease care. In: Bioactive dietary factors and plant extracts in dermatology. https://doi.org/10.1007/978-1-62703-167-7_36
Zhou YX, Xin HL, Rahman K, Wang SJ, Peng C, Zhang H (2015) Portulaca oleracea L.: a review of phytochemistry and pharmacological effects. Biomed Res Int 925631

Natural Product-Based Treatment for Gastritis

8

Kanishka Uthansingh, Kenguruka Sidoine, Swarup Kumar Patnaik, Girish Kumar Pati, and Rabindra Nath Padhy

Abstract

The current chapter describes how various natural products can treat and prevent gastritis. Individual health can be improved by modifying personal lifestyle, including the food and level of exercise. Gastritis is a common illness that affects millions of people worldwide. Numerous drugs, such as nizatidine (Axid AR), cimetidine (Tagamet HB), and famotidine (Pepcid), which neutralize stomach acid, are used to treat gastritis. Many synthetic medications are used to treat gastritis, including those that lower the secretion of acid in the stomach, known as antacids; protecting the mucous tissues lining the stomach and upper small intestine (demulcents); or get rid of *Helicobacter pylori* (*H. pylori*). The current chapter aimed to introduce a safe medicine (or more) of natural origin that can control gastritis without causing negative side effects. Targeting to find new natural anti-gastritis treatments, thorough findings have been continued by different research groups globally. Herbs, medicinal plants, other spices, vegetables, and unprocessed pharmacological substances are thought to be possible sources for treating various illnesses, including gastritis and peptic ulcers. Numerous medicinal

K. Uthansingh (✉)
Department of Gastroenterology and Hepatobiliary Sciences, IMS and SUM Hospital, Siksha 'O' Anusandhan Deemed to be University, Bhubaneswar, Odisha, India

Central Research Laboratory, IMS and SUM Hospital, Siksha 'O' Anusandhan Deemed to be University, Bhubaneswar, Odisha, India

K. Sidoine · R. N. Padhy
Central Research Laboratory, IMS and SUM Hospital, Siksha 'O' Anusandhan Deemed to be University, Bhubaneswar, Odisha, India

S. K. Patnaik · G. K. Pati
Department of Gastroenterology and Hepatobiliary Sciences, IMS and SUM Hospital, Siksha 'O' Anusandhan Deemed to be University, Bhubaneswar, Odisha, India

T. Sharma et al. (eds.), *Natural Products for Antibacterial Drug Development: Recent Advancement of Computational Approach*,
https://doi.org/10.1007/978-981-97-9634-2_8

plants and their secondary metabolites have been identified in the scientific literature as having potential anti-gastric properties. Natural remedies offer promising outcomes and less negative side effects. The objective of the current chapter is to compile information published within the last couple of decades and examine natural remedies used to treat these illnesses and their purposive effects.

Keywords

Gastritis · *H. pylori* · Natural products · Treatment

Abbreviations

CMV	Cytomegalovirus
COX-2	Cyclooxygenase-2
DASH	Dietary approaches to stop hypertension
DGL	Deglycyrrhizinated licorice
EGC	Early gastric cancer
EGCG	Epigallocatechin gallate
EGCG	Epigallocatechin gallate
ESD	Endoscopic submucosal dissection
FGIDs	Functional gastrointestinal disorders
FXR	Farnesoid X receptor
GERD	Gastro esophageal reflex disease
HSV	Herpes simplex virus
IL-1β	Interleukin-1 beta
IL-6	Interleukin-6
iNOS	Inducible nitric oxide synthase
LOX	Lipoxygenase
MAPKs	Mitogen-activated protein kinases
NF-Kb	Nuclear factor-kappa
NSAIDs	Non-steroidal anti-inflammatory drugs
PPIs	Proton pump inhibitors
TLRs	Toll-like receptors
TNF-α	Tumor necrosis factor-alpha

8.1 Introduction

The swelling and redness of the stomach are signs of gastritis. Gastritis is mainly caused by the aerobic *Helicobacter pylori* (*H. pylori*). Gastritis can also result in other digestive disorders like ulcers, gastric tumors, and other bowel problems (Makola et al. 2007). Smoking, abusing certain substances, drinking alcohol, and certain dietary habits can all lead to gastritis. Additionally, the virulence factor of *H. pylori* also causes gastritis; the enzyme produced by *H. pylori*, that is, urease, converts urea into ammonia. This creates a neutral milieu that enables the bacterium to colonize the stomach and duodenum's epithelium and survive in the stomach lining, which causes gastritis (Sokolova and Naumann 2022). There are several symptoms such as abdominal pain, belching, nausea, vomiting, upper GI bleed, and abdominal fullness. These symptoms directly or indirectly associated with the infection of several pathogens such as *Fusobacterium* and *Neisseria* sp. continue to promote malignancy in patients with chronic gastritis even after *H. pylori* is eradicated, based on the results of coculture experiments and human samples (Nikura et al. 2023). There is a possibility of treatment of *H. pylori* infection; however, the eradication failure rate has remained high at a fraction ranging from 5% to 20% with the frequent recurrence of gastritis ulcers after healing using currently available antibiotics. Hence, an alternative mode of eradication of *H. pylori* using natural products such as peel of pomegranate and licorice extracts has been investigated, using ethyl acetate and methanol on the growth and urease activity of *H. pylori* (Bataineh et al. 2023). Moreover, other natural products such as golden spices (turmeric), essential oils, coconut water, apple cider vinegar, *Aloe vera*, other anti-inflammatory foods, licorice, herbal teas (chamomile tea, ginger tea), broccoli, papaya, *Garcinia gardneriana* (medical tree), and honey products help in eradicating *H. pylori* (Wang 2014). Thus, the significance of prebiotics and probiotics, or symbiotic complexes, has been carefully evaluated. The positive findings from another study showed that several microbes are critical for immune system regulation, lipid and carbohydrate metabolism, bile acid biotransformation, and modulation of their central receptors, TGR-5 (membrane Takeda G protein-coupled receptor 5) and FXR (nuclear farnesoid X receptor), which have the necessary immunomodulatory and metabolic functions (Rodríguez-Pastén et al. 2022). Furthermore, biotic organisms have been shown to benefit the host by displacing pathogenic species, enhancing the dysbiosis associated with metabolic syndrome (Met S). Recent research has investigated the beneficial effects of para- and post-probiotics (dead or inactive probiotics) on hepato-gastrointestinal health (Bindels et al. 2015). As mentioned, certain diets, comorbidities, and the use of antibiotics can inhibit the growth of microbiota out of balance. These changes have been connected to the emergence of chronic diseases related to the metabolic disorders mentioned earlier in gastrointestinal disorders such as inflammatory bowel disease, colitis, and gastritis caused by the infection of *H. pylori* (Huang et al. 2021).

Once inflammation develops due to gastritis, the patient may undergo some clinical symptoms such as abdominal pain, bloating, nausea, and indigestion. In the current scenario, the conventional medical treatments for gastritis often involve the use of medications like proton-pump inhibitors (PPIs) or triple therapy regimen for the treatment of *H. pylori*-infected patients; however, some people prefer natural and holistic approaches by applying natural remedies in the form of product-based treatments such as essential oil, coconut water, apple cider vinegar, and turmeric to alleviate gastritis symptoms. Therefore, the current chapter elucidated the treatment modalities of gastritis particularly through natural products.

8.2 Current Methods of Treatment Different Gastritis

The course of treatment and advice for gastritis may differ depending on the underlying cause and the disease severity. Gastritis can be classified as acute gastritis, which lasts for a shorter duration, and chronic gastritis, which lasts for a longer duration. It is recommended to get the advice of a medical professional, especially a gastroenterologist, for an accurate diagnosis for the best course of treatment. Since *H. pylori* is the major cause of gastritis, gastric cancer, and colorectal cancer, the 'test it and treat it' approach is considered best treatment management for gastritis (Shah et al. 2021). The best choice to come out from the different gastritis, peptic ulcer, gastric cancer, and colorectal cancer is the eradication of *H. pylori* rather making the treatment guideline for the treatment of these GI disorders. In the treatment of gastritis by any treatment method, such as natural remedies, allopathic medications, homeopathic therapies, and ayurvedic approaches, the physician should know the disease severity and the pathophysiological conditions for better treatment outcomes as illustrated (Fig. 8.1). There are several causes that cause gastritis, and it is crucial to identify the underlying causes elaborated in the current chapter.

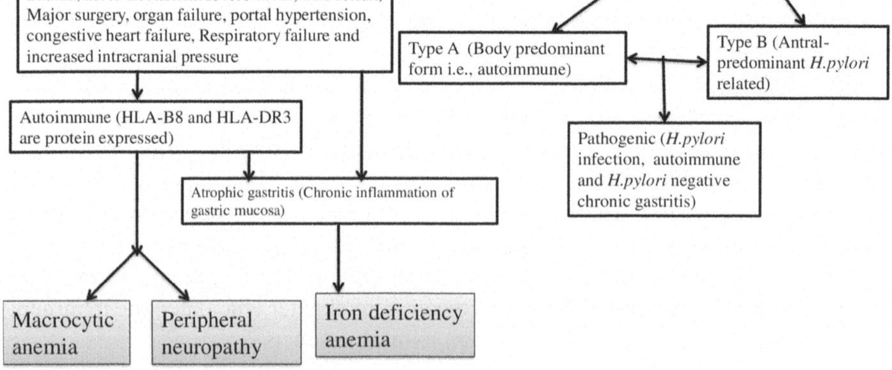

Fig. 8.1 Approaches for the treatment of gastritis according to severity

8.3 Causes of Gastritis

8.3.1 Gastritis by Microbial Infection

When the *H. pylori* bacteria, a common cause of gastritis, is present in the digestive tract, a combination of antibiotics, such as metronidazole (Flagyl), amoxicillin (Amoxil, Augmentin, etc.), or clarithromycin (Biaxin XL), along with acid-suppressing drugs, is used to treat gastritis (Guevara and Cogdill 2020). The antibiotic suggested by the physician usually lasts 7–14 days, during which PPIs are advised to be taken to treat gastritis and infection of *H. Pylori* (Uthansingh et al. 2021). It is usually followed as a precautionary measure to eliminate the infection of *H. pylori* completely (de Boer and Tytgat 2000). Moreover, there are other types of infectious agents responsible for gastritis, including *pyogenic bacteria* that cause phlegmonous gastritis, *mycobacterium tuberculosis* that cause mycobacterial gastritis, and *cytomegalovirus* (CMV) and *herpes simplex virus* (HSV) that cause viral gastritis. Furthermore, *Anisakis, Cryptosporidium, Ascaris lumbricoides, Giardia, Toxoplasma,* and *Schistosoma* cause parasitic gastritis and *Candida, Aspergillus, Mucor, Coccidioides, Histoplasma, Cryptococcus neoformans, Pneumocystis carinii,* and *Torulopsis glabrata* cause fungal gastritis (Pennelli et al. 2020). The stomach is entirely distinct from that of the throat, mouth, or other upper gastrointestinal tract regions. The infection by *H. pylori* may be linked by the stomach microbiome. The relationship between *H. pylori* and the microbiota also influences the course of gastritis. Furthermore, an opportunistic infection called *Candida albicans* causes gastritis in people receiving antibiotics (Mason et al. 2012). Numerous case reports have mentioned gastritis linked to streptococcus. However, the condition was never thoroughly examined. Therefore, investigating gastric pathogens is crucial to learning more about infectious disorders of the stomach. According to the Kyoto classification, the etiologies of acute gastritis include non-transmissible and transmissible agents, as illustrated by Sugano et al. (2015), Lauwers et al. (2010), and Rugge et al. (2024) (Table 8.1).

8.3.2 Gastritis by Medications

If the cause of the gastritis is due to non-steroidal anti-inflammatory drugs (NSAIDs) or other medications, the physician may recommend halting, reducing, or substituting a new therapy for better improvement. Medications may alter the GI physiology in terms of constipation, induced by anticholinergic medications (medications that block acetylcholine) that may cause toxicity and damage the tissue leading to changes in the intestinal microbiota such as antibiotics causing *Clostridium difficile* infection, which can lead to ulcers from the NSAIDs (Kumar et al. 1996; Talley et al. 2001). Drug-induced gastritis is primarily distinguished by the variety and nonspecificity of its presentations; it is identified in adults and older people more often. Such gastritis was probably caused by older age groups using more therapeutic medications like anti-inflammatory or anti-oncologic therapy (Liang and Lauwers 2018).

Table 8.1 Main etiologies associated with gastritis

Aetiology			Identified pathogen
Environmental Host-related	Communicable	Bacteria	*Helicobacter pylori* *Actinomyces* *Mycobacterium avium* *tuberculosis* *Clostridium perfringens* *Staphylococcus aureus* *Pseudomonas aeruginosa* *Treponema pallidum*
		Virus	*Coronaviridae (including SARS-Cov-2)* *Herpes viridae* *Picornaviridae*
		Fugal	*Candida* *Histoplasma capsulatum* *Rhizopusarrhizus*
		Parasite	*Anisakiasis* *Cryptosporium* *Strongyloides stercoralis*
	Noncommunicable (chemical and radiation exposure)	Chemical	Doxycline, iron pill, Kayexalate in sorbitol, lanthanum carbonate, NSAIDs
		Physical	Radiation
	Noncommunicable (immunological)		Autoimmune Lymphocytic, eosinophilic, and collagenous-related
	Noncommunicable (immune-mediated) etiologies		Autoimmune Immune-mediated Allergic Graft-versus-host disease

8.3.3 Gastritis by Acid-Blocking Drugs

The prevalence of acid-related gut disorders, such as peptic ulcers, GERD gastroesophageal reflex disease (GERD), and several others, are rising due to the modern lifestyle. The medications used to treat these illnesses either inhibit the H+, K+, and ATPase enzyme or block the H2 receptor (Rawla et al. 2018). PPIs lower acid by blocking the response of the constituent parts of the cell that produce acid. These medications are used to aid in tissue healing by inhibiting acid secretion. These medications include rabeprazole (Aciphex), pantoprazole (Protonix), lansoprazole (Prevacid), omeprazole (Prilosec), and other prescription and over-the-counter medications (Shin and Sachs 2008). Furthermore, antacids are no longer the primary treatment for most acid-peptic illnesses, histamine 2 antagonists and proton-pump inhibitors (PPIs) still have a place in therapy since they are widely accessible, affordable, and safe for most people (Mejia and Kraft 2009). Understanding the physiology of acid production, the medications targeting gastric acid production, and their efficacy in treating gastritis is crucial.

8.3.4 Gastritis by Stomach Neutralizing Drugs

Antacid medications are an extra prescription regimen used to neutralize preexisting stomach acid or lower the manufacture of new acid to relieve pain quickly. Constipation or diarrhea may occur as adverse effects, depending on the main ingredients. Proton-pump inhibitors (PPIs) and acid blockers are less harmful and more effective (Leung and Graham 2000). The active ingredients of antacids include various calcium, magnesium, and aluminium salts mixed with other materials. Antacids neutralize stomach acid and stop pepsin, a proteolytic enzyme, from working. Each cationic salt has distinct pharmacological properties that indicate the intended clinical use. Antacids are advised for a variety of conditions, including osteoporosis, phosphate binding in chronic renal failure, duodenal and stomach ulcers, stress gastritis, pancreatic insufficiency, non-ulcer dyspepsia, diarrhea caused by bile acid reflux, biliary reflux, and constipation (Salisbury and Terrell 2024). The stomach-neutralizing drugs may bring a lot of side effects on the digestive system of the human body.

8.3.5 Gastritis by Cytoprotective Agents

The capacity of prostaglandins to prevent stomach mucosal damage through different ulcerogenic and necrotizing substances without inhibiting gastric acid output or neutralizing intra-gastric acidity is known as cytoprotection (Tarnawski et al. 1987). The cyclooxygenase inhibitor indomethacin reduced gastritis; however, it never improved much on the gastro-protective action of other medications. As an endogenous product, PG was implicated in mediating the effect of different treatments, such as sofalcone and sucralfate (Szabo 2014). Medications such as sucralfate can help protect the stomach lining from further irritation.

8.3.6 Gastritis by Pain Relievers

There are several pain relievers advised during the treatment of gastritis and stomach ache. Most importantly, there are several drugs that can reduce the chance of developing a peptic ulcer. These specifically include medications known as H2 blockers and PPIs, both of which lower stomach acid production and stomach irritation. The drawback of using these drugs for a few weeks or months is that they frequently result in several GI problems. When treating stomach pain with over-the-counter medications, oral acetaminophen can be efficiently absorbed by the gastrointestinal tract within 30–60 min (Gerriets et al. 2024).

8.3.7 Gastritis by Use of Excessive Proton-Pump Inhibitors

The long-term or high dosage usage of proton-pump inhibitors may raise the risk of hip, wrist, and spine fractures (Lespessailles and Toumi 2022). It is usually recommended to speak with a specialist before using a dosage. Alcohol or smoking: If lifestyle factors like excessive alcohol consumption or smoking are contributing to gastritis, the doctor will advise you to limit or eliminate these habits (Capurso and Lahner 2017). It is widely acknowledged that the most frequent bacterium found in gastritis cases is the infection of *H. pylori*. PPIs are said to block the urease and bioactivity of *H. pylori*, both urease and bioactivity are necessary for the bacteria to survive in an acidic environment.

8.3.8 Gastritis by Certain Unhealthy Diet and Unusual Lifestyle

Eating at irregular times is a significant issue, particularly when it comes to nausea and stomach aches. Epidemiological analysis has shown that circadian disruption substantially increases the likelihood of gastrointestinal symptoms, particularly abdominal pain (Nojkov et al. 2010). Spicy food, acidic, or heavy in fat should be avoided since they might aggravate the gastritis in the stomach layers. Intake of small amount, more frequent meals can ease the strain on the stomach, and limiting or avoiding these foods is advised (Lembong et al. 2019). Moreover, it is also advised to be out of consumption of alcohol and smoke as these can irritate the lining of the stomach. Furthermore, the use of NSAIDs, such as aspirin, ibuprofen, and naproxen, is discouraged. Making long-term lifestyle adjustments is crucial in the event of chronic gastritis in order to avoid recurrence (Bhowmik et al. 2010). Apart from the diet, drug intake and long-term lifestyle adjustments reducing stress through relaxation techniques, physical exercises, or counselling may be recommended if stress is a contributing factor (Xiao et al. 2021).

The specific treatment plan for gastritis may vary depending on individual factors and the cause of gastritis. Always consult with a health-care professional for personalized advice and treatment (Lahner et al. 2018). Additionally, new guidelines or treatment options may have emerged time after time, so it is a good idea to check with a health-care provider for the most current information. There are several diets and life modifications steps for better health, importantly to come out from the common gastrointestinal disease like gastritis. The diet that was recommended by the top 20 remedies for the improvement of gastritis symptoms naturally is illustrated by Veremey (Fig. 8.2) infographics: 61819544 (https://www.dreamstime.com/stock-illustration-infographics-proper-nutrition-gastritis-diseases-digestive-system-high-acidity-eps-image61819544).

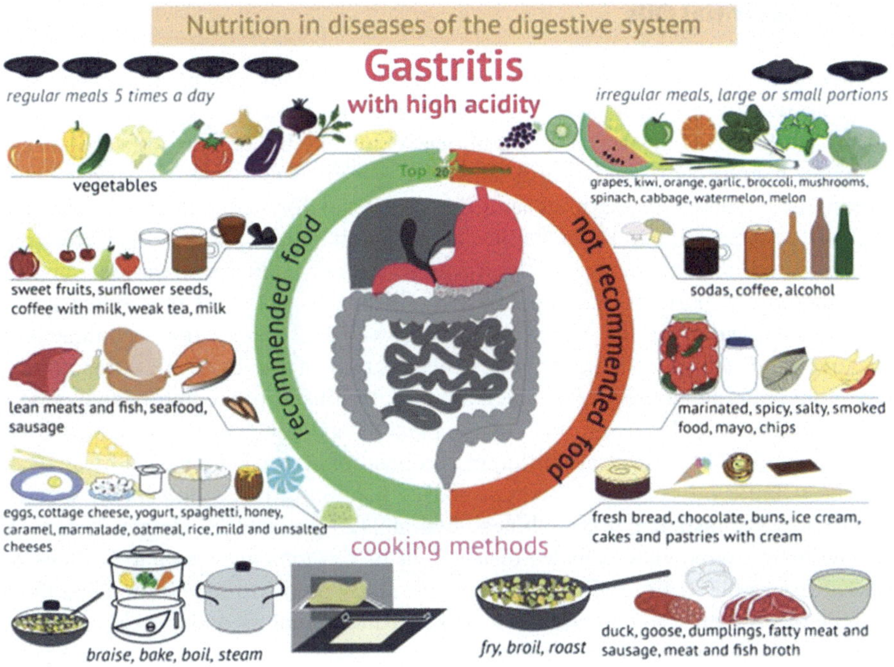

Fig. 8.2 Nutrition recommended for healthy gut microbiome

8.4 Alternative Treatment Modalities for Gastritis

Gastritis is the inflammation of the stomach lining and can be caused by various factors, including infection, medication use, excessive alcohol consumption, stress, and certain autoimmune diseases. Conventional medical treatments for gastritis may involve antacids, proton-pump inhibitors (PPIs), or antibiotics (Hsu et al. 2011). The treatment of chronic gastritis frequently involves the use of either combination or monotherapy recent days. Furthermore, the implication of western medicine through a variety of approaches to manage chronic gastritis, including preventing the production of gastric acid, eliminating *HP*, preventing intestinal metaplasia and dysplasia, and protecting the gastric mucosa has been noticed (Chen et al. 2023). Moreover, some people seek alternative or complementary therapies to manage their symptoms or support their overall digestive health. It is important to note that you should always consult a professional before trying any alternative treatments, especially if you have a severe or chronic condition. Some of the natural products and their potential activities have been studied by different scientists across the globe. These include cranberry (*Vaccinium macrocarpon*), probiotics, honey, green tea (*Camellia sinensis*), garlic (*Allium sativum*), mastic gum (*Pistacia lentiscus*), Turmeric (*Curcuma longa*), licorice (*Glycyrrhiza glabra*), olive leaf extract (*Olea europaea*), and cinnamon (*Cinnamomum verum*) (Shapla et al. 2018).

8.4.1 Essential Oils

Numerous investigations have been conducted to find plants and plant extracts/constituents with gastro-protective and anti-*H. pylori* properties (Vale and Oleastro 2014; Wang 2014). Gastritis is traditionally treated by certain oil substances; *H. pylori* bacteria are less harmful when one use oils such as lemon verbena and lemongrass. Ginger, peppermint, and clove oils aid for better digestion (Thosar et al. 2013). According to Bergonzelli et al., essential oils are unlikely to be effective in vivo anti-*Helicobacter* agents. However, as dietary additives to current supplement therapy, their effects might not be insignificant. Essential oil antioxidant activity varies depending on the source of the oil as well as the extraction technique and solvents employed. Furthermore, the concentration of the main oil components, including phenolic compounds, which are directly linked to the antioxidant action of essential oils, is also determined by the plant's harvesting period (Tongnuanchan and Benjakul 2014). There are different compositional trends identified by analyzing the volatile oil contents of certain leaves belonging to *Cinnamomum* species worldwide. For instance, *trans*-sabinene hydrate, (Z)-β-ocimene, and germacrene A are the main ingredients in hydro-distilled oil of *Cinnamomum tamala* T. Nees & Eberm from India (Mir et al. 2004). Thus, the study on phyto-exploration and their volatile oils are must to be known for their anti-inflammatory and gastro-protective effects through certain identified leaves.

8.4.2 Coconut Water

The monocotyledon palm coconut (*Cocos nucifera L.*), which belongs to the Palmaceae family, is a perennial plant that bears fruit 12–13 times a year for as long as 60–70 years, yielding 30–75 fruits annually (Coulibaly et al. 2023). The antibacterial properties of coconut water and milk may prevent the formation of germs. It increases the strength of the immune system, which helps the healing process (Herron 2016). Coconut water includes both organic and inorganic components that are essential for supporting the body's antioxidant system. It has been successfully used throughout the world to treat cholera, gastroenteritis, and diarrhea in children (Balit et al. 2018; Muhahid et al. 2019). Hence, coconut water is an important juice for the healthy gut microbiota in human.

8.4.3 Apple Cider Vinegar

Apple cider vinegar is a natural product used for different health conditions including gastritis from centuries. Cider vinegar, sometimes referred to as apple cider vinegar, is a kind of vinegar produced from cider. It probably contains some pectin, vitamin B1, B2, and B6, as well as biotin, folic acid, niacin, pantothenic acid, and vitamin C, just like apple juice (Akanksha and Sunita 2017). It acts as antimicrobial agent against *Escherichia coli*, *Staphylococcus aureus*, and *Candida albicans*, downregulating cytokine and microbial protein expression (Yagnik et al. 2018). Furthermore, combining a small amount of honey with vinegar helps soothe the irritable stomach lining (Branson 2017). It is suggested to drink a glass of water with two teaspoons of apple cider vinegar before every meal; moreover, this mixture is also effective in treating stomach aches (Akanksha and Sunita 2017).

8.4.4 Aloe Vera

Aloe vera juice can have soothing properties for the stomach lining. Drinking a small amount (about ¼ cup) of aloe vera juice before meals may help relieve gastritis symptoms. Its anti-inflammatory properties may help reduce and alleviate stomach discomfort. *In addition to stimulating inflammation and increasing vascular permeability, Aloe vera has bradykininase activity, which reduces inflammation. Therefore, the effects of leukocyte adherence following gastric ulcer are induced by 20% acetic acid, and the reduction of inflammatory activity reduces the causes of the inflammatory process* (Eamlamnam et al. 2006). Moreover, *Aloe vera* is able to decrease gastric acid secretion and increase mucus secretion (Brzozowski et al. 2000). In order to reduce costs as an affordable treatment, it is advised to treat chronic ulcers with *Aloe vera* gel for individuals with persistent ulcers, as it could be a helpful and affordable treatment for gastritis (Avijgan et al. 2016).

8.4.5 Use of Anti-Inflammatory Products

Inflammation is a physiological process, which has a crucial role in the defense mechanisms and tissue repair of the body. Knowing how an anti-inflammatory diet affects silent inflammation can take the diet from being a calorie-dense food to being at the forefront of gene-silencing innovation (Sears 2015). Fruits and vegetables rich in vitamin K help relieve the symptoms of gastritis. Furthermore, whole grains like brown rice, oats, whole-wheat bread, beans, nuts, and fish, as well as some spices like turmeric, have properties that can reduce the symptoms of gastritis (Seitz 2021). Hence, it can be stated that there is not just one diet that reduces inflammation. Eating various fruits and vegetables, unsaturated fats, whole grains that have been minimally refined, tea, coffee, herbs, spices, and oily fish are generally emphasized. Popular diet regimens like the *dietary approaches to stop hypertension* (DASH) and Mediterranean diets already highlight various anti-inflammatory items.

8.4.6 Licorice

Specifically, in traditional medicine, it is still commonly used to treat tremors, respiratory infections, gastritis, and peptic ulcers. *G. glabra* root is often used to make a tea that works well as a thirst quencher. Moreover, it has been said that the dried root cleans teeth as well (Armanini et al. 2002). Manufacturing food additives like flavors and sweeteners is the primary industrial application of *G. glabra*. The root is specifically used as a flavoring in chewing gum, candy, baked products, ice cream, soft beverages, and American-style tobacco (Rizzato et al. 2017) Furthermore, deglycyrrhizinated licorice (DGL) supplements or tea can help soothe the stomach lining. It is important to use DGL, as regular licorice can raise blood pressure in cases of low blood pressure (Kulkarni 2017).

8.5 Herbal Products

Plant or botanical products used to treat or prevent disease are known as herbal, botanical, or phytomedicines. Herbal supplements are usually extracted from plants that are meant to be consumed internally. Herbs used in Ayurveda medicine that are known to have digestive benefits can strengthen Agni, reduce inflammation, and improve digestive health in general. Herbs, including licorice, *Aloe vera*, ginger, Indian gooseberry (amla), and fennel (saunf), are frequently used to treat gastritis (https://adyantayurveda.com/ayurveda-for-gastritis). The commonly found herbal products from various herbs used as phytomedicines have drawn attention in past and recent times.

In silico studies, which involve computational methods such as molecular modeling and bioinformatics, have been increasingly used to explore the potential therapeutic effects of phytocompounds on gastritis (Akash et al., 2023; Hanafiah et al., 2024). These provide valuable insights into the molecular interactions between phytocompounds and their targets, helping to elucidate their mechanisms of action. Some examples of in silico studies on phytocompounds such as curcumin (*Curcuma longa*), gingerol (*Zingiber officinale*), quercetin, epigallocatechin gallate (EGCG), and *Aloe vera* polysaccharides and their actions on gastritis have been revealed by different research groups across the globe (Panda et al. 2011).

Furthermore, in silico studies have investigated the molecular interactions between curcumin and various targets involved in inflammation and oxidative stress pathways implicated in gastritis (Zhang et al. 2022a, b; Sarkar et al. 2016). Curcumin may modulate key inflammatory mediators, such as nuclear factor-kappa B (NF-κB), cyclooxygenase-2 (COX-2), and tumor necrosis factor-alpha (TNF-α), thereby exerting anti-inflammatory effects in the stomach lining (Buhrmann et al. 2011; He et al. 2015), as it is known that COX-2 and TNF-α are the key regulating factors for gastric cancer (Uthansingh et al. 2019, 2022). Moreover, to predict the binding affinity and mode of interaction between gingerol and target proteins relevant to gastritis, computational approaches have also been employed. Gingerol may inhibit the activity of pro-inflammatory enzymes like COX-2 and inducible nitric oxide synthase (iNOS), thereby attenuating inflammation in the gastric mucosa revealed by in silico studies (Widyowati et al. 2023). Furthermore, molecular docking simulations have been used to investigate the binding interactions between quercetin and inflammatory mediators implicated in gastritis. In silico studies indicate that quercetin may inhibit the activity of NF-κB and downregulate the expression of pro-inflammatory cytokines, such as interleukin-1 beta (IL-1β) and interleukin-6 (IL-6), leading to anti-inflammatory effects in the stomach lining (Kang et al. 2023; Raut et al. 2023). Furthermore, computational modeling techniques have been utilized to explore the molecular mechanisms underlying the anti-inflammatory and anti-*H. pylori* activities of EGCG in gastritis (Raj et al. 2021; Fagni Njoya et al. 2022). In silico studies suggest that EGCG may disrupt *H. pylori* adhesion to gastric epithelial cells by targeting bacterial adhesins, as well as exerting anti-inflammatory effects through the inhibition of NF-κB signaling pathway (Deng et al. 2024). Apart from the above aspects of molecular modeling by bioinformatics tools that have been employed to predict the potential targets and signaling pathways modulated by *Aloe vera* polysaccharides in the context of gastritis, the effects of phytochemical compounds from aloe species on signaling networks in cancer model has also been revealed (Svitina et al. 2021; Byun et al. 2018). Furthermore, in silico analyses suggest that *Aloe vera* polysaccharides interact with proteins involved in inflammatory signaling cascades, such as mitogen-activated protein kinases (MAPKs) and Toll-like receptors (TLRs), leading to the suppression of gastric inflammation (Choi and Chung 2003). There are different phytocompounds from different plant species, each having specific actions on gastritis, as represented in Table 8.2. Figure 8.3

Table 8.2 Phytocompounds and their action on gastritis

Sl. No	Phytocompound	Source of plant material	Action on gastritis	References
1	Curcumin	Turmeric (*Curcuma longa*)	Anti-inflammatory, antioxidant, gastro-protective properties, may help to inhibit the growth of *H. pylori*	Ghosh et al. (2014), Hewlings and Kalman (2017)
2	Gingerol	Ginger (*Zingiber officinale*)	Anti-inflammatory, anti-ulcer, and antioxidant effects, reduce gastric acid secretion	Hu et al. (2021), Singletary (2023)
3	Boswellic acids	*Boswellia serrata*	Anti-inflammatory, inhibits leukotriene synthesis, reduces gastric inflammation	Cuaz-Pérolin et al. (2020)
4	Chamomile	Chamomile plant	Anti-inflammatory, anti-ulcer, calming effect	Sarkar et al. (2016)
5	Quercetin	Onions, apples, berries	Anti-inflammatory, anti-ulcer, reduces oxidative stress, promotes gastric mucosal healing	Soleimani et al. (2021)
6	Epigallocatechin gallate	Green tea (*Camellia sinensis*)	Anti-inflammatory, antioxidant, protects gastric mucosa, inhibits *H. pylori* growth	Hamzeh-Mivehroud et al. (2020)
7	Allicin	Garlic (*Allium sativum*)	Anti-inflammatory, antibacterial, anti-*H. pylori* activity, promotes ulcer healing	Hasan et al. (2021)
8	Peppermint oil	Peppermint	Antispasmodic, may alleviate symptoms of indigestion	Merat et al. (2010)

(continued)

Table 8.2 (continued)

Sl. No	Phytocompound	Source of plant material	Action on gastritis	References
9	Marshmallow root	Marshmallow	Forms protective layer in the stomach, soothing effect	Zaghlool et al. (2019)
10	*Aloe vera*	*Aloe vera* plant	Anti-inflammatory, may promote healing of stomach lining	Retiu et al. (2021)
11	Licorice root	Licorice plant	Anti-inflammatory, may promote mucus secretion	Fiore et al. (2005)
12	Licorice	*Glycyrrhiza glabra*	Anti-inflammatory, mucosal protection, inhibits *H. pylori* growth, stimulates mucus secretion	Mahady et al. (2021)
13	Berberine	Berberis species	Anti-inflammatory, antimicrobial, gastro-protective, inhibits *H. pylori* growth	Zheng et al. (2020)
14	Catechins	Green tea (*Camellia sinensis*)	Anti-inflammatory, antioxidant, anti-*H. pylori* activity, promotes gastric mucosal repair	Knoops et al. (2021)
15	Silymarin	Milk thistle (*Silybum marianum*)	Anti-inflammatory, antioxidant, protects gastric mucosa, stimulates regeneration of damaged tissue	Sorrentino et al. (2019)

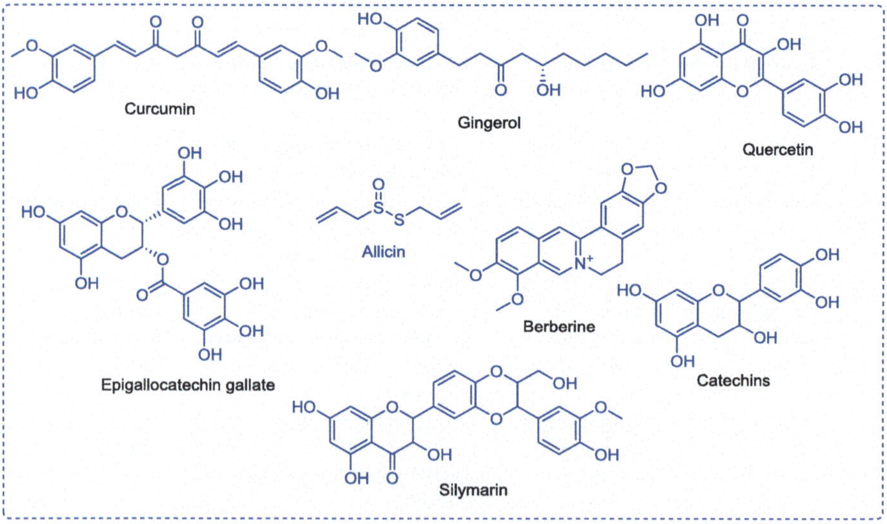

Fig. 8.3 Chemical structure of some phytocompounds inhibitory action against gastritis

represents chemical structure of some phytocompounds with inhibitory action against gastritis.

8.5.1 Chamomile

Herbs like chamomile can ease gastritis symptoms by lowering discomfort, inflammation, and spasms in the digestive tract. The most significant quality of chamomile is its ability to quickly heal gastritis and peptic ulcers. Because of its scent, similar to apples, the Greek terms Chemos and Melos are the source of the English name "chamomile." Chamomile in the form of chamomile tea has anti-inflammatory properties and can help soothe an irritated stomach. Chamomile tea between meals helps alleviate gastritis symptoms (Chauhan and Jaya 2017).

Chamomile (*Matricaria chamomilla*) is renowned for its medicinal properties, particularly its potential therapeutic effects on gastrointestinal disorders such as gastritis. Some key bioactive compounds found in chamomile along with their effects on gastritis are apigenin, luteolin, chamazulene, and bisabolol. Apigenin is a flavonoid abundant in chamomile with potent anti-inflammatory and antioxidant properties. It has been shown to inhibit the production of pro-inflammatory cytokines and mediators involved in gastritis, thereby exerting gastro-protective effects (McKay and Blumberg 2006). Luteolin, another flavonoid present in Chamomile, exhibits anti-inflammatory properties by suppressing NF-κB activation and inhibiting the expression of pro-inflammatory cytokines. It has been suggested to alleviate gastritis symptoms through its anti-inflammatory effects (Nabavi et al. 2015). Chamazulene is a sesquiterpene compound found in chamomile essential oil. It possesses anti-inflammatory and antioxidant properties and has been reported to protect gastric mucosa against injury by inhibiting inflammatory mediators (Avallone et al. 1997). Moreover, bisabolol is a terpene compound found in chamomile oil, known for its anti-inflammatory and gastro-protective effects. It has been reported to inhibit the production of pro-inflammatory cytokines and reduce gastric mucosal damage (Kassi et al. 2004).

8.5.2 Ginger

Stomach pain, nausea, and bloating are some of the symptoms of gastritis that ginger might help with. Gingerols and shogaols, two substances found in ginger, offer anti-inflammatory properties for the body. Ginger helps lessen the chance that stomach acid will rise and enter the esophagus (Thosar et al. 2013). Additionally, ginger contains a lot of compounds and antioxidants that may have a variety of medical uses. Ginger has been shown to be an effective treatment for nausea and vomiting in specific trials. Ginger tea is a standard treatment for numerous digestive problems, including gastritis. Digestion and inflammation both can be reduced by the consumption of ginger. Bioactive substances found in ginger, such as gingerol, have anti-inflammatory and antioxidant activities (Mashhadi et al. 2013). This may lessen inflammation and calm the stomach lining. Ginger can be added as an additional ingredient to food products, capsules, or tea.

There are studies that explored the molecular mechanisms and bioactive compounds of ginger in various contexts, including inflammation and gastrointestinal disorders. Some of the phytochemical bioactive compounds such as gingerol, shogaol, and gingeron have the potential therapeutic effects in various gastrointestinal diseases. Gingerol, the major bioactive compound in ginger, has been extensively studied for its anti-inflammatory properties. In silico studies have investigated its interactions with targets involved in inflammatory pathways, which could be relevant to the treatment of gastritis (Ali et al. 2008). Shogaol, a dehydrated form of gingerol, has shown promising anti-inflammatory and antioxidant activities. Although in silico studies specifically focusing on its effects on gastritis may be limited, its interactions with inflammatory mediators can provide insights into its potential therapeutic effects (Frota et al. 2021). Moreover, zingerone (4-(4-hydroxy-3-methoxyphenyl)-2-butanone), another bioactive compound in ginger, has been investigated for its potential anti-inflammatory effects (Srinivasan 2016). Computational studies can shed light on its interactions with inflammatory targets relevant to gastritis. The cited studies by Ali et al. (2008), Frota et al. (2021), and Srinivasan (2016) may not directly address effects on gastritis by ginger through in silico methods, they provide insights into the potential mechanisms by which ginger bioactive compounds exert their anti-inflammatory and gastro-protective effects, which could be relevant to the treatment of gastritis.

8.5.3 Garcinia Gardneriana

Brazil uses the medicinal tree species *Garcinia gardneriana* to cure gastritis and hepatitis. This has been implicated in phenolic chemicals, primarily 7-epiclusianone, guttiferone-A, and fukugetin, which are responsible for this utilization and have demonstrated multiple biological functions (Rodrigues et al. 2021).

8.5.4 Garlic

Garlic extracts can reduce the growth of *H. pylori*, according to in vitro tests (Ernst 1999). Antifungal and antibacterial agents like garlic may assist in managing gastritis by concentrating on *H. pylori* and lowering gas production in the stomach (Zardast et al. 2016). Therefore, garlic extract may help prevent infected gastritis by *H. pylori*, lowering the risk of stomach cancer.

8.5.5 Broccoli

Broccoli is one of the best home remedies that can be used to get rid of gastritis because it is used in dairy food (Yanaka et al. 2009). One of the molecules in the isothiocyanate group of organosulfur compounds is sulforaphane. It comes from cruciferous plants, including cabbage, Brussels sprouts, and broccoli. In vitro, sulforaphane exhibits a potent bactericidal effect against *H. pylori* (Basaga 1990).

8.6 Other Natural Compounds

8.6.1 Papaya

Oats and papaya are natural foods that are utilized in many countries' traditional medical practices. Papayas are rich in digestive enzymes that aid in digestion. Minerals, beta-glucan fibers, and immune-modulatory and anti-inflammatory probiotics are found in oats. Both ingredients are combined in Caricol®-Gastro, an organic vegan medication for inflammatory bowel illnesses (Weiser et al. 2018). Papaya has a lot of antioxidants and helps in digestion through its enzyme papain (Yogiraj et al. 2014).

8.6.2 Peppermint

Mint has been utilized for decades as a gastrointestinal distress reliever worldwide. Peppermint is a perennial herb that grows well in Europe and North America. Menthol, the active component of peppermint oil, has a variety of applications in both gastroenterology and non-gastroenterology, particularly in functional gastrointestinal disorders (FGIDs) (Scarpellini et al. 2023). Peppermint tea or capsules containing peppermint oil can help relax the muscles of the gastrointestinal tract and relieve symptoms like indigestion and discomfort (Mughal 2022).

8.6.3 Honey

According to research, using honey with triple therapy may hasten the removal of *H. pylori* from gastritis and duodenal ulcer patients. Honey has antimicrobial and anti-inflammatory properties and may help soothe an irritated stomach lining. Raw, unprocessed honey is recommended for its potential therapeutic benefits for gastritis (Mandal and Mandal 2011).

8.6.4 Mānuka Honey

The other products of honey, that is, mānuka honey, have antibacterial properties and help inhibit the growth of *H. pylori*, a bacterium often associated with gastritis. Consuming mānuka honey daily may provide relief from gastritis (Rückriemen et al. 2017; Kumar et al. 2010).

8.6.5 Turmeric

Curcumin, the primary ingredient in turmeric, has antioxidant and anti-inflammatory properties that may help reduce gastritis symptoms. You can use turmeric as a dietary supplement or as a food additive. Black pepper can enhance its absorption when combined with it. It is advised that one can consume turmeric or take pills that contain curcumin (Abd et al. 2021).

8.6.6 Slippery Elm

In North America, slippery elm (*Ulmus fulva*) has long been utilized as a natural treatment for coughing, diarrhea, and gastrointestinal tract diseases. Slippery elm is an herbal remedy that can help coat and soothe the stomach lining. It is advised to take it in the form of capsules or to mix the powder with water and consume it, as recommended long ago (Beveridge et al. 1969). Both slippery elm and licorice are frequently used to treat ulcerative disorders, hyperacidity, and damaged mucous membranes. However, certain groups dedicated to plant preservation, including the United Plant Savers, classify slippery elm as endangered; as a result, alternatives should be considered in clinical treatment (Peterson et al. 2018).

8.6.7 Mastic Gum

Mastic gum, derived from the resin of the mastic tree, has been studied for its potential to inhibit *H. pylori* and reduce inflammation in the stomach. It is available in the form of a supplement (Pachi et al. 2020). Moreover, an alternate treatment option for people with *H. pylori* infection, mainly those resistant to antibiotics, was to add mastic gum to the usual triple-drug regimen. Mastic gum application is beneficial and may lessen the need for antibiotics while lowering the possibility of resistance building (Addissouky et al. 2023).

8.7 Dietary Adjustments

8.7.1 Avoid Trigger Foods

Certain foods and beverages, such as spicy foods, alcohol, caffeine, and acidic items, can exacerbate gastritis. A bland diet with foods like oatmeal, rice, and boiled potatoes are recommended (Li et al. 2020). It is helpful to take minimized meals but frequently to decrease the burden on the stomach (Ananthakrishnan and Xavier 2020). It was found that the nutritional status, gastric mucosa, and gastrointestinal symptoms of individuals with early gastric cancer (EGC) following endoscopic submucosal dissection (ESD) were found to improve after a year of dietary adjustments when following a "traditional food" diet (Zhang et al. 2022a, b).

8.7.2 Hydration

Ingestion of commercial (usual) water daily may be a contributing factor to the occurrence of alterations in the gastric mucosal. Drink plenty of water to help maintain a healthy stomach lining and aid in digestion (Van Hecke et al. 2017). According to the Sidney categorization system, drinking water with a modified pH range (from 8.0 to 10.0) for 5 months could lessen the severity of gastritis. This suggests that changing one's lifestyle could benefit people with gastritis in the Amazon region (Chaves et al. 2020).

8.7.3 Isothiocyanates

The prevalence of gastrointestinal diseases and cancer, in particular, has been rising steadily. These diseases have a significant morbidity and mortality rate as well as a significant financial burden on the health-care system. Natural phytochemicals with specific bioactive qualities in food have demonstrated curative and prophylactic benefits for various diseases. It has been observed that the phytochemical plants such as broccoli, turnips, radishes, brussels sprouts, wasabi, and isothiocyanates are known as ITCs, consumed by humans, have the chemo-preventive effects (Coscueta et al. 2022).

8.7.4 Probiotics

Probiotics improve digestive health by killing harmful bacteria. The composition of probiotics is food-containing bacteria, such as lactobacillus and *Bifidobacterium*, which can help restore the balance of gut flora, which may be beneficial for gastritis patients. Probiotics immediately compete with *H. pylori* and aid in restoring the gut microbial ecosystem; in terms of managing symptoms associated with the pathogenic bacteria, these live microorganisms are somewhat more effective than the conventional triple-antibiotic regimen. They avoid the progression of *H. pylori*, which is causing gastritis in the first place. Consuming probiotic-rich foods like yogurt with live cultures or probiotic supplements can help maintain a healthy gut microbiome, which may aid in preventing or treating gastritis (Amara and Shibl 2015). There are several plant species and their parts have been used for the traditional treatment of gastritis used from ancient to modern days explored by different scientific communities. Few of the species and the mode of using those plants have been illustrated (Table 8.3); moreover, most of the countries across the globe have applied those plant materials for the treatment of gastritis (Fig. 8.4).

Table 8.3 List of medicinal plants used for the treatment of gastritis

Family	Species (codes)	Common names of the plants	Part of the plant used	Mode of application	Purpose	Country/region	References
Acanthaceae	*Dicliptera peruviana* (Lam.) Juss. (1)	Chuncho-chuncho (Q)	Leaves	Infusion	Abdominal pain	Perú (Western)	De-la-Cruz et al. (2007)
	Carlina Acaulis L. (2)	Carolina (I)	Roots	Decoction	Abdominal pain	Italy (South-West)	Vitalini et al. (2013)
	Tagetes elliptica Smith (3)	Chinche (S)	Leaves	Decoction	Abdominal pain	Perú (Canta, Lima)	De-la-Cruz et al. (2007)
	Target filifolia Lag. (4)	Anis Serrano (S)	Aerial parts	Decoction	Abdominal pain	Perú (North-East)	Hammond et al. (1998)
	Vernonia katschyana Scheme (5)	–	Roots	Decoction, cold macerate	Gastritis	Mali	Nergard et al. (2004)
Brassicaceae	*Brassica carinata* A. Braun. (6)	Koza (ET)	Leaves	–	Gastritis	Ethiopia (South-est)	Giday et al. (2010)
	Brassica ca nigra (L) W. D.J.Koch (7)	–	Seeds	–	Abdominal pain	Ethiopia (South-West)	Giday et al. (2010)
Celastraceae	*Maytenusilicifolia* Mart. (8)	Espinhier a santa (P)	Leaves	Infusion	Abdominal pain	Argentina, Brazil, Paraguay	Hatsuko Baggio et al. (2007)
Compositae	*Achilleatomentosa* L. (9)	Milfoil (E)	Aerial parts	Infusion	Abdominal pain	Jordan (Northwest)	AL-qura'n et al. (2009)
	Croton cajucara Benth (10)	Sacasa (P)	Leaves and barks	Infusion	Abdominal pain	Brazil	Hiruma-Lima et al. (2000)
Euphorbiacea	*Jatropha isabelli* Muell. (11)	Yagwarova (G)	Rhizome	Infusion	Abdominal pain	Paraguay, Argentina	Pertino et al. (2007)

Geraniacea	*Geraniummoella* L. (12)	Evra de S. Roberto, Bico de Pinga Amor (P)	Aerial parts, roots	–	Stomach acidity	Portugal, China	Zhang et al. (2013)
	Pelargonium roseum (Andrews) W.T.Aiton (13)	Ceranio (S)	Leaves	Infusion/ decoction	Gastritis	Perú (Western)	Rehecho et al. (2011)
Krameriaceae	*Kramerialappace* (Dombey) Burdet et B. Simpson (14)	Ratanya or Ratiñay (Q)	Roots, stems	Decoction	Abdominal pain	Perú (Northern)	De-la-Cruz et al. (2007)
Laminaceae	*Glechomahederacea* L. (15)	Malvela, redondinha (P)	Aerial parts	–	Gastritis	Portugal, Germany	Wiehle et al. (2009)
	Melissa officinalis L. (16)	Cidreira (P)	Aerial parts	–	Gastritis	Portugal, Perú, Iran	Neves et al. (2009)
	Mentha Piperita L. (17)	Hortelão, Piperita, Pimenta (P); Menta (S)	Green leaves	Infusion	Gastritis	Portugal, Perú, Iran, Argentina	Neves et al. (2009), Marcial et al. (2011)
	Minthostachys mollis (Kunth.) Griseb. (18)	Muña (S)	Leaves	Infusion	Gastritis	Perú (Western)	De-la-Cruz et al. (2007)
	MarrubiumvulgareL. (19)	Mula mujer (S)	Leaves, stems	Infusion	Abdominal pain	Perú (Western)	De-la-Cruz et al. (2007)
Malvaceae	*Althaea rosea* (L.) Cav (20)	Rose mallow (E); Alteia (P)	Aerial parts, roots	Infusion	Gastritis	Jordan, Portugal	Neves et al. (2009)
	Abelmoschusesculentus (L.) Moench (21)	Okra	Fruits	Fresh fruits	Gastritis	Africa, Asia, and America	Benchasri (2012), Lengsfeld et al. (2004)
Paperaceae	*Peperomia galioides* HBK var*gladioides* (22)	Congona (Q)	Aerial parts	Crushed/ juice	Gastric ulcers	Argentina, Perú	Marcial et al. (2011), Hammond et al. (1998)

(continued)

Table 8.3 (continued)

Family	Species (codes)	Common names of the plants	Part of the plant used	Mode of application	Purpose	Country/region	References
Poaceae	*Cynodondactylon* L. (23)	Grama (P)	Dried roots	–	Gastritis	Portugal (Northeast)	Neves et al. (2009), Beltrão et al. (2009)
Polygalaceae	*Polígalapaniculata*Lineau (24)	Barba-de-são-joão, vassourinhabranca or mimosa (P)	Aerial parts	–	Abdominal pain	Brazil, (Eastern coast)	Lapa et al. (2007)
Punicaceae	*Punica granatum* L. (25)	Pomegranate (E)	Fruits	Fresh fruits	Gastritis	Europe, Indo-China, South Africa	Ajaikumar et al. (2005)
Scrophulariaceae	*Calceolaria bicolor* Ruiz & Pav. (26)	Zapatito or globo (S)	Leaves, flowers	Infusion	Abdominal pain	Perú (Western)	De-la-Cruz et al. (2007)
	Calceolarialabata Cav (27)	Globo-globo (S)	Leaves, flowers	Infusion	Abdominal pain	Perú (Western)	

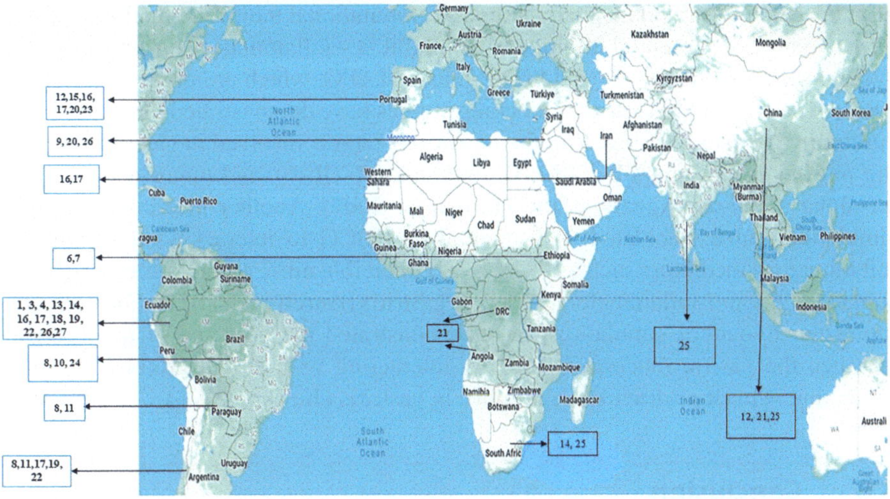

Fig. 8.4 Repartition of natural products in different countries

Numbers in the map specify the codes for the medicinal plant species. Regions are as mentioned:

- Perú (1, 3, 4, 13, 14, 16, 17, 18, 19, 22, 26, 27), Portugal (12, 15, 16, 17, 20, 23); Argentina (8, 11, 17, 19, 22); Brazil (8, 10, 24); Jordan (9, 20, 26); Ethiopia (6,7), Iran (16,17); Paraguay (8, 11), Europe, Indo-China, South Africa (25); Mali (5); Africa, Asia, and America (21); Italy (2); China (12).
- N.B: Red-marked species belong to most of the regions in those countries.

The map (Fig. 8.4) shows the different species found in different geographical locations, along with the codes for these species listed in Table 8.3. For example, in Peru, 12 different types of species identified which are useful for the treatment of gastritis naturally. About 31 different types of plant species identified were retried from Scopus, PubMed, and Web of Science databases explored by different researchers from different countries. Furthermore, the common names of these particular species, as mentioned in column 3, along with the parts of the plant used and their modes of application across the globe, have been mentioned in Table 8.3.

Therefore, consultation with a medical doctor for gastric disorder treatment is crucial before starting any natural treatment for gastritis, especially if an individual have severe or chronic symptoms. Gastritis can have various underlying causes, and the appropriate treatment may vary depending on the specific diagnosis. Additionally, some natural remedies may interact with medications or have various underlying causes, with a health-care provider to ensure safety and effectiveness.

Some specific natural products such as gingerol from ginger, curcumin from turmeric, epigallocatechin gallate (EGCG) from green tea (*Camellia sinesis*), and quercetin from fruits and vegetables along with their therapeutic models of gastritis have been demonstrated in silico method. Computational molecular docking studies have revealed the potential of gingerol to inhibit key inflammatory enzymes such as cyclooxygenase-2 (COX-2) and lipoxygenase (LOX), which are implicated in gastritis (Ali et al. 2008). Molecular docking studies have demonstrated the potential of curcumin to interact with and modulate various targets involved in gastritis, including nuclear factor-kappa B (NF-κB), interleukin-6 (IL-6), and tumor necrosis factor-alpha (TNF-α) (Aggarwal and Harikumar 2009). Another molecular docking simulations have elucidated the potential of EGCG to interact with and inhibit inflammatory mediators such as COX-2 and inducible nitric oxide synthase (iNOS), suggesting its anti-inflammatory effects in gastritis (Aureli et al. 2011). Furthermore, molecular dynamics simulations have demonstrated the potential of quercetin to stabilize the gastric mucosal barrier by interacting with mucin proteins, thereby protecting against gastric mucosal injury in gastritis (Jia et al. 2023).

8.8 Conclusion

These results obtained from different literatures highlighted the potentiality of natural products to treat gastritis and inhibit the *H. pyroli* infection*, and this would be considered a potential tool to understand and treat* gastritis. Moreover, effective natural product treatment or prophylactic use for preventing *gastritis* evolution may be favorable caution for Indians and the global population. According to several epidemiological studies, consuming natural foods, including fruits, vegetables, spices, isoflavones, and quercetin, may reduce the incidence of stomach cancer. Researchers have explored that phytoconstituents have antiulcer properties by targeting inflammatory mediators, various enzymes, and other routes, even though preclinical studies have yielded encouraging results, as discussed in different sections of the current chapter.

The most effective treatment for peptic ulcers is herbal medicine, several allopathic medications, such as NSAIDs, proved detrimental to the gastric mucosa and resulted in duodenal and stomach ulcers. Clinical trials and their pharmacological activity demonstrate that herbal medicine is the primary treatment for peptic ulcers, becoming a global health concern. This treatment has a significant influence on the provision of health care worldwide. *Curcuma longa* and licorice, as active ingredients, can most effectively prevent inflammation in the stomach mucosa and the upper portion of the duodenum produced by bacteria and other causative agents. These ingredients are safe to treat peptic ulcers and contain antiulcer properties.

References

Abd El Hack ME, El Saadony MT, Swelum AA et al (2021) Curcumin, the active substance of turmeric: its effects on health and ways to improve its bioavailability. J Sci Food Agric 101:5747–5762. https://doi.org/10.1002/jsfa.11372

Addissouky TA, Khalil AA, El Agroudy AE (2023) Assessing the efficacy of a modified triple drug regimen supplemented with mastic gum in the eradication of helicobacter pylori infection. Am J Clin Pathol 160:S19–S19

Aggarwal BB, Harikumar KB (2009) Potential therapeutic effects of curcumin, the anti-inflammatory agent, against neurodegenerative, cardiovascular, pulmonary, metabolic, autoimmune and neoplastic diseases. Int J Biochem Cell Bio 41:40–59

Ajaikumar KB, Asheef M, Babu BH et al (2005) The inhibition of gastric mucosal injury by Punica granatum L.(pomegranate) methanolic extract. J Ethnopharmacol 96:171–176. https://doi.org/10.1016/j.jep.2004.09.007

Akanksha S, Sunita M (2017) Study about the nutritional and medicinal properties of apple cider vinegar. Asia Pac J Sci Technol 8:12

Akash S, Bayıl I, Mahmood S et al (2023) Mechanistic inhibition of gastric cancer-associated bacteria helicobacter pylori by selected phytocompounds: a new cutting-edge computational approach. Heliyon 9:e20670

Ali BH, Blunden G, Tanira MO et al (2008) Some phytochemical, pharmacological and toxicological properties of ginger (Zingiber officinale roscoe): a review of recent research. Food Chem Toxicol 46:409–420

Al-Qura'n S (2009) Ethnopharmacological survey of wild medicinal plants in Showbak, Jordan. J Ethnopharmacol 123:45–50. https://doi.org/10.1016/j.jep.2009.02.031

Amara AA, Shibl A (2015) Role of probiotics in health improvement, infection control, and disease treatment and management. Saudi Pharm J 23:107–114. https://doi.org/10.1016/j.jsps.2013.07.001

Ananthakrishnan AN, Xavier RJ (2020) Gastrointestinal diseases. Hunter's Trop Med Emerg Infect Dis 2020:16–26. https://doi.org/10.1016/B978-0-323-55512-8.00003-X

Armanini D, Fiore C, Mattarello MJ (2002) History of the endocrine effects of licorice. Exp Clin Endocrinol Diabet 110:257–261. https://doi.org/10.1055/s-2002-34587

Aureli P, Capurso L, Castellazzi AM et al (2011) Probiotics and health: an evidence-based review. Pharmacol Res 63:366–376

Avallone R, Zanoli P, Corsi L et al (1997) Benzodiazepine-like compounds and GABA in flower heads of Matricaria chamomilla. Phytother Res 10:S177–S179

Avijgan M, Kamran A, Abedini A (2016) Effectiveness of Aloe vera gel in chronic ulcers in comparison with conventional treatments. Iran J Med Sci 41:S30

Balit T, Asae A, Boonyoung P et al (2018) Optimal doses and neuroprotective effects of prolonged treatment with young coconut juice in orchidectomized rats. A preliminary study. Songklanakarin J Sci Technol 40:475–483

Basaga HS (1990) Biochemical aspects of free radicals. Biochem Cell Biol 68:989–998. https://doi.org/10.1139/o90-146

Bataineh SM, Bara'ah A, Mahawreh BS et al (2023) Inhibitory effect of partially purified compounds from pomegranate Peel and licorice extracts on growth and urease activity of *helicobacter pylori*. Jordan J Biol Sci 16:199–206. https://doi.org/10.54319/jjbs/160204

Beltrão JO, Neves AL, de Brito JC et al (2009) Salt removal potential of turfgrasses in golf courses in the Mediterranean Basin. WSEAS Trans Environ Dev 5:394–403

Benchasri S (2012) Okra (Abelmoschus esculentus (L.) Moench) as a valuable vegetable of the world. Ratarstvoipovrtarstvo 49:105–112. https://doi.org/10.5937/ratpov49-1172

Beveridge RJ, Stoddart JF, Szarek WA et al (1969) Some structural features of the mucilage from the bark of Ulmus fulva (slippery elm mucilage). Carbohydr Res 9:429–439. https://doi.org/10.1016/S0008-6215(00)80028-2

Bhowmik D, Chiranjib TK, Pankaj KS (2010) Recent trends of treatment and medication peptic ulcerative disorder. Int J Pharm Tech Res 2:970–980

Bindels LB, Delzenne NM, Cani PD (2015) Towards a more comprehensive concept for prebiotics. Nat Rev Gastroenterol Hepatol 12:303–310. https://doi.org/10.1038/nrgastro.2015.47

Branson S (2017) 101 Amazing uses for apple Cider Vinegar: soothe an upset stomach, get more energy, wash out cat urine and 98 more! Workman Publishing

Brzozowski T, Konturek PC, Konturek SJ et al (2000) Role of gastric acid secretion in progression of acute gastric erosions induced by ischemia-reperfusion into gastric ulcers. Eur J Pharmacol 398:147–158. https://doi.org/10.1016/S0014-2999(00)00287-9

Buhrmann C, Mobasheri A, Busch F et al (2011) Curcumin modulates nuclear factor κB (nf-κB)-mediated inflammation in human tenocytes in vitro: role of the phosphatidylinositol 3-kinase/Akt pathway. J Biol Chem 286:28556–28566

Byun EB, Kim HM, Sung NY et al (2018) Gamma irradiation of aloe-emodin induced structural modification and apoptosis through a ROS-and caspase-dependent mitochondrial pathway in stomach tumor cells. Int J Radiat Biol 94:403–416

Capurso G, Lahner E (2017) The interaction between smoking, alcohol and the gut microbiome. Best Pract Res Clin Gastroenterol 31:579–588. https://doi.org/10.1016/j.bpg.2017.10.006

Chauhan ES, Jaya A (2017) Chamomile an ancient aromatic plant-a review. J Ayu Med Sci 2:251–255. https://doi.org/10.5530/jams.2017.2.26

Chaves JR, de Souza CR, Modesto AA (2020) Effects of alkaline water intake on gastritis and miRNA expression (miR-7, miR-155, miR-135b and miR-29c). Am J Transl Res 12:4043

Chen L, Wei S, He Y et al (2023) Treatment of chronic gastritis with traditional Chinese medicine: pharmacological activities and mechanisms. Pharmaceuticals 16:1308–1323. https://doi.org/10.3390/ph16091308

Choi S, Chung MH (2003) A review on the relationship between Aloe vera components and their biologic effects. Sem Int Med 1:53–62

Coscueta ER, Sousa AS, Reis CA (2022) Phenylethyl isothiocyanate: a bioactive agent for gastrointestinal health. Molecules 27:794. https://doi.org/10.3390/molecules27030794

Coulibaly WH, Camara F, Tohoyessou MG et al (2023) Nutritional profile and functional properties of coconut water marketed in the streets of Abidjan (Côte D'ivoire). Sci Afr 20:e01616. https://doi.org/10.1016/j.sciaf.2023.e0161

Cuaz-Pérolin C, Veutg Cm Angoulvant D et al (2020) Anti-inflammatory and anti-oxidant effects of a combination of Boswellia serrata and Zingiber officinale extracts in TNBS-induced colitis in mice. Phytother Res:2020

de Boer WA, Tytgat GN (2000) Treatment of *helicobacter pylori* infection. Br Med J 320:31–34. https://doi.org/10.1136/bmj.320.7226.31

De-la-Cruz H, Vilcapoma G, Zevallos PA (2007) Ethnobotanical study of medicinal plants used by the Andean people of Canta, Lima, Peru. J Ethnopharmacol 111:284–294. https://doi.org/10.1016/j.jep.2006.11.018

Deng R, Chen X, Zhao S et al (2024) The effects and mechanisms of natural products on helicobacter pylori eradication. Front Cell Infect Microbiol 14:1360852

Eamlamnam K, Patumraj S, Visedopas N et al (2006) Effects of Aloe vera and sucralfate on gastric microcirculatory changes, cytokine levels and gastric ulcer healing in rats. World J Gastroenterol 12:2034. https://doi.org/10.3748/wjg.v12.i13.2034

Ernst E (1999) Is garlic an effective treatment for helicobacter pylori infection. Arch Intern Med 159:2484–2485. https://doi.org/10.1001/archinte.159.20.2484-a

Fagni Njoya ZL, Mbiantcha M, Djuichou Nguemnang SF et al (2022) Anti-helicobacter pylori, anti-inflammatory, and antioxidant activities of trunk bark of Alstonia boonei (Apocynaceae). Biomed Res Int 2022:1

Fiore C, Eisenhut M, Ragazzi E et al (2005) A history of the therapeutic use of liquorice in Europe. J Ethnopharmacol 99:317–324

Frota NF, de Sousa RA, Fuzo CA et al (2021) Alemtuzumab scFv fragments and CD52 interaction study through molecular dynamics simulation and binding free energy. J Mol Gra Model 107:107949

Gerriets V, Anderson J, Patel P et al (2024) Acetaminophen. [Updated 2024 Jan 11]. In: StatPearls [Internet]. Treasure Island (FL): StatPearls Publishing; Available from: https://www.ncbi.nlm.nih.gov/books/NBK482369/

Ghosh SS, Gehr TW, Ghosh S (2014) Curcumin and chronic kidney disease (CKD): major mode of action through stimulating endogenous intestinal alkaline phosphatase. Molecules 19:20139–20156

Giday M, Asfaw Z, Woldu Z (2010) Ethnomedicinal study of plants used by Sheko ethnic group of Ethiopia. J Ethnopharmacol 132:75–85. https://doi.org/10.1016/j.jep.2010.07.046

Guevara B, Cogdill AG (2020) *Helicobacter pylori*: a review of current diagnostic and management strategies. Dig Dis Sci 65:1917–1931. https://doi.org/10.1007/s10620-020-06193-7

Hammond GB, Fernández ID, Villegas LF et al (1998) A survey of traditional medicinal plants from the Callejón de Huaylas, Department of Ancash, Perú. J Ethnopharmacol 61:17–30. https://doi.org/10.1016/s0378-8741(98)00009-9

Hamzeh-Mivehroud M, Moghadam MT, Dastmalchi S et al (2020) Green tea catechins inhibit the endonuclease activity of the helicobacter pylori restriction endonuclease Hpy99I. Helicobacter 25:e12774

Hanafiah A, Abd Aziz SN, Nesran ZN et al (2024) Molecular investigation of antimicrobial peptides against helicobacter pylori proteins using a peptide-protein docking approach. Heliyon 10:e28128

Hasan MM, Hossain MM, Akter R, Rahman MA et al (2021) Beneficial effects of Allium sativum on gastric ulcer induced by aspirin in rats. J Diet Suppl 18:397–411. https://doi.org/10.1080/19390211.2020.1783429

Hatsuko Baggio C, Freitasb C, Martini Otofuji G (2007) Flavonoid-rich fraction of Maytenusilicifolia Mart. ex. Reiss protects the gastric mucosa of rodents through inhibition of both H+,K+-ATPase activity and formation of nitric oxide. J Ethnopharmacol 113:433–440. https://doi.org/10.1016/j.jep.2007.06.015

He Y, Yue Y, Zheng X et al (2015) Curcumin, inflammation, and chronic diseases: how are they linked? Molecules 20:9183–9213

Herron JG (2016) The gut health protocol: a nutritional approach to healing SIBO, intestinal candida, GERD, gastritis, and other gut health issues. Eagle Stock Publishing

Hewlings SJ, Kalman DS (2017) Curcumin: a review of its effects on human health. Food Secur 6:92

Hiruma-Lima CA, Gracioso JS, Rodrıguez JA et al (2000) Gastroprotective effect of essential oil from croton cajucaraBenth (Euphorbiaceae). J Ethnopharmacol 69:229–234. https://doi.org/10.1016/s0378-8741(99)00127-0

Hsu PI, Wu DC, Wu JY et al (2011) Modified sequential helicobacter pylori therapy: proton pump inhibitor and amoxicillin for 14 days with clarithromycin and metronidazole added as a quadruple (hybrid) therapy for the final 7 days. Helicobacter 16:139–145. https://doi.org/10.1111/j.1523-5378.2011.00828.x

Hu ML et al (2021) Ginger extract (Zingiber officinale roscoe) ameliorates alcohol-induced gastric ulcer via inductions of anti-oxidative and anti-inflammatory activities in rats. BMC Complement Med Ther

Huang K, Wu L, Yang Y (2021) Gut microbiota: an emerging biological diagnostic and treatment approach for gastrointestinal diseases. J Gastroenterol Hepatol 5:973–975. https://doi.org/10.1002/jgh3.12659

Jia X, He Y, Li L et al (2023) Pharmacological targeting of gastric mucosal barrier with traditional Chinese medications for repairing gastric mucosal injury. Front Pharmacol 14:1091530

Kang C, Li X, Liu P et al (2023) Quercetin inhibits the activity and function of dendritic cells through the TLR4/IRAK4/NF-κB signaling pathway. Contemp Oncol 27:182–189

Kassi E, Papoutsi Z, Fokialakis N et al (2004) Greek plant extracts exhibit selective estrogen receptor modulator (SERM)-like properties. J Agric Food Chem 52:6956–6961

Knoops M, Schoofs G et al (2021) The role of green tea catechins in prevention of helicobacter pylori-induced diseases: molecular targets, anti-adhesion activity and clinical trials. J Appl Microbiol 131:1125–1137

Kulkarni K (2017) Evaluation of antiulcer activity of DGL (Deglycyrrhizinated liquorice). J Pharm Sci 9:2019–2022

Kumar A, Aitas AT, Hunter AG et al (1996) Sweeteners, dyes, and other excipients in vitamin and mineral preparations. Clin Pediatr 35:443–450. https://doi.org/10.1177/000992289603500903

Kumar KS, Bhowmik D, Biswajit C (2010) Medicinal uses and health benefits of honey: an overview. J Chem Pharm Res 2:385–395

Lahner E, Carabotti M, Annibale B (2018) Treatment of helicobacter pylori infection in atrophic gastritis. World J Gastroenterol 24:2373–2380. https://doi.org/10.3748/wjg.v24.i22.2373
Lapa FD, Freitas CS, Baggio CH et al (2007) Gastroprotective activity of the hydroalcoholic extract obtained from polygala paniculate L. in rats. J Pharm Pharmacol 59:1413–1419. https://doi.org/10.1211/jpp.59.10.0012
Lauwers GY, Fujita H, Nagata K et al (2010) Pathology of non-helicobacter pylori gastritis: extending the histopathologic horizons. J Gastroenterol 45:131–145. https://doi.org/10.1007/s00535-009-0146-3
Lembong E, Kurniati D, Utama GL (2019) Correlation between students diet towards gastritis in faculty of agricultural industrial technology of padjadjaran university. Sci Pap Manag 19:347–356
Lengsfeld C, Titgemeyer F, Faller G, Hensel A (2004) Glycosylated compounds from okra inhibit adhesion of Helicobacter pylori to human gastric mucosa. J Agric Food Chem 52:1495–503. https://doi.org/10.1021/jf030666n
Lespessailles E, Toumi H (2022) Proton pump inhibitors and bone health: an update narrative review. Int J Mol Sci 23:10733. https://doi.org/10.3390/ijms231810733
Leung WK, Graham DY (2000) Clarithromycin for *helicobacter pylori* infection. Expert Opin Pharmacother 1:507–514. https://doi.org/10.1517/14656566.1.3.507
Li Y, Su Z, Li P et al (2020) Association of symptoms with eating habits and food preferences in chronic gastritis patients: a cross-sectional study. J Evid Based Complement Altern Med 2020:1–11. https://doi.org/10.1155/2020/5197201
Liang WY, Lauwers GY (2018) Drug-induced gastritis. In: Carneiro F, Chaves P, Ensari A (eds) Pathology of the gastrointestinal tract: encyclopedia of pathology, pp 194–198. https://doi.org/10.1007/978-3-319-40560-5_1612
Mahady GB, Pendland SL, Chadwick LR (2021) Inhibition of helicobacter pylori in vitro by various berry juices: comparison with other red fruit extracts. Phytother Res 35:2697–2703
Makola D, Peura DA, Crowe SE (2007) *Helicobacter pylori* infection and related gastrointestinal diseases. J Clin Gastroenterol 41:548–558. https://doi.org/10.1097/MCG.0b013e318030e3c3
Mandal MD, Mandal S (2011) Honey: its medicinal property and antibacterial activity. Asian Pac J Trop Biomed 1:154–160. https://doi.org/10.1016/S2221-1691(11)60016-6
Marcial G, Rodríguez C, Medici M et al (2011) New approaches in gastritis treatment. Gastritis and gastric cancer-new insights in gastro-protection. Diag Treat:153–176. https://doi.org/10.5772/23221
Mashhadi NS, Ghiasvand R, Askari G et al (2013) Anti-oxidative and anti-inflammatory effects of ginger in health and physical activity: review of current evidence. Int J Prev Med 4:S36
Mason KL, Erb Downward JR, Falkowski NR et al (2012) Interplay between the gastric bacterial microbiota and Candida albicans during postantibiotic recolonization and gastritis. Infect Immun 80:150–158. https://doi.org/10.1128/iai.05162-11
McKay DL, Blumberg JB (2006) A review of the bioactivity and potential health benefits of chamomile tea (Matricaria recutita L.). Phytother Res An Int J Dev Pharma Toxico Eval Nat Prod Deriv 20:519–530
Mejia A, Kraft WK (2009) Acid peptic diseases: pharmacological approach to treatment. Expert Rev Clin Pharmacol 2:295–314. https://doi.org/10.1586/ecp.09.8
Merat S, Khalili S, Mostajabi P et al (2010) The effect of enteric-coated, delayed-release peppermint oil on irritable bowel syndrome. Dig Dis Sci 55:1385–1390
Mir SR, Ali M, Kapoor R (2004) Chemical composition of essential oil of Cinnamomum tamala Nees et Eberm. Leaves. Flavour Fragr J 19:112–114. https://doi.org/10.1002/ffj.1236
Mughal SS (2022) Peppermint oil, its useful, and adverse effects on human health: a review. J Ayu Integr Med 8:1–4
Mujahid I, Mulyanto A, Khasanah TU (2019) The effectiveness of coconut water in inhibiting shigella sp. bacteria from diarrhea. Medisains 17:8
Nabavi SF, Braidy N, Gortzi O et al (2015) Luteolin as an anti-inflammatory and neuroprotective agent: a brief review. Brain Res Bull 119:1

Nergard CS, Diallo D, Michaelsen TE et al (2004) Isolation, partial characterisation and immunomodulating activities of polysaccharides from Vernonia kotschyana Sch. Bip. Ex Walp. J Ethnopharmacol 91:141–152. https://doi.org/10.1016/j.jep.2003.12.007
Neves JM, Matos C, Moutinho C (2009) Ethnopharmacological notes about ancient uses of medicinal plants in Trás-Os-Montes (northern of Portugal). J Ethnopharmacol 124:270–283. https://doi.org/10.1016/j.jep.2009.04.041
Niikura R, Hayakawa Y, Nagata N et al (2023) Non-helicobacter pylori gastric microbiome modulates prooncogenic responses and is associated with gastric cancer risk. Gastro Hep Adv 2:684–700. https://doi.org/10.1016/j.gastha.2023.03.010
Nojkov B, Rubenstein JH, Chey WD et al (2010) The impact of rotating shift work on the prevalence of irritable bowel syndrome in nurses. Am J Gastroenterol 105:842–847. https://doi.org/10.1038/ajg.2010.48
Pachi VK, Mikropoulou EV, Gkiouvetidis P (2020) Traditional uses, phytochemistry and pharmacology of Chios mastic gum (Pistacia lentiscus var. Chia, Anacardiaceae): a review. J Ethnopharmacol 254:112485
Panda SK, Padhi LP, Mohanty G (2011) Antibacterial activities and phytochemical analysis of Cassia fistula (Linn.) leaf. JAPTR 2:62–67
Pennelli G, Grillo F, Galuppini F et al (2020) Gastritis: update on etiological features and histological practical approach. Pathological 112:153–165. https://doi.org/10.32074/1591-951X-163
Pertino M, Schmeda-Hirschmann G, Rodríguez JA et al (2007) Gastroprotective effect and cytotoxicity of terpenes from the Paraguayan crude drug "yagua rova"(Jatropha isabelli). J Ethnopharmacol 111:553–559. https://doi.org/10.1016/j.jep.2007.01.003
Peterson CT, Sharma V, Uchitel S et al (2018) Prebiotic potential of herbal medicines used in digestive health and disease. J Altern Complement Med 24:656–665. https://doi.org/10.1089/acm.2017.0422
Raj R, Agarwal N, Raghavan S et al (2021) Epigallocatechin gallate with potent anti-helicobacter pylori activity binds efficiently to its histone-like DNA binding protein. ACS Omega 6:3548–3570
Raut BK, Upadhyaya SR, Bashyal J et al (2023) In Silico and in vitro analyses to repurpose quercetin as a human pancreatic α-amylase inhibitor. ACS Omega 8:43617–43631
Rawla P, Sunkara T, Ofosu A et al (2018) Potassium-competitive acid blockers-are they the next generation of proton pump inhibitors? World J Gastrointest Pharmacol Ther 9:63. https://doi.org/10.4292/wjgpt.v9.i7.63
Rehecho S, Uriarte-Pueyo I, Calvo J, Vivas LA, Calvo MI (2011) Ethnopharmacological survey of medicinal plants in nor-Yauyos, a part of the landscape reserve nor-Yauyos-Cochas, Peru. J Ethnopharmacol 133:75–85. https://doi.org/10.1016/j.jep.2010.09.006
Retiu A, Budescu T, Menae I (2021) Aloe Vera extract for stomach acid use safe and effective treatment. Int J Pap Adv Sci Rev 2:59–64
Rizzato G, Scalabrin E, Radaelli M et al (2017) A new exploration of licorice metabolome. Food Chem 221:959–968. https://doi.org/10.1016/j.foodchem.2016.11.068
Rodrigues DA, de Sousa BL, da Silva JG et al (2021) Phytotoxic property of metabolites isolated from Garcinia gardneriana. Comput Biol Chem 92:107460. https://doi.org/10.1016/j.compbiolchem.2021.107460
Rodríguez-Pastén A, Pérez-Hernández N, Añorve-Morga J et al (2022) The activity of prebiotics and probiotics in hepatogastrointestinal disorders and diseases associated with metabolic syndrome. Int J Mol Sci 23:7229. https://doi.org/10.3390/ijms23137229
Rückriemen J, Klemm O, Henle T (2017) Manuka honey (Leptospermum scoparium) inhibits jack bean urease activity due to methylglyoxal and dihydroxyacetone. Food Chem 230:540–546. https://doi.org/10.1016/j.foodchem.2017.03.075
Rugge M, Genta RM, Malfertheiner P et al (2024) IN.: the real-world gastritis initiative–updating the updates. Gut 73:407–441. https://doi.org/10.1136/gutjnl-2023-331164
Salisbury BH, Terrell JM (2024) Antacids. In: Stat Pearls [Internet]. Treasure Island (FL): Stat Pearls Publishing; Available from: https://www.ncbi.nlm.nih.gov/books/NBK526049

Sarkar A, De R, Mukhopadhyay AK (2016) Curcumin as a potential therapeutic candidate for helicobacter pylori associated diseases. World J Gastroenterol 22:2736

Scarpellini E, Broeders B, Schol J et al (2023) The use of peppermint oil in gastroenterology. Curr Pharm Des 29:576–583. https://doi.org/10.2174/1381612829666230328163449

Sears B (2015) Anti-inflammatory diets. J Am Coll Nutr 34:14–21. https://doi.org/10.1080/07315724.2015.1080105

Seitz S (2021) Natural ways to treat Gastrits. Alt Med 1:20–21

Shah SC, Piazuelo MB, Kuipers EJ et al (2021) AGA clinical practice update on the diagnosis and management of atrophic gastritis: expert review. Gastroenterol 161:1325–1332. https://doi.org/10.1053/j.gastro.2021.06.078

Shapla UM, Raihan J, Islam A et al (2018) Propolis: the future therapy against helicobacter pylori-mediated gastrointestinal diseases. J Appl Biomed 16:81–99. https://doi.org/10.1016/j.jab.2017.10.007

Shin JM, Sachs G (2008) Pharmacology of proton pump inhibitors. Curr Gastroenterol Rep 10:528–534. https://doi.org/10.1007/s11894-008-0098-4

Singletary KW (2023) Ginger update: potential health benefits. Nutr Today 58:263–273

Sokolova O, Naumann M (2022) Matrix metalloproteinases in *helicobacter pylori*–associated gastritis and gastric cancer. Int J Mol Sci 23:1883. https://doi.org/10.3390/ijms23031883

Soleimani V, Delghandi PS, Moallem SA et al (2021) Quercetin and quercetin-rich foods in gastrointestinal diseases. J Gastrointest Cancer 52:738–749

Sorrentino R, Angelo G, Di Giacomo A et al (2019) Silymarin exerts anti-inflammatory effects by modulating NF-κB and IL-1β pathways in LPS-activated murine peritoneal macrophages: a network pharmacological study. Drug Des Devel Ther 13:4925–4934

Srinivasan K (2016) Biological activities of red pepper (Capsicum annuum) and its pungent principle capsaicin: a review. Crit Rev Food Sci Nutr 56:1488–1500

Sugano K, Tack J, Kuipers EJ et al (2015) Kyoto global consensus report on helicobacter pylori gastritis. Gut 64:1353–1367. https://doi.org/10.1136/gutjnl-2015-309252

Svitina H, Hamman JH, Gouws C (2021) Molecular mechanisms and associated cell signalling pathways underlying the anticancer properties of phytochemical compounds from aloe species. Exp Ther Med 22:1–23

Szabo S (2014) "Gastric cytoprotection" is still relevant. J Gastroenterol Hepatol 29:124–132. https://doi.org/10.1111/jgh.12735

Talley NJ, Young L, Bytzer P et al (2001) Impact of chronic gastrointestinal symptoms in diabetes mellitus on health-related quality of life. Am J Gastroenterol 96:71–76. https://doi.org/10.1111/j.1572-0241.2001.03350.x

Tarnawski A, Hollander D, Cergely H (1987) Cytoprotective drugs. Focus on essential fatty acids and sucralfate. Scand J Gastroenterol 22:39–43. https://doi.org/10.3109/00365528709090949

Thosar N, Basak S, Bahadure RN et al (2013) Antimicrobial efficacy of five essential oils against oral pathogens: an in vitro study. Eur J Dent 7:S071–S077. https://doi.org/10.4103/1305-7456.119078

Tongnuanchan P, Benjakul S (2014) Essential oils: extraction, bioactivities, and their uses for food preservation. J Food Sci 79:R1231–R1249. https://doi.org/10.1111/1750-3841.12492

Uthansingh K, Sahu MK, Narayan J et al (2019) COX-2 and CYP2C9 polymorphism and the risk of gastric cancer in the population of Odisha. J Gastroenterol Hepatol 34:258–258

Uthansingh K, Kumari R, Pati GK et al (2021) Molecular docking of anti helicobacter pylori antibiotics and proton pump inhibitor: a single center survey. J Pure Appl Microbiol 15:2103–2116. https://doi.org/10.22207/JPAM.15.4.33

Uthansingh K, Pati GK, Parida PK et al (2022) Evaluation of tumor necrosis factor-alpha gene (− 308 G/A,− 238 G/A and− 857 C/T) polymorphisms and the risk of gastric cancer in eastern Indian population. Gastroenterol Insights 13:340–348. https://doi.org/10.3390/gastroent13040034

Vale FF, Oleastro M (2014) Overview of the phytomedicine approaches against helicobacter pylori. World J Gastroenterol 20:5594–5609. https://doi.org/10.3748/wjg.v20.i19.5594

Van Hecke T, Van Camp J, De Smet S (2017) Oxidation during digestion of meat: interactions with the diet and helicobacter pylori gastritis, and implications on human health. Compr Rev Food Sci Food Saf 16:214–233. https://doi.org/10.1111/1541-4337.12248

Vitalini S, Iriti M, Puricelli C et al (2013) Traditional knowledge on medicinal and food plants used in Val San Giacomo (Sondrio, Italy)-an alpine ethnobotanical study. J Ethnopharmacol 145:517–529. https://doi.org/10.1016/j.jep.2012.11.024

Wang YC (2014) Medicinal plant activity on *helicobacter pylori* related diseases. World J Gastroenterol 20:10368–10382. https://doi.org/10.3748/wjg.v20.i30.10368

Weiser FA, Fangl M, Mosgoeller W (2018) Supplementation of Caricol®-gastro reduces chronic gastritis disease associated pain. Neuroendocrinol Lett 39:19–25

Widyowati R, Pratama RR, Sholikhah I et al (2023) Herbal medicine and rheumatic disorders management and prevention. Herb Med Phytochem 1:1–40

Wiehle M, Eusemann P, Thevs N (2009) Root suckering patterns in Populus euphratica (Euphrates poplar, Salicaceae). Trees 23:991–1001. https://doi.org/10.1007/s00468-009-0341-0

Xiao H, Zhao Z, Zhang C et al (2021) Influence of standardized nursing intervention combined with mindfulness stress reduction training on the curative effect, negative emotion, and quality of life in patients with chronic gastritis and gastric ulcer. Evid Based Complement Alternat Med 2021:2131405. https://doi.org/10.1155/2021/2131405

Yagnik D, Serafin V, Shah AJ (2018) Antimicrobial activity of apple cider vinegar against *Escherichia coli*, *Staphylococcus aureus* and *Candida albicans*; downregulating cytokine and microbial protein expression. Sci Rep 8:1732. https://doi.org/10.1038/s41598-017-18618-x

Yanaka A, Fahey JW, Fukumoto A et al (2009) Dietary sulforaphane-rich broccoli sprouts reduce colonization and attenuate gastritis in helicobacter pylori–infected mice and humans. Can Preven Res 2:353–360. https://doi.org/10.1158/1940-6207.CAPR-08-0192

Yogiraj V, Goyal PK, Chauhan CS et al (2014) Carica papaya Linn: an overview. Int J Herb Med 2:01–08

Zaghlool SS, Abo-Seif AA, Rabeh MA et al (2019) Gastro-protective and anti-oxidant potential of Althaea officinalis and solanum nigrum on pyloric ligation/indomethacin-induced ulceration in rats. Antioxidants 8:512

Zardast M, Namakin K, Esmaelian Kaho J et al (2016) Assessment of antibacterial effect ofgarlic in patients infected with *helicobacter pylori* using urease breath test. Avicenna J Phytomed 6:495–501

Zhang XQ, Gu HM, Li XZ et al (2013) Anti-helicobacter pylori compounds from the ethanol extracts of geranium wilfordii. J Ethnopharmacol 147:204–207. https://doi.org/10.1016/j.jep.2013.02.032

Zhang W, Cui N, Ye J et al (2022a) Curcumin's prevention of inflammation-driven early gastric cancer and its molecular mechanism. Chin Herb Med 14:244–253

Zhang Y, Liu C, Ma X et al (2022b) Effect of dietary modification on gastric mucosa, gastrointestinal symptoms and nutritional status of patients with early gastric cancer after endoscopic submucosal dissection surgery: a retrospective cohort study. Front Nutr 9:741630. https://doi.org/10.3389/fnut.2022.741630

Zheng X, Chen T, Zhao L et al (2020) Berberine protects against helicobacter pylori-induced gastric diseases in BALB/c mice via modulation of Nrf2-mediated oxidative stress and suppression of STAT3 activation. Phytother Res 34:1886–1897

Natural Product-Based Treatment for Food Poisoning

9

Nainee Goyal, Anshuman Chandra, Manu Vashistha, Anand Prakash Singh, Vijay Kumar Goel, and Nagendra Singh

Abstract

Food poisoning is a prevalent global health issue that happens when individuals ingest contaminated food that contains many types of pathogens, including bacteria, viruses, and parasites. This chapter offers a thorough examination of foodborne illness, encompassing its categorization, the mechanisms via which it induces sickness, and its substantial implications on global health and the economy, which have been further influenced by recent globalization trends. The progress in technology, namely in the techniques of treating diseases with natural substances, has had a crucial impact on decreasing the impact of food poisoning. In addition, the utilization of in silico tools has significantly enhanced the identification and control of foodborne illnesses. Computational technologies have changed the process of generating antibacterial medications by providing efficient and cost-effective methods for discovering new therapeutic agents. Computational approaches, such as molecular modeling and pharmacophore mapping, expedite the discovery of novel medications that can specifically target

Nainee Goyal and Anshuman Chandra have equally contributed to this work.

N. Goyal · N. Singh (✉)
School of Biotechnology, Gautam Buddha University, Greater Noida, Uttar Pradesh, India

A. Chandra · V. K. Goel
School of Physical Sciences, Jawaharlal Nehru University, New Delhi, India

M. Vashistha
Advanced Instrumentation Research & Facility, Jawaharlal Nehru University, New Delhi, India

A. P. Singh
IIMT College of Engineering, Greater Noida, Uttar Pradesh, India

T. Sharma et al. (eds.), *Natural Products for Antibacterial Drug Development: Recent Advancement of Computational Approach*,
https://doi.org/10.1007/978-981-97-9634-2_9

bacterial illnesses. This chapter also explores the therapeutic potential of natural chemicals in combating foodborne diseases, highlighting their remarkable structural diversity and complex mechanisms of action. Furthermore, the chapter elucidates the mechanisms by which antibacterial medications hinder bacterial physiology, such as by membrane disruption and interference with nucleic acid synthesis. Additionally, a comprehensive analysis of the latest scientific breakthroughs and emerging patterns in utilizing natural products for the management of food poisoning. The integration of computational and experimental methods holds significant promise for expediting the discovery and development of highly effective antibacterial medications, thereby effectively addressing the challenge of managing foodborne illnesses in a rapid and cost-efficient manner.

Keywords

Foodborne diseases · Natural products · Antimicrobial resistance · Computational approaches

Abbreviations

AI	Artificial intelligence
CDC	Centers for Disease Control and Prevention
CNS	Central nervous system
EOs	Essential oils
GC-MS	Gas chromatography-mass spectrometry
HTS	High-throughput screening
LC-MS	Liquid chromatography–mass spectrometry
LC-NMR	Liquid chromatography (LC) and nuclear magnetic resonance (NMR)
LPS	Lipopolysaccharide
MDS	Molecular dynamics simulations
ML	Machine learning
WHO	World Health Organization

9.1 Introduction

Food poisoning, commonly referred to as foodborne illness, is a medical condition caused by ingesting foods contaminated with harmful microbes such as pathogenic bacteria, parasites, viruses, or their toxins (Silva and Domingues 2017). Among all, bacterial pathogens are the most frequent cause of foodborne diseases (Asmat and Khan 2020; Ramos et al. 2020; Vaillant et al. 2005). Pathogen contact can occur at any stage in food production, transport, and preparation process, leading to over 200 foodborne illnesses in millions of people worldwide, making it a significant and wide-ranging concern for public health (Farid et al. 2023; Mead et al. 2000). Pathogens can enter through various routes, such as contaminated water, dairy products, poultry, and soil in farming, as well as through improper food handling and storage by workers and inappropriate cooking methods (Gonçalves-Tenório et al. 2018; Low and Hareeff 2022; 2018–2020 National Advisory Committee on Microbiological Criteria for Foods 2022; Rafiq et al. 2022; Sapp et al. 2023). Therefore, it is crucial to implement comprehensive food safety measurements throughout the food supply chain. Food poisoning has a significant and extensive effect on the overall health of people worldwide, impacting millions of individuals each year across the globe (Afr et al. 2007; Sharif et al. 2018). Geographical, socioeconomical, and cultural boundaries are irrelevant when it comes to this problem, highlighting a universal susceptibility to the dangers of tainted food. World Health Organization (WHO) reports that roughly 10% of the global population becomes sick annually because of consuming contaminated food, resulting in around 420,000 fatalities (Choi et al. 2016; Havelaar et al. 2013). Similar alarming numbers apply to the United States as well. Centers for Disease Control and Prevention (CDC) report that around 48 million individuals contact foodborne diseases each year, leading to lakhs of hospitalizations and causing thousands of fatalities (https://www.cdc.gov).

Food poisoning affects people differently across the world, with underdeveloped nations seeing a disproportionately high rate of illness. Subpar food safety measures, low sanitation, and restricted access to clean water worsen foodborne illnesses. Africa, Southeast Asia, and the Eastern Mediterranean regions are most affected by this issue (Mehlhorn 2016). These regions have the highest rates of occurrence and death, especially among children under the age of 5. This problem is made more difficult by the globalization of supply chain, which increases the probability of global food poisoning outbreaks fatalities (Choi et al. 2016; Havelaar et al. 2013). Furthermore, foodborne infections have enormous economic consequences, resulting in significant expenses for communities. It is anticipated that the annual economic impact in the United States alone will be in the billions of dollars, involving reduced productivity, health-care costs, and disease surveillance operations (https://www.cdc.gov). This condition has a particularly severe financial impact on low- and middle-income countries. Therefore, it can impact people's ability to make a living, put a burden on public health systems, and impede economic progress.

The primary symptoms of food poisoning are diverse, reflecting the broad spectrum of potential causes. In addition to gastrointestinal issues such as nausea, vomiting, abdominal pain, and diarrhea, common symptoms may include fever and dehydration signs (Aljamali et al. 2021; CDC 2019; Gupta and Chaudhary 2022). Depending on the individual's health and immune system as well as the nature of the contaminated food consumed, these symptoms might vary significantly in terms of their severity and duration. Furthermore, a wide range of microorganisms, including Norovirus, *Listeria monocytogenes, Salmonella typhi, Escherichia coli, Clostridium botulinum, Bacillus subtilis, Campylobacter jejuni*, and *Staphylococcus aureus,* cause the majority of foodborne diseases (Scallan et al. 2011). The emergence of pathogens like *E. coli* and *L. monocytogenes* raises significant concern because of their capacity to cause severe health repercussions and elevated mortality rates. Robust food safety regulations and the development of novel antimicrobial drug candidates are required to combat the evolving traits of these pathogenic microbes and lessen their consequences.

Furthermore, the influence of climate change on foodborne infections is becoming a concerning issue, as changing temperatures may affect the occurrence and transmission of bacteria (Dietrich et al. 2023). The worldwide consequences of foodborne illnesses on human health present a complex and diverse problem that necessitates a coordinated effort by governments, industry, and communities globally. Despite progress in understanding and combating foodborne diseases, the ever-changing nature of pathogenic microorganisms, and the interconnectedness of the global food chain demand an unwavering commitment and advancement in research for novel, side effect-free therapeutic alternatives (Ranveer et al. 2024). Also, the spectrum of foodborne microbial illnesses has changed over time, with an alarming increase in cases linked to newly emerging microbes and a rise in antibiotic-resistance (ABR) in pathogens (Fones et al. 2020; Wu-Wu et al. 2023). As a result, researchers have focused their efforts on investigating the antimicrobial activity of natural products as potential novel treatments for foodborne illness (Stan et al. 2021). Natural products provide a variety of unexplored chemotypes with therapeutic potential due to their complex chemical structures, ensuring they are promising candidates for foodborne disease treatments (do Prado et al. 2022; Grout et al. 2020; He et al. 2022; Ibrahim et al. 2021; Jackson et al. 2011; Rodrigues et al. 2016; Zhou et al. 2023). These natural substances also exhibit target-specific activity, higher bioavailability, and less toxicity. Furthermore, natural products have wide pharmacophore patterns and a physiologically significant molecular framework that favors compound-target protein interactions (Koch et al. 2005). Thus, utilizing cutting-edge computational approaches for drug development, researchers have examined natural products as effective pharmacological agents, with an emphasis on foodborne infections (Grout et al. 2020; He et al. 2022; Ibrahim et al. 2021; Jackson et al. 2011; Sarno et al. 2021; Zhou et al. 2023).

Research and development advancements combined with technology, particularly computational techniques, had led to a considerable improvement in the treatment of a variety of foodborne illnesses, including those caused by microorganisms (Sharma et al. 2024; Wang et al. 2021a, b). In addition, computational approaches have revolutionized the pharmaceutical industry by enabling the identification of new therapeutic molecules and the optimization of existing drugs (Sadybekov and Katritch 2023). Computational-aided drug discovery relies on several critical technologies to generate natural products as antimicrobial therapies. It mainly involved high-throughput screening (HTS) to conduct screening of millions of compounds, molecular docking and simulation to predict binding affinity and solvent effect, quantitative structure-activity relationship (QSAR) models to identify toxicity and biological activity, machine learning and artificial intelligence (AI)-based programs to predict ADME properties and improve drug efficacy, systems biology to identify drug effects, and pharmacogenomics to minimize adverse effects of natural compounds. Collectively, which provide valuable insights for developing fast, personalized, efficient, and cost-effective therapeutics, particularly in the context of treating food poisoning (Quazi 2022). As computational power and analytical techniques continue to evolve, which can expect further innovations that will enhance our ability to combat foodborne pathogens especially using natural products as antimicrobials. The incorporation of these technologies represents a substantial advancement towards more efficient, tailored, and prompt therapies for food poisoning and other conditions. This chapter will provide an understanding of foodborne illnesses, their causative agents, and their treatment based on natural products by applying computer-aided drug discovery approaches.

9.2 Agents Causing Food Poisoning

Foodborne disease causative agents may cause a wide range of symptoms, from minor gastrointestinal discomfort to severe life-threatening disorders (Bintsis 2017). These contaminants are often of two types: biological and non-biological agents. Certain toxins secreted by bacteria or other pathogens, some poisonous substances, and physical agents are example of non-biological agents, while bacteria, viruses, parasites, and other microbes are biological agents as shown in Table 9.1 (Kumar 2020). Thorough knowledge of these causative agents, including their characteristics, behaviors, and associated clinical manifestations, is essential for identifying and developing treatments using natural products. Moreover, such understanding is fundamental to both preventing outbreaks and facilitating more precise diagnosis and efficient treatment methods, ultimately reducing the incidence and severity of foodborne illnesses.

Foodborne bacteria: Bacterial pathogens are the predominant causative agent behind foodborne illnesses, which come in a wide range of shapes, properties, and types, each associated with specific dietary sources and exhibiting particular symptoms. Certain foodborne bacteria are spore-forming in nature, so they are extremely heat-resistant, such as *C. botulinum, C. perfringens, B. subtilus*, and *B. cereus* (Andryukov et al. 2020). Furthermore, *C. botulinum*, the botulism causing agent, is extremely deadly despite being less prevalent (Rawson et al. 2023). While most foodborne bacteria thrive at an ideal growing temperature ranging from 20 °C to 40 °C, like *B. cereus, Salmonella, C. jejuni,* and *E. coli.*, *C. jejuni* thrives in poultry, unpasteurized milk, and water-based sources and causes bloody diarrhea, fever, and cramping in the abdomen (Lopes et al. 2021). However, some pathogens (known as psychrotrophs) can grow in a refrigerator or at temperatures below 10 °C, including *L. monocytogenes* and *Yersinia enterocolitica. L. monocytogenes* is especially concerning due to its symptoms: fever, muscle aches, and nausea, which can be more severe in pregnant women, newborns, and those with compromised immune systems (Osek et al. 2022).

Foodborne viruses: Viral particles can only multiply within other living cells. They cannot, therefore, endure for extended periods of time without their host. Foodborne illnesses have been linked to over 100 different types of enteric viruses; the most prevalent ones, noroviruses and hepatitis A, are significant in the landscape of foodborne illnesses, each linked to specific transmission vectors and exhibiting

Table 9.1 Common causative agent of food borne infections, including their characteristics, and associated clinical manifestations

Category	Causative agent	Common foods involved	Symptoms	References
Bacteria	*Salmonella* spp.	Poultry, eggs, dairy products, and produce	Diarrhea, fever, abdominal cramps, vomiting	Ehuwa et al. (2021)
	Escherichia coli (*E. coli*)	Undercooked beef, raw milk, fresh produce	Diarrhea, abdominal pain, vomiting	Ramos et al. (2020)
	Listeria monocytogenes	Unpasteurized dairy products, ready-to-eat meats	Fever, muscle aches, nausea, diarrhea	Eallonardo et al. (2023)
	Campylobacter jejuni	Poultry, unpasteurized milk, contaminated water	Diarrhea, fever, abdominal pain	Bai et al. (2021)
	Clostridium botulinum	Improperly canned or preserved foods	Blurred vision, slurred speech, difficulty swallowing	Rawson et al. (2023)
Viruses	Norovirus	Contaminated water, raw produce, uncooked foods	Diarrhea, vomiting, nausea, stomach pain	Graziano et al. (2019)
	Hepatitis A	Raw shellfish, contaminated water	Fatigue, nausea, stomach pain, jaundice	Fiore (2004)
Parasites	*Giardia lamblia*	Contaminated water, fresh produce	Diarrhea, bloating, stomach cramps, nausea	Porter et al. (1990)
	Cryptosporidium	Contaminated water, fresh produce	Watery diarrhea, stomach cramps, nausea	Zhang et al. (2022)
Toxins	*Staphylococcus aureus*	Improperly stored or handled meats, salads, dairy products	Nausea, vomiting, diarrhea	Bencardino et al. (2021)
	Bacillus cereus	Rice, pasta, pastries	Nausea, vomiting, diarrhea	Tirloni et al. (2022)

distinct symptom profiles. Extremely contagious, norovirus is frequently associated with outbreaks that originate from contaminated water, raw produce, and undercooked food items (Hardstaff et al. 2018). The symptoms include acute gastroenteritis, characterized by intense stomach discomfort, diarrhea, and vomiting, which are disruptive and can lead to severe dehydration. Hepatitis A, on the other hand, is often transmitted through raw shellfish or water tainted with the virus (Sattar et al. 2000). It has a direct impact on the liver, resulting in systemic symptoms like fatigue, nausea, abdominal pain, and jaundice. Both viruses highlight the critical need for novel therapeutics to contain outbreaks and protect community health.

Foodborne parasites: Parasites derive nourishment and protection from host organisms; several parasites have been identified as causing agents of foodborne illness. Moreover, *Taenia solium, Echinococcus granulosus, Echinococcus multilocularis*, and *Toxoplasma gondii* are the most significant contributors to foodborne illness based on global impact, mortality, and morbidity, particularly through their contamination of water and fresh produce (Gabriël et al. 2023). These parasites thrive in environments where water treatment and sanitation are inadequate, making them common culprits in outbreaks within such regions. The transmission of these parasites is typically through ingestion of contaminated water or eating fresh produce that has not been properly cleaned.

Foodborne toxins: Sometimes foodborne illnesses are caused by non-biological agents like toxins produced by bacteria rather than the bacteria themselves (Schmidt et al. 1969). *S. aureus* is one prominent example, known for secreting toxins in foods that are improperly stored or handled, such as meats, salads, and dairy products (Otto 2014). These toxins can trigger rapid onset symptoms including nausea, vomiting, and diarrhea, demonstrating their potency and the swift response of the human body to these substances. A nonbacterial agent that can cause foodborne illness is paralytic shellfish toxin, which occurs due to the ingestion of mussels, clams, or scallops that have consumed toxic dinoflagellates (Todd 2014).

9.3 The Historical Perspective

Foodborne diseases have been a constant companion on our evolutionary journey. Thus, natural products have been critical to therapeutic practices and the mainstay of foodborne illnesses for many civilizations since antiquity (Chunarkar-Patil et al. 2024). The evolution of natural compounds into medications for foodborne diseases reflects the broader trajectory of medical advancement, from rudimentary understandings of compounds to today's precise, evidence-based effects (Fig. 9.1). This journey demonstrates how cultural, scientific, and technological knowledge have shaped natural products to combat the numerous pathogens responsible for foodborne illnesses. Natural products have a long history of being used as source of active substances with health advantages, whether from aquatic or terrestrial creatures.

These therapeutics, which contain compounds derived from plants and minerals, have been widely utilized worldwide, particularly in India, China, Egypt, Greece, and Sumer, for more than 5000 years (Nasim et al. 2022). For instance, the ancient Egyptians treated digestive disorders using coriander, garlic, and other herbs presumed to have antimicrobial qualities, a primitive form of combating bacterial infections without an understanding of the pathogens involved (Metwaly et al. 2021). China and India are frequently regarded as the motherlands where natural products are used (Afr et al. 2007). Around 80–90% of people on the planet still perform conventional therapeutic practices, and drugs obtained from plants offer

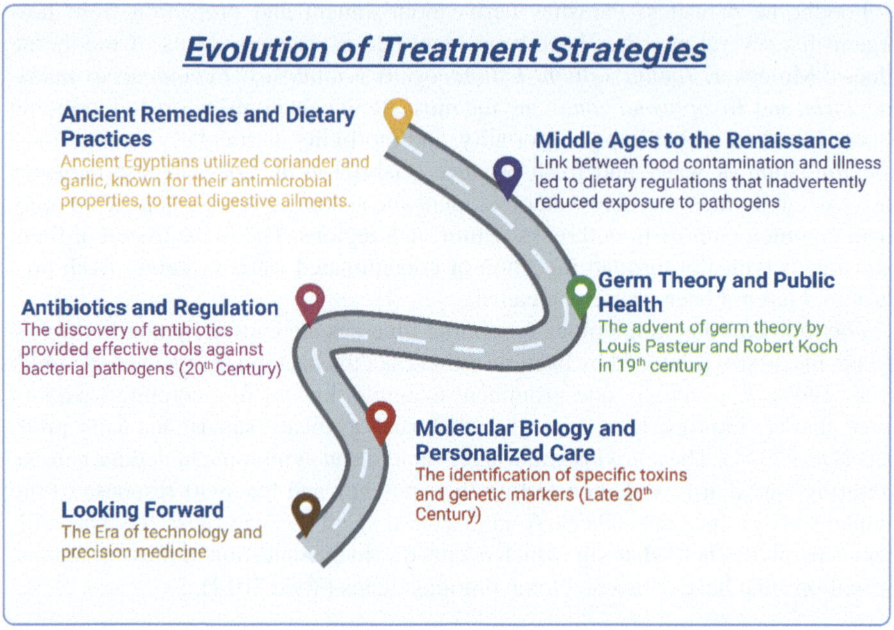

Fig. 9.1 Evolution of treatment strategies. From ancient to precision medicine

substantial therapeutic potential (Singh et al. 2023a, b). Several secondary metabolites from plants have been identified so far, exhibiting a unique structure and set of pharmacological characteristics (Gad et al. 2013). The age of drug discovery for natural products officially began in 1803 with morphine isolation from *Papaver somniferum,* the first naturally occurring, pure plant-derived molecule (Krishnamurti and Rao 2016). Furthermore, people have been employing natural products to prevent and treat infectious illnesses in recent years, of which secondary metabolites are the most potent and promising compounds (Newman and Cragg 2016). Notably, over half of the therapeutics authorized by the Food and Drug Administration (FDA) have been derived from natural products or their derivatives (Patridge et al. 2016).

The inclusion of natural products into modern pharmacology represents a significant breakthrough in medicinal research. While traditional pharmaceutics served as the initial impetus for the use of natural products, their incorporation into mainstream medicine has required rigorous scientific validation and optimization. Natural products like secondary metabolites that are included in pharmacology mark a shift in medical practice from anecdotal and empirical to evidence-based. The utilization of plant-based natural drug compounds and healthcare products is currently experiencing a global upsurge. In addition, it highlighted how natural compounds could be a plentiful source of potential drug candidates capable of treating a wide range of intricate illnesses, including various foodborne diseases.

9.4 The Importance and Potential of Natural Products in Food Poisoning

It is commonly known that nature has selected and refined over millions of years to create chemical scaffolds and molecules rich in biological activity (Dias et al. 2012). Due to the development of antimicrobial resistance in foodborne pathogens, researchers have spent the past few years observing and identifying natural products' antimicrobial and therapeutic potential for better treatment alternatives. Plant natural products are phytochemicals that are well known for their contribution to plant defensive mechanisms against predators and pathogens (Che and Zhang 2019). Additionally, scientists have a significant degree of interest in natural compounds because of their wide range of advantageous qualities.

The remarkably varied yet unique physical and chemical space, from simple organic molecules to complex ring structures and more stereogenic centers, each equipped with diverse functional groups, directly contributes to the variety of bioactivates that are observed in natural products. The structural diversity of natural compounds is a vast and unexplored resource that offers a rich spectrum of potential therapeutics that reflect an intricate history of evolutionary adaptations (Scott and Piel 2019). These natural compounds are mainly products of secondary metabolic processes in various organisms, including plants, fungi, bacteria, and marine species. This structural diversity has evolved through natural selection over millions of years, enabling organisms to navigate ecological challenges such as deterring predators, combating infections, and facilitating symbiotic relationships. Furthermore, pharmacology is increasingly using new, advanced techniques such as LC-MS, GC-MS, and LC-NMR to isolate, analyze, and conduct biochemical and metabolomic studies of natural compounds (Chopra and Dhingra 2021).

Therefore, this structural complexity makes natural compounds highly valuable in pharmacology; their ability to act as biocatalysts, activate or block receptors, and interact with DNA offers unique opportunities for therapeutic application (Sheldon and Woodley 2018). Furthermore, functional characterization of such catalysts is another important area of research in the natural compound-based drug discovery field. Natural compounds often work in ways that are distinct from synthetic drugs. For example, they may interact with multiple targets simultaneously or evade traditional resistance mechanisms found in pathogenic microorganisms and tumor cells (Dzobo 2022). However, the ability to function as natural metabolites, on the other hand, may give natural products an edge over synthetic ones; promising therapeutic substances may have metabolite-like properties (Katrachanca and Koleske 2017). Another property of natural compounds, particularly plant-derived molecules, is that they are less expensive, easily procurable, and have a lesser extent of adverse side effects. The fact that natural compounds have many properties underscores their potential to serve as blueprints for developing new therapeutics, providing novel mechanisms that could address unmet medical needs and enhance the efficacy of treatment regimens against foodborne diseases. As research continues to explore and harness this chemical and structural diversity, natural compounds are likely to play an increasingly prominent role in drug innovation.

Researchers have discovered several natural compounds with potent antibacterial and antiviral properties to combat foodborne illnesses. Novel synthetic molecules are created by structurally altering natural compounds, which are also very efficient.

9.5 Natural Product Involves in Foodborne Disease Treatment

The preferred source of medications has historically been natural products derived from plants and animals, particularly as antimicrobial and anticancer drugs. However, modern medicine has surpassed traditional therapeutics as the means of treatment for most of the recent diseases. But due to the HT screening of large synthetic compounds and combinatorial libraries failing to create significant new pharmacological agents between 1995 and 2005, there was a renewed interest in studying how to mine plants and microorganisms for potent natural therapeutic products (Katz and Baltz 2016). Currently, several plant-derived natural compounds are used as prescribed medications in developed nations like the United Kingdom, France, and China (Ji et al. 2017). Studies on naturally occurring compounds have mostly concentrated on the several aspects listed below.

9.5.1 Polyphenols

Plant physiology depends heavily on polyphenols, which are naturally occurring micronutrients. These natural substances, which have over 8000 distinct phenolic structures, are widely found in many plant components, such as fruits, vegetables, grains, leaves, roots, and flowers (Abbas et al. 2017). Polyphenols are identified by their hydroxyl groups and aromatic rings. Polyphenols are well known for their potent antibacterial activity that effectively hinders the growth of numerous harmful microbes. These are typically found in abundance in various sources, including wine, green tea, grapes, vegetables, berries, and coffee.

Furthermore, the antimicrobial activity of polyphenols can aid in the fight against antibiotic resistance. Globally, polyphenols are becoming more widely acknowledged and accepted as a natural product-based treatment against foodborne diseases as knowledge of their therapeutic potential and protective function in health increases. In addition to being known for their phytochemical and antioxidant qualities, polyphenols are also believed to have anti-inflammatory benefits. Numerous studies have explored their role in preventing and treating conditions related to oxidative stress, inflammation, and pollutant-induced cell damage. Furthermore, humans frequently include various vegetables and drinks high in polyphenols in their diets. Polyphenols, which include phenolic acids, flavonoids, lignans, and stilbenes, have various health benefits as they exhibit neuroprotective, cardioprotective, antioxidant, anticarcinogenic, and antimicrobial properties (Zhang and Tsao 2016).

Certain polyphenols are known to be resistant to foodborne viruses. Some functional elements, such as licoflavonol, glyasperin, glyasperin D, and methoxyisoliquiritigenin obtained from the roots of *Glycyrrhiza uralensis,* have demonstrated their resistance activity against rotaviruses (Kwon et al. 2010). In another example, black raspberry extracts, which are rich in caffeic acid, gallic acid, ellagic acid, cyanidin-3-glucose, and quercetin, have a deleterious effect on human norovirus (Zhang et al. 2012). Polyphenols exhibit antiviral effects by inhibiting viral replication, binding with their receptors, and/or causing structural damage to foodborne viruses. Additionally, polyphenols are recognized to regulate bacterial, fungal, and chemical contamination in food. They typically work against foodborne bacteria by altering the characteristics of the bacterial cell membrane (Plumed-Ferrer et al. 2013). By modifying the morphology of the cell membrane, polyphenols, including rutin, ellagic acid, and gallic acid, can also have an antifungal impact (Saravanan and Parimelazhagan 2014). Meanwhile, polyphenols exhibit chemical contaminant inhibition; for instance, epicatechin decreases lipid oxidation by trapping carbonyl molecules to reduce the amount of acrylamide in meals (Liu et al. 2015).

9.5.2 Essential Oils

Essential oils (EOs) are aromatic liquids derived from various parts of plants, including flowers, buds, seeds, leaves, twigs, bark, herbs, wood, fruits, and roots. EOs can be extracted through methods like expression, fermentation, enfleurage, or extraction. Around 300 of the approximately 3000 known EOs are employed commercially, primarily in the flavor, fragrance, and medicinal industries. Even though it has been recognized that several EOs have antibacterial qualities, current "green" consumer enthusiasm has spurred new scientific research on these compounds. They can be extracted through methods like expression, fermentation, enfleurage, or extraction, but steam distillation is the most used method for commercial production. Beyond their antibacterial effects, EOs and their components have demonstrated antiviral, antifungal, antitoxigenic, antiparasitic, and insecticidal properties. These diverse characteristics may be linked to the natural roles these compounds play in plants (Pandey et al. 2023b; Sánchez-González et al. 2011).

Differences in the aroma profile suggest variations in the composition of EOs obtained through solvent extraction compared to distillation, which may also affect their antimicrobial properties. This is supported by studies showing that EOs extracted with hexane exhibit stronger antimicrobial activity than those obtained by steam distillation. Due to their volatile nature, EOs require storage in airtight containers and kept in the dark to maintain their composition over time. Numerous studies have detailed the composition of various EOs. Comprehensive compositional analysis is typically conducted using gas chromatography and mass spectrometry. EOs can contain over 60 individual compounds with major constituents often comprising up to 85% of the EO, while others are present in trace amounts. Phenolic compounds primarily contribute to the antibacterial properties of EOs.

The composition of EOs from a particular plant species can vary depending on the harvesting season and geographical location. This variability can be attributed, in part, to the formation of antibacterial compounds from their precursors. For instance, p-Cymene and g-terpinene are precursors to carvacrol and thymol in *Origanum* and *Thymus* species. EOs harvested from herbs during or shortly after flowering generally exhibit the most potent antimicrobial activity. Additionally, enantiomers of EO components can display varying degrees of antimicrobial activity. Moreover, EOs derived from different parts of the same plant can exhibit significant compositional differences. For example, the EO from coriander seeds differs markedly from that of cilantro, derived from the plant's immature leaves (Pandey et al. 2023b; Sánchez-González et al. 2011).

9.5.3 Proteins

Proteins play an important role in the treatment of foodborne diseases. Some proteins, such as lactoferrin, lysozyme, and certain antimicrobial peptides, are known to have antibacterial activity and are found in foods or produced by the body. Natural antimicrobials could be added post-processing as a bio-preservative measure to reduce the risk of bacterial contamination in dairy milk. One such protein with potential bio-preservative properties is lactoferrin, a naturally occurring protein found in mammalian excretory fluids like saliva, tears, and milk. Lactoferrin concentrations vary across mammalian species and body fluids, with bovine milk containing relatively low levels. Extensive studies have highlighted lactoferrin's diverse properties, including antiviral, antifungal, anti-inflammatory, and antimicrobial effects. Lactoferrin's antimicrobial actions were examined, it can bind to the lipid A portion of the LPS layer in Gram-negative bacteria, target teichoic acid in Gram-positive bacteria, and sequester iron to limit its availability for pathogens. The antibacterial efficacy against Gram-positive bacteria is commonly attributed to the lysozyme's enzymatic breakdown of peptidoglycans within the bacterial cell wall. However, Gram-negative bacteria have an additional barrier in a lipidic outer membrane, which shields their peptidoglycans. Despite this obstacle, lysozyme still exhibits some level of activity against Gram-negative bacteria, albeit weakly (Gruden and Ulrih 2021).

The gut serves as the primary battleground for many significant bacterial pathogens transmitted through food. Symptoms associated with foodborne illnesses include diarrhea, vomiting, fever, and headache, although severe cases can lead to neurological complications or even fatalities. Many of these diseases are linked to disturbances in the gastrointestinal microbiota; Probiotics have garnered significant attention in both the food industry and nutritional research as a promising solution for addressing common gastrointestinal disorders and combating enteric infections, particularly in the treatment of diarrhea. The main mechanism by which probiotics promote host health is through physical interaction with the human gastrointestinal tract's epithelium. When probiotic microorganisms adhere to the intestinal wall, they enhance mucus production and strengthen the intestinal barrier. The ability to adhere is considered one of the most efficient functions of probiotic strains. This capacity of probiotic bacteria to attach to epithelial cells or mucus is a proposed mechanism for protecting the host organism from invasion and preventing the adherence of pathogens. The intestinal barrier serves as a protective mechanism that prevents the entry or retention of harmful agents. Recent research suggests that probiotic bacteria may enhance the functions of the cellular barrier and modulate cytoskeletal structures in the intestinal mucosa, potentially inhibiting the pathogenic activity.

9.5.4 Alkaloid

Berberine, a natural isoquinoline alkaloid, is increasing attention due to its diverse therapeutic effects on conditions such as cancer, diabetes, hyperlipidemia, cardiovascular diseases, and central nervous system (CNS) disorders (Imenshahidi and Hosseinzadeh 2019). Traditionally, berberine has been extensively utilized in treating bacterial diarrhea and gastroenteritis. Furthermore, berberine seems to be effective in inhibiting the excessive electrolyte secretion induced by cholera toxin. Moreover, it exhibits anti-inflammatory properties, inhibits gastrointestinal motility, improves impaired gastrointestinal function, and diminishes intestinal secretion and exudation. Berberine's ability to relax intestinal smooth muscle and prolong the retention of intestinal contents aids in the complete digestion and absorption of nutrients, thereby addressing functional diarrhea. Hence, berberine demonstrates notable efficacy in managing infectious and noninfectious forms of secretory or exudative diarrhea. Several mechanisms have been proposed where berberine improves colitis by inhibiting the growth of Gram-negative intestinal bacteria such as *E. coli*, *Klebsiella pneumoniae*, and *Proteus mirabilis*. Besides this, berberine has been seen to show some positive effects on gut microbiota, such as *Bifidobacterium adolescentis* and *Lactobacillus acidophilus* (Imenshahidi and Hosseinzadeh 2019).

9.5.5 Organic Sulfur Compounds

Food poisoning, often caused by food contamination with pathogenic bacteria such as *Salmonella* sp., *E. coli*, and *S. aureus*, can be effectively mitigated by the antibacterial properties of organosulfur compounds. Allicin, a volatile compound containing sulfur, can inhibit these illness-causing bacteria. It is primarily found in white garlic (*Allium sativum* L.) but can also be found in other Allium species like elephant garlic (*A. ampeloprasum* L.), wild garlic (*A. ursinum* L.), field garlic (*A. vineale* L.), and alpine leek (*A. victorialis* L.). Allicin has a broad antibacterial range. Allicin and other sulfur-containing chemicals achieve this through multiple mechanisms. They affect biofilm development, bacterial enzyme systems, and cell activity, which is important since biofilms protect bacteria from standard cleaning methods and allow them to stay on food surfaces and processing equipment. Food preservation with allicin-rich extracts is also noteworthy (Marchese et al. 2016).

Food can be preserved naturally by inhibiting microbial development with these extracts, prolonging shelf life and minimizing bacterial contamination. Natural substances are desirable in the food sector because consumers prefer products without synthetic ingredients. Organosulfur compounds also protect against food poisoning due to their anti-inflammatory and antioxidant capabilities. These chemicals prevent bacterial infections and reduce foodborne disease symptoms by lowering inflammation and oxidative stress.

9.6 Action Mechanism of Natural Compounds as Antibacterial

Selective toxicity is an essential characteristic for antimicrobial drugs, indicating their capability to specifically target and hinder the growth of microbial agents while sparing the host from harm (Fig. 9.2). Currently, antibacterial medications dominate clinical usage due to the wider array of distinct targets available in prokaryotic cells, offering greater potential for achieving selective toxicity compared to fungi, parasites, and viruses. Various classes of antibacterial agents inhibit steps in peptidoglycan biosynthesis, achieving bacterial cells more vulnerable to osmotic lysis. Consequently, antibacterials that focus on cell wall biosynthesis exert a bactericidal effect. Since human cells lack the capacity to produce peptidoglycan, this mechanism exemplifies exceptional selective toxicity (Nina Parker et al. 2018). Few natural products demonstrate antibacterial activity by binding to lipopolysaccharide (LPS), a key component of the outer membrane in Gram-negative bacteria. This binding disrupts the protective barrier function of LPS, rendering the bacteria more susceptible to environmental stresses (Johnson et al. 2008). Similarly, few natural

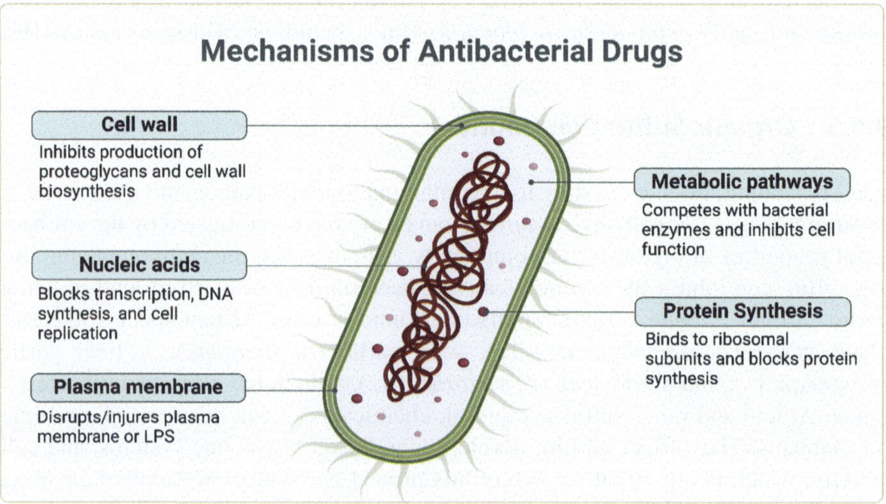

Fig. 9.2 Mechanism of antibacterials

products exhibit antibacterial activity by inducing morphological damage to the cell membrane, compromising membrane permeability, and disrupting membrane integrity (Wang et al. 2021a, b).

9.6.1 Cell Wall Degradation

The outer membranes of bacteria serve as protective barriers, enabling the bacteria to grow in harsh environments. Meanwhile, the cytoplasmic membrane (inner membrane) controls the intake of solutes and transportation of proteins and other large molecules into the cell. Changes to these membranes can have detrimental effects on crucial bacterial functions such as nutrient uptake and ion transport, impacting bacterial integrity. As membrane fluidity decreases, fluorescence polarization undergoes corresponding changes, providing insights into the antioxidant's impact on membrane properties (Ali Raza Naqvi et al. 2019).

The significance of antioxidants in combating bacterial inflammations is increasingly acknowledged, particularly as they demonstrate synergy with existing antibacterial agents against resistant bacterial strains. Unlike synthetic agents, the structural diversity of natural products prevents bacteria from replicating their functional components. Additionally, these specific scaffolds may be crucial for facilitating the uptake of flavonoids into bacterial cells. For instance, polyhydroxylated flavonoids exhibit more potent antibacterial activity compared to mono-hydroxylated or non-hydroxylated counterparts (Nina Parker et al. 2018).

The pathogenicity of Gram-negative bacteria is closely tied to their LPS layer, which reduces sensitivity to antibacterial agents. As a result, antibacterial agents tend to exhibit greater efficacy against Gram-positive bacteria. Catechins, compounds found in tea, interact with and target various bacterial enzymes, membrane proteins, and fatty acid synthase. These have been reported to alter membrane fluidity, particularly in the well-organized phosphatidylcholine and phosphatidylethanolamine bilayers that constitute bacterial membranes. In addition, the antibacterial effects of catechins result from their interactions with oxygen, genes, and cell membranes. Epigallocatechin gallate (EGCG) demonstrates potent antibacterial function by disrupting bacterial membranes and causing leakage. EGCG binds directly to the peptidoglycan of *S. aureus*, affecting cell integrity and reducing the cell's ability to withstand high osmotic pressure and ionic strength. It also induces morphological changes and oxidative stress in Gram-negative bacteria (Kitichalermkiat et al. 2019).

Other flavonoids, such as epigallocatechin, myricetin, and quercetin, similarly damage membrane proteins, disrupt cytoplasmic coagulation, alter fatty acid and phospholipid constituents, inhibit energy formation and metabolism, affect RNA and DNA production, and impede protein translocation. Epicatechin gallate (ECG) significantly alters the physical properties of phosphatidylcholine (PC) and phosphatidylethanolamine (PE) bilayers, leading to membrane leakage. Therefore, PE is responsible for catechins' antibacterial properties at the membrane level (Ali Raza Naqvi et al. 2019).

Catechins are renowned for their remarkable affinity to partition into lipid bilayers composed of diverse compositions (Tsuchiya 2001). This characteristic has led to the hypothesis that catechins' partitioning within bacterial cell membranes reduces lipid packing and promotes lateral expansion of the membrane, consequently diminishing its barrier activity (Fig. 9.3).

The antibacterial mechanisms of catechins can be described as follows:

(a) The lateral expansion/membrane disruption hypothesis proposes that catechins intercalate into the lipid bilayer of bacterial membranes. This intercalation increases the spacing between lipids, leading to membrane destabilization and enhanced permeability (Fig. 9.3a).
(b) In the hydrogen peroxide generation mechanism, react with dissolved oxygen to generate hydrogen peroxide and hydroxyl radicals. These reactive oxygen species later induce lipid oxidation and damage DNA and proteins within bacterial cells (Fig. 9.3b).

Furthermore, ferulic and gallic acids interact with cell membranes in various bacteria. These interactions lead to reductions in surface negative charges, formation of membrane pores, leakage of intracellular components, and alterations in hydrophobicity, all contributing to their antibacterial effects (Akbari et al. 2019). p-coumaric acid, found in apple, pear, and quinoa, exhibits antibacterial activity by disrupting membrane permeability and inducing pore formation in bacteria such as *E. coli*, *S. dysenteriae*, and *S. typhimurium* (Lou et al. 2012).

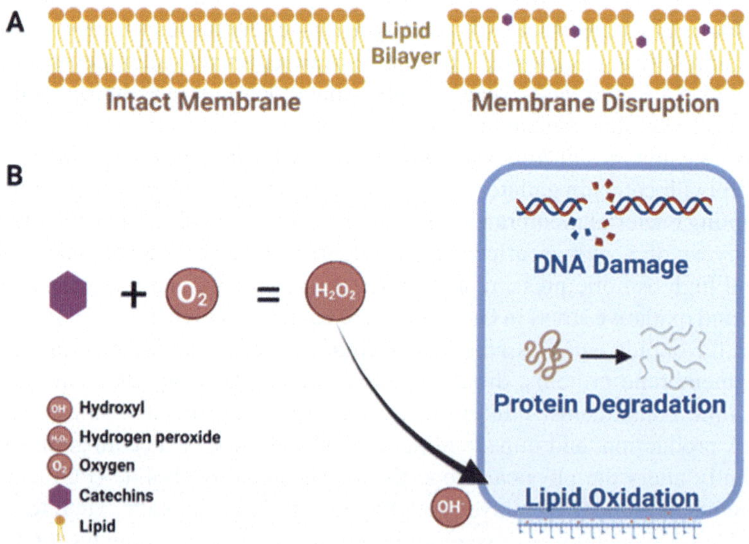

Fig. 9.3 Antibacterial mechanisms of catechins. (**a**) The lateral expansion/membrane disruption hypothesis and (**b**) the hydrogen peroxide generation mechanism

9.6.2 Nucleic Acid Synthesis Interference

DNA Gyrase Inhibitors The B rings of few natural product have the ability to inhibit DNA and RNA synthesis. DNA gyrase, classified as a type II isomerase, is pivotal in converting DNA into a negative super helical form post-replication. Additionally, it facilitates the separation of chromosomes in daughter cells connected in a loop structure. Consequently, DNA gyrase serves as a prime target for antibacterial agents. Comprising two catalytic subunits, DNA gyrase includes GyrA, responsible for DNA breakage and recombination, and GyrB, housing an ATP-binding site (Oblak et al. 2007).

p-coumaric acid, a phenolic acid, disrupts bacterial cell membranes and triggers cell death by binding to DNA (Lou et al. 2012). Coumarin and cyclothialidine inhibit DNA gyrase by preventing the binding of ATP to the GyrB subunit, exhibiting antibacterial effects (Oblak et al. 2007; Plaper et al. 2003). Quercetin, catechin, and epigallocatechin gallate have been investigated and found to possess antibacterial properties by inhibiting bacterial DNA gyrase activity and impeding nucleic acid synthesis (Plaper et al. 2003). For example, trimethoprim-sulfonamides combine the antibacterial effects of trimethoprim with those of sulfonamides. Their efficacy stems from the synergistic action in impeding folate pathway in pathogen. When administered together, they exhibit a broad spectrum of activity against susceptible bacterial infections, targeting both Gram-negative and Gram-positive bacteria.

The antibacterial activity of flavonoids may arise from their capacity to form complexes with proteins through various nonspecific forces, including hydrogen bonding and hydrophobic effects, as well as through covalent bond formation (Kumar and Pandey 2013). When polyphenols bind to proteins, they can become sequestered into soluble or insoluble complexes, altering the function of both the polyphenol and the protein (Brudzynski and Maldonado-Alvarez 2015). This binding can result in modifications to the protein structure, including blocked amino acids or conformational transitions, leading to changes in protein solubility, hydrophobicity, thermal stability, and isoelectric point. Studies have demonstrated that naturally occurring polyphenols, such as condensed tannins, can inhibit various digestive enzymes, including α-glycosidase, α-amylase, lipase, pepsin, trypsin, and chymotrypsin, thereby affecting nutrient availability and modulating microbiota composition (Sun et al. 2020). Additionally, polyphenols can bind to essential bacterial proteins, such as adhesins, enzymes, and cell envelope transport proteins, leading to their inactivation and exerting an antimicrobial effect.

Kaempferol-3-rutinoside (nicotiflorin) has been shown to inhibit *Streptococcus mutans* by targeting the protein sortase A, a membrane enzyme crucial for bacterial adhesion and invasion of host cells (Yang and Hinner 2015). Curcumin was found to inhibit purified sortase A from *S. mutans* at a half-maximal inhibitory concentration, and it was also observed to release the Pac protein into the supernatant, thereby reducing *S. mutans* biofilm formation (Hu et al. 2013). Also, sortase enzymes, also known as cysteine transpeptidases, are utilized by Gram-positive bacteria to anchor proteins, such as glycoproteins, to the cell surface and attach them to the cross-bridge peptide of the cell wall, making them essential virulence factors. Morin, myricetin, and quercetin have shown strong inhibitory activity against both sortase A and B from *S. aureus* (Hu et al. 2013; Kang et al. 2006).

9.6.3 Modulation of Bacterial Metabolism

Disruption of Energy Production Numerous facets of cellular metabolism are centered on producing and consuming ATP, which serves as the universal energy currency connecting anabolic and catabolic processes. ATP facilitates essential cellular activities such as motile contraction, phosphorylation, and active transport of nutrients. In bacteria, membrane-bound F1F0 ATP synthase is responsible for ATP production via oxidative or photophosphorylation pathways (Chinnam et al. 2010).

Studies have shown that morin, baicalein, silibinin, and epicatechin completely inhibit ATPase activity, while hesperidin, chrysin, kaempferol, diosmin, apigenin, genistein, and rutin exert partial inhibition, reducing activity by approximately 40%–60%. It was found that resveratrol, piceatannol, quercetin, quercitrin, and quercetin-3-β-D glucoside inhibited *E. coli* ATP synthase to varying extents. Among these compounds, piceatannol demonstrated the most potent inhibition, while the inhibition by other compounds ranged from approximately 20% residual activity for quercetin to around 60% residual activity for quercitrin or resveratro l (Ahmad et al. 2015; Chinnam et al. 2010; Oblak et al. 2007).

Enzyme Inhibition Trimethoprim, a dihydropyrimidine antimicrobial and antiparasitic agent, serves as the prototype for a class of nonsulfonamide drugs that target dihydrofolate reductase in bacterial and protozoal cells (Watson et al. 2007). Originally introduced as an antimalarial agent, trimethoprim is now predominantly used as an antibacterial agent, particularly with sulfamethoxazole. Trimethoprim and its related agent, 2,4-diaminopyrimidine, act as competitive inhibitors of dihydrofolate reductase, an enzyme crucial for the reduction of dihydrofolate to tetrahydrofolate utilizing NADPH. This enzyme plays a vital role in cells for the biosynthesis of purines, pyrimidines, and certain amino acids. Notably, trimethoprim leads to disrupting folate metabolism in bacteria which lead to cell death. Trimethoprim is currently exclusively utilized as an antibacterial agent. It effectively treats initial uncomplicated urinary tract infections caused by susceptible strains of *E. coli, P. mirabilis, Klebsiella* sp., and *Enterobacter* sp. Additionally, it is valuable in the treatment of *Pneumocystis carinii* pneumonia in AIDS patients when combined with sulfamethoxazole (Watson et al. 2007).

Interference with Protein Synthesis Tetracyclines penetrate the outer membrane of Gram-negative enteric bacteria through the OmpF and OmpC porin channels which include magnesium ions and act as positively charged cation-tetracycline coordination complexes. The Donnan potential across the outer membrane attracts these complexes, which causes them to accumulate in the periplasm. The metal ion-tetracycline combination is probably break in this area, releasing uncharged tetracycline molecules that can diffuse through the inner membrane's lipid bilayer because they are mildly lipophilic (Ian and Marilyn 2001).

In Gram-positive bacteria, it is believed that the lipophilic form of tetracyclines is transported across the cytoplasmic membrane. The transport of tetracyclines via this membrane relies on energy use and is facilitated by the difference in pH across the membrane, known as the proton motive force. Upon entering the cytoplasm, tetracycline molecules have the potential to form chelates because of the elevated internal pH and larger concentrations of divalent metal ions in comparison to the external surroundings. The most probable scenario is that the active drug species that binds to the ribosome combines magnesium and tetracycline (Ian and Marilyn 2001).

9.7 Computational Approaches in Antibacterial Drug Discovery

The battle against bacterial pathogens that cause foodborne illnesses has entered a crucial phase, largely due to the rapid emergence of antibiotic-resistant strains (Ali et al. 2020; Rahman et al. 2018; Zenu and Bekele 2024). Traditional methods of drug discovery are increasingly proving inadequate in keeping pace with the evolving threat. As a result, there is a growing emphasis on computational approaches, which offer a more efficient and precise method for identifying and developing new antibacterial agents (Geronikaki. 2022; Mohamed et al. 2022; Ortmayr et al. 2022; Straniero et al. 2020). These approaches integrate advanced algorithms, data analysis, and molecular biology techniques to expedite and enhance the drug discovery process. Figure 9.4 depicts different computational strategies in drug discovery.

9.7.1 Molecular Modeling and Rational Drug Design

Molecular modeling and rational drug design represent cornerstone methodologies in computational pharmacology (Ramírez 2016). These techniques involve the use of computer-assisted design tools to conceptualize and optimize molecules that can interact with specific biological targets, such as proteins essential to bacterial survival. By understanding the structural and functional landscape of these targets, researchers can devise potent inhibitors that are tailor-made to block bacterial pathways or deactivate crucial enzymes (Adelusi et al. 2022; Darby et al. 2023; Reeve et al. 2015; Stogios and Savchenko 2020). Molecular modeling can be performed using homology modeling. Homology modeling, also known as comparative modeling, is an indispensable computational technique in drug design used when the three-dimensional structure of a target protein is unavailable through experimental

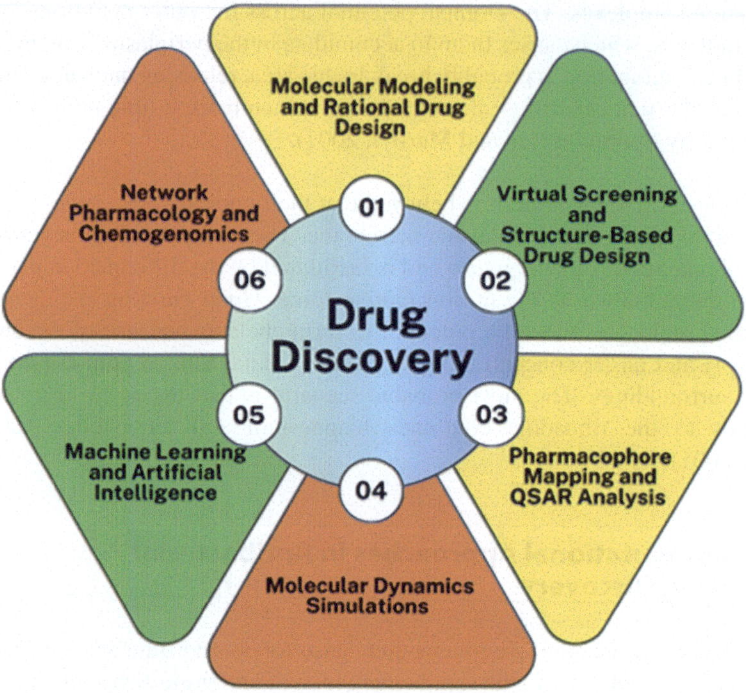

Fig. 9.4 Computational strategies in drug discovery

methods like X-ray crystallography or NMR spectroscopy (Jalily Hasani and Barakat 2017). Homology modeling is crucial in drug discovery, providing insights into the structural characterization of drug targets, facilitating virtual screening, and enabling the rational design of new drugs by showing how compounds/ligands interact at the molecular level (Hameduh et al. 2020). For example, a recent study reported cinnamic aldehyde and 1, 2-epoxycyclododecane as lead molecule against drug-resistant *Listeria monocytogenes* by modeling and targeting the enzymes of a secretion (Sec) pathway (Skariyachan et al. 2015). Thus, homology modeling not only enhances the efficacy of potential drugs but also minimizes off-target effects, thereby streamlining the drug development process (Muhammed and Aki-Yalcin 2019).

9.7.2 Virtual Screening and Molecular Docking

Virtual screening and molecular docking are critical computational techniques in the discovery of new drugs, particularly in identifying compounds that could potentially have therapeutic effects against specific targets, such as bacterial enzymes or receptors (Hossain et al. 2015). These methods allow researchers to efficiently screen vast libraries of compounds, dramatically accelerating the early stages of drug discovery and reducing the reliance on more time-consuming and costly experimental assays (Alam et al. 2022). Virtual screening involves the automated evaluation of large collections of chemical compounds using computer programs to identify those that are most likely to bind to a drug target, typically a protein involved in disease progression (Li et al. 2023). Virtual screening has been reported to identify natural and drug like compounds specific to pathogens responsible for food borne diseases (Nizami et al. 2020; Skariyachan et al. 2015).

Molecular docking simulates the binding process between a drug molecule and a target protein, predicting how the molecular structures will align with each other, the kind of interactions they will form, and the strength and stability of these interactions. Molecular docking provides a "score" that estimates the affinity and binding stability between the compound and the target, aiding in the identification of the most promising candidates for further testing. One of the major advantages of molecular docking and virtual screening is their ability to explore thousands to millions of compounds in a relatively less time, which is a huge enhancement over traditional methods (Chauhan et al. 2023; Nizami et al. 2020; Shukla and Singh 2019). By prioritizing compounds that are more likely to succeed in real-world tests, these techniques can significantly reduce the resources required for later drug development stages (Chauhan et al. 2023). These techniques not only streamline the drug discovery process but also enhance the precision with which researchers can predict and molecular interactions, leading to more effective and targeted therapies. Recent studies identified a few natural compounds, such as amentoflavone, cardamom, baicalein, and barbatimão, as well as extracts of *A. altilis*, *Chlorella* spp., and *Alhagi graecorum*, as potential leads in the fight against foodborne diseases (Ahmad et al. 2020; Gupta et al. 2023; Mishra et al. 2023; Tingting et al. 2022).

9.7.3 Pharmacophore Mapping and QSAR Analysis

Pharmacophore mapping and QSAR analysis are sophisticated computational techniques that play a pivotal role in the field of pharmacology, particularly in the design and development of antibacterial agents (Tyagi et al. 2021). These methods are instrumental in identifying and optimizing compounds that can effectively target and inhibit bacterial pathogens, addressing the urgent need for new treatments in the face of rising antibiotic resistance. Pharmacophore mapping is a process used to define the essential features necessary for a molecule to bind to a specific biological target and exert its therapeutic effect (Meshram et al. 2017). A pharmacophore model represents a spatial arrangement of these critical chemical features such as hydrogen bond acceptors, hydrogen bond donors, aromatic rings, and hydrophobic regions that are crucial for molecular recognition and interaction with a biological receptor. This model does not represent a real molecule or a real group of atoms; rather, it is an idealized combination of the features necessary to ensure biological activity (Meshram et al. 2017; Tyagi et al. 2021). QSAR analysis extends the concept of pharmacophore mapping by employing statistical methods to correlate the chemical properties or molecular descriptors of compounds with their observed biological activities (Dudek et al. 2006). These descriptors may include structural, physical, chemical, and even quantum mechanical properties (Winkler 2002). QSAR models allow the prediction of the biological activity of new compounds based on their molecular descriptors (Muhammad et al. 2018; Winkler 2002). By establishing quantitative relationships between the chemical structure and biological activity, QSAR models can predict how modifications to chemical structures might impact the effectiveness of a compound, enabling the optimization of lead compounds into safer and more effective drugs (Kwon et al. 2019).

In the context of antibacterial drug discovery, pharmacophore models and QSAR analyses have been crucial in the identification and optimization of novel inhibitors targeting key bacterial enzymes essential for pathogen survival and virulence (Badaoui et al. 2023; Mkhayar et al. 2023; Thakur et al. 2006; Wang et al. 2018). One notable application has been in the development of novel beta-lactamase inhibitors. Beta-lactamases are enzymes produced by bacteria that confer resistance to beta-lactam antibiotics like penicillin and cephalosporins. Using pharmacophore mapping, scientists have identified key features that are critical for binding to the active site of various beta-lactamase enzymes (Chen et al. 2013; Verma and Hansch 2008; Wang et al. 2020). By providing a deeper understanding of the molecular basis of drug-receptor interactions, pharmacophore mapping and QSAR analysis contribute to targeted and effective development of antibacterial therapies, promising a brighter future in the battle against bacterial infections (Badaoui et al. 2023).

9.7.4 Molecular Dynamics Simulations

MDS is a fundamental tool in the field of drug discovery and optimization, providing deep insights into the behavior of ligand/substrate molecules within biological systems (Ahmed et al. 2023). These simulations involve computationally solving the equations of motion for atoms and molecules, enabling researchers to observe the dynamic evolution of these entities in a simulated biological environment. This approach allows for a detailed understanding of the interactions between natural molecules and their target sites on a molecular level, which is critical for the development of effective and stable drugs (AlRawashdeh and Barakat 2024). By visualizing how natural molecules interact with their biological targets, molecular dynamics simulations help researchers predict the stability and efficacy of potential candidates under different physiological conditions (Liu et al. 2018; Nizami et al. 2020; Singh et al. 2023a, b; Sivasankar et al. 2020; Trevisan et al. 2020; Xu et al. 2022). These simulations mimic the natural environment of the body, accounting for factors such as temperature, pressure, and the presence of various biomolecules. This enables the simulation of compound behavior in conditions that closely resemble those in human cells, providing predictive insights that are crucial for drug formulation and development. These simulations are critical for understanding how compound interact within various biological environments, thereby guiding decisions on dosage and delivery methods (Choi et al. 2018). This includes predicting potential interactions with biomolecules and other compounds, which is crucial for minimizing adverse drug reactions and enhancing patient safety (Kopaczynska et al. 2016). Furthermore, simulations can show how altering a side chain on a drug molecule improves its ability to fit into the binding pocket of a target protein, thereby increasing efficacy and reducing the likelihood of binding to unintended targets (Ahmed et al. 2023; Liu et al. 2018). Optimization also extends to the improvement of drug solubility and bioavailability. By observing how drugs interact with solvents in the simulation, researchers can modify chemical groups to enhance solubility, ensuring that drugs can be absorbed more efficiently when administered (Defelipe et al. 2018). In addition to enhancing drug properties, molecular dynamics simulations assist in overcoming challenges such as drug resistance. By simulating drug interactions with mutant strains of target proteins, researchers can design drugs that retain efficacy against resistant forms of diseases, a common issue in the treatment of conditions like cancer and infectious diseases (Amusengeri et al. 2022; Chen et al. 2024). This proactive approach in drug design is crucial for staying ahead of rapidly evolving pathogens.

9.7.5 Machine Learning and Artificial Intelligence

Machine learning (ML) and AI are changing how we identify and produce natural antibacterial agents, especially for food poisoning (Deng et al. 2021; Kumari et al. 2023; Mu et al. 2024; Pandey et al. 2023a). Computers use massive volumes of data to find patterns and predictions that humans cannot. This capability revolutionizes pharmaceuticals by delivering natural approaches to fight bacterial infections. ML algorithms can search vast natural libraries for antibacterial characteristics. By analyzing vast datasets containing structural information, biological activity, and phytochemical features, ML can predict natural product potency against specific bacteria (Pandey et al. 2023a; Zisanur Rahman et al. 2022). Rapid identification and development of efficient antibacterials can prevent food poisoning, making this critical for foodborne pathogens (Deng et al. 2021). AI models can predict how natural substances interact with bacterial cells and human tissues, helping to develop safe and effective antibacterial treatments (Sahayasheela et al. 2022). AI can model the interaction dynamics between natural compound and bacterial enzymes or cell membranes to predict its efficacy and resistance (Sahayasheela et al. 2022). Deep learning, a subset of ML, helps analyze multidimensional genomes, proteomics, and metabolomics data (Nicora et al. 2020). These data points can help determine which chemicals are efficient against specific microorganisms. Deep learning can examine bacterial genetic sequences to predict natural agent susceptibility, enabling tailored foodborne pathogen therapy (Chen et al. 2023; Martorell-Marugán et al. 2019). AI in drug discovery accelerates the slow, expensive drug development cycle. As these technologies advance, their use in generating food poisoning antibacterial drugs should improve. AI could also personalize therapy for bacterial illnesses and improve food safety by tailoring medicines to genetics, lifestyle, and circumstances (Martorell-Marugán et al. 2019).

9.7.6 Network Pharmacology and Chemogenomics

Rapidly increasing network pharmacology and chemogenomics are essential for creating targeted food poisoning treatments (Adnan et al. 2023; Ho et al. 2018). These fields use systems-based approaches to investigate how natural compounds can fight foodborne infections, maximizing the discovery and development of antibacterial medicines. Network pharmacology studies biological network connections to understand how natural substances affect infections and hosts. This method is useful for determining how natural antibacterial drugs disrupt complex bacterial metabolic and pathogenicity networks (Hopkins 2008). Identifying dual insights improve food poisoning treatment without affecting the host's microbiota. Chemogenomics studies chemical compound-gene target interactions across the genome (Bredel and Jacoby 2004). This allows the screening of several natural compounds to determine their activity and identify bacterial genomic targets. Researchers can identify natural compounds that are effective against specific bacterial strains by studying how these compounds interact with bacterial genes. Targeted food poisoning treatments require precision since various bacteria require different methods. Network pharmacology and chemogenomics provide a solid framework for natural antibacterial drug discovery (Chandrasekaran 2019; Chandrasekaran et al. 2016). Since this integrated method can swiftly find promising natural substances and predict and limit adverse effects by understanding their broad interactions, it can speed up treatment development and make therapies safer (Bağcıoğlu et al. 2019; Xu et al. 2022).

9.8 Future Perspectives and Conclusion

The global socioeconomic situation has been severely impacted by the introduction of new infections and the increasing prevalence of ABR, leading to a significant rise in foodborne illness incidence. Throughout history, natural products have served as the primary means of protecting against foodborne infections due to their affordability, efficacy, and low toxicity. Moreover, the extensive range of forms and intricate biological functionalities render natural products a considerably untapped reservoir for the pharmaceutical sector. Researchers have thus far evaluated numerous plant-based chemicals to determine their effect on inhibiting microorganisms and their role in causing disease in the host organism. The main natural compounds that showed antimicrobial properties were cinnamic amides, flavonoids, terpenoids, chalcones, and tanshinones. Various traditional and experimental studies have emphasized the capacity of natural substances to inhibit foodborne microorganisms. These studies not only showcase the healing capabilities of natural substances but also emphasize the complex process of finding new drugs. Simultaneously, the incorporation of computational methods like HTS, molecular docking, and molecular dynamic simulations in drug discovery is transforming the fight against foodborne diseases. These strategies expedite the rapid identification and improvement of potential natural products by predicting their biological activity and molecular interactions.

This interdisciplinary relationship accelerates the discovery of novel natural medicinal molecules and offers innovative approaches to tackle the challenges posed by emerging pathogens and antibiotic resistance. These computational techniques have the ability to accurately predict pharmaceutical results in real time, enabling timely decision-making in the drug development process. With advancements in computer power and increasingly advanced algorithms, these models will have the capability to effectively evaluate intricate biological datasets. This will enable the discovery of a diverse array of natural products and their derivatives. It is crucial to produce new natural chemicals that can target specific bacteria responsible for foodborne diseases, and computer-aided medication design is essential in achieving this objective. Moreover, progress in computer technology will expedite the analysis of vast repositories of natural compounds with significant potential for antibacterial activities. Consequently, this will streamline the process and increase the pool of viable therapy options that may be evaluated in clinical trials. Integrating genomes and proteomics will enhance the customization of medications. By acquiring a deeper understanding of the genetic and molecular characteristics of infections and their interactions with various natural chemicals, it becomes possible to create treatments that are more effective and have fewer negative consequences. This method ensures that medications are both precise and harness the unique properties of natural chemicals, which often have complex structures that can function as potent antibacterial agents.

Furthermore, the field of drug development will gain significant benefits by utilizing collaborative and open-source platforms, which will allow professionals from all over the world to openly exchange data and experimental discoveries. The formation of this global partnership is essential for pooling resources and expediting the drug development process, specifically in the pursuit of powerful natural antibacterial agents. Open-source platforms can level the playing field by providing equitable access to drug development tools. This enables researchers with limited financial resources to participate in and benefit from global initiatives that target the fight against foodborne diseases.

Although there have been significant strides in creating antimicrobial medications derived from natural sources, there are still some challenges that remain. The challenges encompass issues pertaining to formulation, solubility, and resistance that have arisen throughout the development of antimicrobial medications. These challenges underscore the crucial necessity for the field to persist in its efforts to innovate and adapt. Ultimately, the simultaneous utilization of computational and experimental methods enhances the investigation and utilization of natural items in the advancement of antibacterial therapies. By improving these approaches and utilizing their collaborative potential, we can greatly enhance the likelihood of finding new and powerful antibacterial medications. This presents a promising opportunity for enhanced control and treatment of foodborne illnesses, especially at this critical moment in the advancement of the pharmaceutical industry.

Acknowledgments The figures are generated by Biorender.com and Authors used LLMs to increase the readability of the scientific data. All authors take full responsibility for the presented content.

References

2018–2020 National Advisory Committee on Microbiological Criteria for Foods (2022) Microbiological testing by industry of ready-to-eat foods under FDA's jurisdiction for pathogens (or appropriate indicator organisms): verification of preventive controls. J Food Prot 85(11):1646–1666. https://doi.org/10.4315/JFP-22-143

Abbas M, Saeed F, Anjum FM, Afzaal M, Tufail T, Bashir MS, Ishtiaq A, Hussain S, Suleria HAR (2017) Natural polyphenols: an overview. Int J Food Prop 20(8):1689–1699. https://doi.org/10.1080/10942912.2016.1220393

Adelusi TI, Oyedele AQK, Boyenle ID, Ogunlana AT, Adeyemi RO, Ukachi CD, Idris MO, Olaoba OT, Adedotun IO, Kolawole OE, Xiaoxing Y, Abdul-Hammed M (2022) Molecular modeling in drug discovery. Inform Med Unlocked 29. https://doi.org/10.1016/j.imu.2022.100880

Adnan M, Siddiqui AJ, Noumi E, Hannachi S, Ashraf SA, Awadelkareem AM, Snoussi M, Badraoui R, Bardakci F, Sachidanandan M, Patel M, Patel M (2023) Integrating network pharmacology approaches to decipher the multi-target pharmacological mechanism of microbial biosurfactants as novel green antimicrobials against Listeriosis. Antibiotics 12(1). https://doi.org/10.3390/antibiotics12010005

Afr S, Cam JT, Traditional AJ (2007) Review 4(3):319–337

Ahmad Z, Laughlin TF, Kady IO (2015) Thymoquinone inhibits Escherichia coli ATP synthase and cell growth. PLoS One 10(5). https://doi.org/10.1371/journal.pone.0127802

Ahmad MN, Karim NU, Normaya E, Mat Piah B, Iqbal A, Ku Bulat KH (2020) Artocarpus altilis extracts as a food-borne pathogen and oxidation inhibitors: RSM, COSMO RS, and molecular docking approaches. Sci Rep 10(1). https://doi.org/10.1038/s41598-020-66488-7

Ahmed M, Maldonado AM, Durrant JD (2023) From byte to bench to bedside: molecular dynamics simulations and drug discovery. BMC Biol 21(1). https://doi.org/10.1186/s12915-023-01791-z

Akbari R, Hakemi-Vala M, Pashaie F, Bevalian P, Hashemi A, Bagheri KP (2019) Highly synergistic effects of melittin with conventional antibiotics against multidrug-resistant isolates of acinetobacter baumannii and pseudomonas aeruginosa. Microb Drug Resist 25(2). https://doi.org/10.1089/mdr.2018.0016

Alam Z, Bappy MNI, Sultana A, Laskar FS, Miah K, Zinnah KMA, Saha S (2022) In-silico exploration of plant metabolites as potential remedies of norovirus. Advances in Virology 2022. https://doi.org/10.1155/2022/8905962

Ali A, Parisi A, Conversano MC, Iannacci A, D'Emilio F, Mercurio V, Normanno G (2020) Food-borne bacteria associated with seafoods: a brief review. J Food Qual Hazards Control 7(1). https://doi.org/10.18502/JFQHC.7.1.2446

Aljamali NM, Najim MM, Alabbasy AJ (2021) Review on food poisoning (types, causes, symptoms, diagnosis, treatment). Glob Acad J Pharm Drug Res 3(4)

AlRawashdeh S, Barakat KH (2024) Applications of molecular dynamics simulations in drug discovery. Methods Mol Biol 2714. https://doi.org/10.1007/978-1-0716-3441-7_7

Amusengeri A, Khan A, Tastan Bishop Ö (2022) The structural basis of mycobacterium tuberculosis RpoB drug-resistant clinical mutations on rifampicin drug binding. Molecules 27(3). https://doi.org/10.3390/molecules27030885

Andryukov BG, Karpenko AA, Lyapun IN (2020) Learning from nature: bacterial spores as a target for current technologies in medicine. Sovrem Tehnol Med 12(3):105–122. https://doi.org/10.17691/stm2020.12.3.13

Asmat TM, Khan A (2020) Campylobacter-review on its significance as a foodborne pathogen. Pak-Euro J Med Life Sci 3(3):131–137

Badaoui H, Koubi Y, Moukhliss Y, El Khatabi K, Elbouhi M, Ouabane M, Choukrad M, Jana MA, Sbai A, Lakhlifi T, Bouachrine M (2023) Cytotoxic activity of new thioglycosides of the arylamino-1,3,4-thiadiazole and glycosyl-1,3,4-thiadiazole-triazole conjugates against Escherichia coli: QSAR study. Turk Comp Theo Chem 7(1). https://doi.org/10.33435/tcandtc.1093987

Bağcıoğlu M, Fricker M, Johler S, Ehling-Schulz M (2019) Detection and identification of bacillus cereus, bacillus cytotoxicus, bacillus thuringiensis, bacillus mycoides and bacillus weihenstephanensis via machine learning based FTIR spectroscopy. Front Microbiol 10(APR). https://doi.org/10.3389/fmicb.2019.00902

Bai Y, Lin XH, Zhu JH, Cui SH, Guo LX, Yan SF, Wang W, Xu J, Li FQ (2021) Quantification of cross-contamination of campylobacter jejuni during food preparation in a model kitchen in China. J Food Prot 84(5). https://doi.org/10.4315/JFP-20-280

Bencardino D, Amagliani G, Brandi G (2021) Carriage of Staphylococcus aureus among food handlers: an ongoing challenge in public health. Food Control 130. https://doi.org/10.1016/j.foodcont.2021.108362

Bintsis T (2017) Foodborne pathogens. AIMS Microbiol 3(3):529–563. https://doi.org/10.3934/microbiol.2017.3.529

Bredel M, Jacoby E (2004) Chemogenomics: an emerging strategy for rapid target and drug discovery. Nat Rev Genet 5(4). https://doi.org/10.1038/nrg1317

Brudzynski K, Maldonado-Alvarez L (2015) Polyphenol-protein complexes and their consequences for the redox activity, structure and function of honey. A current view and new hypothesis – a review. Pol J Food Nutr Sci 65(2). https://doi.org/10.1515/pjfns-2015-0030

CDC (2019) Signs and {Symptoms} of {Food} {Poisoning}. *Centers for Disease Control and Prevention*

Chandrasekaran S (2019) Predicting drug interactions from chemogenomics using INDIGO. Methods Mol Biol 1888. https://doi.org/10.1007/978-1-4939-8891-4_13

Chandrasekaran S, Cokol-Cakmak M, Sahin N, Yilancioglu K, Kazan H, Collins JJ, Cokol M (2016) Chemogenomics and orthology-based design of antibiotic combination therapies. Mol Syst Biol 12(5). https://doi.org/10.15252/msb.20156777

Chauhan SS, Jamal T, Singh A, Sehrawat A, Parthasarathi R (2023) Structure-based virtual screening. In: Cheminformatics, QSAR and machine learning applications for novel drug development. https://doi.org/10.1016/B978-0-443-18638-7.00016-5

Che CT, Zhang H (2019) Plant natural products for human health. Int J Mol Sci 20(4):2–5. https://doi.org/10.3390/ijms20040830

Chen J, Yu R, Shen B, Xu Y, Liu Y, Zheng H, Yao W (2013) Docking-based 3D-QSAR modeling of the inhibitors of IMP metallo-β-lactamase. Med Chem Res 22(4). https://doi.org/10.1007/s00044-012-0172-1

Chen C, Wang J, Pan D, Wang X, Xu Y, Yan J, Wang L, Yang X, Yang M, Liu GP (2023) Applications of multi-omics analysis in human diseases. MedComm 4(4). https://doi.org/10.1002/mco2.315

Chen X, Chang Y, Ye M, Wang Z, Wu S, Duan N (2024) Rational Design of a Robust G-Quadruplex aptamer as an inhibitor to alleviate listeria monocytogenes infection. ACS Appl Mater Interfaces 16(13). https://doi.org/10.1021/acsami.4c00496

Chinnam N, Dadi PK, Sabri SA, Ahmad M, Kabir MA, Ahmad Z (2010) Dietary bioflavonoids inhibit Escherichia coli ATP synthase in a differential manner. Int J Biol Macromol 46(5). https://doi.org/10.1016/j.ijbiomac.2010.03.009

Choi S, Puligundla P, Mok C (2016) Who estimates of the global burden or foodborne diseases. Ann Microbiol 66(2)

Choi SB, Yap BK, Choong YS, Wahab H (2018) Molecular dynamics simulations in drug discovery. In: Encyclopedia of bioinformatics and computational biology: ABC of bioinformatics, vol 1–3. https://doi.org/10.1016/B978-0-12-809633-8.20154-4

Chopra B, Dhingra AK (2021) Natural products: a lead for drug discovery and development. Phytother Res 35(9):4660–4702. https://doi.org/10.1002/ptr.7099

Chunarkar-Patil P, Kaleem M, Mishra R, Ray S, Ahmad A, Verma D, Bhayye S, Dubey R, Singh HN, Kumar S (2024) Anticancer drug discovery based on natural products: from computa-

tional approaches to clinical studies. Biomedicines 12(1):1–35. https://doi.org/10.3390/biomedicines12010201
Darby EM, Trampari E, Siasat P, Gaya MS, Alav I, Webber MA, Blair JMA (2023) Molecular mechanisms of antibiotic resistance revisited. Nat Rev Microbiol 21(5). https://doi.org/10.1038/s41579-022-00820-y
Defelipe LA, Arcon JP, Modenutti CP, Marti MA, Turjanski AG, Barril X (2018) Solvents to fragments to drugs: MD applications in drug design. Molecules 23(12). https://doi.org/10.3390/molecules23123269
Deng X, Cao S, Horn AL (2021) Emerging applications of machine learning in food safety. Annu Rev Food Sci Technol 12. https://doi.org/10.1146/annurev-food-071720-024112
Dias DA, Urban S, Roessner U (2012) A historical overview of natural products in drug discovery. Meta 2(2):303–336. https://doi.org/10.3390/metabo2020303
Dietrich J, Hammerl J-A, Johne A, Kappenstein O, Loeffler C, Nöckler K, Rosner B, Spielmeyer A, Szabo I, Richter MH (2023) Impact of climate change on foodborne infections and intoxications. J Health Monit 8(Suppl 3):78–92. https://doi.org/10.25646/11403
do Prado FG, Pagnoncelli MGB, de Melo Pereira GV, Karp SG, Soccol CR (2022) Fermented soy products and their potential health benefits: a review. Microorganisms 10(8). https://doi.org/10.3390/microorganisms10081606
Dudek A, Arodz T, Galvez J (2006) Computational methods in developing quantitative structure-activity relationships (QSAR): a review. Comb Chem High Throughput Screen 9(3). https://doi.org/10.2174/138620706776055539
Dzobo K (2022) The role of natural products as sources of therapeutic agents for innovative drug discovery. Comprehen Pharmacol 2(January):408–422. https://doi.org/10.1016/B978-0-12-820472-6.00041-4
Eallonardo SJ, Wang Y, Freitag NE (2023) Listeria monocytogenes. In: Molecular medical microbiology, 3rd edn. https://doi.org/10.1016/B978-0-12-818619-0.00127-1
Ehuwa O, Jaiswal AK, Jaiswal S (2021) Salmonella, food safety and food handling practices. Food Secur 10(5). https://doi.org/10.3390/foods10050907
Eifert JD, National Advisory Committee on Microbiological Criteria for Foods (2022) Microbiological testing by industry of ready-to-eat foods under FDA's jurisdiction for pathogens (or appropriate indicator organisms): verification of preventive controls. J Food Prot 85(11). https://doi.org/10.4315/JFP-22-143
Farid N, Waheed A, Motwani S (2023) Synthetic and natural antimicrobials as a control against food borne pathogens: a review. Heliyon 9(6). https://doi.org/10.1016/j.heliyon.2023.e17021
Fiore AE (2004) Hepatitis a transmitted by food. Clin Infect Dis 38(5). https://doi.org/10.1086/381671
Fones HN, Bebber DP, Chaloner TM, Kay WT, Steinberg G, Gurr SJ (2020) Threats to global food security from emerging fungal and oomycete crop pathogens. Nat Food 1(6):332–342. https://doi.org/10.1038/s43016-020-0075-0
Gabriël S, Dorny P, Saelens G, Dermauw V (2023) Foodborne parasites and their complex life cycles challenging food safety in different food chains. Food Secur 12(1):1–27. https://doi.org/10.3390/foods12010142
Gad HA, El-Ahmady SH, Abou-Shoer MI, Al-Azizi MM (2013) Application of chemometrics in authentication of herbal medicines: a review. Phytochem Anal 24(1):1–24. https://doi.org/10.1002/pca.2378
Geronikaki. (2022) Antibacterial and antiviral agents: computational insight. Curr Top Med Chem 22(32). https://doi.org/10.2174/1568026623222122112027
Gonçalves-Tenório A, Nunes Silva B, Rodrigues V, Cadavez V, Gonzales-Barron U (2018) Prevalence of pathogens in poultry meat: a meta-analysis of European published surveys. Food Secur 7(5). https://doi.org/10.3390/foods7050069
Graziano VR, Wei J, Wilen CB (2019) Norovirus attachment and entry. Viruses 11(6). https://doi.org/10.3390/v11060495
Grout L, Baker MG, French N, Hales S (2020) A review of potential public health impacts associated with the global dairy sector. GeoHealth 4(2). https://doi.org/10.1029/2019GH000213

Gruden Š, Ulrih NP (2021) Diverse mechanisms of antimicrobial activities of lactoferrins, lactoferricins, and other lactoferrin-derived peptides. Int J Mol Sci 22(20). https://doi.org/10.3390/ijms222011264

Gupta AK, Chaudhary A (2022) Food poisoning: causes, its effects and control. INWASCON Technol Mag 4. https://doi.org/10.26480/itechmag.04.2022.59.61

Gupta N, Bhattacharya S, Urbanová K, Dutta A, Hazra AK, Fernández-Cusimamani E, Leuner O (2023) Systematic analysis of antimicrobial activity, phytochemistry, and in silico molecular interaction of selected essential oils and their formulations from different Indian spices against foodborne bacteria. Heliyon 9(12). https://doi.org/10.1016/j.heliyon.2023.e22480

Hameduh T, Haddad Y, Adam V, Heger Z (2020) Homology modeling in the time of collective and artificial intelligence. Comput Struct Biotechnol J 18. https://doi.org/10.1016/j.csbj.2020.11.007

Hardstaff JL, Clough HE, Lutje V, McIntyre KM, Harris JP, Garner P, O'Brien SJ (2018) Foodborne and food-handler norovirus outbreaks: a systematic review. Foodborne Pathog Dis 15(10):589–597. https://doi.org/10.1089/fpd.2018.2452

Havelaar AH, Cawthorne A, Angulo F, Bellinger D, Corrigan T, Cravioto A, Gibb H, Hald T, Ehiri J, Kirk M, Lake R, Praet N, Speybroeck N, de Silva N, Stein C, Torgerson P, Kuchenmüller T (2013) WHO initiative to estimate the global burden of foodborne diseases. Lancet 381. https://doi.org/10.1016/s0140-6736(13)61313-6

He H, Peng S, Song X, Jia R, Zou Y, Li L, Yin Z (2022) Protective effect of isoflavones and triterpenoid saponins from pueraria lobata on liver diseases: a review. Food Sci Nutr 10(1). https://doi.org/10.1002/fsn3.2668

Ho TT, Tran QT, Chai CL (2018) The polypharmacology of natural products. Future Med Chem 10(11). https://doi.org/10.4155/fmc-2017-0294

Hopkins AL (2008) Network pharmacology: the next paradigm in drug discovery. Nat Chem Biol 4(11). https://doi.org/10.1038/nchembio.118

Hossain MM, Roy PK, Jannatul Mosnaz ATM, Shakil SK, Hasan MM, Prodhan SH (2015) Structural analysis and molecular docking of potential ligands with chorismate synthase of listeria monocytogenes: a novel antibacterial drug target. Indian J Biochem Biophys 52(1)

Hu P, Huang P, Chen MW (2013) Curcumin reduces Streptococcus mutans biofilm formation by inhibiting sortase a activity. Arch Oral Biol 58(10). https://doi.org/10.1016/j.archoralbio.2013.05.004

Ian C, Marilyn R (2001) Tetracycline antibiotics: mode of action, applications, molecular biology, and epidemiology of bacterial resistance. Microbiol Mole. https://doi.org/10.1128/MMBR.65.2.232

Ibrahim SA, Ayivi RD, Zimmerman T, Siddiqui SA, Altemimi AB, Fidan H, Esatbeyoglu T, Bakhshayesh RV (2021) Lactic acid bacteria as antimicrobial agents: food safety and microbial food spoilage prevention. Food Secur 10(12). https://doi.org/10.3390/foods10123131

Imenshahidi M, Hosseinzadeh H (2019) Berberine and barberry (Berberis vulgaris): a clinical review. Phytother Res 33(3). https://doi.org/10.1002/ptr.6252

Jackson AL, Kulchaiyawat C, Sullivan GA, Sebranek JG, Dickson JS (2011) Use of natural ingredients to control growth of Clostridium perfringens in naturally cured frankfurters and hams. J Food Prot 74(3). https://doi.org/10.4315/0362-028X.JFP-10-365

Jalily Hasani H, Barakat K (2017) Homology modeling: an overview of fundamentals and tools. Int Rev Modell Simul 10(2). https://doi.org/10.15866/iremos.v10i2.11412

Ji S, Fattahi A, Raffel N, Hoffmann I, Beckmann MW, Dittrich R, Schrauder M (2017) Antioxidant effect of aqueous extract of four plants with therapeutic potential on gynecological diseases; semen persicae, Leonurus cardiaca, Hedyotisdiffusa, and Curcuma zedoaria. Eur J Med Res 22(1):1–8. https://doi.org/10.1186/s40001-017-0293-6

Johnson BJ, Delehanty JB, Lin B, Ligler FS (2008) Immobilized proanthocyanidins for the capture of bacterial lipopolysaccharides. Anal Chem 80(6). https://doi.org/10.1021/ac7024128

Kang SS, Kim JG, Lee TH, Oh KB (2006) Flavonols inhibit sortases and sortase-mediated Staphylococcus aureus clumping to fibrinogen. Biol Pharm Bull 29(8). https://doi.org/10.1248/bpb.29.1751

Katrachanca SM, Koleske AJ (2017) 乳鼠心肌提取 HHS public access. Physiol Behav 176(5):139–148. https://doi.org/10.1038/nchembio.180.Quantifying

Katz L, Baltz RH (2016) Natural product discovery: past, present, and future. J Ind Microbiol Biotechnol 43(2–3):155–176. https://doi.org/10.1007/s10295-015-1723-5

Kitichalermkiat A, Kurahachi M, Nonaka A, Nakayama M, Shimatani K, Shigemune N, Tsugukuni T, Hitomi J, Sato J, Sonoda T, Masuda Y, Honjoh K et al (2019) Effects of epigallocatechin Gallate on viability and cellular proteins of Staphylococcus aureus. Food Sci Technol Res 25(2). https://doi.org/10.3136/fstr.25.277

Koch MA, Schuffenhauer A, Scheck M, Wetzel S, Casaulta M, Odermatt A, Ertl P, Weldmann H (2005) Charting biologically relevant chemical space: a structural classification of natural products (SCONP). Proc Natl Acad Sci USA 102(48):17272–17277. https://doi.org/10.1073/pnas.0503647102

Kopaczynska M, Schulz A, Fraczkowska K, Kraszewski S, Podbielska H, Fuhrhop JH (2016) Selective condensation of DNA by aminoglycoside antibiotics. Eur Biophys J 45(4). https://doi.org/10.1007/s00249-015-1095-9

Krishnamurti C, Rao SSCC (2016) The isolation of morphine by serturner. Indian J Anaesth 60(11):861–862. https://doi.org/10.4103/0019-5049.193696

Kumar A (2020) Food poisoning: causes, precautions, diagnosis and treatment: a brief review. World J Biol Biotechnol 5(1):33. https://doi.org/10.33865/wjb.005.01.0287

Kumar S, Pandey AK (2013) Chemistry and biological activities of flavonoids: an overview. Sci World J 2013. https://doi.org/10.1155/2013/162750

Kumari S, Venkatesh VG, Tan FTC, Bharathi SV, Ramasubramanian M, Shi Y (2023) Application of machine learning and artificial intelligence on agriculture supply chain: a comprehensive review and future research directions. Ann Oper Res. https://doi.org/10.1007/s10479-023-05556-3

Kwon HJ, Kim HH, Ryu YB, Kim JH, Jeong HJ, Lee SW, Chang JS, Cho KO, Rho MC, Park SJ, Lee WS (2010) In vitro anti-rotavirus activity of polyphenol compounds isolated from the roots of Glycyrrhiza uralensis. Bioorg Med Chem 18(21):7668–7674. https://doi.org/10.1016/j.bmc.2010.07.073

Kwon S, Bae H, Jo J, Yoon S (2019) Comprehensive ensemble in QSAR prediction for drug discovery. BMC Bioinform 20(1). https://doi.org/10.1186/s12859-019-3135-4

Li Z, Zhang S, Meng W, Zhang J, Zhang D (2023) Screening and activity analysis of α-glucosidase inhibitory peptides derived from Coix seed Prolamins using bioinformatics and molecular docking. Food Secur 12(21). https://doi.org/10.3390/foods12213970

Liu Y, Wang P, Chen F, Yuan Y, Zhu Y, Yan H, Hu X (2015) Role of plant polyphenols in acrylamide formation and elimination. Food Chem 186:46–53. https://doi.org/10.1016/j.foodchem.2015.03.122

Liu X, Shi D, Zhou S, Liu H, Liu H, Yao X (2018) Expert opinion on drug discovery molecular dynamics simulations and novel drug discovery molecular dynamics simulations and novel drug discovery. Expert Opin Drug Discov 13(1)

Lopes GV, Ramires T, Kleinubing NR, Scheik LK, Fiorentini ÂM, Padilha da Silva W (2021) Virulence factors of foodborne pathogen campylobacter jejuni. Microb Pathog 161(April). https://doi.org/10.1016/j.micpath.2021.105265

Lou Z, Wang H, Rao S, Sun J, Ma C, Li J (2012) P-Coumaric acid kills bacteria through dual damage mechanisms. Food Control 25(2). https://doi.org/10.1016/j.foodcont.2011.11.022

Low YL, Hareeff MM (2022) W074 prevalence of acute gastroenteritis (age) pathogens infection among pre-school children in Malaysia. ClinicaChimica Acta 530. https://doi.org/10.1016/j.cca.2022.04.812

Marchese A, Barbieri R, Sanches-Silva A, Daglia M, Nabavi SF, Jafari NJ, Izadi M, Ajami M, Nabavi SM (2016) Antifungal and antibacterial activities of allicin: a review. Trends Food Sci Technol 52. https://doi.org/10.1016/j.tifs.2016.03.010

Martorell-Marugán J, Tabik S, Benhammou Y, del Val C, Zwir I, Herrera F, Carmona-Sáez P (2019) Deep learning in omics data analysis and precision medicine. Computat Biol. https://doi.org/10.15586/computationalbiology.2019.ch3

Mead PS, Slutsker L, Dietz V, McCaig LF, Bresee JS, Shapiro C, Griffin PM, Tauxe RV (2000) Food-related illness and death in the United States. J Environ Health 62(7):9–18
Mehlhorn H (2016) Food-borne disease burden epidemiology reference group. In: Encyclopedia of parasitology, pp 1068–1069. https://doi.org/10.1007/978-3-662-43978-4_3884
Meshram RJ, Baladhye VB, Gacche RN, Karale BK, Gaikar RB (2017) Pharmacophore mapping approach for drug target identification: a chemical synthesis and in silico study on novel thiadiazole compounds. J Clin Diagn Res 11(5). https://doi.org/10.7860/JCDR/2017/22761.9925
Metwaly AM, Ghoneim MM, Eissa IH, Elsehemy IA, Mostafa AE, Hegazy MM, Afifi WM, Dou D (2021) Traditional ancient Egyptian medicine: a review. Saudi J Biol Sci 28(10):5823–5832. https://doi.org/10.1016/j.sjbs.2021.06.044
Mishra T, Kidie E, Vuppu S (2023) Microalgae-based disinfectant formulation for aseptic processing of Ethiopian ingredient-sourced functional bread and its molecular docking analysis to reduce hypernatremia. Mol Biotechnol. https://doi.org/10.1007/s12033-023-00970-1
Mkhayar K, Haloui R, Daoui O, Elkhattabi K, Chtita S, Elkhattabi S (2023) In silico studies of 2-aryloxy-1,4- naphthoquinone derivatives as antibacterial agents against Escherichia coli using 3D-QSAR, ADMET properties, molecular docking, and molecular dynamics. Chem Collect 47. https://doi.org/10.1016/j.cdc.2023.101060
Mohamed AM, Abbas WA, Khedr GE, Abass W, Allam NK (2022) Computational and experimental elucidation of the boosted stability and antibacterial activity of ZIF-67 upon optimized encapsulation with polyoxometalates. Sci Rep 12(1). https://doi.org/10.1038/s41598-022-20392-4
Mu W, Kleter GA, Bouzembrak Y, Dupouy E, Frewer LJ, Radwan Al Natour FN, Marvin HJP (2024) Making food systems more resilient to food safety risks by including artificial intelligence, big data, and internet of things into food safety early warning and emerging risk identification tools. Compr Rev Food Sci Food Saf 23(1). https://doi.org/10.1111/1541-4337.13296
Muhammad U, Uzairu A, Ebuka Arthur D (2018) Review on: quantitative structure activity relationship (QSAR) modeling. J Anal Pharm Res 7(2). https://doi.org/10.15406/japlr.2018.07.00232
Muhammed MT, Aki-Yalcin E (2019) Homology modeling in drug discovery: overview, current applications, and future perspectives. Chem Biol Drug Des 93(1). https://doi.org/10.1111/cbdd.13388
Naqvi SAR, Nadeem S, Komal S, Naqvi SAA, Mubarik MS, Qureshi SY, Ahmad S, Abbas A, Zahid M, Raza SS, Aslam N (2019) Antioxidants: natural antibiotics. Antioxidants. https://doi.org/10.5772/intechopen.84864
Nasim N, Sandeep IS, Mohanty S (2022) Plant-derived natural products for drug discovery: current approaches and prospects. Nucleus (India) 65(3):399–411. https://doi.org/10.1007/s13237-022-00405-3
Newman DJ, Cragg GM (2016) Natural products as sources of new drugs from 1981 to 2014. J Nat Prod 79(3):629–661. https://doi.org/10.1021/acs.jnatprod.5b01055
Nicora G, Vitali F, Dagliati A, Geifman N, Bellazzi R (2020) Integrated multi-omics analyses in oncology: a review of machine learning methods and tools. Front Oncol 10. https://doi.org/10.3389/fonc.2020.01030
Nizami B, Tan W, Arias-Moreno X (2020) In silico identification of novel PrfA inhibitors to fight listeriosis: a virtual screening and molecular dynamics studies. J Mol Graph Model 101. https://doi.org/10.1016/j.jmgm.2020.107728
Oblak M, Kotnik M, Solmajer T (2007) Discovery and development of ATPase inhibitors of DNA gyrase as antibacterial agents. Curr Med Chem 14(19). https://doi.org/10.2174/092986707781368414
Ortmayr K, de la Cruz Moreno R, Zampieri M (2022) Expanding the search for small-molecule antibacterials by multidimensional profiling. Nat Chem Biol 18(6). https://doi.org/10.1038/s41589-022-01040-4
Osek J, Lachtara B, Wieczorek K (2022) Listeria monocytogenes – how this pathogen survives in food-production environments? Front Microbiol 13(April):1–21. https://doi.org/10.3389/fmicb.2022.866462

Otto M (2014) Staphylococcus aureus toxins. Curr Opin Microbiol 17(1):32–37. https://doi.org/10.1016/j.mib.2013.11.004

Pandey VK, Srivastava S, Dash KK, Singh R, Mukarram SA, Kovács B, Harsányi E (2023a) Machine learning algorithms and fundamentals as emerging safety tools in preservation of fruits and vegetables: a review. PRO 11(6). https://doi.org/10.3390/pr11061720

Pandey VK, Tripathi A, Srivastava S, Dar AH, Singh R, Farooqui A, Pandey S (2023b) Exploiting the bioactive properties of essential oils and their potential applications in food industry. Food Sci Biotechnol 32(7). https://doi.org/10.1007/s10068-023-01287-0

Parker N, Schneegurt M, Tu A-HT, Forster BM (2018) Micobiology. Angew Chem Int Ed 6(11):951–952

Patridge E, Gareiss P, Kinch MS, Hoyer D (2016) An analysis of FDA-approved drugs: natural products and their derivatives. Drug Discov Today 21(2). https://doi.org/10.1016/j.drudis.2015.01.009

Plaper A, Golob M, Hafner I, Oblak M, Šolmajer T, Jerala R (2003) Characterization of quercetin binding site on DNA gyrase. Biochem Biophys Res Commun 306(2). https://doi.org/10.1016/S0006-291X(03)01006-4

Plumed-Ferrer C, Väkeväinen K, Komulainen H, Rautiainen M, Smeds A, Raitanen JE, Eklund P, Willför S, Alakomi HL, Saarela M, Von Wright A (2013) The antimicrobial effects of wood-associated polyphenols on food pathogens and spoilage organisms. Int J Food Microbiol 164(1):99–107. https://doi.org/10.1016/j.ijfoodmicro.2013.04.001

Porter JDH, Gaffney C, Heymann D, Parkin W (1990) Food-borne outbreak of Giardia lamblia. Am J Public Health 80(10). https://doi.org/10.2105/AJPH.80.10.1259

Quazi S (2022) Artificial intelligence and machine learning in precision and genomic medicine. Med Oncol 39(8):1–18. https://doi.org/10.1007/s12032-022-01711-1

Rafiq K, Islam MR, Siddiky NA, Samad MA, Chowdhury S, Hossain KMM, Rume FI, Hossain MK, Mahbub-E-Elahi ATM, Ali MZ, Rahman M, Amin MR, Masuduzzaman M, Ahmed S, Ara Rumi N, Hossain MT (2022) Antimicrobial resistance profile of common foodborne pathogens recovered from livestock and poultry in Bangladesh. Antibiotics 11(11). https://doi.org/10.3390/antibiotics11111551

Rahman MA, Rahman AKMA, Islam MA, Alam MM (2018) Antimicrobial resistance of Escherichia coli isolated from milk, beef and chicken meat in Bangladesh. Bangladesh J Vet Med 15(2). https://doi.org/10.3329/bjvm.v15i2.35525

Ramírez D (2016) Computational methods applied to rational drug design. Open Med Chem J 10(1). https://doi.org/10.2174/1874104501610010007

Ramos S, Silva V, Dapkevicius MDLE, Caniça M, Tejedor-Junco MT, Igrejas G, Poeta P (2020) Escherichia coli as commensal and pathogenic bacteria among food-producing animals: health implications of extended spectrum β-lactamase (ESBL) production. Animals 10(12):2239

Ranveer SA, Dasriya V, Ahmad MF, Dhillon HS, Samtiya M, Shama E, Anand T, Dhewa T, Chaudhary V, Chaudhary P, Behare P, Ram C, Puniya DV, Khedkar GD, Raposo A, Han H, Puniya AK (2024) Positive and negative aspects of bacteriophages and their immense role in the food chain. NPJ Sci Food 8(1). https://doi.org/10.1038/s41538-023-00245-8

Rawson AM, Dempster AW, Humphreys CM, Minton NP (2023) Pathogenicity and virulence of clostridium botulinum. Virulence 14(1):1–28. https://doi.org/10.1080/21505594.2023.2205251

Reeve SM, Lombardo MN, Anderson AC (2015) Understanding the structural mechanisms of antibiotic resistance sets the platform for new discovery. Future Microbiol 10(11). https://doi.org/10.2217/fmb.15.78

Rodrigues T, Reker D, Schneider P, Schneider G (2016) Counting on natural products for drug design. Nat Chem 8(6):531–541. https://doi.org/10.1038/nchem.2479

Sadybekov AV, Katritch V (2023) Computational approaches streamlining drug discovery. Nature 616(7958):673–685. https://doi.org/10.1038/s41586-023-05905-z

Sahayasheela VJ, Lankadasari MB, Dan VM, Dastager SG, Pandian GN, Sugiyama H (2022) Artificial intelligence in microbial natural product drug discovery: current and emerging role. Nat Prod Rep 39(12). https://doi.org/10.1039/d2np00035k

Sánchez-González L, Vargas M, González-Martínez C, Chiralt A, Cháfer M (2011) Use of essential oils in bioactive edible coatings: a review. Food Eng Rev 3(1). https://doi.org/10.1007/s12393-010-9031-3
Sapp AC, Nane GF, Amaya MP, Niyonzima E, Hategekimana JP, VanSickle JJ, Gordon RM, Havelaar AH (2023) Estimates of disease burden caused by foodborne pathogens in contaminated dairy products in Rwanda. BMC Public Health 23(1). https://doi.org/10.1186/s12889-023-15204-x
Saravanan S, Parimelazhagan T (2014) In vitro antioxidant, antimicrobial and anti-diabetic properties of polyphenols of Passiflora ligularis Juss. Fruit pulp. Food Sci Human Wellness 3(2):56–64. https://doi.org/10.1016/j.fshw.2014.05.001
Sarno E, Pezzutto D, Rossi M, Liebana E, Rizzi V (2021) A review of significant European foodborne outbreaks in the last decade. J Food Prot 84(12). https://doi.org/10.4315/JFP-21-096
Sattar SA, Tetro J, Bidawid S, Farber J (2000) Foodborne spread of hepatitis a: recent studies on virus survival, transfer and inactivation. Can J Infect Dis 11(3):159–163. https://doi.org/10.1155/2000/805156
Scallan E, Hoekstra RM, Angulo FJ, Tauxe RV, Widdowson MA, Roy SL, Jones JL, Griffin PM (2011) Foodborne illness acquired in the United States-major pathogens. Emerg Infect Dis 17(1):7–15. https://doi.org/10.3201/eid1701.P11101
Schmidt RH, Goodrich RM, Archer DL, Schneider KR (1969) General overview of the causative agents of foodborne illness. Edis 2003(6). https://doi.org/10.32473/edis-fs099-2003
Scott TA, Piel J (2019) The hidden enzymology of bacterial natural product biosynthesis. Nat Rev Chem 3(7):404–425. https://doi.org/10.1038/s41570-019-0107-1
Sharif MK, Javed K, Nasir A (2018) Foodborne illness: threats and control. In: Foodborne diseases. Elsevier Inc. https://doi.org/10.1016/B978-0-12-811444-5.00015-4
Sharma A, Lysenko A, Jia S, Boroevich KA, Tsunoda T (2024) Advances in AI and machine learning for predictive medicine. J Hum Genet:1–11. https://doi.org/10.1038/s10038-024-01231-y
Sheldon RA, Woodley JM (2018) Role of biocatalysis in sustainable chemistry. Chem Rev 118(2):801–838. https://doi.org/10.1021/acs.chemrev.7b00203
Shukla R, Singh TR (2019) Virtual screening, pharmacokinetics, molecular dynamics and binding free energy analysis for small natural molecules against cyclin-dependent kinase 5 for Alzheimer's disease. J Biomol Struct Dyn. https://doi.org/10.1080/07391102.2019.1571947
Silva F, Domingues FC (2017) Antimicrobial activity of coriander oil and its effectiveness as food preservative. Crit Rev Food Sci Nutr 57(1):35–47. https://doi.org/10.1080/10408398.2013.847818
Singh M, Lo SH, Dubey R, Kumar S, Chaubey KK, Kumar S (2023a) Plant-derived natural compounds as an emerging antiviral in combating COVID-19. Indian J Microbiol 63(4):429–446. https://doi.org/10.1007/s12088-023-01121-5
Singh PP, Jaiswal AK, Raghuvanshi TS, Prakash B (2023b) Insights into the antimicrobial efficacy of Coleus aromaticus essential oil against food-borne microbes: biochemical and molecular simulation approaches. Food Chem Toxicol 182. https://doi.org/10.1016/j.fct.2023.114111
Sivasankar C, Jha NK, Singh SR, Murali A, Shetty PH (2020) Molecular evaluation of quorum quenching potential of vanillic acid against yersinia enterocolitica through transcriptomic and in silico analysis. J Med Microbiol 69(11). https://doi.org/10.1099/jmm.0.001261
Skariyachan S, Pachiappan A, Joy J, Bhaduri R, Aier I, Vasist KS (2015) Investigating the therapeutic potential of herbal leads against drug resistant listeria monocytogenes by computational virtual screening and in vitro assays. J Biomol Struct Dyn 33(12). https://doi.org/10.1080/07391102.2015.1004110
Stan D, Enciu AM, Mateescu AL, Ion AC, Brezeanu AC, Stan D, Tanase C (2021) Natural compounds with antimicrobial and antiviral effect and Nanocarriers used for their transportation. Front Pharmacol 12(September):1–25. https://doi.org/10.3389/fphar.2021.723233
Stogios PJ, Savchenko A (2020) Molecular mechanisms of vancomycin resistance. Protein Sci 29(3). https://doi.org/10.1002/pro.3819
Straniero V, Sebastián-Pérez V, Hrast M, Zanotto C, Casiraghi A, Suigo L, Zdovc I, Radaelli A, De Giuli Morghen C, Valoti E (2020) Benzodioxane-Benzamides as antibacterial agents: com-

putational and SAR studies to evaluate the influence of the 7-substitution in FtsZ interaction. ChemMedChem 15(2). https://doi.org/10.1002/cmdc.201900537

Sun L, Wang Y, Miao M (2020) Inhibition of α-amylase by polyphenolic compounds: substrate digestion, binding interactions and nutritional intervention. Trends Food Sci Technol 104. https://doi.org/10.1016/j.tifs.2020.08.003

Thakur A, Thakur M, Khadikarc PV (2006) QSAR study on inhibition of E. Coli by sulfonamides. ARKIVOC 2006(14). https://doi.org/10.3998/ark.5550190.0007.e12

Tingting W, Tianqi F, Xinyu W, Can Z, Xue S, Xuming D, Jianfeng W (2022) Amentoflavone attenuates listeria monocytogenes pathogenicity through an LLO-dependent mechanism. Br J Pharmacol 179(14). https://doi.org/10.1111/bph.15827

Tirloni E, Stella S, Celandroni F, Mazzantini D, Bernardi C, Ghelardi E (2022) Bacillus cereus in dairy products and production plants. Food Secur 11(17). https://doi.org/10.3390/foods11172572

Todd ECD (2014) Foodborne diseases: overview of biological hazards and foodborne diseases. Encyclopedia of Food Safety 1(January):221–242. https://doi.org/10.1016/B978-0-12-378612-8.00071-8

Trevisan DAC, da Silva PV, Farias ABP, Campanerut-Sá PAZ, Ribeiro TDVR, Faria DR, de Mendonça PSB, de Mello JCP, Seixas FAV, Mikcha JMG (2020) Antibacterial activity of Barbatimão (Stryphnodendronadstringens) against Staphylococcus aureus: in vitro and in silico studies. Lett Appl Microbiol 71(3). https://doi.org/10.1111/lam.13317

Tsuchiya H (2001) Stereospecificity in membrane effects of catechins. Chem Biol Interact 134(1). https://doi.org/10.1016/S0009-2797(00)00308-2

Tyagi R, Singh A, Chaudhary KK, Yadav MK (2021) Pharmacophore modeling and its applications. In: Bioinformatics: methods and applications. https://doi.org/10.1016/B978-0-323-89775-4.00009-2

Vaillant V, De Valk H, Baron E, Ancelle T, Colin P, Delmas MC, Dufour B, Pouillot R, Le Strat Y, Weinbreck P, Jougla E, Desenclos JC (2005) Foodborne infections in France. Foodborne Pathog Dis 2(3). https://doi.org/10.1089/fpd.2005.2.221

Verma RP, Hansch C (2008) Combating the threat of anthrax: a quantitative structure – activity relationship approach. Mol Pharm 5(5). https://doi.org/10.1021/mp8000149

Wang D, Shi J, Xiong Y, Hu J, Lin Z, Qiu Y, Cheng J (2018) A QSAR-based mechanistic study on the combined toxicity of antibiotics and quorum sensing inhibitors against Escherichia coli. J Hazard Mater 341. https://doi.org/10.1016/j.jhazmat.2017.07.059

Wang X, Gao Y, Yu Y, Yang Y, Wang G, Sun L, Niu X (2020) Design of dipicolinic acid derivatives as New Delhi metallo-β-lactamase-1 inhibitors using a combined computational approach. J Biomol Struct Dyn 38(11). https://doi.org/10.1080/07391102.2019.1663262

Wang R, Zhang Y, Jia Y, Zhang M, Huang Y, Li C, Li K (2021a) Persimmon oligomeric Proanthocyanidins exert antibacterial activity through damaging the cell membrane and disrupting the energy metabolism of Staphylococcus aureus. ACS Food Science and Technology 1(1). https://doi.org/10.1021/acsfoodscitech.0c00021

Wang W, Ye Z, Gao H, Ouyang D (2021b) Computational pharmaceutics – a new paradigm of drug delivery. J Control Release 338(March):119–136. https://doi.org/10.1016/j.jconrel.2021.08.030

Watson M, Liu JW, Ollis D (2007) Directed evolution of trimethoprim resistance in Escherichia coli. FEBS J 274(10). https://doi.org/10.1111/j.1742-4658.2007.05801.x

Winkler DA (2002) The role of quantitative structure--activity relationships (QSAR) in biomolecular discovery. Brief Bioinform 3(1). https://doi.org/10.1093/bib/3.1.73

Wu-Wu JWF, Guadamuz-Mayorga C, Oviedo-Cerdas D, Zamora WJ (2023) Antibiotic resistance and food safety: perspectives on new technologies and molecules for microbial control in the food industry. Antibiotics 12(3). https://doi.org/10.3390/antibiotics12030550

Xu L, Zhou Y, Xu J, Xu X, Lu G, Lv Q, Wei L, Deng X, Shen X, Feng H, Wang J (2022) Anti-inflammatory, antioxidant and anti-virulence roles of atractylodin in attenuating listeria monocytogenes infection. Front Immunol 13. https://doi.org/10.3389/fimmu.2022.977051

Yang NJ, Hinner MJ (2015) Getting across the cell membrane: an overview for small molecules, peptides, and proteins. In: Methods in molecular biology, vol 1266, Clifton. https://doi.org/10.1007/978-1-4939-2272-7_3

Zenu F, Bekele T (2024) Major food-borne zoonotic bacterial pathogens of livestock origin: review. Foods and Raw Materials 12(1). https://doi.org/10.21603/2308-4057-2024-1-595

Zhang H, Tsao R (2016) Dietary polyphenols, oxidative stress and antioxidant and anti-inflammatory effects. Curr Opin Food Sci 8:33–42. https://doi.org/10.1016/j.cofs.2016.02.002

Zhang XF, Dai YC, Zhong W, Tan M, Lv ZP, Zhou YC, Jiang X (2012) Tannic acid inhibited norovirus binding to HBGA receptors, a study of 50 Chinese medicinal herbs. Bioorganic and Medicinal Chemistry 20(4):1616–1623. https://doi.org/10.1016/j.bmc.2011.11.040

Zhang K, Fu Y, Li J, Zhang L (2022) Public health and ecological significance of rodents in cryptosporidium infections. One Health 14. https://doi.org/10.1016/j.onehlt.2021.100364

Zhou Y, Wang D, Yan W (2023) Treatment effects of natural products on inflammatory bowel disease in vivo and their mechanisms: based on animal experiments. Nutrients 15(4). https://doi.org/10.3390/nu15041031

Zisanur Rahman ASM, Liu C, Sturm H, Hogan AM, Davis R, Hu P, Cardona ST (2022) A machine learning model trained on a high-throughput antibacterial screen increases the hit rate of drug discovery. PLoS Comput Biol 18(10). https://doi.org/10.1371/journal.pcbi.1010613

Multidrugs Resistance: Concepts, Mechanisms, and Recent Advances

10

Diptikanta Acharya, Sagarika Satapathy, Sandhyarani Patra, Goutam Jha, Somanath Sahoo, and T. Gayatri

Abstract

In the current scenario, bacterial infections are most common and affect different organs of humans. Antibiotics have been used to inhibit bacterial infections to save many lives. The extensive use of antibiotics can cause antibiotic resistance. Recently, extensive research and approaches have been developed to combat the antibiotic resistance of bacteria. There are many drug delivery strategies and materials having antimicrobial potentials that are now emerging methodologies. The nano-based targeted drug delivery of antibiotics gives a big scope to decrease the resistance of bacteria. This chapter emphasizes the recent progress and advancement of different approaches to inhibit multidrug-resistant bacteria and recapitulate the modern antimicrobial materials and drug delivery systems.

Keywords

Bacterial infections · Antibiotics · Drug resistance · Nanomaterial · Drug delivery

Abbreviations

AgNPs	Silver nanoparticles
CNTs	Carbon nanotubes

D. Acharya (✉) · S. Satapathy · S. Patra · G. Jha · S. Sahoo
Department of Biotechnology, School of Sciences, GIET University, Gunupur, Odisha, India
e-mail: dkacharya@giet.edu

T. Gayatri
Department of Biotechnology, School of Sciences, OP Jindal University, Raigarh, Chhattisgarh, India

T. Sharma et al. (eds.), *Natural Products for Antibacterial Drug Development: Recent Advancement of Computational Approach*,
https://doi.org/10.1007/978-981-97-9634-2_10

DDSs	Drug delivery systems
ELDT	Electroluminescent power therapy
EPIs	Efflux pump inhibitors
F-MWNTs	Multi-walled carbon nanotubes with functionalized walls
GARDP	Global Antibiotic Research and Development Partnership
GLASS	Global Antimicrobial Surveillance System
HGT	Horizontal gene transfer
IACG	Interagency Coordination Group on Antimicrobial Resistance
ODEs	Ordinary differential equations
PACT	Photodynamic antimicrobial chemotherapy
ROS	Reactive oxygen species
TIR	Translation initiation region
TM	transmembrane

10.1 Introduction

Microbes are ubiquitous in the biosphere, and their presence affects the environment. The effect of microorganisms may be beneficial or harmful. They are essential to maintain our balanced ecosystem. The bacteria that cause infections and diseases are abundant worldwide. There is a big public health consequence from these bacterial illnesses. When pathogenic bacteria enter our body and produce disease, it is called bacterial infection. Bacterial infections and diseases are caused by multiple reasons. An organism's infectivity initially establishes how many people will contract it relative to exposed and vulnerable people (Gupta et al. 2017). Second, an infectious organism's pathogenicity measures its capacity to spread illness. Certain strains of pathogenic bacteria enable them to overcome the body's defenses and exploit its resources in order to spread illness. The ability to spread the disease is due to traits like toxin production and invasiveness, referred to as virulence. When a bacterial agent spreads, the presence or absence of disease largely depends on the host body's age, nutritional status, genetic variation, length of exposure to the organism, and concurrent medical conditions (Doron and Gorbach 2008).

One of the main social health issues of the twenty-first century is antimicrobial resistance (AMR), which imposes a major threat for prevention and control of infections caused by harmful bacteria, parasites, viruses, and fungi, now some of them are resistant to the common medications used to treat them (Prestinaci et al. 2015). Because microbes are becoming more resistant to antibiotics, the issue of AMR is particularly critical. Harmful bacteria that cause serious infections have grown resistant to new antibiotic that strike the market over a number of times. Multidrug-resistant organisms are found mainly in hospitals and long-term care facilities. Bacteria that resist with more than one antibiotic are called multidrug-resistant organisms (Llor and Bjerrum 2014). They often affect people who are immunologically weak, older, or very ill and can cause chronic infections. Multidrug resistance has been associated with a higher risk of invasive infection, higher frequency of spreading, and increased risk of mortality as compared to infections caused by susceptible strains. Prolonged period of antibiotic consumption and hospitalization is a major cause of resistance towards antibiotics (Ventola 2015).

The multidrug resistance has been emerged as a public health concern due to the rampant use of antibiotics, leading to the development of multidrug resistance. The use of combinational multidrugs sometimes develops resistance. Bacteria develops resistance to multiple drugs and make them ineffective which leads to a prolonged period of sickness, a significant risk of death, and increased health-care costs. Researchers are continually striving to discover and develop the potential new antibiotics with novel mode of action to overcome the resistance problem (Nikaido 2009).

Bacteria can alter the cell structure, biocatalytic activity of enzymes, and target site of the drugs rendering them less susceptible to the effectiveness of drugs. The changing of bacterial cell membrane, cell wall can prevent the entry of drugs into the cell, sometimes bacterial enzymes can chemically modify the drug structure so that limiting the activity of potential therapeutics. The protective bacterial biofilm community enhance resistance to antimicrobial agents and antibiotics. Bacteria reducing the concentration of medications by using bacterial pump system and can be actively removed from cells (Egorov et al. 2018).

10.2 Common Bacterial Infection

A bacterial infection happens when pathogenic microorganisms enter into our body and make us sick. Certain bacteria are good for humans or harmless, but others can cause illness. These harmful bacteria can grow and interfere with our biological system once they get inside of our body. The specific bacterial pathogenicity causing the infection can determine how mild or severe it is and how it affects our body. Different bacteria have a different tendency to target the locations. The symptom of a bacterial infection might differ based on the body part and the type of infection (Finlay and Falkow 1989).

Potential bacterial infection and its symptoms:

- Fever: A bacterial infection can typically raise body temperature, resulting in fever. In order to combat the infection, the body naturally reacts (Islam et al. 2021).
- Pain and inflammation: Pain, itching, redness, swelling, and soreness at the site of infection are all possible side effects of infections (Chiu 2018).
- Coughing and sneezing: Symptoms of bacterial respiratory infections include congestion, sneezing, and coughing. The throat, sinuses, or lungs may be impacted by these infections (Oates et al. 1990).
- Gastrointestinal symptoms: The digestive tract can become infected with bacteria, which can cause symptoms like cramps, diarrhea, nausea, and vomiting. Bacterial food poisoning is a common case (Kolsin et al. 2018).
- Fatigue: Bacterial infections can sometimes be accompanied by nonspecific symptoms such as extreme exhaustion or tiredness. Patient may feel exhausted as a result of their body's immunological reaction to the infection (Cortes Rivera et al. 2019).

Following are the common types of bacterial infections that are given in the Table 10.1.

Table 10.1 Common bacterial infections and their resistance to antibiotics

Bacterial infection	Name of the species	Resistant pattern	Characteristics of bacterial infection	References
Food poisoning	*Salmonella* sp.	Ampicillin, chloramphenicol, cotrimoxazole, ciprofloxacin	Raw or undercooked eggs, poultry, pork, and dairy products are frequently contaminated with this bacterium. Contamination may occur during production or by cross-contaminating cooked and raw ingredients	Bintsis (2017)
	Escherichia coli	Cephalosporins and fluoroquinolones	Severe food poisoning can be brought on by specific strains of *Escherichia coli*, including O157:H7. Undercooked ground beef, unpasteurized milk or juice, tainted water, and vegetables are some of the things that might cause contamination	Ingerson-Mahar and Reid (2011)
	Campylobacter sp.	Fluoroquinolones, macrolides	Unpasteurized milk, contaminated water, and raw or undercooked poultry can all contaminate meals. Additionally, cross-contamination may happen when raw meat comes into contact with other foods while being prepared	Ehuwa et al. (2021)
Bacterial respiratory infections	Pneumonia: *Streptococcus pneumoniae* *Staphylococcus aureus*	Chloramphenicol, rifampin, methicillin, erythromycin, tetracycline Penicillin, chloramphenicol, erythromycin, trimethoprim-sulfamethoxazole	An infection that inflames one or both of the lungs' air sacs is known as pneumonia. *Streptococcus pneumoniae* is frequently the culprit behind bacterial pneumonia; however, it can also be caused by *Haemophilus influenzae* and *Staphylococcus aureus*	Yamaguchi et al. (2009)

	Bronchitis: *Haemophilus influenzae Streptococcus pneumoniae*	Beta-lactams, chloramphenicol, trimethoprim, tetracycline Chloramphenicol, rifampin, methicillin, erythromycin, tetracycline	Bacterial bronchitis is an infection that makes the bronchial tubes, which carry air to and from the lungs, inflamed. Common microorganisms that cause it include *Haemophilus influenzae* and *Streptococcus pneumoniae*	Nowicki and Murray (2020)
	Sinusitis: *Moraxella catarrhalis Haemophilus influenzae*	Penicillin, ampicillin Beta-lactams, chloramphenicol, trimethoprim, tetracycline	The air-filled spaces around the nasal passages called sinuses become inflamed due to an illness called sinusitis. Bacteria like *Haemophilus influenzae*, *Moraxella catarrhalis*, or *Streptococcus pneumoniae* can cause bacterial sinusitis	Meltzer et al. (2004)
	Strep throat: *Streptococcus pyogenes*	Spiramycin	The bacterium that causes strep throat is *streptococcus pyogenes*, also referred to as group A *streptococcus*. It results in tonsil swelling, dysphagia, and sore throats by inflaming and infecting the throat	Ibrahim et al. (2016)
	Whooping cough (pertussis): *Bordetella pertussis*	Erythromycin	The bacterium responsible for whooping cough is *Bordetella pertussis*. A strong coughing fit, a "whooping" sound made during inhalation, and trouble breathing are all symptoms of this extremely contagious respiratory tract infection.	Wang et al. (2014)
	Diphtheria: *Corynebacterium diphtheriae*	Fosfomycin	Diphtheria is caused by the bacteria *Corynebacterium diphtheriae*. Breathing difficulties, sore throats, and fever are caused by this dangerous respiratory illness that leaves a thick, grayish coating in the throat	Kandi and Vaish (2019)
	Tuberculosis: *Mycobacterium tuberculosis*	Rifampicin, isoniazid, fluoroquinolone, ethambutol, pyrazinamide	Mycobacterium tuberculosis, which infects the lungs, is the cause of tuberculosis (TB), a lower respiratory tract infection. Due to an increase in drug-resistant strains, it continues to be a significant killer	Seung et al. (2015)

(continued)

Table 10.1 (continued)

Bacterial infection	Name of the species	Resistant pattern	Characteristics of bacterial infection	References
Bacterial skin infections	Impetigo: *Staphylococcus aureus* *Streptococcus pyogenes*	Penicillin, chloramphenicol, erythromycin, trimethoprim-sulfamethoxazole Spiramycin	The bacteria *streptococcus pyogenes* or *Staphylococcus aureus* are the main culprits behind the highly contagious skin illness. Red sores or blisters that burst, leak, and create a crust color of honey are frequently seen in young children and babies	Anthony et al. (1976)
	Cellulitis: *Streptococcus or staphylococcus bacteria*	Chloramphenicol, rifampin, methicillin, erythromycin, penicillin, chloramphenicol, erythromycin, trimethoprim-sulfamethoxazole	Deeper skin layers and underlying tissues are susceptible to cellulitis, a bacterial infection. Bacteria from the *streptococcus* or *staphylococcus* species are frequently the cause. The affected area may feel heated, swollen, red, and painful	Raff and Kroshinsky (2016)
	Folliculitis: *Staphylococcus aureus*	Penicillin, chloramphenicol, erythromycin, trimethoprim-sulfamethoxazole	When bacteria, usually *Staphylococcus aureus*, infiltrate hair follicles, the condition known as folliculitis develops. It usually affects the face, scalp, or places that come in contact with friction, like the thighs or buttocks. It causes little red pimples or pustules surrounding hair follicles	Miró and Sánchez (2012)
	Carbuncles: *Staphylococcus aureus*	Penicillin, chloramphenicol, erythromycin, trimethoprim-sulfamethoxazole	Carbuncles are collections of linked boils. These are more serious infections that might result in excruciating pain and a hard, swollen tumor that is red. *Staphylococcus aureus* germs are also frequently responsible for carbuncles	Lin et al. (2018)

	Methicillin-resistant *Staphylococcus aureus* infections: *Staphylococcus aureus*	Erythromycin, trimethoprim-sulfamethoxazol, chloramphenicol, penicillin	A form of *Staphylococcus aureus* known as MRSA has developed resistance to numerous antibiotics. A number of skin illnesses, including as cellulitis, boils, abscesses, and surgical wound infections, can be brought on by MRSA	Elward et al. (2009)
Urinary tract infections	*Escherichia coli*	Cephalosporins and fluoroquinolones	Bacteria can enter and multiply in any region of the urinary system, but they most frequently do so in the bladder and urethra, causing these infections. Urine that smells strongly, contains blood, and causes a burning feeling when peeing are among the symptoms	Zhou et al. (2023)

10.3 Multidrug Resistance

A microorganism's ability to give resistance against various antimicrobials that were previously effective against them is known as antimicrobial resistance. Multidrug resistance (MDR) is the term used to describe this type of antimicrobial resistance to numerous medications. Microorganisms have been found to exhibit a variety of resistance mechanisms (Chinemerem Nwobodo et al. 2022). These include acquired resistance from other species, genetic mutations, and innate resistance in some microorganisms against a specific antibiotic. From an evolutionary perspective, bacteria alter the effects of antibiotics either by acquiring foreign DNA by horizontal gene transfer (HGT) or by altering chromosomal genes, which leads to the acquisition of genes that determine resistance. According to research conducted by the World Health Organization, the bacteria are showing high rate of antibiotic resistance to drugs such as cephalosporin and fluoroquinolones (Table 10.2) (Davies and Davies 2010; Munita and Arias 2016).

Antibiotic-resistant bacteria may possess one or more genes from acquired, intrinsic, or adaptive sources. A bacteria's innate ability to resist particular antibiotic classes is referred to as "intrinsic resistance" because these medications are found in their chromosomal genes without being acquired through gene mutation. The term "acquired resistance" refers to the evolutionary process by which a previously susceptible bacterium develops resistance as a result of chromosomal gene mutation acquisition or external genetic material acquisition by horizontal gene transfer (HGT) (C Reygaert 2018; Salam et al. 2023). HGT is based on three primary mechanisms: conjugation, transposition, and transformation. The acquired resistance can be either transient or permanent, and it is often passed on via a plasmid obtained by conjugation. Both in humans and livestock, bacteria can acquire adaptive resistance when their growth is impacted by subinhibitory antibiotic doses in conjunction with particular environmental signals such as growth hormones, nutrition, stress, pH, and ion concentrations. When the inciting signals are removed, adaptive resistance typically returns to its initial state (Lopatkin et al. 2016; Michaelis and Grohmann 2023).

10.4 Problem Associated with MDR

MDR causes barriers to disease control by increasing resistance capability in microorganisms, which lowers treatment efficacy, prolongs the patient's infection period, and increases treatment costs. The development of resistance in microorganisms to

Table 10.2 List of bacteria having high rate of antibiotic resistance (Breijyeh et al. 2020)

Name of the species	Antibiotic-resistant
Neisseria gonorrhoeae	Cephalosporin
Klebsiella pneumoniae	Cephalosporin and carbapenems
Nontyphoidal salmonella	Fluoroquinolones
Shigella sp.	Fluoroquinolones

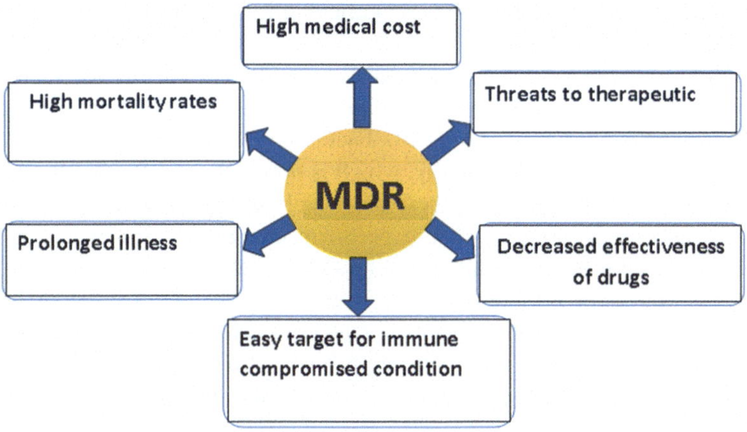

Fig. 10.1 Problems associated with multidrug resistance

commercially accessible medications enable them to adapt to costly therapies. Variations in the resistance profiles of fungal and bacterial infections, along with the standard of public hygiene, significantly influence the efficacy of antimicrobial drugs (Kunz Coyne et al. 2022; Hetta et al. 2023). Figure 10.1 represents problems associated with multidrug resistance.

10.5 National and Global Scenario of MDR

One major economic burden on the world economy is multidrug resistance. In 2018, the World Health Organization's Global Antimicrobial Surveillance System (GLASS) released a list of resistant bacteria (including *E. coli*, *K. pneumoniae*, *S. aureus*, and *S. pneumoniae*) extensively distributed across several nations. A significant problem with antibacterial resistance is faced by nations including Brazil, Nigeria, Bangladesh, Thailand, Nepal, and Pakistan (Zhen et al. 2019; Pulingam et al. 2022). This resistance is spread via various sources such as humans, pollution, and poultry. India is still battling against ancient problems like drug-resistant cholera pathogens, malaria, and tuberculosis, even though there are substantial obstacles in diagnosing and treating emerging multidrug-resistance (MDR) organisms. Factors such as malnourishment, impoverishment, lack of education, and crowded living conditions exacerbate the situation. Antibiotic resistance is predicted to kill about 3 million people by 2050 across continents such as Asia, Africa, and North America (Taneja and Sharma 2019; Iskandar et al. 2022).

The main goals of researchers and scientists in reducing the prevalence of antibiotic resistance are raising public awareness and educating the public. In addition to GLASS, the World Health Organization collaborates with the Global Antibiotic Research and Development Partnership (GARDP) and the Interagency Coordination Group on Antimicrobial Resistance (IACG) to enhance and develop antibiotics and execute a worldwide action plan aimed at alleviating the current situation (Muteeb et al. 2023).

According to the evaluation carried out by the Centers for Disease Control and Prevention (CDC), there occur millions of deaths worldwide due to antibiotic resistance. The analysis predicts that by 2050, bacterial resistance may lead to the death of 10 million people per year globally, which is more than cancer deaths. The conventional antibiotics uses because these are widely utilized in clinical settings and have been thoroughly covered in the pertinent literature (Murray et al. 2022). Instead, we concentrate on a number of alternative strategies for dealing with drug-resistant bacteria; specific strategies for addressing the urgent issue of bacterial resistance, such as diverse antimicrobial materials and drug delivery systems are outlined. Lastly, we go over the possible problem with bacterial medication resistance and look at the current developments (Yao et al. 2023).

10.6 Role of Antibiotic and Associated Problem

10.6.1 Traditional Antimicrobial Drugs

Medicinal plants serve as a promising source of potential new antibiotics to treat infections caused by pathogens that are resistant to antibiotics. Around 25% and 50% of currently available pharmaceuticals are derived from plants. This is an attempt to address the growing threat to public health as well as prevent the world from becoming a post-antibiotic era (Murugaiyan et al. 2022). In traditional medicine, plant species possess certain physiologically active chemicals called phytochemicals, utilized for pharmacological and therapeutic for both humans and animals. Traditionally, therapeutic plant extracts were considered as safe when used at appropriate dosages (Nasim et al. 2022). These extracts also lacking the side effects, which add a potential significance in the advancement of drug discovery research and could support the efficacy of conventional therapy. Through the process of extracting, isolating, and identifying plants that contain phytochemicals, research and development (R&D) in the pharmaceutical industry has historically discovered new therapies. On the health-care system, this will be a positive outcome (Thomford et al. 2018).

Many studies on the antibacterial activity of specific plants have been published by researchers in relation to the use of medicinal plants against multidrug-resistant (MDR) diseases. This demonstrates that the creation of novel antibiotics with active anti-MDR properties can be facilitated by the use of medicinal plants. Out of the 350,000 vascular plant species that science has identified, only about 26,000, or 7%, have been recorded and documented for use in medicine, according to recent estimates. Thus, it is undeniable that the demand for medical plant products around the world is extremely at the top level which can lead to endanger the existence of well-known species (Vaou et al. 2021). Table 10.3 lists common medicinal plants with antimicrobial activity.

Table 10.3 Common medicinal plants with antimicrobial activity (Cowan 1999)

Common name	Scientific name	Compound	Activity
Alfalfa	*Medicago sativa*	Tannins	Gram-positive organisms
Aloe	*Aloe barbadensis*, *Aloe vera*	Latex	*Corynebacterium*, *Salmonella*, *Streptococcus*, *S. aureus*
Mountain tobacco	*Arnica montana*	Helanins	General
Barberry	*Berberis vulgaris*	Berberine	Bacteria, protozoa
Black pepper	*Piper nigrum*	Piperine	Fungi, *Lactobacillus*, *Micrococcus*, *E. coli*, *E. faecalis*
Blueberry	*Vaccinium* spp.	Fructose	*E. coli*
Ceylon cinnamon	*Cinnamomum verum*	Essential oils, others	General
Chapparal	*Larrea tridentata*	Nordihydroguaiaretic acid	Skin bacteria
Eucalyptus	*Eucalyptus globulus*	Tannin	Bacteria, viruses
Hemp	*Cannabis sativa*	β-Resercyclic acid	Bacteria and viruses
Hops	*Humulus lupulus*	Lupulone, humulone	General
Legume	*Millettia thonningii*	Alpinumisoflavone	*Schistosoma*
Lemon	*Aloysia triphylla*	Essential oil	*Ascaris*
Green tea	*Camellia sinensis*	Catechin	*Shigella*, *Vibrio*, *S. mutans*
Olive oil	*Olea europaea*	Hexanal	General
Papaya	*Carica papaya*	Latex	General
Quinine	*Cinchona* sp.	Quinine	*Plasmodium* spp.
Rosemary	*Rosmarinus officinalis*	Essential oil	General

10.7 Active Antimicrobial Compounds

Compounds that are designed to combat infections and have microbicidal properties against bacteria, fungus, and viruses include small molecules, macromolecules, polymers, ceramics, metals, and composites. Although antimicrobial materials have the ability to suppress or eradicate bacteria on their surface or in their immediate surroundings, they suffer from several clinical drawbacks, including low antimicrobial activity, problems with microbial resistance, and difficulties operating in a changing environment (Balasubramaniam et al. 2021; Agarwalla et al. 2023). Therefore, to meet the needs in biomedicine, active efficient antimicrobial materials are required against bacteria and biofilms (Vaou et al. 2021). The following are some examples of some significant antimicrobial material:

10.7.1 Nanoparticles

A nanoparticle-based coating is now very popular due to their long-term durability and enhanced properties. Silver nanoparticles (AgNPs) exhibit considerable toxicity towards a wide variety of microorganisms while lesser toxic towards the human cells (Ferdous and Nemmar 2020). The antibacterial and self-cleaning properties of silver (Ag) nanoparticles are widely known. Numerous natural and synthetic textiles have been treated with metal oxide nanoparticles and other metals, including copper, gold, titanium, tin, and zinc, in addition to silver (Gulati et al. 2022). The antibacterial activity of the functionalized CuO-NPs-coated textile materials against Gram-positive and Gram-negative bacteria can be determined by three processes: the release of copper ions, the direct interaction of CuO-NPs with bacteria, and the production of reactive oxygen species (Román et al. 2020). To create capped ZnO-NPs, along with the seed extract an additional natural capping agent have been employed. These NPs showed UV protection and antibacterial activity in cotton textiles (Pandimurugan and Thambidurai 2017).

10.7.2 Natural Compounds

Natural materials have long been recognized for their antibacterial properties. Natural products are vital to current medication research initiatives because of their antibacterial properties.Numerous antimicrobial substances that are found in nature can originate from microorganisms, minerals, plants, or animals, among other sources (Dias et al. 2012). Numerous plant and animal products possess antimicrobial properties because they contain a range of active ingredients, secondary metabolites, or phytochemicals (in the case of plants), such as proteins and polypeptides, quinones, coumarins, polyphenols and phenolic compounds, enzymes, lysozymes, phagocytic cells, and numerous other organic constituents (Rahman et al. 2021). Milk has antimicrobial effects because it contains a variety of proteins and peptides, including certain lysozymes, lactoferrin, immunoglobulins, and the lactoperioxidase system. There is a well-known therapeutic significance to the black meat and blood of the Kadaknath breed of poultry. A variety of bacteria, including *B. subtilis* and *P. asoginosa*, were shown to be susceptible to the antibacterial properties of cow dung and urine concentrate (Kazimierska and Kalinowska-Lis 2021; Singh et al. 2023). Table 10.4 lists phytocompounds with inhibitory action against MDR.

10.7.3 Inhibitors as Adjuvants

When faced with a situation like current antibiotics are resistant to infection, continuous work is necessary to enhance the effectiveness of antibiotics against pathogenic microbes (Mancuso et al. 2021). To improve the efficacy of antibiotics, it can be combined with non-antibiotic substances known as antibiotic adjuvants that target bacteria. Adjuvants, such as inhibitors, are frequently used to increase antibiotic

Table 10.4 List of phytocompounds inhibitory action against MDR

Compound	Inhibitory action	Structure	References
Acetosyringone	*S. cerevisiae* (depolarization of the bacterial cell membrane)	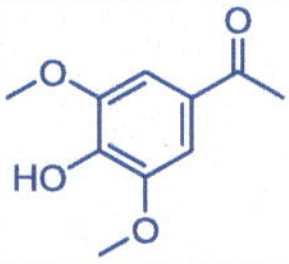	Epand et al. (2010)
Agasyllin	*Salmonella* sp. (cell membrane disruption)	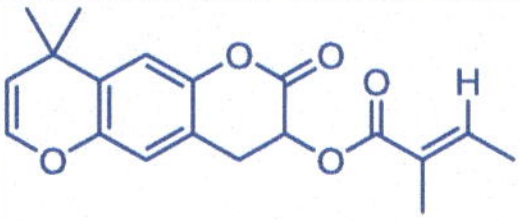	Villanueva et al. (2022)
	Enterobacter cloacae (cell membrane disruption)		Mosby et al. (2023)
Alfa-Bisabolol	*Propionibacterium acnes* (disrupting cell membrane integrity)	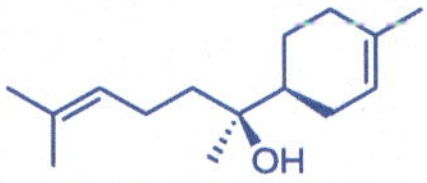	Ryu ct al. (2015)
Andrographolide	*Streptococcus pneumoniae* (protein and DNA synthesis)	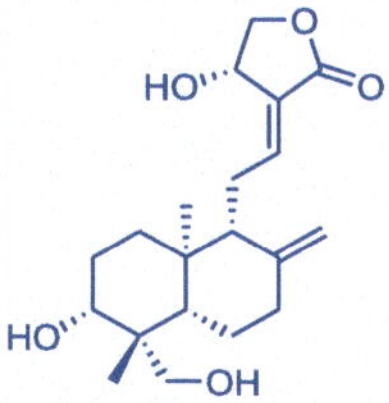	Briggs et al. (2021)
Baicalein	*S. typhimurium* (inhibition of efflux pump)	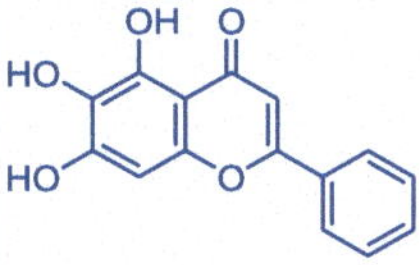	Anbazhagan et al. (2019)
Carvacrol	*Streptococcus pyogenes* (disrupting cell membrane integrity, inhibition of efflux pump)	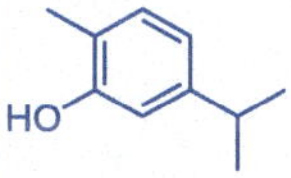	Brouwer et al. (2023)
Chlorogenic acid	*Escherichia coli* (interacting with some crucial enzymes)	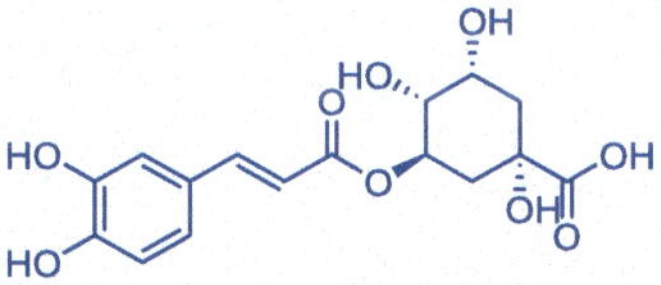	Bhattacharyya et al. (2016)
Citronellol	*Trichophyton rubrum* (deteriorating membrane integrity)	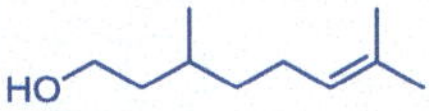	Pereira et al. (2011)
Conessine	*Micrococcus letues* (inhibition of efflux pump)	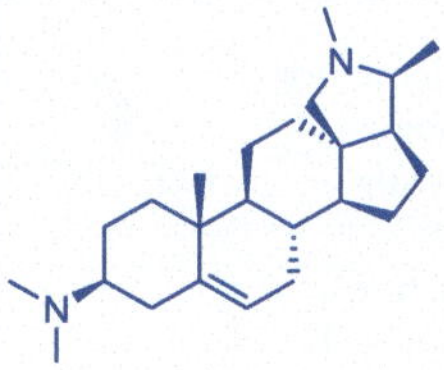	Sharma et al. (2019)

(continued)

Table 10.4 (continued)

Compound	Inhibitory action	Structure	References
Curcumin	*Shigella dysenteriae* (damaging the cell membranes)	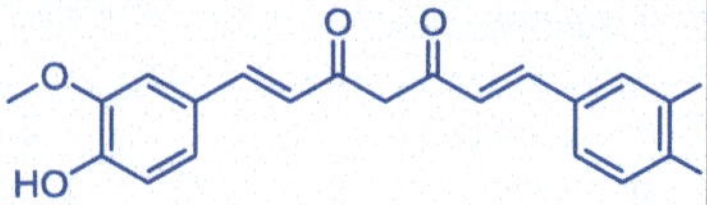	Liang et al. (2022)
Daphnetin	*Shewanella putrefaciens* (cell membrane disruption, type III secretion inactivation)	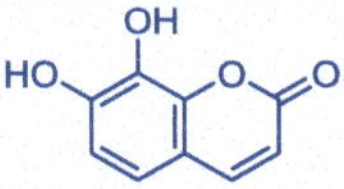	Suganya et al. (2022)
Esculetin	*Ralstonia pseudosolanacearum* (cell membrane disruption, type III secretion inactivation)	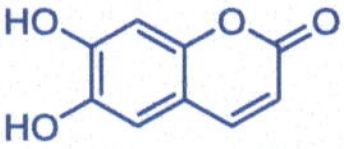	Yang et al. (2016)
Farnesol	*Lactobacillus* spp. (disrupting cell membrane integrity)		Wang et al. (2017)
Luteolin	*Listeria monocytogenes* (cell membrane disruption, DNA gyrase inhibition, type III secretion inactivation, interacting with some crucial enzymes)	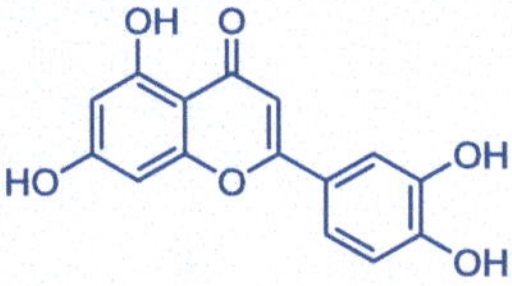	Lechner et al. (2008)
Menthol	*C. albicans* (disrupting cell membrane integrity)	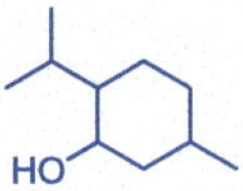	Zore et al. (2022)
Novobiocin	*S. gallinarum* (inhibition of DNA topoisomerase type II DNA gyrase)	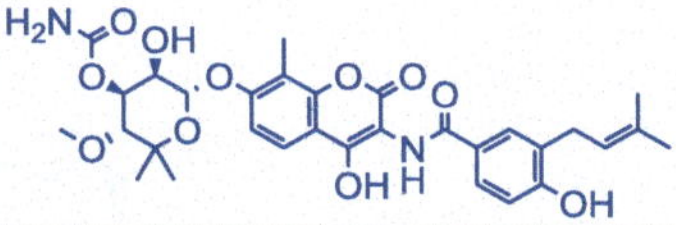	(Tse-Dinh 2007)
Piperine	*Staphylococcus aureus* and *Bacillus subtilis* (inhibition of efflux pump)	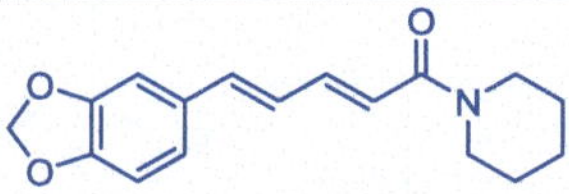	Mirza et al. (2011)
Quercetin	*Aspergillus fumigatus* (inhibition of efflux pump, interacting with some crucial enzymes such as lactamase, and cell membrane disruption)	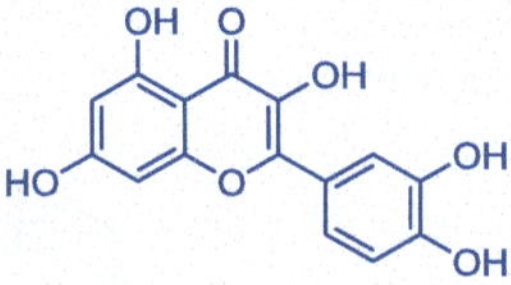	Siriwong et al. (2016)

(continued)

Table 10.4 (continued)

Compound	Inhibitory action	Structure	References
Sanguinarine	*Serratia marcescens* (inhibiting replication and transcription)	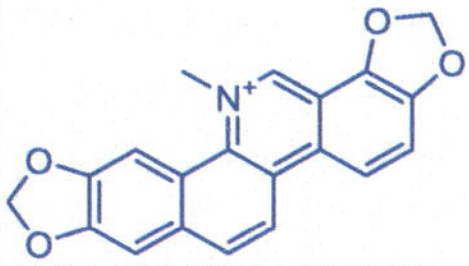	Fu et al. (2021)
Taxifolin	*Helicobacter pylori* (interacting with some crucial enzymes)	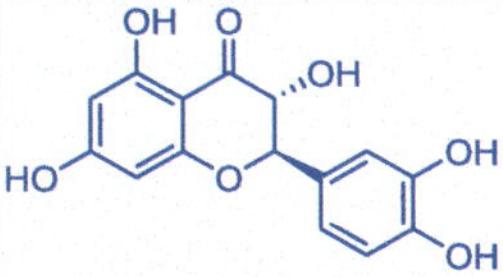	Liu et al. (2023)
Thymol	*B. cereus* (disrupting cell membrane integrity, inhibition of efflux pump)	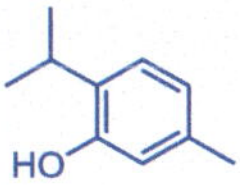	Miladi et al. (2016)

efficacy (Kumar et al. 2023). β-lactamase inhibitors when used in conjunction with β-lactam antibiotic, it inhibits the enzyme activity by attaching themselves to the enzyme's active site. This prevents the antibiotic's lactam ring from hydrolyzing and preserves the antibiotic's structural integrity (Bush and Bradford 2016). Inhibitors that target drug-resistant bacteria's enzymes and efflux pumps can still be employed to investigate potential therapeutic targets for antidrug-resistant bacteria, even after resistance has developed (Zhang et al. 2024).

10.7.4 Efflux Pump Inhibitors (EPIs)

Bacterial resistance, both acquired and inherent, depends on efflux pumps. Research is now being done to identify and inhibit EPIs from increasing the effectiveness of already available antibiotics. There have been a lot of efforts over the last 20 years to find new EPIs activity (Lamut et al. 2019). In MRSA, the multidrug efflux pump NorA is encoded chromosomally. The transmembrane (TM) α-helices that comprise NorA are arranged into two 6-TM bundles, known as the N- and C-terminal domains. This resembles a potential substrate-binding pocket (Brawley et al. 2022). By analyzing the substrate binding pocket's structure about NorA, synthetic antigen-binding fragment Fabs are designed. The substrate binding site of NorA received the Fab ring in order to prevent the pumping of antibiotic drugs, hence resolving the issue of bacterial resistance (Palazzotti et al. 2023). When the Fab ring of peptide combined with the antibiotic norfloxacin, it suppressed both NorA and MRSA growth. It was shown that when the Fab CDRH3 loop was inserted into NorA's substrate binding pocket, it reduces antibiotics' efflux (Brawley et al. 2022).

10.7.5 Photosensitive Materials

Photosensitive materials are used to induce cellular and microbiological inactivation and this process is termed as photodynamic antimicrobial chemotherapy (PACT). The three components of PACT light, a photosensitizer, and oxygen work together to cause oxidative damage. Because of PACT's great spatiotemporal selectivity, noninvasiveness, low drug-resistance rate, and suitability for treating localized infections with its penetrating laser, it has drawn a lot of attention from several antimicrobial fields to design it (Oyim et al. 2021; Hampden-Martin et al. 2021). Electroluminescent power therapy (ELDT) is a novel antimicrobial strategy that has gained attention recently. Under an electric field, it uses photosensitizers and EL materials in nano-assemblies to produce reactive oxygen species (ROS) in real time. In essence, the photosensitizers are excited by the fluorescence that EL molecules emit, resulting in the production of singlet oxygen (1O2), which causes oxidative damage (Chandna et al. 2020; Leanse et al. 2023).

10.8 Action of Antibiotics

10.8.1 Disruption of Cell Membrane

The microorganism has a selectively permeable plasma membrane which performs active transportation of nutrients and generates ATP. Throughout active transportation, the integral transporter proteins, the cytoplasmic gradient ions, macro and micro molecules transportation are carried out. Fatty acids serve as the building blocks to make the plasma membranes of bacteria that are either synthesized within the cell (Stillwell 2013). Antimicrobials agents mainly target membrane phospholipids and the metabolic stages of fatty acid production. One of the bactericidal antibiotics used to kill Gram-negative bacteria like Pseudomonas is polymyxin B (Ayoub Moubareck 2020). This antibiotic is composed of detergent-like peptides with hydrophilic and lipophilic groups, which disturbs the membrane's phosphatidylethanolamine (Laverty et al. 2011). Valinomycin is another antibiotic which interferes with the potential of the cell membrane by creating holes in the cellular membrane, this in turn promotes oxidative phosphorylation. β-lactam antibiotics are frequently utilized to treat wound and soft skin infections, it works by depolarizing the membrane, which causes potassium ions to be released from the cytoplasm into the extracellular space (Kimbrell et al. 2011).

10.8.2 Inhibition of Cell Wall Synthesis

Many antibiotics function by preventing the production of bacterial cell walls. These substances are commonly known as β-lactam antibiotics and are distinguished by a lactam ring in their molecular structure. This antibiotic act by inhibiting the synthesis of peptidoglycan layer of bacterial cell wall (Bush and Bradford

2016). Glycopeptides and antibiotics of class β-lactams are examples of substances that hinder the formation of cell walls. Monobactams, carbapenem, cephalosporins, and penicillin are B-lactam antibiotics. The most widely used antibiotics in this class are glycopeptides, such as teicoplanin and vancomycin (Sarkar et al. 2017). The only antibiotic in this class that is sold commercially is aztreonam, which is found in penicillin, which also include ampicillin, oxacillin, cephalosporins, cefpodoxime, and cefepime. The components of carbapenem are ertapenem, imipenem, and meropenem. The most popular class of antibiotics for treating both Gram-positive and Gram-negative infections are those that impede the synthesis of cell walls (Laws et al. 2019).

10.8.3 Inhibition of Nucleic Acid Synthesis

In the creation of nucleic acids one important component is the synthesis of folic acid, also known as folate. A number of antibiotics competitively impede this mechanism, including sulfonamides. Sulfonamides can block folic acid from developing because they imitate para-aminobenzoic acid, an intermediary molecule that is converted to folic acid by an enzyme. Due to their structural resemblances, para-aminobenzoic acid and sulfonamide compete with one another for the enzyme that converts the intermediate to folic acid. This reversible process inhibits but does not kill the bacteria, which can be stopped by removing the chemical (Fernández-Villa et al. 2019; Shahzad et al. 2020).

10.8.4 Inhibition of Protein Synthesis

Protein is produced by the ribosome's tRNA from mRNA following the process of translation. The binding of formyl methionine tRNA-aminoacyl-tRNA (fMet-tRNAfMet), initiation factor 2 (IF-2) and the translation initiation region (TIR) of mRNA to the P site of the 30S subunit of the ribosome, as well as the creation of the 70S complex due to the release of IF-2, all serve to initiate translation. In order to translate the polypeptide sequence, the tRNA needs to bind with the A, P, and E site during the initiation, elongation, and termination. When the antibiotic like tetracycline is employed, it binds with the 30S ribosomal subunit. Hence the binding of the tRNA to the A-site on the mRNA-ribosome complex blocks leading to the death of bacteria (Laursen et al. 2005; Julián et al. 2011). Figure 10.2 depicts action of antibiotics on bacteria.

10.9 Antibiotic-Resistance Mechanism

The world's persistent misuse and abuse of conventional antimicrobial medications is the cause of the alarming rise of bacterial resistance and tolerance. The WHO listed antibiotic resistance as one of the top 10 threats to global health in 2019. The

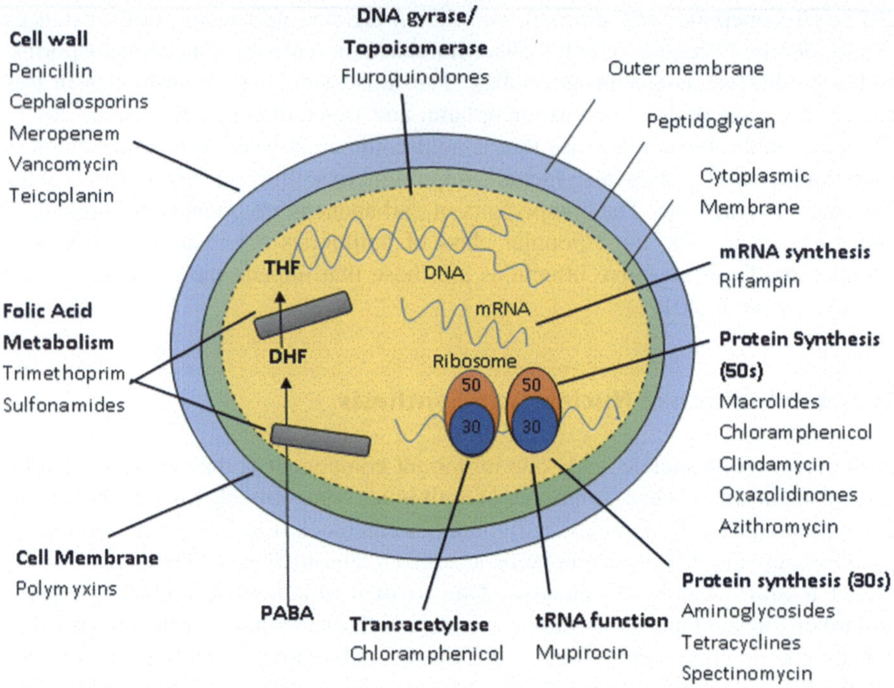

Fig. 10.2 Action of antibiotics on bacteria

current situation is very worrying due to the rise in multidrug-resistance bacteria and the lack of innovative antibiotics that can eradicate them. The Gram-positive bacteria that constitute a major concern to public health, comprising vancomycin-resistant, *Enterococcus faecium*, methicillin-resistant *Staphylococcus aureus*, and medication-resistant *Streptococcus pneumonia*. It has been observed that acquired genomes or genetic changes cause bacteria to develop this kind of resistance (Ayukekbong et al. 2017; Salam et al. 2023).

10.9.1 Genetic Basis of Antimicrobial Resistance

Bacteria can become resistant to antibiotics through a variety of genetic pathways. These methods alter the bacterial cells specific properties through biochemical modifications, which often make the organism sensitive to antibiotics and lead to resistance (Kapoor et al. 2017). Drug resistance can result from a variety of biochemical changes, such as the synthesis of inactive enzymes, drug efflux pumps that are activated to deliberately remove the drug from cells, changes to cell-wall proteins that prevent drug uptake, and alterations to the protein, enzyme, or receptor that a drug target (Li and Nikaido 2009).

Mutation and the acquisition of new genetic material are two important genetic processes that lead to antibiotic resistance. Due to their high rate of replication and single chromosome genome, bacteria are particularly vulnerable to mutation through which an organism's genetic composition can permanently change (Uddin et al. 2021). The acquisition of new genetic material also occurs spontaneously in bacteria, this lead to develop resistance in bacteria. They are prokaryotic organisms and have little DNA pieces called plasmids that reside outside of chromosomes in their cells. Because bacteria's genetic material is free to roam around inside their cells, they are susceptible to gene transfer, which frequently entails the spread of plasmids (Hawkey 1998; Coculescu 2009).

10.9.2 Gram-Positive Bacteria's Antibiotic-resistance Mechanism

Gram-positive and Gram-negative bacteria have slightly different cell wall compositions because the former have an LPS layer while the latter have not. Gram-positive bacteria's cytoplasmic membrane is thickly coated with peptidoglycan, which protects them from damaging environmental conditions. It is commonly known that there are two primary morphologic forms of Gram-positive bacteria: cocci, such as *Staphylococcus*, and bacilli, such as *Bacillus*. Moreover, the lengthy anionic polymers known as teichoic acids are present in Gram-positive bacteria (Costerton et al. 1974; Silhavy et al. 2010). Capsular polysaccharides covalently linked to peptidoglycan, along with the membrane proteins that help other molecules enter and exit. There are two main ways that antibiotic resistance in Gram-positive bacteria is brought about. These are either the enzymatic degradation of antibiotic by the production of β-lactamase, or by decreasing affinity and sensitivity of their target site, such as penicillin-binding protein (PBP) either creating mutations in native PBP genes or absorbing exogenous DNA (Navarre and Schneewind 1999; Eberhardt et al. 2012).

10.9.3 Gram-Negative Bacteria's Antibiotic-resistance Mechanism

Gram-positive bacteria's outer membrane serves as a defense and barrier, setting them apart from Gram-negative bacteria. The outer membrane's inner leaflets contain lipopolysaccharides, while the outside leaflets contain phospholipids. Repeating units of NAM and NAG make up the peptidoglycan that makes up the bacterial cell wall (Silhavy et al. 2010). This helps to regulate the shape of the cell. Phospholipids play a variety of functions in the second layer, which is found in the inner membrane. These functions include transport, biosynthesis, and structure. It facilitates the separation of sister chromosomes and acts as a site for DNA anchoring (Strahl and Thorner 2007).

Antibiotic resistance in bacteria has been associated with their outer membrane, which includes β-lactase, quinilons, colistins, and other medications. For antibiotics to reach their intended target, they must be able to penetrate the cell wall. Resistance

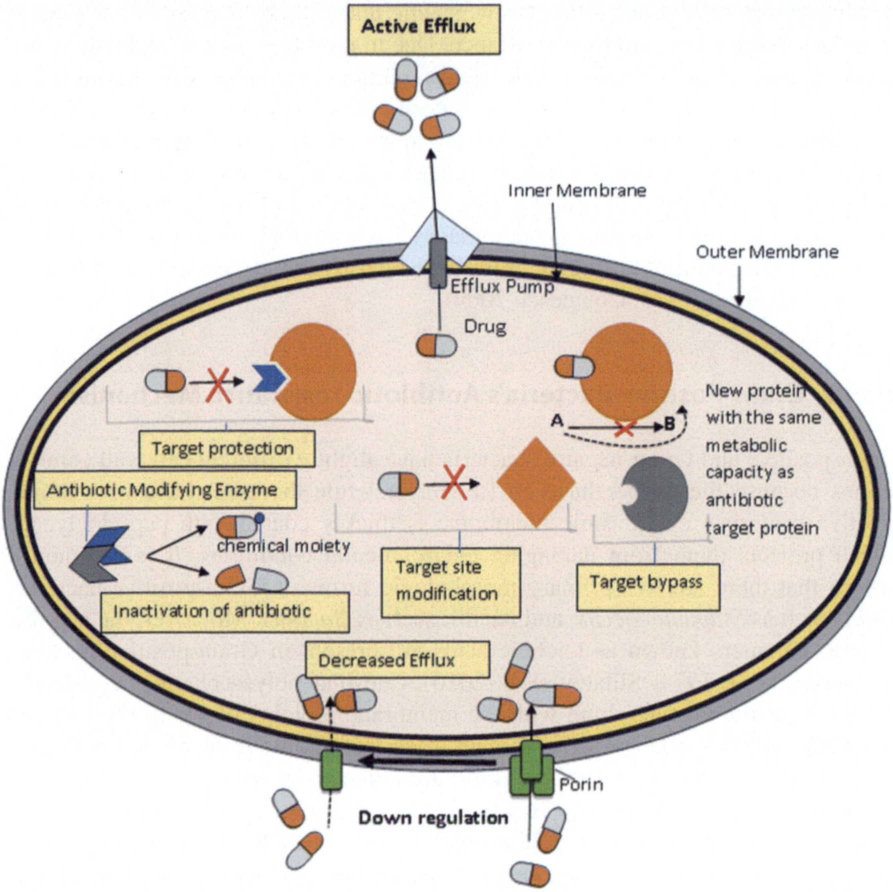

Fig. 10.3 Mechanism of antibiotic resistance

may arise as a result of any alteration or mutation to the outer membrane. Since Gram-positive bacteria do not have this layer, antibiotic resistance in Gram-negative bacteria is higher than in Gram-positive bacteria. Gram-negative bacteria's primary source of antibiotic resistance is antibiotic-inactivating enzymes. These enzymes can be produced by chromosomal gene alterations, mobility genetic elements carrying resistance genes, plasmids, and aminoglycosides that modify enzymes (Delcour 2009; Bharadwaj et al. 2022). Figure 10.3 shows mechanism of antibiotic resistance.

10.10 Approaches Used for Drug Delivery

The systematic antibiotic treatment possesses certain drawbacks such as antibiotic resistance, adverse effects, and limited absorption. Antibiotic drug delivery systems (DDSs) are a beneficial technique to prevent antibiotic resistance and boost the

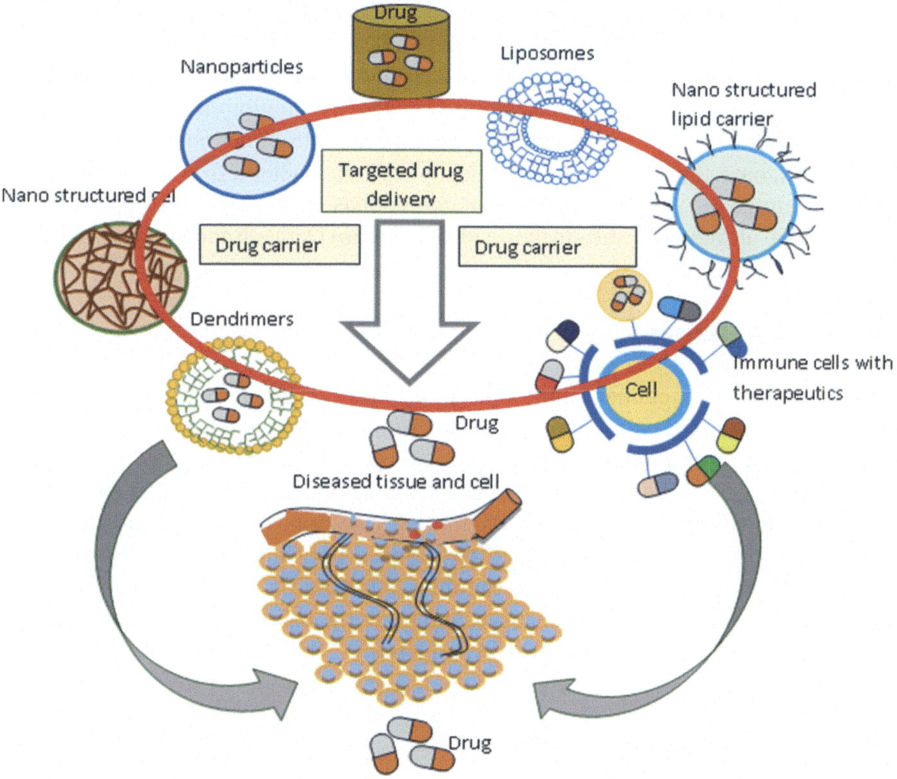

Fig. 10.4 Various approaches used in drug delivery

biodistribution and bioavailability of new drugs, thereby increasing their half-life (Pham et al. 2019). A "Trojan horse" tactic has been suggested by scientists for the creation of DDSs. In order to accomplish effective medication delivery, the fundamental component of this method is the combination of antimicrobial agents with various types of carriers, such as polymers, self-assembled peptides, erythrocytes, liposomes, and exosomes. Additionally, to release medication at the infection site, DDSs can be directed by external inputs to target the particular microenvironment connected to the diseased tissue (Ibarra et al. 2020; Huang et al. 2024). Figure 10.4 shows various approaches used in drug delivery.

10.10.1 Carbon-Based Nano-delivery Systems

The carbon-based nano materials with strong antibacterial include graphene oxide nanoparticles, fullerenes, and carbon nanotubes (CNTs). Because of their unique structural features and physicochemical qualities, Carbon-based nanomaterials are biocompatible and eco-friendly in nature. When bacteria directly exposed to carbon-based nanostructures, this can damage their shape, metabolic processes, and integrity of their cell membranes (Mahor et al. 2021). When used as carriers of common antibiotics, functionalized carbon-based nanomaterials can decrease antibiotic resistance, increase antibiotic bioavailability, and deliver antibiotics precisely where they are needed. Multiwalled carbon nanotubes with functionalized walls, or F-MWNTs, have the capacity to biologically separate bacteria from their surroundings, encourage the synthesis of harmful compounds, and subject cells to oxidative stress, which ultimately results in the death of bacteria (Debnath and Srivastava 2021).

10.10.2 Liposomes

When self-assembling amphiphilic lipid molecules come together, liposomal vesicles with a diameter ranging from 20 nm to 10 μm and a bilayer structure are formed. Hydrophilic and amphiphilic compounds are present in the core of partitioned hydrophobic lipid bilayers, which enables the selective loading of medications. Liposomes have certain advantages like low toxicity, non-immunogenicity, and biofilm matrix–cell membrane tumorgenicity. Hence liposome-based antibiotic drug carriers can efficiently lower the systemic administration dose with less harmful side effects, prevent bacterial resistance, and boost the local concentration of antibiotics in inflammatory areas (Lombardo and Kiselev 2022; Giordani et al. 2023).

10.10.3 Erythrocytes

Red blood cells or erythrocytes have the ability to conceal nanoparticles. Antigens, medications, enzymes, and peptides can all be delivered in vivo via erythrocyte membrane nano-vesicles. Thus, in recent years, erythrocyte membrane-encapsulated nanoparticles have become a promising avenue for the administration of nanodrugs because of its high targeting, a long circulation cycle, and excellent biocompatibility (Xia et al. 2019). The virulence of MRSA infections can be countered by using an erythrocyte membrane-coated nanogel (RBC-nanogel) technology that combines reactive and antiviral antibiotic delivery for the treatment of MRSA infection. The infections are brought on by numerous pore-forming toxins (PFTs) that the bacteria secrete. It was demonstrated that the RBC-nanogels could efficiently neutralize MRSA-associated toxins, and that this action encouraged macrophages to pick up the germs. When the RBC-nanogels entered the cell, they released the medication more quickly, leading to more potent suppression of bacteria (Zhang et al. 2017).

10.10.4 Self-Assembled Peptide Drug Delivery Systems

Even while each of the delivery methods discussed above has many benefits, the complex in vivo environment can lead to a number of issues regarding in vivo microenvironment's degradation and their capacity to penetrate. As self-assembly is more stable and biocompatible than free peptides, it is an effective technique for preparing and delivering polymeric materials. Peptide-based self-assembled nanostructures are already showing great potential in the field of antibacterial DDS (Ezike et al. 2023). In order to overcome the difficulties in creating peptide-based nanomaterials ex vivo for use in biomedical applications, scientists have recently put forth a novel tactic called in vivo self-assembly. The phospholipid membrane is displaced and forms a cleft when the peptide attaches itself to the bacterial membrane. The introduction of nano-AMPs causes the lysis and demise of bacteria (Habibi et al. 2016).

10.10.5 In Silico Approaches in MDR

The interaction between protein motifs of pathogen with phytochemical ligands through screening techniques provides the identity of antibacterial property, stability, and safety prospects of phytochemicals, in silico molecular docking study provides the thorough approaches about the mathematical modelling, data analysis techniques, and computer simulation to understand the complete mechanism of multidrug resistance and also identify potential strategies to overcome from it (Turabi et al. 2023). A computational method called molecular docking is used to determining the way to identify the orientation of one molecule to a second when they form a stable complex. The interaction between potential drug and target molecules (e.g., enzymes or receptors) and its resistant mechanism can be determined by molecular docking. Large libraries of compounds and its effectiveness against MDR can be determined through computational screening. The identification of essential structural and chemical features of a pharmacophore molecule, its binding target and biological activity can be determined by pharmacophore modelling also it is used to design a new drug, modify existing drugs to better one which target the resistant pathogens (Meng et al. 2011). Thorough understandings of interaction between various components like genes, proteins and metabolites can be determined through system biology. Network analysis tools and techniques like pathway analysis and network modelling can be used to understand the complex interaction between drugs and targets and identify key role of genes and pathways involved in drug resistance. To understand the molecular mechanisms of multidrug resistance, data mining techniques can be used to extract meaningful information from large scale data (e.g., genomics, transcriptomics, and proteomics). Based on the structural features of compounds, QSAR analysis can be used to predict the activity of new potent compounds against multidrug resistant pathogens allowing for the rational design of novel therapeutics (Zou et al. 2013). Ordinary differential equations (ODEs) or agent-based models, can be used to simulate the dynamics of drug resistance within the populations of pathogens by using mathematical models. Researchers can gain insights into the mechanism of multidrug resistance, identity the novel drug targets, and optimize the treatment regimens to combat the multidrug resistance more effectively by using the in silico approaches (Lavi et al. 2012; Youlden et al. 2022).

10.11 Conclusion

It has been said that the discovery and use of antibiotics represent, one of the greatest scientific and technological achievements of the twentieth century. Millions of individuals who were suffering from bacterial illnesses have been spared death. Potential antibiotic development and their diverse bactericidal mechanisms is now acting as a very potent weapon in the fight against mortality. Up until this point, antibiotics have been a necessary anti-infection medication. But the abuse of antibiotics has made bacterial resistance worse, which increases the risk to humans as well. The complicated mechanism of resistance resulting from the use of different medicines and antibacterial processes causes the emergence of multiresistant bacteria. For this reason, the investigation and study of the mechanisms underlying bacterial resistance and the bactericidal effects of antibiotics is very necessary.

The majority of research being done right now is aimed at developing antibiotic substitutes. Although antimicrobial materials are generally straightforward, there are certain issues that must be taken into consideration, such as the toxicity and metabolism of metal nanoparticles. Antibiotic drug delivery systems have been developed with several advantages. The nanoscale benefits, including the improved drug-specific targeting, decreased toxicity, and longer in vivo circulation, all of which led to greater efficacy of antibiotics. However, cutting-edge research and innovative technology will be applied in clinical settings for effective antibacterial treatment against multidrug resistant bacteria.

References

Agarwalla A, Ahmed W, Al-Marzouqi AH et al (2023) Characteristics and key features of antimicrobial materials and associated mechanisms for diverse applications. Molecules 28:8041. https://doi.org/10.3390/molecules28248041

Anbazhagan PV, Thavitiki PR, Varra M et al (2019) Evaluation of efflux pump activity of multidrug-resistant *salmonella* Typhimurium isolated from poultry wet markets in India. Infect Drug Resist 12:1081–1088. https://doi.org/10.2147/IDR.S185081

Anthony BF, Kaplan EL, Wannamaker LW, Chapman SS (1976) The dynamics of streptococcal infections in a defined population of children: serotypes associated with skin and respiratory Infections1. Am J Epidemiol 104:652–666. https://doi.org/10.1093/oxfordjournals.aje.a112344

Ayoub Moubareck C (2020) Polymyxins and bacterial membranes: a review of antibacterial activity and mechanisms of resistance. Membranes (Basel) 10:181. https://doi.org/10.3390/membranes10080181

Ayukekbong JA, Ntemgwa M, Atabe AN (2017) The threat of antimicrobial resistance in developing countries: causes and control strategies. Antimicrob Resist Infect Control 6:47. https://doi.org/10.1186/s13756-017-0208-x

Balasubramaniam B, Prateek RS et al (2021) Antibacterial and antiviral functional materials: chemistry and biological activity toward tackling COVID-19-like pandemics. ACS Pharmacol Transl Sci 4:8–54. https://doi.org/10.1021/acsptsci.0c00174

Bharadwaj A, Rastogi A, Pandey S et al (2022) Multidrug-resistant bacteria: their mechanism of action and prophylaxis. Biomed Res Int 2022:1–17. https://doi.org/10.1155/2022/5419874

Bhattacharyya S, Bershtein S, Yan J et al (2016) Transient protein-protein interactions perturb E. coli metabolome and cause gene dosage toxicity. elife 5:10.7554/eLife.20309

Bintsis T (2017) Foodborne pathogens. AIMS Microbiol 3:529–563. https://doi.org/10.3934/microbiol.2017.3.529
Brawley DN, Sauer DB, Li J et al (2022) Structural basis for inhibition of the drug efflux pump NorA from Staphylococcus aureus. Nat Chem Biol 18:706–712. https://doi.org/10.1038/s41589-022-00994-9
Breijyeh Z, Jubeh B, Karaman R (2020) Resistance of gram-negative bacteria to current antibacterial agents and approaches to resolve it. Molecules 25:1340. https://doi.org/10.3390/molecules25061340
Briggs NS, Bruce KE, Naskar S et al (2021) The pneumococcal divisome: dynamic control of Streptococcus pneumoniae cell division. Front Microbiol 12. https://doi.org/10.3389/fmicb.2021.737396
Brouwer S, Rivera-Hernandez T, Curren BF et al (2023) Pathogenesis, epidemiology and control of group a streptococcus infection. Nat Rev Microbiol 21:431–447. https://doi.org/10.1038/s41579-023-00865-7
Bush K, Bradford PA (2016) β-Lactams and β-lactamase inhibitors: an overview. Cold Spring Harb Perspect Med 6:a025247. https://doi.org/10.1101/cshperspect.a025247
Chandna S, Thakur NS, Kaur R, Bhaumik J (2020) Lignin–bimetallic Nanoconjugate doped pH-responsive hydrogels for laser-assisted antimicrobial photodynamic therapy. Biomacromolecules 21:3216–3230. https://doi.org/10.1021/acs.biomac.0c00695
Chinemerem Nwobodo D, Ugwu MC, Oliseloke Anie C et al (2022) Antibiotic resistance: the challenges and some emerging strategies for tackling a global menace. J Clin Lab Anal 36:e24655. https://doi.org/10.1002/jcla.24655
Chiu IM (2018) Infection, pain, and itch. Neurosci Bull 34:109–119. https://doi.org/10.1007/s12264-017-0098-1
Coculescu B-I (2009) Antimicrobial resistance induced by genetic changes. J Med Life 2:114–123
Cortes Rivera M, Mastronardi C, Silva-Aldana C et al (2019) Myalgic encephalomyelitis/chronic fatigue syndrome: a comprehensive review. Diagnostics 9:91. https://doi.org/10.3390/diagnostics9030091
Costerton JW, Ingram JM, Cheng KJ (1974) Structure and function of the cell envelope of gram-negative bacteria. Bacteriol Rev 38:87–110. https://doi.org/10.1128/br.38.1.87-110.1974
Cowan MM (1999) Plant products as antimicrobial agents. Clin Microbiol Rev 12:564–582. https://doi.org/10.1128/CMR.12.4.564
Davies J, Davies D (2010) Origins and evolution of antibiotic resistance. Microbiol Mol Biol Rev 74:417–433. https://doi.org/10.1128/MMBR.00016-10
Debnath SK, Srivastava R (2021) Drug delivery with carbon-based nanomaterials as versatile Nanocarriers: Progress and prospects. Front Nanotechnol 3. https://doi.org/10.3389/fnano.2021.644564
Delcour AH (2009) Outer membrane permeability and antibiotic resistance. Biochim Biophys Acta Proteins Proteomics 1794:808–816. https://doi.org/10.1016/j.bbapap.2008.11.005
Dias DA, Urban S, Roessner U (2012) A historical overview of natural products in drug discovery. Meta 2:303–336. https://doi.org/10.3390/metabo2020303
Doron S, Gorbach SL (2008) Bacterial infections: overview. In: International encyclopedia of public health. Elsevier, pp 273–282
Eberhardt A, Hoyland CN, Vollmer D et al (2012) Attachment of capsular polysaccharide to the cell wall in *Streptococcus pneumoniae*. Microb Drug Resist 18:240–255. https://doi.org/10.1089/mdr.2011.0232
Egorov AM, Ulyashova MM, Rubtsova MY (2018) Bacterial enzymes and antibiotic resistance. Acta Nat 10:33–48
Ehuwa O, Jaiswal AK, Jaiswal S (2021) Salmonella, food safety and food handling practices. Food Secur 10:907. https://doi.org/10.3390/foods10050907
Elward AM, McAndrews JM, Young VL (2009) Methicillin-sensitive and methicillin-resistant Staphylococcus aureus: preventing surgical site infections following plastic surgery. Aesthet Surg J 29:232–244. https://doi.org/10.1016/j.asj.2009.01.010

Epand RF, Pollard JE, Wright JO et al (2010) Depolarization, bacterial membrane composition, and the antimicrobial action of Ceragenins. Antimicrob Agents Chemother 54:3708–3713. https://doi.org/10.1128/AAC.00380-10

Ezike TC, Okpala US, Onoja UL et al (2023) Advances in drug delivery systems, challenges and future directions. Heliyon 9:e17488. https://doi.org/10.1016/j.heliyon.2023.e17488

Ferdous Z, Nemmar A (2020) Health impact of silver nanoparticles: a review of the biodistribution and toxicity following various routes of exposure. Int J Mol Sci 21:2375. https://doi.org/10.3390/ijms21072375

Fernández-Villa D, Aguilar MR, Rojo L (2019) Folic acid antagonists: antimicrobial and Immunomodulating mechanisms and applications. Int J Mol Sci 20:4996. https://doi.org/10.3390/ijms20204996

Finlay BB, Falkow S (1989) Common themes in microbial pathogenicity. Microbiol Rev 53:210–230. https://doi.org/10.1128/mr.53.2.210-230.1989

Fu Y, Liu W, Liu M et al (2021) *In vitro* anti-biofilm efficacy of sanguinarine against carbapenem-resistant *Serratia marcescens*. Biofouling 37:341–351. https://doi.org/10.1080/08927014.2021.1919649

Giordani S, Marassi V, Zattoni A et al (2023) Liposomes characterization for market approval as pharmaceutical products: analytical methods, guidelines and standardized protocols. J Pharm Biomed Anal 236:115751. https://doi.org/10.1016/j.jpba.2023.115751

Gulati R, Sharma S, Sharma RK (2022) Antimicrobial textile: recent developments and functional perspective. Polym Bull 79:5747–5771. https://doi.org/10.1007/s00289-021-03826-3

Gupta A, Gupta R, Singh RL (2017) Microbes and environment. In: Principles and applications of environmental biotechnology for a sustainable future. Springer Singapore, Singapore, pp 43–84

Habibi N, Kamaly N, Memic A, Shafiee H (2016) Self-assembled peptide-based nanostructures: smart nanomaterials toward targeted drug delivery. Nano Today 11:41–60. https://doi.org/10.1016/j.nantod.2016.02.004

Hampden-Martin A, Fothergill J, El Mohtadi M et al (2021) Photodynamic antimicrobial chemotherapy coupled with the use of the photosensitizers methylene blue and temoporfin as a potential novel treatment for Staphylococcus aureus in burn infections. Access Microbiol 3. https://doi.org/10.1099/acmi.0.000273

Hawkey PM (1998) The origins and molecular basis of antibiotic resistance. BMJ 317:657–660. https://doi.org/10.1136/bmj.317.7159.657

Hetta HF, Ramadan YN, Al-Harbi AI et al (2023) Nanotechnology as a promising approach to combat multidrug resistant bacteria: a comprehensive review and future perspectives. Biomedicines 11:413. https://doi.org/10.3390/biomedicines11020413

Huang Y, Guo X, Wu Y et al (2024) Nanotechnology's frontier in combatting infectious and inflammatory diseases: prevention and treatment. Signal Transduct Target Ther 9:34. https://doi.org/10.1038/s41392-024-01745-z

Ibarra LE, Beaugé L, Arias-Ramos N et al (2020) Trojan horse monocyte-mediated delivery of conjugated polymer nanoparticles for improved photodynamic therapy of glioblastoma. Nanomedicine 15:1687–1707. https://doi.org/10.2217/nnm-2020-0106

Ibrahim J, Eisen JA, Jospin G et al (2016) Genome analysis of streptococcus pyogenes associated with pharyngitis and skin infections. PLoS One 11:e0168177. https://doi.org/10.1371/journal.pone.0168177

Ingerson-Mahar M, Reid A (2011) FAQ: E. Coli: good, bad, & deadly

Iskandar K, Murugaiyan J, Hammoudi Halat D et al (2022) Antibiotic discovery and resistance: the chase and the race. Antibiotics 11:182. https://doi.org/10.3390/antibiotics11020182

Islam MA, Kundu S, Alam SS et al (2021) Prevalence and characteristics of fever in adult and paediatric patients with coronavirus disease 2019 (COVID-19): a systematic review and meta-analysis of 17515 patients. PLoS One 16:e0249788. https://doi.org/10.1371/journal.pone.0249788

Julián P, Milon P, Agirrezabala X et al (2011) The Cryo-EM structure of a complete 30S translation initiation complex from Escherichia coli. PLoS Biol 9:e1001095. https://doi.org/10.1371/journal.pbio.1001095

Kandi V, Vaish R (2019) Diphtheria or streptococcal pharyngitis: a case report highlighting the diagnostic dilemma in the post-vaccination era. Cureus 11:e6190. https://doi.org/10.7759/cureus.6190

Kapoor G, Saigal S, Elongavan A (2017) Action and resistance mechanisms of antibiotics: a guide for clinicians. J Anaesthesiol Clin Pharmacol 33:300. https://doi.org/10.4103/joacp.JOACP_349_15

Kazimierska K, Kalinowska-Lis U (2021) Milk proteins – their biological activities and use in cosmetics and dermatology. Molecules 26:3253. https://doi.org/10.3390/molecules26113253

Kimbrell JB, Hite JR, Skala KN et al (2011) Direct binding of halide ions by valinomycin. Supramol Chem 23:782–789. https://doi.org/10.1080/10610278.2011.593627

Kolsin JM, Lopman BA, Payne DC et al (2018) Evaluating previous antibiotic use as a risk factor for acute gastroenteritis among children in Davidson County, Tennessee, 2014–2015. J Pediatr Infect Dis Soc 7:e86–e91. https://doi.org/10.1093/jpids/piy044

Kumar V, Yasmeen N, Pandey A et al (2023) Antibiotic adjuvants: synergistic tool to combat multi-drug resistant pathogens. Front Cell Infect Microbiol 13. https://doi.org/10.3389/fcimb.2023.1293633

Kunz Coyne AJ, El Ghali A, Holger D et al (2022) Therapeutic strategies for emerging multidrug-resistant Pseudomonas aeruginosa. Infect Dis Ther 11:661–682. https://doi.org/10.1007/s40121-022-00591-2

Lamut A, Peterlin Mašič L, Kikelj D, Tomašič T (2019) Efflux pump inhibitors of clinically relevant multidrug resistant bacteria. Med Res Rev 39:2460–2504. https://doi.org/10.1002/med.21591

Laursen BS, Sørensen HP, Mortensen KK, Sperling-Petersen HU (2005) Initiation of protein synthesis in bacteria. Microbiol Mol Biol Rev 69:101–123. https://doi.org/10.1128/MMBR.69.1.101-123.2005

Laverty G, Gorman SP, Gilmore BF (2011) The potential of antimicrobial peptides as biocides. Int J Mol Sci 12:6566–6596. https://doi.org/10.3390/ijms12106566

Lavi O, Gottesman MM, Levy D (2012) The dynamics of drug resistance: a mathematical perspective. Drug Resist Updat 15:90–97. https://doi.org/10.1016/j.drup.2012.01.003

Laws M, Shaaban A, Rahman KM (2019) Antibiotic resistance breakers: current approaches and future directions. FEMS Microbiol Rev 43:490–516. https://doi.org/10.1093/femsre/fuz014

Leanse LG, Marasini S, dos Anjos C, Dai T (2023) Antimicrobial resistance: is there a 'light' at the end of the tunnel? Antibiotics 12:1437. https://doi.org/10.3390/antibiotics12091437

Lechner D, Gibbons S, Bucar F (2008) Plant phenolic compounds as ethidium bromide efflux inhibitors in mycobacterium smegmatis. J Antimicrob Chemother 62:345–348. https://doi.org/10.1093/jac/dkn178

Li X-Z, Nikaido H (2009) Efflux-mediated drug resistance in bacteria. Drugs 69:1555–1623. https://doi.org/10.2165/11317030-000000000-00000

Liang Z, Liu X, Qin Z et al (2022) Photodynamic inactivation of Shigella flexneri by curcumin. LWT 153:112491. https://doi.org/10.1016/j.lwt.2021.112491

Lin H-S, Lin P-T, Tsai Y-S et al (2018) Interventions for bacterial folliculitis and boils (furuncles and carbuncles). Cochrane Database Syst Rev. https://doi.org/10.1002/14651858.CD013099

Liu Y, Shi X, Tian Y et al (2023) An insight into novel therapeutic potentials of taxifolin. Front Pharmacol 14. https://doi.org/10.3389/fphar.2023.1173855

Llor C, Bjerrum L (2014) Antimicrobial resistance: risk associated with antibiotic overuse and initiatives to reduce the problem. Ther Adv Drug Saf 5:229–241. https://doi.org/10.1177/2042098614554919

Lombardo D, Kiselev MA (2022) Methods of liposomes preparation: formation and control factors of versatile Nanocarriers for biomedical and Nanomedicine application. Pharmaceutics 14:543. https://doi.org/10.3390/pharmaceutics14030543

Lopatkin AJ, Sysoeva TA, You L (2016) Dissecting the effects of antibiotics on horizontal gene transfer: analysis suggests a critical role of selection dynamics. BioEssays 38:1283–1292. https://doi.org/10.1002/bies.201600133

Mahor A, Singh PP, Bharadwaj P et al (2021) Carbon-based nanomaterials for delivery of biologicals and therapeutics: a cutting-edge technology. C (Basel) 7:19. https://doi.org/10.3390/c7010019
Mancuso G, Midiri A, Gerace E, Biondo C (2021) Bacterial antibiotic resistance: the Most critical pathogens. Pathogens 10:1310. https://doi.org/10.3390/pathogens10101310
Meltzer EO, Hamilos DL, Hadley JA et al (2004) Otolaryngol Head Neck Surg 131:1–62. https://doi.org/10.1016/j.otohns.2004.09.067
Meng X-Y, Zhang H-X, Mezei M, Cui M (2011) Molecular docking: a powerful approach for structure-based drug discovery. Curr Comput Aided-Drug Des 7:146–157. https://doi.org/10.2174/157340911795677602
Michaelis C, Grohmann E (2023) Horizontal gene transfer of antibiotic resistance genes in biofilms. Antibiotics 12:328. https://doi.org/10.3390/antibiotics12020328
Miladi H, Zmantar T, Chaabouni Y et al (2016) Antibacterial and efflux pump inhibitors of thymol and carvacrol against food-borne pathogens. Microb Pathog 99:95–100. https://doi.org/10.1016/j.micpath.2016.08.008
Miró EM, Sánchez NP (2012) Cutaneous manifestations of infectious diseases. In: Atlas of dermatology in internal medicine. Springer, New York, pp 77–119
Mirza ZM, Kumar A, Kalia NP et al (2011) Piperine as an inhibitor of the MdeA efflux pump of Staphylococcus aureus. J Med Microbiol 60:1472–1478. https://doi.org/10.1099/jmm.0.033167-0
Mosby CA, Edelmann MJ, Jones MK (2023) Murine norovirus interaction with Enterobacter cloacae leads to changes in membrane stability and packaging of lipid and metabolite vesicle content. Microbiol Spectr 11. https://doi.org/10.1128/spectrum.04691-22
Munita JM, Arias CA (2016) Mechanisms of antibiotic resistance. Microbiol Spectr 4. https://doi.org/10.1128/microbiolspec.VMBF-0016-2015
Murray CJL, Ikuta KS, Sharara F et al (2022) Global burden of bacterial antimicrobial resistance in 2019: a systematic analysis. Lancet 399:629–655. https://doi.org/10.1016/S0140-6736(21)02724-0
Murugaiyan J, Kumar PA, Rao GS et al (2022) Progress in alternative strategies to combat antimicrobial resistance: focus on antibiotics. Antibiotics 11:200. https://doi.org/10.3390/antibiotics11020200
Muteeb G, Rehman MT, Shahwan M, Aatif M (2023) Origin of antibiotics and antibiotic resistance, and their impacts on drug development: a narrative review. Pharmaceuticals 16:1615. https://doi.org/10.3390/ph16111615
Nasim N, Sandeep IS, Mohanty S (2022) Plant-derived natural products for drug discovery: current approaches and prospects. Nucleus 65:399–411. https://doi.org/10.1007/s13237-022-00405-3
Navarre WW, Schneewind O (1999) Surface proteins of gram-positive bacteria and mechanisms of their targeting to the Cell Wall envelope. Microbiol Mol Biol Rev 63:174–229. https://doi.org/10.1128/MMBR.63.1.174-229.1999
Nikaido H (2009) Multidrug resistance in bacteria. Annu Rev Biochem 78:119–146. https://doi.org/10.1146/annurev.biochem.78.082907.145923
Nowicki J, Murray MT (2020) Bronchitis and pneumonia. In: Textbook of natural medicine. Elsevier, pp 1196–1201
Oates JA, Wood AJJ, Douglas RG (1990) Prophylaxis and treatment of influenza. N Engl J Med 322:443–450. https://doi.org/10.1056/NEJM199002153220706
Oyim J, Omolo CA, Amuhaya EK (2021) Photodynamic antimicrobial chemotherapy: advancements in porphyrin-based photosensitize development. Front Chem 9. https://doi.org/10.3389/fchem.2021.635344
Palazzotti D, Felicetti T, Sabatini S et al (2023) Fighting antimicrobial resistance: insights on how the *staphylococcus aureus*NorA efflux pump recognizes 2-Phenylquinoline inhibitors by supervised molecular dynamics (SuMD) and molecular docking simulations. J Chem Inf Model 63:4875–4887. https://doi.org/10.1021/acs.jcim.3c00516

Pandimurugan R, Thambidurai S (2017) UV protection and antibacterial properties of seaweed capped ZnO nanoparticles coated cotton fabrics. Int J Biol Macromol 105:788–795. https://doi.org/10.1016/j.ijbiomac.2017.07.097

Pereira FDO, Wanderley PA, Viana FAC, Lima RBD, Sousa FBD, Lima EDO (2011) Growth inhibition and morphological alterations of Trichophyton rubrum induced by essential oil from Cymbopogon winterianus Jowitt ex Bor. Braz J Microbiol 42:233–242. https://doi.org/10.1590/S1517-83822011000100029

Pham T, Loupias P, Dassonville-Klimpt A, Sonnet P (2019) Drug delivery systems designed to overcome antimicrobial resistance. Med Res Rev 39:2343–2396. https://doi.org/10.1002/med.21588

Prestinaci F, Pezzotti P, Pantosti A (2015) Antimicrobial resistance: a global multifaceted phenomenon. Pathog Glob Health 109:309–318. https://doi.org/10.1179/2047773215Y.0000000030

Pulingam T, Parumasivam T, Gazzali AM et al (2022) Antimicrobial resistance: prevalence, economic burden, mechanisms of resistance and strategies to overcome. Eur J Pharm Sci 170:106103. https://doi.org/10.1016/j.ejps.2021.106103

Raff AB, Kroshinsky D (2016) Cellulitis. JAMA 316:325. https://doi.org/10.1001/jama.2016.8825

Rahman MM, Rahaman MS, Islam MR et al (2021) Multifunctional therapeutic potential of Phytocomplexes and natural extracts for antimicrobial properties. Antibiotics 10:1076. https://doi.org/10.3390/antibiotics10091076

Reygaert WC (2018) An overview of the antimicrobial resistance mechanisms of bacteria. AIMS Microbiol 4:482–501. https://doi.org/10.3934/microbiol.2018.3.482

Román LE, Gomez ED, Solís JL, Gómez MM (2020) Antibacterial cotton fabric functionalized with copper oxide nanoparticles. Molecules 25:5802. https://doi.org/10.3390/molecules25245802

Ryu S, Han HM, Song PI et al (2015) Suppression of Propionibacterium acnes infection and the associated inflammatory response by the antimicrobial peptide P5 in mice. PLoS One 10:e0132619. https://doi.org/10.1371/journal.pone.0132619

Salam MA, Al-Amin MY, Salam MT et al (2023) Antimicrobial resistance: a growing serious threat for global public health. Healthcare 11:1946. https://doi.org/10.3390/healthcare11131946

Sarkar P, Yarlagadda V, Ghosh C, Haldar J (2017) A review on cell wall synthesis inhibitors with an emphasis on glycopeptide antibiotics. Medchemcomm 8:516–533. https://doi.org/10.1039/C6MD00585C

Seung KJ, Keshavjee S, Rich ML (2015) Multidrug-resistant tuberculosis and extensively drug-resistant tuberculosis. Cold Spring Harb Perspect Med 5:a017863. https://doi.org/10.1101/cshperspect.a017863

Shahzad S, Qadir MA, Ahmed M et al (2020) Folic acid-sulfonamide conjugates as antibacterial agents: design, synthesis and molecular docking studies. RSC Adv 10:42983–42992. https://doi.org/10.1039/D0RA09051D

Sharma A, Gupta V, Pathania R (2019) Efflux pump inhibitors for bacterial pathogens: from bench to bedside. Indian J Med Res 149:129. https://doi.org/10.4103/ijmr.IJMR_2079_17

Silhavy TJ, Kahne D, Walker S (2010) The bacterial cell envelope. Cold Spring Harb Perspect Biol 2:a000414. https://doi.org/10.1101/cshperspect.a000414

Singh A, Duche RT, Wandhare AG et al (2023) Milk-derived antimicrobial peptides: overview, applications, and future perspectives. Probiotics Antimicrob Proteins 15:44–62. https://doi.org/10.1007/s12602-022-10004-y

Siriwong S, Teethaisong Y, Thumanu K et al (2016) The synergy and mode of action of quercetin plus amoxicillin against amoxicillin-resistant Staphylococcus epidermidis. BMC Pharmacol Toxicol 17:39. https://doi.org/10.1186/s40360-016-0083-8

Stillwell W (2013) Membrane transport. In: An introduction to biological membranes. Elsevier, pp 305–337

Strahl T, Thorner J (2007) Synthesis and function of membrane phosphoinositides in budding yeast, Saccharomyces cerevisiae. Biochim Biophys Acta 1771:353–404. https://doi.org/10.1016/j.bbalip.2007.01.015

Suganya T, Packiavathy IASV, Aseervatham GSB et al (2022) Tackling multiple-drug-resistant bacteria with conventional and complex phytochemicals. Front Cell Infect Microbiol 12. https://doi.org/10.3389/fcimb.2022.883839

Taneja N, Sharma M (2019) Antimicrobial resistance in the environment: the Indian scenario. Indian J Med Res 149:119. https://doi.org/10.4103/ijmr.IJMR_331_18

Thomford N, Senthebane D, Rowe A et al (2018) Natural products for drug discovery in the 21st century: innovations for novel drug discovery. Int J Mol Sci 19:1578. https://doi.org/10.3390/ijms19061578

Tse-Dinh Y-C (2007) Exploring DNA topoisomerases as targets of novel therapeutic agents in the treatment of infectious diseases. Infect Disord Drug Targets 7:3–9. https://doi.org/10.2174/187152607780090748

Turabi KS, Sannakki JA, Aich J et al (2023) Computational screening of phytochemicals for antibacterial drug discovery. In: Phytochemistry, computational tools and databases in drug discovery. Elsevier, pp 213–243

Uddin TM, Chakraborty AJ, Khusro A et al (2021) Antibiotic resistance in microbes: history, mechanisms, therapeutic strategies and future prospects. J Infect Public Health 14:1750–1766. https://doi.org/10.1016/j.jiph.2021.10.020

Vaou N, Stavropoulou E, Voidarou C et al (2021) Towards advances in medicinal plant antimicrobial activity: a review study on challenges and future perspectives. Microorganisms 9:2041. https://doi.org/10.3390/microorganisms9102041

Ventola CL (2015) The antibiotic resistance crisis: part 1: causes and threats. Pharm Ther 40:277–283

Villanueva JA, Crooks AL, Nagy TA et al (2022) Salmonella enterica infections are disrupted by two small molecules that accumulate within phagosomes and differentially damage bacterial inner membranes. MBio 13:e0179022. https://doi.org/10.1128/mbio.01790-22

Wang K, Bettiol S, Thompson MJ et al (2014) Symptomatic treatment of the cough in whooping cough. Cochrane Database Syst Rev 2014:CD003257. https://doi.org/10.1002/14651858.CD003257.pub5

Wang F, Liu Z, Zhang D, Niu X (2017) In vitro activity of farnesol against vaginal lactobacillus spp. Eur J Obstet Gynecol Reprod Biol 212:25–29. https://doi.org/10.1016/j.ejogrb.2017.03.010

Xia Q, Zhang Y, Li Z et al (2019) Red blood cell membrane-camouflaged nanoparticles: a novel drug delivery system for antitumor application. Acta Pharm Sin B 9:675–689. https://doi.org/10.1016/j.apsb.2019.01.011

Yamaguchi M, Minamide Y, Terao Y et al (2009) Nrc of Streptococcus pneumoniae suppresses capsule expression and enhances anti-phagocytosis. Biochem Biophys Res Commun 390:155–160. https://doi.org/10.1016/j.bbrc.2009.09.099

Yang L, Ding W, Xu Y et al (2016) New insights into the antibacterial activity of Hydroxycoumarins against Ralstonia solanacearum. Molecules 21:468. https://doi.org/10.3390/molecules21040468

Yao J, Zou P, Cui Y et al (2023) Recent advances in strategies to combat bacterial drug resistance: antimicrobial materials and drug delivery systems. Pharmaceutics 15:1188. https://doi.org/10.3390/pharmaceutics15041188

Youlden G, McNeil HE, Blair JMA et al (2022) Mathematical modelling highlights the potential for genetic manipulation as an adjuvant to counter efflux-mediated MDR in salmonella. Bull Math Biol 84:56. https://doi.org/10.1007/s11538-022-01011-9

Zhang Y, Zhang J, Chen W et al (2017) Erythrocyte membrane-coated nanogel for combinatorial antivirulence and responsive antimicrobial delivery against Staphylococcus aureus infection. J Control Release 263:185–191. https://doi.org/10.1016/j.jconrel.2017.01.016

Zhang L, Tian X, Sun L et al (2024) Bacterial efflux pump inhibitors reduce antibiotic resistance. Pharmaceutics 16:170. https://doi.org/10.3390/pharmaceutics16020170

Zhen X, Lundborg CS, Sun X et al (2019) Economic burden of antibiotic resistance in ESKAPE organisms: a systematic review. Antimicrob Resist Infect Control 8:137. https://doi.org/10.1186/s13756-019-0590-7

Zhou Y, Zhou Z, Zheng L et al (2023) Urinary tract infections caused by Uropathogenic Escherichia coli: mechanisms of infection and treatment options. Int J Mol Sci 24:10537. https://doi.org/10.3390/ijms241310537

Zore G, Thakre A, Abdulghani M et al (2022) Menthol inhibits Candida albicans growth by affecting the membrane integrity followed by apoptosis. Evid Based Complement Alternat Med 2022:1–10. https://doi.org/10.1155/2022/1297888

Zou J, Zheng M-W, Li G, Su Z-G (2013) Advanced systems biology methods in drug discovery and translational biomedicine. Biomed Res Int 2013:1–8. https://doi.org/10.1155/2013/742835

Approaches of Bioinformatics in Antibacterial Drug Development

11

Smriti Dewangan and Varsha Rawat

Abstract

In the face of rising antibiotic resistance and a dwindling pipeline of novel antibacterial agents, bioinformatics has emerged as a critical tool in the search for effective antibacterial drugs. Bioinformatics harnesses the power of computational tools and data-driven analysis to expedite the discovery, design, and development of antibacterial agents. One prominent application is in genomics, where comparative genomics, metagenomics, and functional genomics are employed to elucidate bacterial genomics associated with resistance mechanisms, virulence factors, and potential drug targets. This information aids in the identification of essential bacterial proteins and pathways that can be exploited for drug development. Bioinformatics is crucial in high-throughput screening, efficiently evaluating large chemical libraries against multiple bacterial targets. It employs virtual screening, molecular docking, and simulations to predict interactions, aiding in prioritizing drug candidates. Additionally, bioinformatics significantly contributes to rational drug design by identifying and validating drug targets. Computational methods like structure-based drug design and pharmacophore modeling enhance the development of better antibacterial agents, while metagenomics aids in finding new natural products with antibacterial properties. Collaborative research and data sharing have facilitated the creation of databases and tools essential for antibacterial drug development, expediting the process by providing key information on bacterial genomes, resistance, and chemicals.

S. Dewangan (✉) · V. Rawat
MATS School of Pharmacy, MATS University, Raipur, Chhattisgarh, India
e-mail: smritidewangan88@gmail.com

T. Sharma et al. (eds.), *Natural Products for Antibacterial Drug Development: Recent Advancement of Computational Approach*,
https://doi.org/10.1007/978-981-97-9634-2_11

Keywords

Antibiotic resistance · Antibacterial agents · Bioinformatics · Computational tools · Data-driven analysis

Abbreviations

ADME/T	Absorption, Distribution, Metabolism, Excretion, and Toxicity
ARDB	Antibiotic Resistance Genes Database
ARG-ANNOT	Antibiotic resistance gene annotation
ARLG	Antibacterial Resistance Leadership Group
CARB-X	Combating antibiotic-resistant bacteria biopharmaceutical accelerator
CARD	Comprehensive Antibiotic Resistance Database
ENA	European Nucleotide Archive
GARDP	Global Antibiotic Research and Development Partnership
GenoList	Genome Database for Enteric Bacteria
IMG	Integrated Microbial Genomes
IMMPACT	Initiative on Methods, Measurement, and Pain Assessment in Clinical Trials
JPIAMR	European Commission Joint Programming Initiative on Antimicrobial Resistance
NCBI	National Center for Biotechnology Information
NIAID	National Institute of Allergy and Infectious Diseases
PATRIC	Pathosystems Resource Integration Center
RefSeq	Reference Sequence

11.1 Introduction

11.1.1 Background and Significance of Antibacterial Drug Development

The development of antibacterial drugs is a critical and ongoing endeavor in medicine and pharmaceutical research. Antibacterial drugs, also known as antibiotics, are pivotal in treating bacterial infections, a significant cause of morbidity and mortality worldwide (Kon and Rai 2016). Understanding the background and significance of antibacterial drug development is essential to appreciate the importance of this field. Antibiotics have revolutionized medicine since the discovery of penicillin by Alexander Fleming in 1928. The subsequent development of antibiotics like streptomycin, tetracycline, and erythromycin marked a new era in health care, as they effectively treated various bacterial infections (Sagar et al. 2019). However, over the decades, bacteria have evolved resistance mechanisms against many antibiotics, making it increasingly challenging to combat infections. This necessitates continuous efforts to develop new antibacterial drugs. Antibiotic resistance has become a global health crisis. Bacteria can acquire resistance to antibiotics through genetic mutations or horizontal gene transfer, rendering once-effective drugs ineffective (Uddin et al. 2021). Multidrug-resistant bacteria, often referred to as "superbugs," are a significant threat to public health, as they cause infections that are difficult, and sometimes impossible, to treat with existing antibiotics. This underscores the urgency of developing new antibacterial agents (Medina and Pieper 2016). Despite the growing need for new antibiotics, the development pipeline has slowed. Many pharmaceutical companies have shifted their focus to more profitable therapeutic areas, as developing antibacterial drugs is complex, time-consuming, and financially challenging. This has resulted in a shortage of novel antibiotics to combat emerging resistant strains (Miethke et al. 2021). Antibacterial drug development is essential for preserving the gains made in modern medicine. Routine surgical procedures, cancer treatments, organ transplants, and even childbirth depend on the availability of effective antibiotics to prevent or treat bacterial infections. Without reliable antibacterial drugs, these medical procedures become riskier and potentially life-threatening (Terreni et al. 2021). Emerging infectious diseases, such as pandemics like COVID-19, highlight the interconnectedness of global health. Bacterial infections can exacerbate these public health crises, making the development of new antibiotics a priority for patient care and global health security (Cameron et al. 2022). The field of antibacterial drug development is witnessing innovative approaches, including the exploration of new drug classes, combination therapies, and the development of antibacterials with novel mechanisms of action. Researchers are also focusing on precision medicine to tailor treatments to individual patients, minimizing the development of resistance. Figure 11.1 represents an overview of the key milestones and eras in antibacterial drug development from 1910 to the present.

11.1.2 Role of Bioinformatics in Drug Discovery

Bioinformatics is pivotal in advancing antibacterial drug discovery through its multifaceted contributions. By harnessing computational tools and techniques, bioinformatics enables the analysis of bacterial genomes to identify essential genes and potential drug targets. It facilitates the prediction of three-dimensional protein structures, allowing for the virtual screening of compound libraries and the selection of promising drug candidates (Wooller et al. 2017). Additionally, bioinformatics aids in exploring existing drugs for new antibacterial applications through drug repurposing efforts. Systems biology and pathway analysis provide a comprehensive understanding of bacterial physiology, which is crucial for designing interventions that disrupt specific pathways (Vila et al. 2020). Bioinformatics also contributes to predicting and monitoring antibiotic resistance, guiding the development of effective drugs against resistant strains (Swaminathan et al. 2016). In essence, integrating diverse biological data and computational analyses accelerates the entire drug discovery process, from target identification to developing effective antibacterial agents. Bioinformatics helps in various functional domains, such as drug discovery and vaccine development, prophylaxis, and studying gene function, as mentioned in Fig. 11.2.

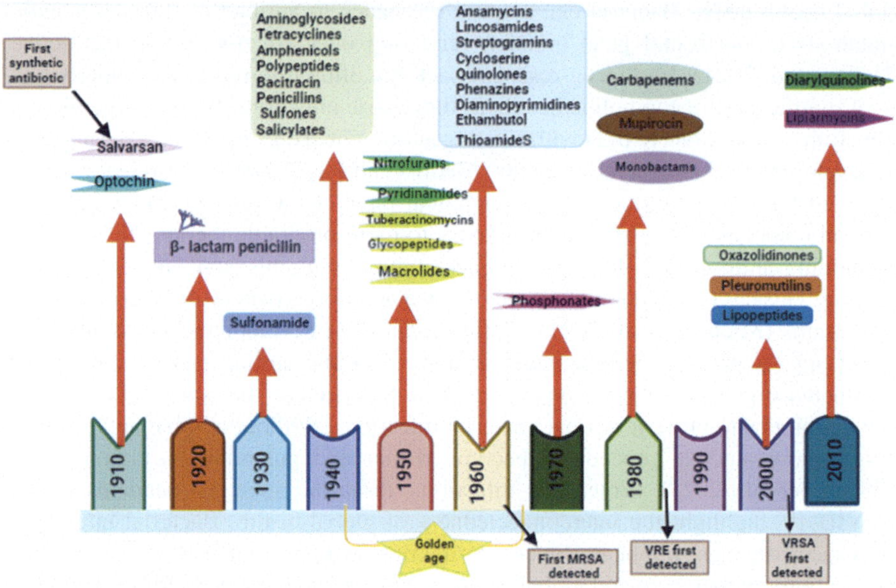

Fig. 11.1 Timeline study of antibacterial drug

11.2 Genomic Approaches

In antibacterial drug discovery, genomics plays a crucial role in unraveling the intricate landscape of bacterial genetics, and bioinformatics serves as the key facilitator in harnessing this wealth of genomic information.

11.2.1 Role of Genomics in Understanding Bacterial Genetics

Genome sequencing, a fundamental aspect of genomics, involves decoding the entire genetic blueprint of bacteria, encompassing genes, regulatory elements, and non-coding regions (Monaghan and Barrett 2006). This comprehensive genomic data offers a foundation for various analytical approaches. Gene identification becomes possible, enabling researchers to annotate individual genes and understand their roles in bacterial traits like virulence, antibiotic resistance, and metabolism (Bansal 2008). Comparative genomics allows for the exploration of genetic variations between bacterial strains, shedding light on evolutionary adaptations and unique genetic features. Functional annotation, another genomic avenue, aids in predicting the functions of bacterial genes, contributing to our understanding of their biological significance (Sivashankari and Shanmughavel 2007).

Moreover, genomics provides insights into the evolutionary history of bacterial species, tracing the gain or loss of genes over time (Loman and Pallen 2015). The study of mobile genetic elements, facilitated by genomic data, uncovers the

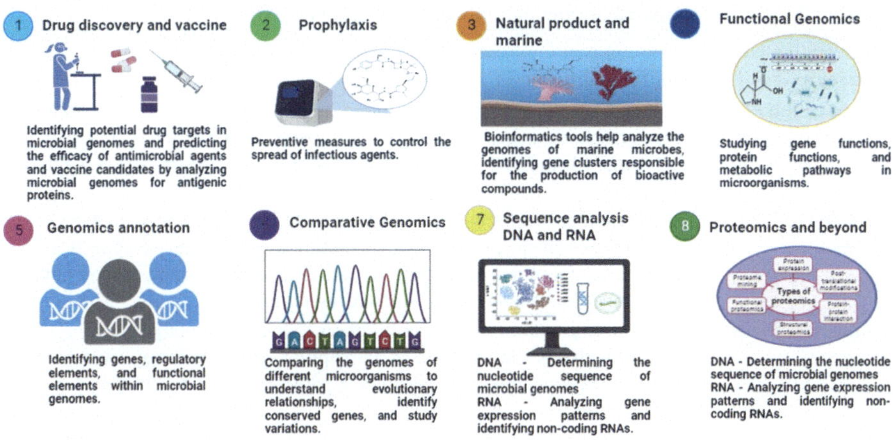

Fig. 11.2 Applications of bioinformatics in various approaches

presence of elements like plasmids and transposons, which contribute to bacterial evolution and the spread of antibiotic resistance (Broaders et al. 2013). The identification of antibiotic-resistance genes, virulence factors, and mechanisms of pathogenicity through genomic analysis is critical for developing strategies to combat bacterial infections. Additionally, genomics delves into gene regulation mechanisms, unveiling transcription factors, operons, and regulatory elements that govern bacterial responses to environmental changes (Chuan et al. 2022).

Functional genomics techniques, including transcriptomics, proteomics, and metabolomics, offer a deeper understanding of gene expression, protein production, and metabolic pathways in bacteria (Franzosa et al. 2015). This holistic view of genetic and physiological responses is invaluable in drug discovery efforts. Lastly, applying genomic approaches to study bacterial communities in complex environments, known as metagenomics, provides insights into the genetic diversity and functional potential of microbial populations in diverse ecosystems, such as the human gut (Zhong et al. 2021). In essence, the role of bioinformatics in antibacterial drug discovery lies in harnessing the power of genomics, employing computational tools to analyze, interpret, and leverage the vast array of genomic data for the identification of drug targets, understanding bacterial biology, and ultimately developing effective antibacterial agents.

11.2.2 Comparative Genomics in Identifying Drug Targets

Comparative genomics is a powerful approach for identifying potential drug targets in various organisms, including bacteria, viruses, and eukaryotes. This method involves comparing the genomes of different species to find conserved genes or proteins that are essential for the survival or pathogenicity of the target organism

Table 11.1 Role of comparative genomics in identifying drug target

Comparative genomics	Identifying drug target
Identification of conserved genes	By comparing the genomes of different organisms, researchers can identify highly conserved genes across species. These conserved genes are likely to be essential for the basic functions of the organism and represent potential drug targets
Orthology analysis	Comparative genomics can help identify orthologous genes, which are genes in different species that share a common evolutionary origin. If a gene is essential in one species, its orthologous in other species may have similar functions and could be viable drug targets
Pathway analysis	Comparative genomics allows researchers to analyze metabolic and signaling pathways across species. Disrupting critical pathways in a pathogenic organism while sparing host pathways can be a strategy for drug development
Virulence factors	In the case of pathogens, comparative genomics can be used to identify genes that are specific to the pathogen and contribute to its virulence. Targeting these virulence factors can disrupt the pathogen's ability to cause disease
Antibiotic targets	In the context of bacteria, comparative genomics can help identify essential genes for bacterial survival. These genes could be targeted with antibiotics inhibit the growth of pathogenic bacteria selectively
Novel targets	Comparative genomics can reveal previously unknown or uncharacterized genes that are unique to the target organism. These novel genes may have essential functions and could serve as promising drug targets
Drug-resistance mechanisms	Comparative genomics can also be used to understand the genetic basis of drug resistance in pathogens. By identifying the genes or mutations responsible for resistance, researchers can develop strategies to overcome resistance and design more effective drugs.
Host–pathogen interactions	In the case of infectious diseases, comparing the genomes of both the host and the pathogen can provide insights into the interactions between the two. This knowledge can help identify targets for disrupting the pathogen's ability to infect and survive in the host
Functional genomics	Beyond sequence comparison, functional genomics approaches can validate potential drug targets by studying gene expression, protein-protein interactions, and the impact of gene knockdown or knockout on the organism's viability.
Genomic epidemiology	Comparative genomics can also be applied in the context of tracking and understanding the evolution of drug-resistant strains in infectious diseases, which is critical for public health efforts

(Shahbaaz et al. 2016). Table 11.1 describes how comparative genomics can be applied to identify drug targets.

11.2.3 Metagenomics for Discovering Novel Antibacterial Agents

Metagenomics is a powerful approach for discovering novel antibacterial agents by studying the genetic material of entire microbial communities in various environments, including soil, water, and the human microbiome. This approach can lead to the identification of previously unknown compounds and pathways with antibacterial properties (Dos Santos et al. 2017; Schmieder and Edwards 2012). Figure 11.3 describes the application of metagenomics in discovering novel antibacterial drugs.

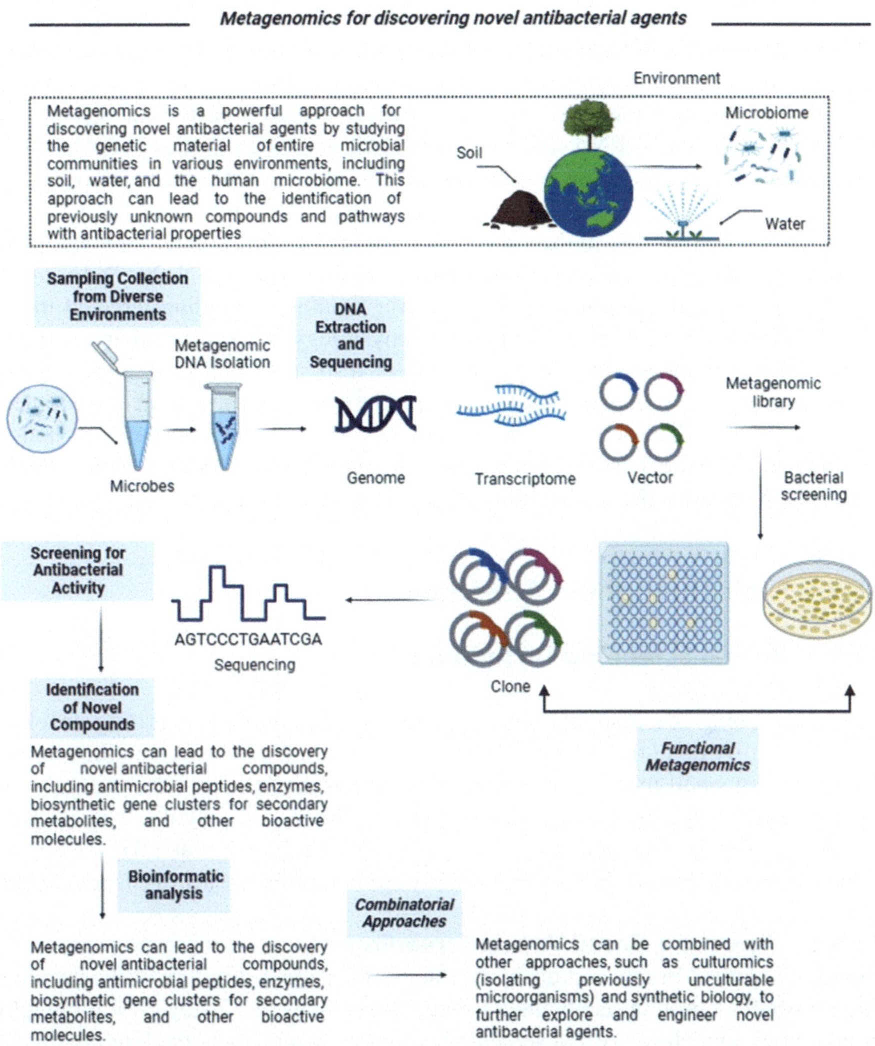

Fig. 11.3 Metagenomics for discovering novel antibacterial agents

11.2.4 Functional Genomics and Its Applications

Functional genomics, a crucial field in biological research, delves into unraveling the mysteries of gene functions and the broader genome landscape. Its primary goal is to answer the fundamental question, "what do genes do?" (Kampourakis 2017). Through a combination of experimental and computational techniques, functional genomics explores gene regulation, RNA and protein functions, and their contributions to diverse biological processes. This multidimensional approach leads to various applications, such as gene function annotation, gene expression profiling, identification of regulatory elements, and proteomics (Bunnik and Le Roch 2013). High-throughput functional genomics screens systematically investigate gene roles, while technologies like RNA interference (RNAi) and CRISPR/Cas9 enable selective gene manipulation. Pathway analysis unveils intricate biological networks, and functional genomics extends its reach to non-coding RNA molecules. Beyond the lab, it plays a pivotal role in drug discovery, cancer genomics, microbiology, plant biology, neurobiology, and evolutionary biology (Montalbano et al. 2017). In cancer research, functional genomics identifies driver mutations, guiding targeted therapies. In microbiology, it unravels microbial physiology and pathogenicity, while in plant biology, it sheds light on growth, stress responses, and crop improvement (Hawkes et al. 2021). In neuroscience, it explores genes' roles in brain development and neurological disorders, and in evolutionary biology, it contributes to understanding the genetic basis of adaptations across species (Wajman et al. 2015). Figure 11.4 specifies the applicability of genomics and its applications.

11.3 High-Throughput Screening

11.3.1 Virtual Screening Techniques

Virtual screening is a computational approach used in antibacterial drug development to identify potential drug candidates by simulating and analyzing the interactions between small molecules and bacterial targets, such as proteins or enzymes. This approach helps prioritize compounds for further experimental testing, significantly reducing the time and cost of drug discovery (Ballester et al. 2012). Here are some virtual screening techniques commonly used in antibacterial drug development:

11.3.1.1 Structure-Based Virtual Screening

Docking Studies In this approach, the three-dimensional structure of the bacterial target (usually a protein) is obtained through techniques like X-ray crystallography or homology modeling. Virtual screening software then predicts the binding modes and affinities of small molecules to the target. The goal is to identify compounds that can bind and inhibit the target's activity (Liu et al. 2013). Fig. 11.5 describes the various steps in molecular docking.

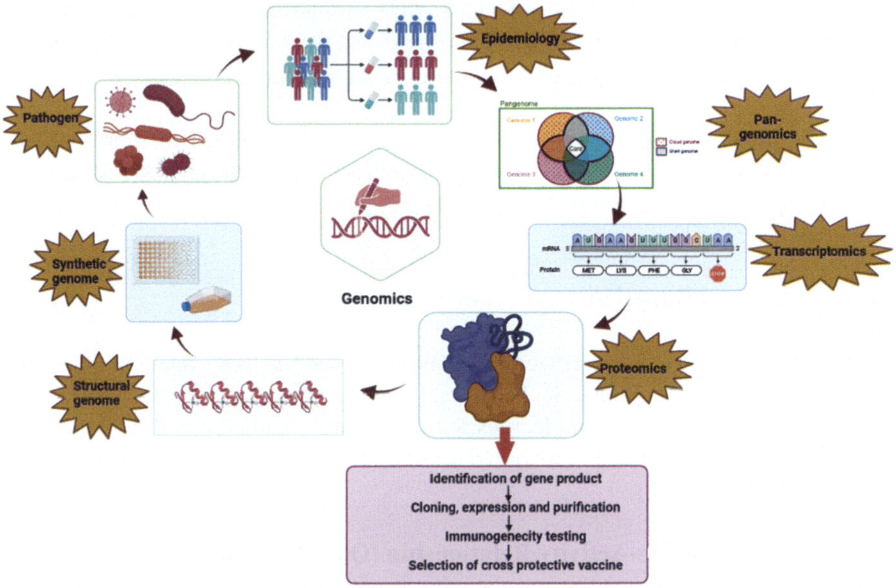

Fig. 11.4 Functional genomics and its applications

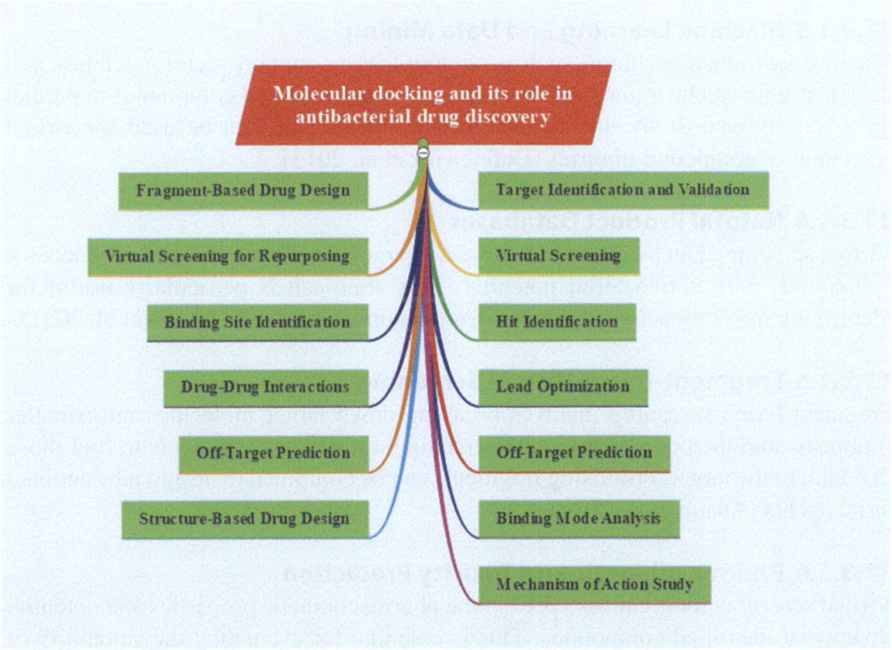

Fig. 11.5 Molecular docking and its role in antibacterial drug discovery

Molecular Dynamics (MD) Simulations MD simulations are used to study the dynamic behavior of the protein-ligand complex over time. This can provide insights into the stability and binding affinity of the interaction (Naik et al. 2022).

Pharmacophore Modeling Pharmacophore models are created to represent the essential features required for a compound to interact with the target. Virtual screening is then performed to identify compounds that match the pharmacophore features (Mathpal et al. 2021).

11.3.1.2 Ligand-Based Virtual Screening

Similarity Searching

In this approach, a known antibacterial compound with activity against a specific target is used as a query molecule. Virtual screening software searches a database of chemical compounds to find molecules with structural similarity to the query molecule. This method assumes that similar molecules have similar biological activity (Lian et al. 2021).

Quantitative Structure-Activity Relationship (QSAR) Models

QSAR models are developed to correlate the structural features of known antibacterial compounds with their biological activity. These models can be used to predict the activity of new compounds (Singh et al. 2012).

11.3.1.3 Machine Learning and Data Mining

Machine learning algorithms, such as random forests, support vector machines, and deep learning, can be trained on large datasets of antibacterial compounds to predict their activity against specific targets. These models can then be used for virtual screening of compound libraries (Butkiewicz et al. 2013).

11.3.1.4 Natural Product Databases

Virtual screening can be employed to search large databases of natural products for compounds with antibacterial potential. This approach is particularly useful for identifying novel bioactive molecules from natural sources (Atanasov et al. 2021).

11.3.1.5 Fragment-Based Virtual Screening

Fragment-based screening involves breaking down larger molecules into smaller fragments and then computationally screening a library of fragments to find those that bind to the target. Promising fragments can be combined to design new antibacterial agents (Austin et al. 2016).

11.3.1.6 Pharmacokinetic and Toxicity Prediction

Virtual screening tools can also predict the pharmacokinetic properties and potential toxicity of identified compounds. This is essential for evaluating the suitability of compounds for further development (Kumari et al. 2022).

11.3.1.7 Resistance Mechanism Prediction

Some virtual screening approaches focus on predicting the likelihood of bacterial resistance to antibacterial agents. Understanding resistance mechanisms can guide the development of more effective drugs (Roca et al. 2015).

11.3.1.8 Network Pharmacology

Network pharmacology uses a systems biology approach to analyze the interactions between drugs, targets, and biological pathways. It can help identify combinations of compounds that target multiple points in bacterial cellular processes (Leung et al. 2013).

11.3.1.9 Adaptive Strategies

Virtual screening can be combined with adaptive strategies where the virtual screening process is adjusted based on the results of initial rounds, allowing for the exploration of a broader chemical space (Puentes et al. 2020).

11.3.2 Molecular Docking and Its Role in Drug Discovery

Molecular docking is a computational technique used in drug discovery to predict and analyze the binding interactions between a small molecule (ligand) and a biological target, typically a protein (receptor). This approach plays a crucial role in various stages of drug discovery, including target identification, lead compound optimization, and the design of potential drug candidates. Here's an overview of molecular docking and its role in drug discovery (Stanzione et al. 2021).

11.3.2.1 Target Identification and Validation

Molecular docking can screen a large database of potential ligands against a known or modeled protein target. By predicting the binding affinity and mode of interaction, researchers can identify lead compounds for further investigation (Hossain et al. 2015).

11.3.2.2 Hit Identification

Once potential binding ligands are identified, molecular docking helps rank and select the most promising candidates. Hit identification is a crucial step in drug discovery, as it determines which compounds will progress to the next stages (Lans et al. 2020).

11.3.2.3 Lead Optimization

Molecular docking assists in the optimization of lead compounds by predicting how modifications to the ligand structure can improve binding affinity and selectivity. This iterative process helps refine lead compounds into more potent drug candidates (Omar et al. 2018).

11.3.2.4 Binding Mode Analysis

Molecular docking provides insights into the binding mode of ligands within the target's binding site. This information can reveal key interactions, such as hydrogen bonds, hydrophobic contacts, and electrostatic interactions, guiding further compound optimization (Abdolmaleki et al. 2018).

11.3.2.5 Structure-Based Drug Design

The 3D structural information generated by molecular docking is valuable for structure-based drug design. It allows medicinal chemists to design new compounds with optimal interactions with the target and predict how chemical modifications affect binding (Edache et al. 2020).

11.3.2.6 Mechanism of Action Study

Molecular docking can be used to investigate the mechanisms of drug action. It helps elucidate how a drug affects the biological target, essential for understanding its efficacy and potential side effects (Kabir and Muth 2022).

11.3.2.7 Off-Target Prediction

Molecular docking can predict potential off-target drug interactions, helping assess their safety and potential side effects. This is particularly important in minimizing unintended biological effects (Zheng et al. 2018).

11.3.2.8 Drug–Drug Interactions

Molecular docking can be used to study potential interactions between different drugs, helping to predict possible adverse drug–drug interactions in polypharmacy scenarios (Vilar and Hripcsak 2017).

11.3.2.9 Binding Site Identification

In cases where the binding site of a target is unknown, molecular docking can help predict potential binding sites on the protein, aiding in target identification and structure determination (Pulido et al. 2016).

11.3.2.10 Virtual Screening for Repurposing

Molecular docking can be applied to repurpose existing drugs for new therapeutic indications by assessing their binding to different targets (Luo et al. 2016).

11.3.2.11 Fragment-Based Drug Design

Molecular docking can be used in fragment-based drug design by examining the binding of smaller fragments to the target and guiding the assembly of these fragments into larger, more potent compounds (Selvaraj 2019).

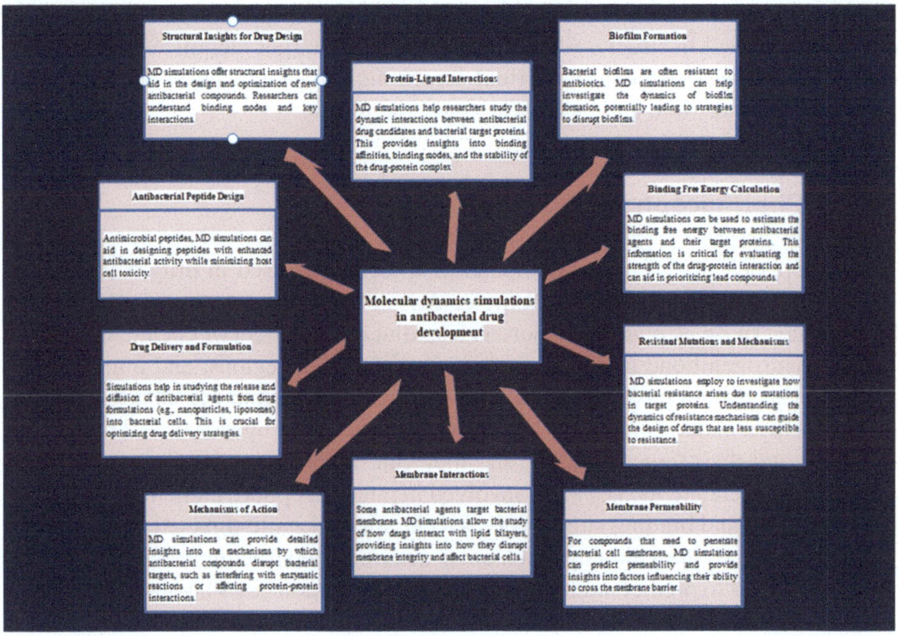

Fig. 11.6 Molecular dynamics simulations in antibacterial drug development

11.3.3 Molecular Dynamics Simulations in Antibacterial Drug Development

Molecular dynamics (MD) simulations are a powerful computational tool used in antibacterial drug development to understand the interactions between antibacterial agents and their target proteins as well as to investigate the behavior of bacterial membranes and other molecular components (Matamoros-Recio et al. 2021). Figure 11.6 describes the key roles and applications of MD simulations in antibacterial drug development.

11.4 Rational Drug Design

11.4.1 Structure-Based Drug Design

Structure-based drug design (SBDD) is a drug discovery approach that relies on the three-dimensional structures of biological targets (such as proteins or nucleic acids) to design and optimize drug candidates. This approach aims to understand the molecular interactions between a potential drug and its target at the atomic or molecular level, which can provide valuable insights for the rational design of new therapeutic agents. (Bruch et al. 2020). Techniques like X-ray crystallography, NMR spectroscopy, and cryo-electron microscopy enable the determination of protein structures at an atomic level (Kaur et al. 2021). The basic key steps and concepts involved in structure-based drug design are target identification and validation, structure determination, virtual screening, hit identification and optimization, lead optimization, and biological assay and testing.

11.4.2 Pharmacophore Modeling

Pharmacophore modeling is a computational technique used in drug discovery to identify the essential features of a molecular structure that are responsible for its biological activity. A pharmacophore is a spatial arrangement of chemical features such as hydrogen bond donors, acceptors, hydrophobic regions, and ionic centers that are critical for a molecule to interact with its biological target. Pharmacophore models help in understanding the key interactions between ligands (small molecules or drugs) and their target proteins, and they can be used for virtual screening and drug design. Here are the key steps involved in pharmacophore modeling. The identification of ligand binding sites involves techniques like site-directed mutagenesis, while virtual screening employs molecular docking and structure-based pharmacophore modeling to predict interactions between small molecules and target proteins. Fragment-based drug design, quantitative structure-activity relationship (QSAR) studies, and computational chemistry methods such as molecular dynamics simulations contribute to refining lead compounds (Pushkaran et al. 2019).

11.4.3 In Silico ADME/Tox Prediction

Considering ADME/T properties, iterative optimization ensures the enhancement of potency, selectivity, and pharmacokinetic characteristics. Case studies and learning from successful structures guide the design process, showcasing the precision and efficiency of SBDD in drug discovery. Pharmacophore modeling, a vital component of rational drug design, focuses on identifying essential features for ligand binding. It involves defining pharmacophores, collecting data from experimental studies, generating pharmacophore models, and validating them through experimental testing. Virtual screening and lead optimization based on structure–activity relationship studies are integral in applying pharmacophore models to design novel compounds with improved properties. In silico ADME/Tox prediction plays a crucial role by using computational methods to anticipate the pharmacokinetic and toxicological properties of potential drug candidates (Wang and Yang 2022). This includes predicting absorption, distribution, metabolism, excretion, and toxicity, and incorporating machine learning and predictive modeling. The integration of in silico ADME/Tox predictions into the drug design workflow helps prioritize and optimize candidates early in the process, reducing the risk of failures in later development stages (Alqahtani 2017). Together, these rational approaches significantly enhance the efficiency and success rate of drug discovery efforts. Figure 11.7 shows high-throughput screening and rational drug design.

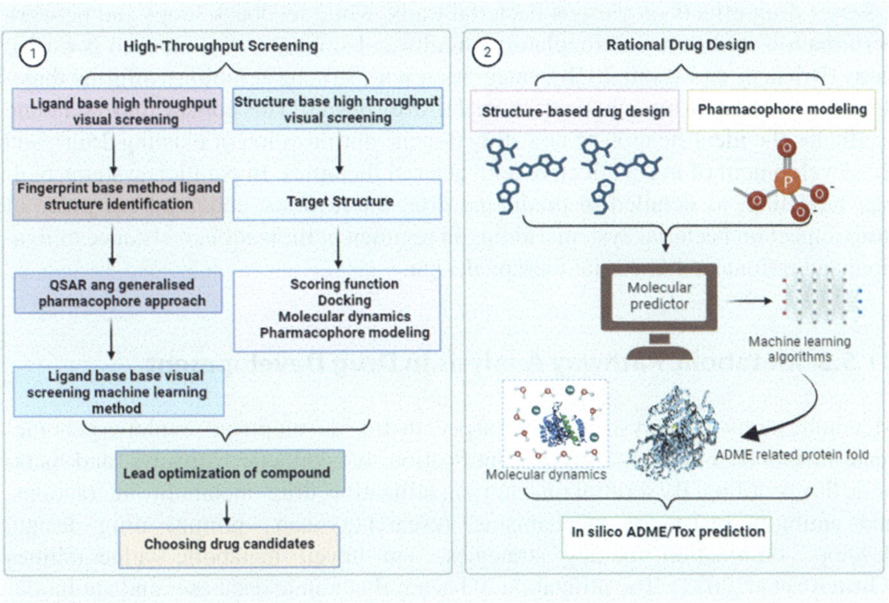

Fig. 11.7 High-throughput screening and rational drug design

11.5 Systems Biology and Pathway Analysis

11.5.1 Understanding Bacterial Pathways Related to Drug Response

Understanding bacterial pathways related to drug response is crucial for developing effective antibiotics and managing bacterial infections. The bacterial pathway in the context of drug response is an iterative process involving experimental and computational techniques.

11.5.2 Systems Biology Modeling for Predicting Drug Interactions

In the realm of antibacterial drug development, systems biology and pathway analysis serve as indispensable tools, offering a comprehensive understanding of bacterial pathways in response to drug interventions (Farha et al. 2021). These approaches unveil key genes and dynamic regulatory patterns by integrating genomic and transcriptomic data. Pathway mapping, including biochemical and metabolic analyses, pinpoints potential drug targets and illuminates impacts on bacterial metabolism (Kozlowska et al. 2014). Exploring protein–protein interaction networks identifies critical nodes, while functional genomics employs gene knockout studies and CRISPR-based approaches to discern essential pathways. Phenotypic profiling assesses drug effects on various bacterial traits, while feedback loops and network perturbation studies unveil regulatory circuits and simulate drug-induced perturbations (Brioschi and Banfi 2018). Integration with structural biology informs three-dimensional protein arrangements crucial to drug design. This holistic understanding facilitates the identification of new drug targets, optimization of existing drugs, and the development of more effective antibacterial therapies. In parallel, systems biology modeling, as detailed in predicting drug interactions, enhances our grasp of drug impact on bacterial systems, aiding in regimen optimization, resistance mitigation, and rational antibacterial therapy design.

11.5.3 Metabolic Pathway Analysis in Drug Development

Metabolic pathway analysis, a pivotal aspect in drug development, explores genome-scale metabolic modeling, target identification, biosynthetic pathways, and metabolic flux profiling. By scrutinizing nutrient utilization, drug–metabolite interactions, and antibiotic-resistance mechanisms, researchers can optimize drug design, develop combination therapy strategies, and unveil metabolic vulnerabilities (Ortmayr et al. 2022). The utilization of bacterial genomic databases and antibiotic-resistance databases further accelerates drug development by providing a wealth of genomic information, enabling the identification of potential drug targets, understanding bacterial pathogenicity, and designing effective antibiotics. These databases compile information on resistance genes, mechanisms, and associated genomic data, facilitating the surveillance of resistance trends, guiding treatment strategies, and contributing to the fight against drug-resistant bacteria (Uddin et al. 2021).

11.6 Data Resources and Database

11.6.1 Bacterial Genomic Databases

Bacterial genomic databases are repositories or collections of genomic information related to various bacterial species. These databases play a crucial role in advancing research in microbiology, infectious diseases, and biotechnology. They typically contain annotated genomic sequences, information about genes and their functions, comparative genomics data, and other relevant metadata. Some important bacterial genomic databases are listed here.

11.6.1.1 National Center for Biotechnology Information (NCBI)

NCBI provides a comprehensive collection of bacterial genomes and related information. The Microbial Genomes database includes complete and draft genome sequences, along with annotation data and tools for comparative analysis (Elbourne et al. 2023).

European Nucleotide Archive (ENA)
The ENA hosts genomic data for bacteria, archaea, and other microorganisms. It provides access to raw sequence data, assembled genomes, and associated metadata (Mukhopadhyay et al. 2023).

Integrated Microbial Genomes (IMG)
IMG, developed by the Joint Genome Institute (JGI), provides a platform for the comparative analysis of microbial genomes. It includes tools for functional annotation, pathway analysis, and exploration of microbial communities (Markowitz et al. 2008).

PATRIC (Pathosystems Resource Integration Center
PATRIC is a bacterial bioinformatics resource center that focuses on bacterial pathogens. It provides access to annotated genomes, tools for comparative genomics, and data related to virulence factors and antimicrobial resistance (Gillespie et al. 2011).

RefSeq
The NCBI's Reference Sequence (RefSeq) database includes annotated bacterial genomes, providing standardized and curated information about genes, transcripts, and proteins (O'Leary et al. 2016).

Genome Database for Enteric Bacteria (GenoList)
GenoList offers genomic information for enteric bacteria, including *Escherichia coli* and *Salmonella* species. It provides tools for genome browsing, annotation, and comparative analysis (Cossart 2023).

BacMap
BacMap is a visualized database of bacterial genomes that allows users to explore the genomic organization of various bacteria. It provides graphical representations of gene arrangements and genomic structures (Cruz et al. 2012).

11.6.2 Antibiotic-Resistance Database

The antibiotic-resistance database collects information on antibiotic-resistance genes, their associated proteins, and the genomic context in which they are found (Xavier et al. 2016). These databases are crucial tools for researchers and healthcare professionals working to understand, monitor, and combat antibiotic resistance, a global health concern. Some notable examples of antibiotic-resistance databases are as follows:

11.6.2.1 CARD (Comprehensive Antibiotic-Resistance Database)
CARD is a comprehensive and regularly updated database that provides curated information on antibiotic-resistance genes, their proteins, and associated resistance

phenotypes. It includes tools for predicting resistance based on genomic data (Alcock et al. 2020).

ResFinder

ResFinder, developed by the Center for Genomic Epidemiology, allows users to identify acquired antibiotic-resistance genes in bacterial genomes and plasmids. It covers many bacterial species (Florensa et al. 2022).

ARG-ANNOT

ARG-ANNOT is a database for antibiotic-resistance gene annotation. It provides information on resistance genes, including their nucleotide and protein sequences, and is designed to be used with genome annotation tools (Wei et al. 2019).

ARDB (Antibiotic-Resistance Genes Database)

ARDB is a database that compiles information on antibiotic-resistance genes and their associated proteins. It includes data on resistance mechanisms, gene families, and functional annotations (Gupta et al. 2020).

Resfams

Resfams is a database that focuses on classifying antibiotic-resistance gene families. It uses profile hidden Markov models to classify genes based on protein domain architecture (Boolchandani et al. 2019).

MegaRes

MegaRes is a database that combines information on antibiotic-resistance genes from various sources. It includes data on resistance genes, their genetic context, and metadata related to their discovery (Doster et al. 2020).

Ecogene-AR

Ecogene-AR is part of the larger EcoGene project and provides information on antibiotic-resistance genes in *Escherichia coli*. It includes data on gene function, regulation, and genomic context (Zhou and Rudd 2012).

11.6.3 Chemical Compound Libraries and Repositories

Chemical compound libraries and repositories serve as valuable resources for researchers engaged in antibacterial drug development, providing a foundation for experimental screening, virtual testing, and the identification of potential drug candidates (Miethke et al. 2021). Table 11.2 describes the different screening techniques and the method involved.

Table 11.2 Various screening techniques and their procedure

Experimental screening	
High-throughput screening	Diverse compound libraries: Bioinformatics tools curate and manage diverse chemical libraries containing small molecules for experimental screening against bacterial targets
Structure-based virtual screening	Docking and binding studies: Using computational methods to virtually screen chemical libraries against bacterial protein structures Identification of potential binders: Predicting the binding affinity and mode of interaction of compounds with bacterial targets
Fragment-based drug design	Fragment libraries: Compiling collections of small molecular fragments that can be used as building blocks for drug design Growing fragments: Expanding and optimizing initial fragment hits to design more potent antibacterial compounds.
Natural product databases	Natural compound repositories: Aggregating information on natural products with antibacterial properties Source of novel antibiotics: Exploring natural compounds as potential leads for new antibacterial drugs
Anti-infective compound libraries	Specialized collections: Curating compound libraries specifically designed for anti-infective drug discovery Focused screening: Prioritizing compounds with known or predicted antibacterial activity
Machine learning and predictive modeling	Quantitative structure-activity relationship (QSAR): Building models to predict the biological activity of compounds based on their chemical structure Virtual screening filters: Implementing machine learning algorithms to prioritize compounds with a higher likelihood of antibacterial activity
Chemoinformatics and data mining	Chemical structure analysis: Utilizing chemoinformatics tools to analyze the chemical structures of compounds in libraries Similarity searches: Identifying structurally similar compounds to known antibacterial agents for lead identification
ADME/T profiling	In silico ADME/T predictions: Integrating bioinformatics tools to predict the absorption, distribution, metabolism, excretion, and toxicity (ADME/T) properties of compounds Optimizing drug-like properties: Selecting compounds with favorable ADME/T profiles for further development

11.7 Case Studies

Bioinformatics has played a crucial role in various stages of antibacterial drug development, from target identification to drug design and optimization. While specific examples may vary, here are a few instances where bioinformatics has contributed to successful antibacterial drug development:

11.7.1 Examples of Successful Antibacterial Drug Development Using Bioinformatics

Several successful antibacterial drugs have been developed using bioinformatics, which leverages computational tools to identify novel drug targets and optimize drug design. Notable examples of novel antibiotics shown in Table 11.3.

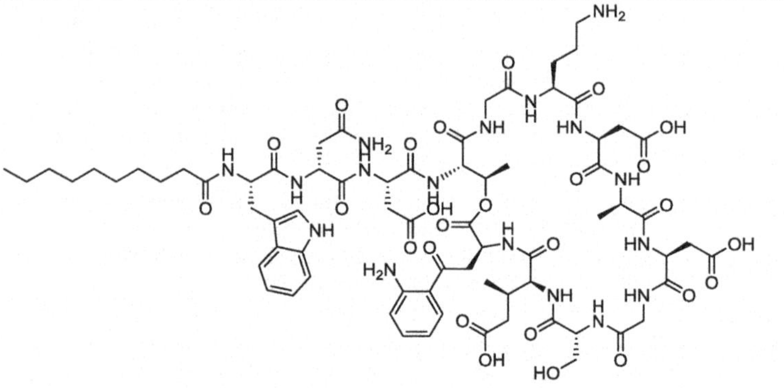

1. Daptomycin

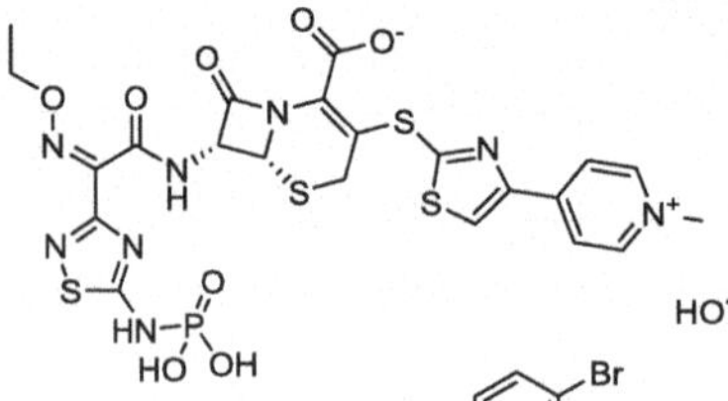

2. Ceftaroline

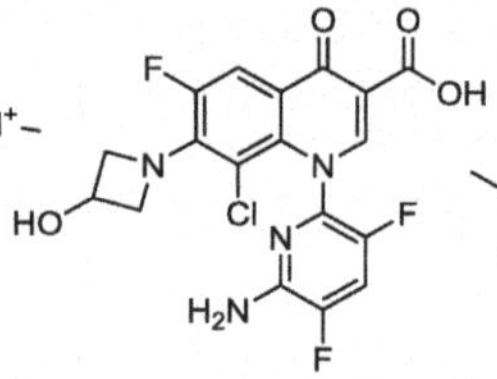

3. Delafloxacin

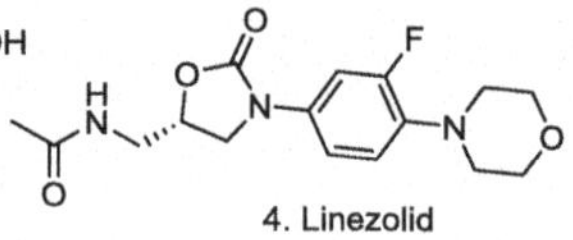

4. Linezolid

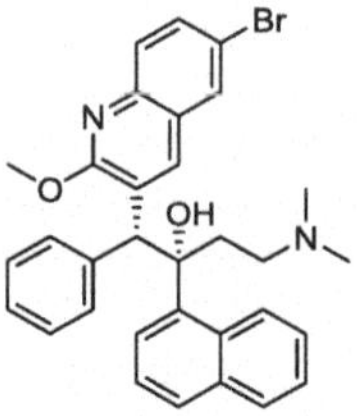

5. Bedaquiline

In these case studies, bioinformatics approaches, including genomics, comparative genomics, structural bioinformatics, and virtual screening, have been pivotal in target identification, understanding resistance mechanisms, and optimizing drug candidates. These examples showcase the integration of computational methods with experimental approaches to achieve successful antibacterial drug development.

Table 11.3 Examples of successful antibacterial drug development using bioinformatics

Antibiotics	Bioinformatics role	Outcome	Refs.
Daptomycin (1) (Cubicin)	Gram-positive bacteria, including methicillin-resistant *Staphylococcus aureus* (MRSA)	Bioinformatics tools were used for genomic analysis to identify essential genes and potential drug targets. Comparative genomics helped in understanding the genetic basis of antibiotic resistance. Additionally, molecular modeling and docking studies facilitated the design of daptomycin, a lipopeptide antibiotic effective against a broad spectrum of gram-positive bacteria	Ma et al. (2018)
Ceftaroline (2) (Teflaro)	Broad-spectrum cephalosporin effective against gram-positive and gram-negative bacteria, including MRSA	Bioinformatics contribution: Genomic analysis and comparative genomics were employed to identify essential bacterial targets. Structural bioinformatics played a role in optimizing ceftaroline to enhance its activity against MRSA and other resistant strains	Gajdács (2019)
Delafloxacin (3) (Baxdela)	Broad-spectrum antibiotic effective against gram-positive and gram-negative bacteria, including MRSA	Genomic and comparative genomics analyses facilitated the identification of conserved bacterial targets. Structural bioinformatics and molecular modeling guided the optimization of delafloxacin, enhancing its efficacy against diverse bacterial strains while minimizing off-target effects	Singh et al. (2012)
Linezolid (4) (Zyvox)	Gram-positive bacteria, including MRSA and vancomycin-resistant *Enterococcus faecium* (VRE).	Genomic and comparative genomics analyses played a vital role in identifying bacterial targets for linezolid. Bioinformatics tools were employed to predict potential drug-resistance mechanisms. Virtual screening and molecular docking were instrumental in the design and optimization of linezolid as a protein synthesis inhibitor	Bender et al. (2016)
Bedaquiline (5) (Sirturo)	*Mycobacterium tuberculosis*, the bacterium causing tuberculosis	Bioinformatics tools identified mycobacterial genes linked to drug resistance, while comparative genomics unveiled unique targets for new drugs. Structural bioinformatics guided the design and optimization of bedaquiline, a pioneering ATP synthase inhibitor addressing drug-resistant tuberculosis	Guo et al. (2021)

11.7.2 Challenges and Limitations in Bioinformatics-Driven Drug Development

Bioinformatics-driven drug development in antibacterials faces several challenges and limitations. Data quality and variability issues, including incomplete genomic information and experimental data variability, hinder accurate target identification and understanding of the bacterial genetic landscape. The rapid evolution of drug resistance in bacteria poses a significant challenge, as predicting resistance based solely on genomic data is complex and may lack sufficient predictive power. The limited functional information derived from genomic data, such as incomplete pathway understanding and annotations, further complicates drug development efforts. Overreliance on in silico predictions introduces challenges in validation, leading to false positives or negatives (McDermott and Davis 2021). The computational resource intensity of bioinformatics analyses, especially for smaller research groups, is a barrier to implementation. Integrating multi-omics data and addressing ethical and legal concerns about data privacy, ownership, and sharing also present formidable challenges. The unpredictable and dynamic nature of bacterial systems and the complexity of host–pathogen interactions add layers of intricacy to drug development. Despite these challenges, ongoing advancements in bioinformatics, alongside experimental validation and interdisciplinary collaboration, offer promise for overcoming these limitations and driving progress in developing effective antibacterial therapeutics. Addressing these challenges will be essential for the successful translation of bioinformatics-driven approaches into tangible therapeutic solutions (Willaert 2018).

11.8 Collaborative Research and Data Sharing

11.8.1 Importance of Collaborative Efforts

Collaborative efforts in antibacterial drug development, specifically through collaborative research and data sharing, are crucial for several reasons. Firstly, the interdisciplinary nature of antibacterial drug development requires expertise from various fields, including genomics, bioinformatics, medicinal chemistry, microbiology, and clinical research. Collaborative endeavors bring together diverse experts, fostering a holistic approach to drug discovery. Secondly, collaborative research accelerates drug discovery by pooling resources, including funding, infrastructure, and expertise. This not only reduces redundancy but also ensures the efficient use of resources, leading to accelerated discoveries. Thirdly, collaboration enables access to diverse datasets, including genomic, phenotypic, and clinical data, enhancing the quality of analyses in bioinformatics-driven drug development (Lawrence and Jeyakumar 2013). Moreover, global collaboration provides a comprehensive perspective on antibacterial drug resistance and development, addressing infectious disease challenges on a global scale. Collaborative studies also allow for cross-validation of findings, strengthening the robustness and reliability of research outcomes (Infectious Diseases Society of America 2011). The integrated solutions developed through collaborative efforts address complex challenges such as antibiotic resistance and host–pathogen interactions. Additionally, collaboration facilitates translational research, speeding up the translation of promising drug candidates from the laboratory to clinical practice. It optimizes clinical trial design by considering diverse patient populations and treatment strategies. Collaborative projects contribute to capacity building by promoting skill and knowledge transfer and providing training opportunities for the next generation of scientists in the field. Furthermore, resource optimization is achieved through shared infrastructure and access to specialized tools. Lastly, community engagement and advocacy efforts are enhanced through collaborative initiatives, involving stakeholders and advocating for increased funding and support for antibacterial drug research at national and international levels. In essence, collaborative efforts in antibacterial drug development create a synergistic environment that harnesses diverse expertise, accelerates the drug discovery, and addresses global challenges associated with bacterial infections and antibiotic resistance.

11.8.2 Data Sharing Initiatives in Antibacterial Drug Discovery

Data sharing initiatives in antibacterial drug discovery play a pivotal role in addressing the global challenge of antibiotic resistance. Notable initiatives, such as the Global Antibiotic Research and Development Partnership (GARDP), focus on developing new antibiotics and sharing data on resistance patterns and clinical trial outcomes to inform research (Dougan et al. 2019). The Combating antibiotic-resistant bacteria biopharmaceutical accelerator (CARB-X) accelerates antibacterial R&D by supporting collaborative projects with a strong emphasis on data sharing (Alm and Gallant 2020). The Antibiotic Resistance (AMR) Benchmark transparently assesses pharmaceutical companies' efforts, promoting data sharing on research pipelines and access plans (Cantón et al. 2022). The Pew Charitable Trusts advocate for policies addressing antibiotic resistance and collaborating with researchers, encouraging data sharing on antibiotic use and resistance patterns (Luepke et al. 2017). Initiatives like the Initiative on Methods, Measurement, and Pain Assessment in Clinical Trials (IMMPACT) and the European Commission Joint Programming Initiative on Antimicrobial Resistance (JPIAMR) promote standardized methods and data sharing to improve clinical trials and research collaboration (Renwick et al. 2016). The Antibacterial Resistance Leadership Group (ARLG), funded by the National Institute of Allergy and Infectious Diseases (NIAID), supports clinical research and data sharing to improve our understanding of antibiotic-resistance mechanisms (Suh et al. 2022). These initiatives collectively emphasize the importance of collaborative efforts and data sharing in advancing antibacterial drug discovery, contributing to the development of effective antibiotics and strategies to combat antibiotic resistance on a global scale.

11.9 Future Directions

11.9.1 Emerging Trends and Technologies in Bioinformatics

Several emerging trends and technologies were shaping the field of bioinformatics. The field is dynamic, and new developments may have occurred since then. Here are some trends and technologies that were noteworthy at that time (Arabnia and Tran 2018).

11.9.2 Potential Breakthroughs in Antibacterial Drug Development

It is important to note that breakthroughs in antibacterial drug development are likely to emerge from a combination of these and other innovative approaches. The landscape of research is dynamic, and ongoing efforts are essential to stay ahead of the evolving challenges posed by bacterial infections and resistance. There are many important approaches or their combinations which can lead to a successful antibacterial drug development. Some of the important approaches are phage therapy advancements, host-targeted therapies, antimicrobial peptides, combination therapies and drug synergies, CRISPR-based approaches, nanotechnology innovations, immunomodulatory agents, AI and machine learning in drug discovery, antibiotic adjuvants and potentiators, drug repurposing, and synthetic biology approaches (Murugaiyan et al. 2022).

11.10 Conclusion

Bioinformaticians play a crucial role across various aspects of the bioinformatics field, contributing to tasks like data analysis, interpretation, mapping, and modeling. To grasp the future landscape, it is vital to recognize key influencers, including major corporations, biotech and pharmaceutical firms, as well as start-ups pioneering innovative technologies. The emergence of blockchain technology introduces another influential dimension, as it has the capacity to transform the future of bioinformatics by offering a decentralized and secure platform for data storage.

References

Abdolmaleki A, Shiri F, Ghasemi JB (2018) Use of molecular docking as a decision-making tool in drug discovery. In: Molecular docking for computer-aided drug design. Academic Press, pp 229–243

Alcock BP, Raphenya AR, Lau TT, Tsang KK, Bouchard M, Edalatmand A, Huynh W, Nguyen AL, Cheng AA, Liu S, Min SY (2020) CARD 2020: antibiotic resistome surveillance with the comprehensive antibiotic resistance database. Nucleic Acids Res 48(D1):D517–D525. https://doi.org/10.1093/nar/gkz935

Alm RA, Gallant K (2020) Innovation in antimicrobial resistance: the CARB-X perspective. ACS Infect Dis 6(6):1317–1322. https://doi.org/10.1021/acsinfecdis.0c00026

Alqahtani S (2017) In silico ADME-Tox modeling: Progress and prospects. Expert Opin Drug Metab Toxicol 13(11):1147–1158. https://doi.org/10.1080/17425255.2017.1389897

Arabnia HR, Tran QN (eds) (2018) Emerging trends in applications and infrastructures for computational biology, bioinformatics, and systems biology: systems and applications. Elsevier

Atanasov AG, Zotchev SB, Dirsch VM, Supuran CT (2021) Natural products in drug discovery: advances and opportunities. Nat Rev Drug Discov 20(3):200–216. https://doi.org/10.1038/s41573-020-00114-z

Austin MJ, Hearnshaw SJ, Mitchenall LA, McDermott PJ, Howell LA, Maxwell A, Searcey M (2016) A natural product inspired fragment-based approach towards the development of novel anti-bacterial agents. Med Chem Commun 7(7):1387–1391. https://doi.org/10.1039/C6MD00229C

Ballester PJ, Mangold M, Howard NI, Robinson RL, Abell C, Blumberger J, Mitchell JB (2012) Hierarchical virtual screening for the discovery of new molecular scaffolds in antibacterial hit identification. J R Soc Interface 9(77):3196–3207. https://doi.org/10.1098/rsif.2012.0569

Bansal AK (2008) Role of bioinformatics in the development of new antibacterial therapy. Expert Rev Anti-Infect Ther 6(1):51–65. https://doi.org/10.1586/14787210.6.1.51

Bender JK, Fleige C, Klare I, Fiedler S, Mischnik A, Mutters NT, Dingle KE, Werner G (2016) Detection of a cfr (B) variant in German enterococcus faecium clinical isolates and the impact on linezolid resistance in enterococcus spp. PLoS One 11(11):e0167042. https://doi.org/10.1371/journal.pone.0167042

Boolchandani M, D'Souza AW, Dantas G (2019) Sequencing-based methods and resources to study antimicrobial resistance. Nat Rev Genet 20(6):356–370. https://doi.org/10.1038/s41576-019-0108-4

Brioschi M, Banfi C (2018) The application of gene silencing in proteomics: from laboratory to clinic. Expert Rev Proteomics 15(9):717–732. https://doi.org/10.1080/14789450.2018.1521275

Broaders E, Gahan CG, Marchesi JR (2013) Mobile genetic elements of the human gastrointestinal tract: potential for spread of antibiotic resistance genes. Gut Microbes 4(4):271–280. https://doi.org/10.4161/gmic.24627

Bruch EM, Petrella S, Bellinzoni M (2020) Structure-based drug design for tuberculosis: challenges still ahead. Appl Sci 10(12):4248. https://doi.org/10.3390/app10124248

Bunnik EM, Le Roch KG (2013) An introduction to functional genomics and systems biology. Adv Wound Care 2(9):490–498. https://doi.org/10.1089/wound.2012.0379

Butkiewicz M, Lowe EW, Mueller R, Mendenhall JL, Teixeira PL, Weaver CD, Meiler J (2013) Benchmarking ligand-based virtual high-throughput screening with the PubChem database. Molecules 18(1):735–756. https://doi.org/10.3390/molecules18010735

Cameron A, Esiovwa R, Connolly J, Hursthouse A, Henriquez F (2022) Antimicrobial resistance as a global health threat: the need to learn lessons from the COVID-19 pandemic. Global Pol 13(2):179–192. https://doi.org/10.1111/1758-5899.13049

Cantón R, Akova M, Langfeld K, Torumkuney D (2022) Relevance of the consensus principles for appropriate antibiotic prescribing in 2022. J Antimicrob Chemother 77(1):i2–i9. https://doi.org/10.1093/jac/dkac211

Chuan J, Belov A, Cloutier M, Li X, Khan IU, Chen W (2022) Comparative genomics analysis and virulence-related factors in novel Aliarcobacter faecis and Aliarcobacter lanthieri species identified as potential opportunistic pathogens. BMC Genomics 23(1):471. https://doi.org/10.1186/s12864-022-08663-w

Cossart P (2023) Raising a bacterium to the rank of a model system: the listeria paradigm. Ann Rev Microbiol 77:1–22. https://doi.org/10.1146/annurev-micro-110422-112841

Cruz J, Liu Y, Liang Y, Zhou Y, Wilson M, Dennis JJ, Stothard P, Van Domselaar G, Wishart DS (2012) BacMap: an up-to-date electronic atlas of annotated bacterial genomes. Nucleic Acids Res 40(D1):D599–D604. https://doi.org/10.1093/nar/gkr1105

Dos Santos DF, Istvan P, Quirino BF, Kruger RH (2017) Functional metagenomics as a tool for identification of new antibiotic resistance genes from natural environments. Microb Ecol 73:479–491. https://doi.org/10.1007/s00248-016-0866-x

Doster E, Lakin SM, Dean CJ, Wolfe C, Young JG, Boucher C, Belk KE, Noyes NR, Morley PS (2020) MEGARes 2.0: a database for classification of antimicrobial drug, biocide and metal resistance determinants in metagenomic sequence data. Nucleic Acids Res 48(D1):D561–D569. https://doi.org/10.1093/nar/gkz1010

Dougan G, Dowson C, Overington J, Participants NG (2019) Meeting the discovery challenge of drug-resistant infections: progress and focusing resources. Drug Discov Today 24(2):452–461. https://doi.org/10.1016/j.drudis.2018.11.001

Edache EI, Uzairu A, Mamza PA, Shallangwa GA (2020) A comparative QSAR analysis, 3D-QSAR, molecular docking and molecular design of iminoguanidine-based inhibitors of HemO: a rational approach to antibacterial drug design. J Pharm Sci 4:21. https://doi.org/10.31248/JDPS2020.036

Elbourne LD, Wilson-Mortier B, Ren Q, Hassan KA, Tetu SG, Paulsen IT (2023) TransAAP: an automated annotation pipeline for membrane transporter prediction in bacterial genomes. Microb Genom 9(1):15. https://doi.org/10.1099/mgen.0.000927

Farha MA, French S, Brown ED (2021) Systems-level chemical biology to accelerate antibiotic drug discovery. Acc Chem Res 54(8):1909–1920. https://doi.org/10.1021/acs.accounts.1c00011

Florensa AF, Kaas RS, Clausen PT, Aytan-Aktug D, Aarestrup FM (2022) ResFinder-an open online resource for identification of antimicrobial resistance genes in next-generation sequencing data and prediction of phenotypes from genotypes. Microb Genom 8(1):1–10. https://doi.org/10.1099/mgen.0.000748

Franzosa EA, Hsu T, Sirota-Madi A, Shafquat A, Abu-Ali G, Morgan XC, Huttenhower C (2015) Sequencing and beyond: integrating molecular 'omics' for microbial community profiling. Nat Rev Microbiol 13(6):360–372. https://doi.org/10.1038/nrmicro3451

Gajdács M (2019) The continuing threat of methicillin-resistant Staphylococcus aureus. Antibiotics 8(2):52. https://doi.org/10.3390/antibiotics8020052

Gillespie JJ, Wattam AR, Cammer SA, Gabbard JL, Shukla MP, Dalay O, Driscoll T, Hix D, Mane SP, Mao C, Nordberg EK (2011) PATRIC: the comprehensive bacterial bioinformatics resource with a focus on human pathogenic species. Infect Immun 79(11):4286–4298. https://doi.org/10.1128/IAI.00207-11

Guo H, Courbon GM, Bueler SA, Mai J, Liu J, Rubinstein JL (2021) Structure of mycobacterial ATP synthase bound to the tuberculosis drug bedaquiline. Nature 589(7840):143–147. https://doi.org/10.1038/s41586-020-3004-3

Gupta CL, Tiwari RK, Cytryn E (2020) Platforms for elucidating antibiotic resistance in single genomes and complex metagenomes. Environ Int 138:105667. https://doi.org/10.1016/j.envint.2020.105667

Hawkes CV, Kjøller R, Raaijmakers JM, Riber L, Christensen S, Rasmussen S, Christensen JH, Dahl AB, Westergaard JC, Nielsen M, Brown-Guedira G (2021) Extension of plant phenotypes by the foliar microbiome. Annu Rev Plant Biol 72:823–846. https://doi.org/10.1146/annurev-arplant-080620-114342

Hossain MM, Roy PK, Mosnaz ATM, Shakil SK, Hasan MM, Prodhan SH (2015) Structural analysis and molecular docking of potential ligands with chorismate synthase of Listeria monocytogenes: a novel antibacterial drug target

Infectious Diseases Society of America (IDSA) (2011) Combating antimicrobial resistance: policy recommendations to save lives. Clin Infect Dis 52(5):S397–S428. https://doi.org/10.1093/cid/cir153

Kabir A, Muth A (2022) Polypharmacology: the science of multi-targeting molecules. Pharmacol Res 176:106055. https://doi.org/10.1016/j.phrs.2021.106055

Kampourakis K (2017) Making sense of genes. Cambridge University Press, pp 1–10. https://doi.org/10.1017/9781316422939

Kaur H, Jakob RP, Marzinek JK, Green R, Imai Y, Bolla JR, Agustoni E, Robinson CV, Bond PJ, Lewis K, Maier T (2021) The antibiotic darobactin mimics a β-strand to inhibit outer membrane insertase. Nature 593(7857):125–129. https://doi.org/10.1038/s41586-021-03455-w

Kon K, Rai M (eds) (2016) Antibiotic resistance: mechanisms and new antimicrobial approaches. Academic Press

Kozlowska J, Vermeer LS, Rogers GB, Rehnnuma N, Amos SB, Koller G, McArthur M, Bruce KD, Mason AJ (2014) Combined systems approaches reveal highly plastic responses to antimicrobial peptide challenge in Escherichia coli. PLoS Pathog 10(5):e1004104. https://doi.org/10.1371/journal.ppat.1004104

Kumari R, Rathi R, Pathak SR, Dalal V (2022) Structural-based virtual screening and identification of novel potent antimicrobial compounds against YsxC of Staphylococcus aureus. J Mol Struct 1255:132476. https://doi.org/10.1016/j.molstruc.2022.132476

Lans I, Anoz-Carbonell E, Palacio-Rodríguez K, Aínsa JA, Medina M, Cossio P (2020) In silico discovery and biological validation of ligands of FAD synthase, a promising new antimicrobial target. PLoS Comput Biol 16(8):e1007898. https://doi.org/10.1371/journal.pcbi.1007898

Lawrence R, Jeyakumar E (2013) Antimicrobial resistance: a cause for global concern. BMC Proc 7(3):1–14. https://doi.org/10.1186/1753-6561-7-S3-S1

Leung EL, Cao ZW, Jiang ZH, Zhou H, Liu L (2013) Network-based drug discovery by integrating systems biology and computational technologies. Brief Bioinform 14(4):491–505. https://doi.org/10.1093/bib/bbs043

Lian X, Xia Z, Li X, Karpov P, Jin H, Tetko IV, Xia J, Wu S (2021) Anti-MRSA drug discovery by ligand-based virtual screening and biological evaluation. Bioorg Chem 114:105042. https://doi.org/10.1016/j.bioorg.2021.105042

Liu C, He G, Jiang Q, Han B, Peng C (2013) Novel hybrid virtual screening protocol based on molecular docking and structure-based pharmacophore for discovery of methionyl-tRNAsynthetase inhibitors as antibacterial agents. Int J Mol Sci 14(7):14225–14239. https://doi.org/10.3390/ijms140714225

Loman NJ, Pallen MJ (2015) Twenty years of bacterial genome sequencing. Nat Rev Microbiol 13(12):787–794. https://doi.org/10.1038/nrmicro3565

Luepke KH, Suda KJ, Boucher H, Russo RL, Bonney MW, Hunt TD, Mohr JF III (2017) Past, present, and future of antibacterial economics: increasing bacterial resistance, limited antibiotic pipeline, and societal implications. Pharma J Hum Pharmacol Drug Ther 37(1):71–84. https://doi.org/10.1002/phar.1868

Luo H, Mattes W, Mendrick DL, Hong H (2016) Molecular docking for identification of potential targets for drug repurposing. Curr Top Med Chem 16(30):3636–3645. https://doi.org/10.2174/1568026616666160530181149

Ma Z, Lasek-Nesselquist E, Lu J, Schneider R, Shah R, Oliva G, Pata J, McDonough K, Pai MP, Rose WE, Sakoulas G (2018) Characterization of genetic changes associated with daptomycin nonsusceptibility in Staphylococcus aureus. PLoS One 13(6):e0198366. https://doi.org/10.1371/journal.pone.0198366

Markowitz VM, Ivanova N, Anderson I, Lykidis A, Mavromatis K, Szeto E, Palaniappan K, Chen IM, Chu K, Grechkin Y, Kyrpides NC (2008) Using IMG: comparative analysis with the integrated microbial genomes system

Matamoros-Recio A, Franco-Gonzalez JF, Forgione RE, Torres-Mozas A, Silipo A, Martín-Santamaría S (2021) Understanding the antibacterial resistance: computational explorations in bacterial membranes. ACS Omega 6(9):6041–6054. https://doi.org/10.1021/acsomega.0c05590

Mathpal D, Masand M, Thomas A, Ahmad I, Saeed M, Zaman GS, Kamal M, Jawaid T, Sharma PK, Gupta MM, Kumar S (2021) Pharmacophore modeling, docking and the integrated use of a ligand-and structure-based virtual screening approach for novel DNA gyrase inhibitors: synthetic and biological evaluation studies. RSC Adv 11(55):34462–34478. https://doi.org/10.1039/D1RA05630A

McDermott PF, Davis JJ (2021) Predicting antimicrobial susceptibility from the bacterial genome: a new paradigm for one health resistance monitoring. J Vet Pharmacol Ther 44(2):223–237. https://doi.org/10.1111/jvp.12913

Medina E, Pieper DH (2016) Tackling threats and future problems of multidrug-resistant bacteria. In: How to overcome the antibiotic crisis: facts, challenges, technologies and future perspectives, p 3. https://doi.org/10.1007/82_2016_492

Miethke M, Pieroni M, Weber T, Brönstrup M, Hammann P, Halby L, Arimondo PB, Glaser P, Aigle B, Bode HB, Moreira R (2021) Towards the sustainable discovery and development of new antibiotics. Nat Rev Chem 5(10):726–749. https://doi.org/10.1038/s41570-021-00313-1

Monaghan RL, Barrett JF (2006) Antibacterial drug discovery-then, now and the genomics future. Biochem Pharmacol 71(7):901–909. https://doi.org/10.1016/j.bcp.2005.11.023

Montalbano A, Canver MC, Sanjana NE (2017) High-throughput approaches to pinpoint function within the noncoding genome. Mol Cell 68(1):44–59. https://doi.org/10.1016/j.molcel.2017.09.017

Mukhopadhyay A, Choudhury S, Kumar M (2023) Metaviromic analyses of DNA virus community from sediments of the N-Choe stream, North India. Virus Res 330:199110. https://doi.org/10.1016/j.virusres.2023.199110

Murugaiyan J, Kumar PA, Rao GS, Iskandar K, Hawser S, Hays JP, Mohsen Y, Adukkadukkam S, Awuah WA, Jose RA, Sylvia N (2022) Progress in alternative strategies to combat antimicrobial resistance: focus on antibiotics. Antibiotics 11(2):200. https://doi.org/10.3390/antibiotics11020200

Naik SR, Bharadwaj P, Dingelstad N, Kalyaanamoorthy S, Mandal SC, Ganesan A, Chattopadhyay D, Palit P (2022) Structure-based virtual screening, molecular dynamics and binding affinity calculations of some potential phytocompounds against SARS-CoV-2. J Biomol Struct Dyn 40(15):6921–6938. https://doi.org/10.1080/07391102.2021.1891969

O'Leary NA, Wright MW, Brister JR, Ciufo S, Haddad D, McVeigh R, Rajput B, Robbertse B, Smith-White B, Ako-Adjei D, Astashyn A (2016) Reference sequence (RefSeq) database at NCBI: current status, taxonomic expansion, and functional annotation. Nucleic Acids Res 44(D1):D733–D745. https://doi.org/10.1093/nar/gkv1189

Omar FA, Abelrasoul M, Sheha MM, Hassan HY, Ibrahiem YM (2018) Synthesis, antibacterial activity and molecular docking of substituted naphthyridines as potential DNA gyrase inhibitors. Chem Select 3(9):2604–2612. https://doi.org/10.1002/slct.201800108

Ortmayr K, de la Cruz Moreno R, Zampieri M (2022) Expanding the search for small-molecule antibacterials by multidimensional profiling. Nat Chem Biol 18(6):584–595. https://doi.org/10.1038/s41589-022-01040-4

Puentes PR, Henao MC, Torres CE, Gómez SC, Gómez LA, Burgos JC, Arbeláez P, Osma JF, Muñoz-Camargo C, Reyes LH, Cruz JC (2020) Design, screening, and testing of non-rational peptide libraries with antimicrobial activity: in silico and experimental approaches. Antibiotics 9(12):854. https://doi.org/10.3390/antibiotics9120854

Pulido MR, García-Quintanilla M, Gil-Marqués ML, McConnell MJ (2016) Identifying targets for antibiotic development using omics technologies. Drug Discov Today 21(3):465–472. https://doi.org/10.1016/j.drudis.2015.11.014

Pushkaran AC, Biswas R, Mohan CG (2019) Impact of target-based drug design in anti-bacterial drug discovery for the treatment of tuberculosis. In: Structural bioinformatics: applications in preclinical drug discovery process, pp 307–346. https://doi.org/10.1007/978-3-030-05282-9_10

Renwick MJ, Simpkin V, Mossialos E & World Health Organization (2016) Targeting innovation in antibiotic drug discovery and development: The need for a One Health-One Europe-One World Framework. World Health Organization. Regional Office for Europe

Roca I, Akova M, Baquero F, Carlet J, Cavaleri M, Coenen S, Cohen J, Findlay D, Gyssens I, Heure OE, Kahlmeter G (2015) The global threat of antimicrobial resistance: science for intervention. New Microb New Infect 6:22–29. https://doi.org/10.1016/j.nmni.2015.02.007

Sagar S, Kaistha S, Das AJ, Kumar R (2019) Antibiotic resistant bacteria: a challenge to modern medicine. Springer Singapore. https://doi.org/10.1007/978-981-13-9879-7

Schmieder R, Edwards R (2012) Insights into antibiotic resistance through metagenomic approaches. Future Microbiol 7(1):73–89. https://doi.org/10.2217/fmb.11.135

Selvaraj C (2019) Molecular modeling and drug design techniques in microbial drug discovery. In: Essentials of bioinformatics, volume II: in Silico life sciences: medicine, pp 185–231. https://doi.org/10.1007/978-3-030-18375-2_11

Shahbaaz M, Bisetty K, Ahmad F, Hassan I (2016) Current advances in the identification and characterization of putative drug and vaccine targets in the bacterial genomes. Curr Top Med Chem 16(9):1040–1069. https://doi.org/10.2174/1568026615666150825143307

Singh N, Chaudhury S, Liu R, AbdulHameed MD, Tawa G, Wallqvist A (2012) QSAR classification model for antibacterial compounds and its use in virtual screening. J Chem Inf Model 52(10):2559–2569. https://doi.org/10.1021/ci300336v

Sivashankari S, Shanmughavel P (2007) Comparative genomics-a perspective. Bioinformation 1(9):376. https://doi.org/10.6026/97320630001376

Stanzione F, Giangreco I, Cole JC (2021) Use of molecular docking computational tools in drug discovery. Prog Med Chem 60:273–343. https://doi.org/10.1016/bs.pmch.2021.01.004

Suh GA, Lodise TP, Tamma PD, Knisely JM, Alexander J, Aslam S, Barton KD, Bizzell E, Totten KM, Campbell JL, Chan BK (2022) Considerations for the use of phage therapy in clinical practice. Antimicrob Agents Chemother 66(3):e02071–e02021. https://doi.org/10.1128/aac.02071-21

Swaminathan S, Sundaramurthi JC, Palaniappan AN, Narayanan S (2016) Recent developments in genomics, bioinformatics and drug discovery to combat emerging drug-resistant tuberculosis. Tuberculosis 101:31–40. https://doi.org/10.1016/j.tube.2016.08.002

Terreni M, Taccani M, Pregnolato M (2021) New antibiotics for multidrug-resistant bacterial strains: latest research developments and future perspectives. Molecules 26(9):2671. https://doi.org/10.3390/molecules26092671

Uddin TM, Chakraborty AJ, Khusro A, Zidan BR, Mitra S, Emran TB, Dhama K, Ripon MK, Gajdács M, Sahibzada MU, Hossain MJ (2021) Antibiotic resistance in microbes: history, mechanisms, therapeutic strategies and future prospects. J Infect Public Health 14(12):1750–1766. https://doi.org/10.1016/j.jiph.2021.10.020

Vila JO, Moreno-Morales J, Ballesté-Delpierre C (2020) Current landscape in the discovery of novel antibacterial agents. Clin Microbiol Infect 26(5):596–603. https://doi.org/10.1016/j.cmi.2019.09.015

Vilar S, Hripcsak G (2017) The role of drug profiles as similarity metrics: applications to repurposing, adverse effects detection and drug-drug interactions. Brief Bioinform 18(4):670–681. https://doi.org/10.1093/bib/bbw012

Wajman JR, Bertolucci PH, Mansur LL, Gauthier S (2015) Culture as a variable in neuroscience and clinical neuropsychology: a comprehensive review. Dement Neuropsychol 9(3):203. https://doi.org/10.1590/1980-57642015dn93000002

Wang Z, Yang B (2022) General strategies for rational design and discovery of multitarget drugs. In: Polypharmacology: principles and methodologies. Springer International Publishing, pp 677–736. https://doi.org/10.1007/978-3-031-04998-9_18

Wei Z, Wu Y, Feng K, Yang M, Zhang Y, Tu Q, Wang J, Deng Y (2019) ARGA, a pipeline for primer evaluation on antibiotic resistance genes. Environ Int 128:137–145. https://doi.org/10.1016/j.envint.2019.04.030

Willaert RG (2018) Micro-and nanoscale approaches in antifungal drug discovery. Fermentation 4(2):43. https://doi.org/10.3390/fermentation4020043

Wooller SK, Benstead-Hume G, Chen X, Ali Y, Pearl FM (2017) Bioinformatics in translational drug discovery. Biosci Rep 37(4):BSR20160180. https://doi.org/10.1042/BSR20160180

Xavier BB, Das AJ, Cochrane G, De Ganck S, Kumar-Singh S, Aarestrup FM, Goossens H, Malhotra-Kumar S (2016) Consolidating and exploring antibiotic resistance gene data resources. J Clin Microbiol 54(4):851–859. https://doi.org/10.1128/JCM.02717-15

Zheng M, Zhao J, Cui C, Fu Z, Li X, Liu X, Ding X, Tan X, Li F, Luo X, Chen K (2018) Computational chemical biology and drug design: facilitating protein structure, function, and modulation studies. Med Res Rev 38(3):914–950. https://doi.org/10.1002/med.21483

Zhong C, Chen C, Wang L, Ning K (2021) Integrating pan-genome with metagenome for microbial community profiling. Comput Struct Biotechnol J 19:1458–1466. https://doi.org/10.1016/j.csbj.2021.02.021

Zhou J, Rudd KE (2012) EcoGene 3.0. Nucleic Acids Res 41(D1):D613–D624. https://doi.org/10.1093/nar/gks1235

Antibiotic Drug Development Methods 12

Jublee Jasmine and Saswat S. Mohapatra

Abstract

Antimicrobial resistance (AMR) is one of the biggest health-care challenges in recent times. With the rising incidence of infections caused by drug-resistant pathogens and the declining antibiotic discovery pipeline, we are staring at what is being called a "post-antibiotic era." The antimicrobial resistance problem is also referred to as the silent pandemic that requires urgent intervention. Traditionally, antibiotic discovery is based on finding natural compounds from a variety of microorganisms in nature and optimizing their chemical properties for clinical use. However, with the rapid progress in genomics and proteomics methods, the entire process of drug development has seen an overhaul. In this approach, finding the suitable drug target is the most significant step against which synthetic or semisynthetic compounds could be developed and validated using bioinformatics tools before proceeding with the in vitro characterization. Moreover, with the rapid progress in artificial intelligence and machine learning tools, drug development methods have taken a leap forward that may significantly reduce the time required from discovery to the development process. In this chapter, we have discussed the traditional antibiotic discovery and development methods and have highlighted the recent achievements in the area aided by omics, artificial intelligence, and machine learning methods.

J. Jasmine (✉)
P.G. Department of Biosciences, College of Pharmaceutical Sciences, Berhampur, Odisha, India

S. S. Mohapatra
Molecular Microbiology Lab, P.G. Department of Biotechnology, Berhampur University, Berhampur, Odisha, India

T. Sharma et al. (eds.), *Natural Products for Antibacterial Drug Development: Recent Advancement of Computational Approach*,
https://doi.org/10.1007/978-981-97-9634-2_12

Keywords

Antimicrobial resistance · Antibiotics · Antibacterial drugs · Omics · Machine learning · Artificial intelligence · High-throughput screening

Abbreviations

AMR	Antimicrobial resistance
ARGs	Antibiotic resistance genes
BGCs	Biosynthetic gene clusters
CLSI	Clinical and Laboratory Standards Institute
dNDP	Deoxynucleoside diphosphates
HPLC	High-performance liquid chromatography
HPLC:MS	High performance liquid chromatography mass spectrometry
HPLC-HRMS	High-performance liquid chromatography-high resolution mass spectrometry
IMG-ABC	Integrated Microbial Genomes Atlas
MIBiG	Minimum Information about a Biosynthetic Gene Cluster
MIC	Minimal inhibitory concentration
MoA	Mode of action
NDPs	Ribonucleoside-5-diphosphates
NPs	Natural products
NRPS	Non-ribosomal peptide synthase
PC	Paper chromatography
PDA	Polymer polydiacetylene
PG	Peptidoglycan
PHMMs	Profile Hidden Markov Models
PKS	Polyketide synthase
PTMs	Post-translational modifications
RiPPs	Ribosomally synthesized and post-translationally modified peptides
RNR	Ribonucleotide reductase
SMBP	Secondary Metabolite Bioinformatics Portal
SMILES	Simplified molecular input line entry system
TCSs	Two-component systems
TLC	Thin-layer chromatography
UDP-GlcNAc	UDP-N-acetylglucosamine

12.1 Antibiotics and Antimicrobial Resistance (AMR): An Introduction

Antibiotics were among the most important classes of pharmaceuticals and major medical innovations of the previous century. Antibiotics are referred to be "wonder medicines" because they are effective against bacterial infections (Uddin et al. 2021), and their discovery and eventual therapeutic use represent a medical marvel (Watkins and Bonomo 2016). Unquestionably, antibiotics have helped humanity combat bacterial infections and saved millions of lives. It also has been used as a prophylactic agent in animal and poultry feeds. However, the threat of incurable illnesses has been present since the start of the twenty-first century, and the frequency of infections brought on by multidrug-resistant (MDR) bacteria is rising globally. Microbes undergo genetic modifications to lessen the fatal effect of antibiotics, making antimicrobial resistance (AMR) an inevitable phenomenon. AMR is an evolutionary outcome to evade deadly selection pressure presented by antibiotics. The evolution of drug resistance is fuelled by the population size of bacterial pathogens and their exposure to antibiotics. Bacteria and other microorganisms adapt with time. Their primary goal is to proliferate, live, and spread quickly. Microbes adapt to their environment and change in ways that ensure their survival. Genetic changes may occur if something, like an antibiotic, prevents the bacteria from growing, making them resistant to the drug and enabling them to live (Munita and Arias 2016). Drug resistance is a natural process that microorganisms go through. Still, several factors control the complex etiology of antibiotic resistance. AMR and related bacterial persistence have significantly increased because of improper waste management, environmental transmission, misuse and abuse of antibiotics in health care and agriculture, inaccurate diagnosis and prescription of antibiotics, loss of patient sensitivity, and self-medication.

It is alarming to see the growth of MDR pathogens as a major threat to global public health. A 2019 World Health Organization (WHO) report estimates that at least annually 700,000 deaths are caused by drug-resistant pathogens. If the problem of AMR is not handled properly, this number might rise to as high as 10 million by 2050 (O'Neill 2016; Tacconelli et al. 2018; Murray et al. 2022). Therefore, urgent R&D efforts are required to develop new antimicrobial agents to control infections caused by drug-resistant bacteria, as the existing treatment regime appears to be unsuccessful when using the antibiotics that are already available. Even though antibiotics have helped advance several medical specialties, including their prophylactic use during surgeries, still antibiotic resistance poses a significant challenge to global health-care systems (Dodds 2017).

The effectiveness of current antibiotics is being undermined by AMR in clinically relevant microorganisms, leading to alarming rates of morbidity and mortality worldwide. A worldwide health emergency has arisen due to the rise and dissemination of resistant pathogens and the dearth of novel drugs (Lushniak 2014; Martens and Demain 2017). In the United States, annually 2.8 million infections resulting in 35,000 deaths are caused by drug-resistant pathogens as per the estimates of the Center for Disease Control and Prevention. Over 50,000 babies in India are at risk of dying from sepsis due to bacteria resistant to standard antibiotics. Moreover, in India, every 9 min a child dies from an illness caused by antibiotic-resistant bacteria (Subramaniam and Girish 2020). The available data also raises the possibility that the remedy could exacerbate the original issue, since antibiotics have been demonstrated to seriously harm the gut microbiota, decreasing species diversity, and promoting the rise and dissemination of antibiotic resistance genes (ARGs). Novel approaches in antibiotic discovery are necessary because most antibiotics undergoing clinical trials are analogs of previously approved medications for which resistance mechanisms have already surfaced. The fact that developing novel antibiotics is a labor-intensive, costly, and sometimes unsuccessful process that can take more than decades and hundreds of millions of dollars, only serves to exacerbate this problem. Fourteen new antibiotics were developed and authorized between 2014 and 2019. The likelihood of new medications treating infectious diseases being successful was estimated to be 25.2% based on a scan of around 186,000 clinical trials including over 21,000 chemicals. This chance fell to just 19.1% for orphan pharmaceuticals, or medications that cure uncommon infectious disorders (Melo et al. 2021).

The pharmaceutical industry has high-cost intensive antibiotic development; however, the rising rate of AMR has led to reduced returns on the investment. The creation of novel antibiotics and research into antibiotics has already been shelved by several pharmaceutical corporations (Miethke et al. 2021). Even if the situation is dire, several novel technologies have the power to improve things. Additionally, various scientific developments may facilitate the discovery and development of novel antibiotics. There is an immediate need for new antibiotics, which needs to be taken up by academics to focus on the initial stages of antibiotic design and optimization. An urgent requirement for purposeful investment in innovative therapeutic alternatives to prevent AMR is necessary to both meet unmet patient demand and offset the rising financial burden on global healthcare. To ensure a consistent flow of novel drug candidates into the development process, scientists ought to concentrate on employing hit identification and hit-to-lead optimization initiatives. The initial stages of drug development must be strengthened to fight antibiotic resistance as these are essential for finding and confirming novel therapeutics. In the future, it will likely be required to prioritize the development of effective drugs to treat all bacterial infections globally, and success will likely require a combination of approaches.

12.1.1 Emergence of Antimicrobial Resistance (AMR)

The rise of antibiotic resistance is a complex problem that has become the driving force behind new antibiotic development (Theuretzbacher 2011). Penicillinase enzyme was detected in *E. coli* as early as 1940, even before penicillin was widely used. In 1942, a penicillinase assay also detected penicillin resistance in a few *Staphylococcus aureus* clinical strains. Reports of resistance appeared soon after each new antibiotic was introduced to the market. This pattern started to appear many decades later in a variety of bacterial infections. The threat of resistance will remain, giving each antibiotic in use an expiration date, unless more innovative antibiotics become available and improved control mechanisms are implemented. Since antibiotic resistance can develop quickly and more than 70% of bacterial pathogens are already resistant to at least one antibiotic, they have become a major cause of concern (Economou and Gousia 2015).

Based on strains of bacterium, antibiotic class, and geographic area, the prevalence of AMR varies across the world. As per the European Antimicrobial Resistance Surveillance Network (EARS-Net) report, *Acinetobacter* spp. (1.7%), *Pseudomonas aeruginosa* (5.6%), *Streptococcus pneumoniae* (5.3%), *Enterococcus faecium* (4.5%), *Escherichia coli* (44.2%), *S. aureus* (20.6%), *Klebsiella pneumoniae* (11.3%), and *Enterococcus faecalis* (6.8%) were the most prevalent antibiotic-resistant bacterial species (de Kraker et al. 2011). According to data from 22 nations' Global Antimicrobial Surveillance System (GLASS), 500,000 individuals with documented bacterial illnesses exhibit widespread AMR (de Kraker et al. 2011). The most commonly found resistant species were *K. pneumoniae*, *S. aureus*, *E. coli*, and *S. pneumoniae*. In countries that are part of GLASS, penicillin resistance reached 51 percent (de Kraker et al. 2011). Resistance to ciprofloxacin, a drug frequently used against urinary tract infections (UTIs), ranged from 8.4% to 92.9% for *E. coli* and from 4.1 to 79.4% for *K. pneumoniae*. Data on methicillin-resistant *S. aureus* (MRSA) bloodstream infections from 25 countries, regions, and zones, as well as data on *E. coli* bloodstream infections from 49 countries, were received by GLASS in 2019. MRSA had a median incidence of 12.11% (IQR 6.4–26.4), while *E. coli* with a median rate of resistance to third-generation cephalosporins was 36% (IQR 15.2–63) (de Kraker et al. 2011). The recent antituberculosis medication statistics indicate that MDR and rifampicin-resistant TB will affect 3.5% of active and 18% of previously treated patients globally (Migliori et al. 2020). Approximately 558,000 cases of MDR/RR-TB were recorded globally in 2017, resulting in 230,000 fatalities (Ballestero et al. 2020).

12.1.2 Molecular Mechanisms of Antibiotic Resistance

Several molecular mechanisms are involved in the development of antibiotic resistance (Fig. 12.1). Enzymes that break down or alter the antibiotic molecule mediate the inactivation of antibiotics (Darby et al. 2023). Antibiotic-modifying enzymes add different functional groups to the antibiotic so that it cannot bind to its intended target (Darby et al. 2023). Target site modification brings about changes in the antibiotic target resulting in reduced binding. This leads to modification in the binding site of the enzyme or mutations in the gene encoding the antibiotic target (Cook and Wright 2022; Darby et al. 2023). Target bypass occurs when a different protein fulfills the role of the drug target, making the original target redundant. Reduced inflow is caused by modifications to the bacterial membrane structure, resulting in the reduced number of porins, and other transmembrane proteins required for the entry of substances, including antibiotics, into the cell. Efflux pumps enable the active efflux of the drug reducing its accumulation inside the cell (Blair et al. 2015;

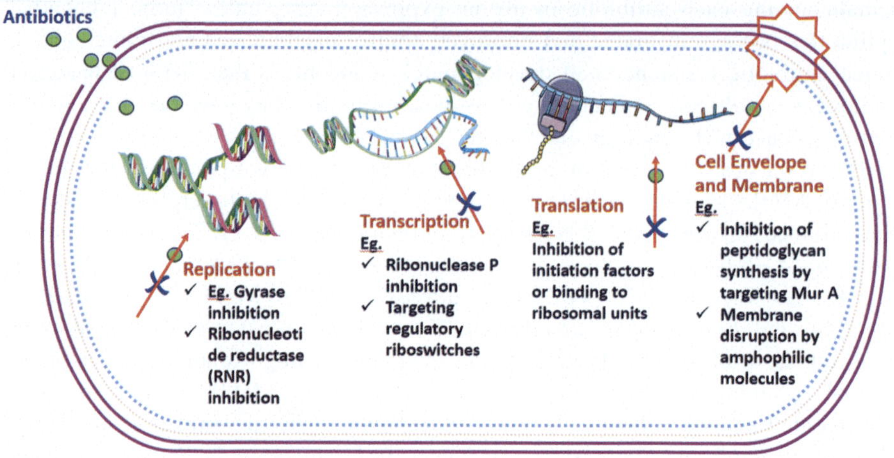

Fig. 12.1 Common antibiotic targets and the inhibition of the cellular processes by the antibiotics (see the text for details). The nucleic acid icons were used from https://smart.servier.com/

Piddock 2017). The process of physically attaching a target protection protein to the antibiotic target to free it from antibiotic-mediated inhibition is known as target protection and has been implicated in the development of antibiotic resistance (Wilson et al. 2020). The different mechanisms involved in developing antibiotic resistance are summarized in Fig. 12.1.

12.2 Strategies and Assays for Antibacterial Screening and Discovery

Actinomycetes are well-known producers of secondary metabolites that include antibiotics. Most of the currently used antibiotics were initially isolated from these bacterial species. Most of the time, a group of genes known as biosynthetic gene clusters (BGCs) in the bacterial genome code for the proteins that are needed for the synthesis, transport, and resistance of secondary metabolites, including antibiotics. Data gathered from the verified BGC databases, and matching sequences of protein family domains (Pfams) has helped in mining bacterial genomes. On average, 20 BGCs can be found in an actinomycete genome. A popular approach used in the initial stages of antibiotic development is whole-cell screenings, where cells are exposed to chemicals derived from these bacterial species. This goes beyond bioinformatics tools that facilitate the discovery of BGCs (Rütten et al. 2022). Target-based strategies have been developed more recently. Several genome-mining technologies and the conventional (phenotypic) approach, described in the following sections, are used in the detection and characterization of antibacterial agents (Rütten et al. 2022). An overview of the different drug development approaches is presented in Fig. 12.2.

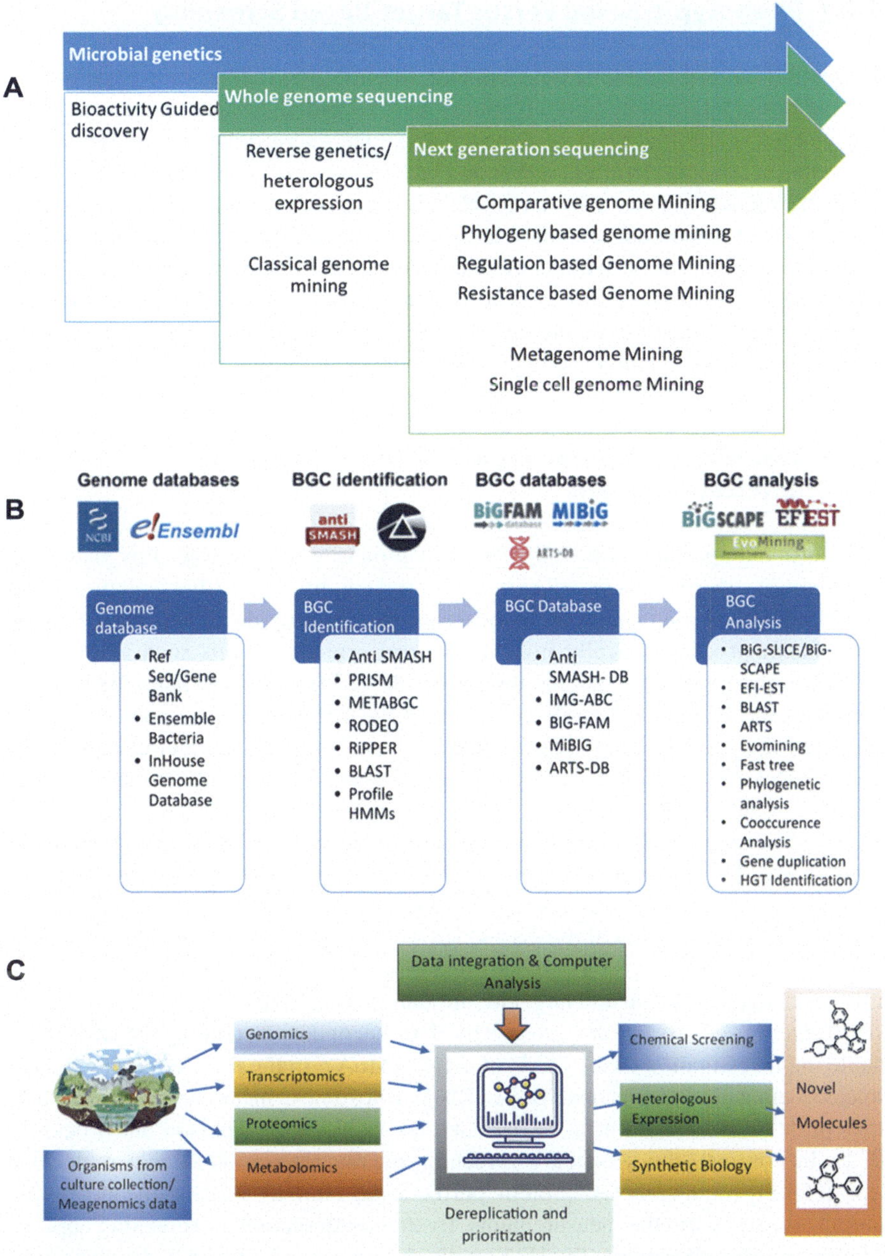

Fig. 12.2 Genome mining in the discovery and development of novel antimicrobial agents. (**a**) Evolution of genome-mining tools for novel secondary metabolites discovery. (**b**) General workflow and examples of common computational tools for natural product discovery. (**c**) Advanced automated genome-mining approaches

12.2.1 Phenotypic-Based Versus Target-Based Screening

In the initial days of antibiotic development, in vitro growth inhibition experiments were the principal method whose readouts were recorded as bacterial growth or no growth (Katz and Baltz 2016; Moloney 2016). One major drawback of employing old whole-cell phenotypic tests is that mechanisms of action were typically discovered later, frequently years after approval. With the development of high-throughput screening (HTS) methods, genomics, and bioinformatics, drug screening approaches changed from phenotypic to molecular target-based platforms, making it possible to identify and validate significant disease-related targets (Lewis 2013). In a target-based approach, a drug candidate interacts in vitro in a system devoid of cells with a specified or validated target such as an enzyme or receptor.

12.2.2 Traditional Approaches in Antibiotic Discovery and Development

The traditional approaches include the cultivation of strain, followed by screening of the culture supernatants or cell extracts for bioactivity, and further isolation of the compound which is an essential downstream step of drug development. The drug development process typically involves several screening steps such as primary, secondary, and even more specific ones are followed for characterizing the isolated compound to elucidate its mode of action. Agar diffusion assays and whole-cell assays are the primary screening methods that are simple, fast, and cheaper ones used to check the antibiotic activity in a sample either containing culture supernatant or cell extract. This is especially necessary for isolating and testing actinomycetes, or other novel potential producers, for the presence of bioactive compounds. The compound is put through additional characterization once the isolate's product inhibits bacterial test strains and dereplication results (using, e.g., HPLC, HPLC-MS, HPLC-HRMS, and chemical databases) strongly suggest a new entity. Since crude extracts are combinations of chemicals that obstruct several downstream processes in structure determination and target identification, this frequently calls for purification. Target-based assays are mostly used in secondary screening because they enable the investigation of the mode of action (MoA). In the past, screening assays targeted cell wall components, DNA, RNA, ribosomes, metabolic enzymes, and other proteins (Kohanski et al. 2010; Grzelak et al. 2016; Hudson and Lockless 2022). In contrast, the secondary screens typically require additional molecular steps (cloning, mutagenesis, protein purification, etc.) as well as the necessary equipment to perform the measurements. Even though secondary screening appears to be more difficult technically, HTS methods can ease the process. Furthermore, with unspecific primary screening methods, it is challenging to gain useful evidence for deciphering the mechanisms of action of antimicrobial drugs.

12.2.2.1 Primary Screening

Traditionally, the procedure of antibiotic discovery begins with the finding of an antibiotic producer such as actinomycetes from various ecosystems. Primarily, soil samples are extracted, diluted, and plated on selective solid media to isolate microorganisms. The plates are evaluated for promising candidates such as actinomycetes which are further propagated. This is done typically using a variety of media to investigate the biosynthetic potential of the bacterial isolates. Actinomycetes are evaluated in agar diffusion assays using agar with mycelium, culture supernatants, and extracts.

Agar Diffusion Methods

Alexander Fleming developed the agar diffusion test to detect antibiotic activity after observing that *S. aureus*, which had been streaked on the same plate, did not grow around a fungal colony (*Penicillium notatum*). In light of this finding, he filled a Petri dish with a test solution after partially removing an agar strip, creating a narrow channel (medium containing penicillin). Then he streaked various bacterial strains across the channel using another portion of the agar plate. The more susceptible strains showed less growth near the channel with penicillin. The same fundamental principle was used later in drug discovery and diagnostics processes. Various modifications have been made to the agar diffusion test, which consists of the agar disc diffusion assay (also known as Kirby-Bauer antibiotic susceptibility tests), the agar plug-based diffusion assay, the agar hole-based (well) diffusion assay, and the "direct" agar diffusion assay (Rütten et al. 2022).

Direct Agar Diffusion Assay

The goal of the direct agar diffusion assay is to screen for antibiotic-producing strains by directly exposing them to a test strain (indicator strain) on a solid medium (Singh et al. 2016; Das et al. 2018). The producer is first spotted and cultured under ideal conditions (incubation at 28–37 °C for 5–14 days for Actinomycetes sp.). An indicator strain with a specific optical density is then applied on top of the agar plate. A suitable solid medium should be used for the agar diffusion assay. It is crucial that antibiotic production is guaranteed, and that both the strains can grow on agar. A rich and transparent medium, the International Streptomyces Project-2 agar (ISP-2, Difco Laboratories), is frequently chosen for the purpose. This improves the likelihood of producing antibiotics and makes the growth inhibition zones easier to see and assess.

This method's concept states that the competition between a target pathogen (trigger) during the producer's serial transfer and exposure to the pathogen activates the antibiotic producer. It is anticipated that the rate of mutation will rise due to competition. The presence of a halo or a wider zone of inhibition, respectively, indicates that a mutation has either triggered the biosynthesis of antibiotics or increased production relative to the precursor strain. Using this system, an MRSA strain (N315) was neutralized and a *Streptomyces clavuligerus* strain was evolved (Charusanti et al. 2012). The evolved strain of *S. clavuligerus* inhibited the pathogen by producing holomycin, unlike the unevolved strain.

In this method of overlaying the agar plates, the suspension washes away the mycelium, and the producer strain often gets mixed up over the agar plate among a few actinomycetes leading to difficulties. This is very common among the sporulating producer strains or those that are fast-growing. Instead of overlaying the whole plate with the indicator strain rather than carefully plating them reduces this problem. Furthermore, using soft agar with the indicator strain for overlaying helps simplify the process for producers that are not washed away but rather stuck in the agar (e.g., *Streptomyces* strains).

Agar Plug-Based Diffusion Assay

In this method, the antibiotic-producing strains are grown on an agar medium, and plugs are punched out of it using a sterile cork borer or a micropipette tip. The plugs are placed onto fresh agar plates having an indicator strain. One advantage of this assay is that the producer can be grown independently on different media. It is therefore possible to test different antibiotic production (Mazza 1983; Elleuch et al. 2010; Saleem et al. 2015; Ganesan et al. 2017; Sarmiento-Vizcaíno et al. 2021). A microplate-based system was recently developed that can be used to screen Actinomycete strains (Ortlieb et al. 2021). In this method, agar was distributed in a microplate with a removable bottom, and actinomycete spore suspensions were added to the solid medium to create agar plugs. Following incubation, the bottom was taken out, and a special tool was used to push the agar plugs containing the bacteria out of the wells. The plugs are placed on Petri dishes containing the indicator strain to assay the bioactivity. This assay requires "custom-made" equipment, but the process is straightforward and allows the investigators to conduct a medium-throughput screening. Because the producers are grown apart from the indicator on different agar plates, the users have greater flexibility and need not worry about the medium, unlike in the "direct" agar diffusion assay.

Agar Well Diffusion Method

The agar plug-based diffusion assays and the agar well diffusion method, also known as the agar hole-based diffusion assay, are comparable. In the first approach, the agar punch device is used to obtain plugs from the producer; however, in the second it is necessary to punch holes in test agar plates (diameter of 6–8 mm) (Souza et al. 2004; Chaudhary et al. 2013; Saleem et al. 2015; Singh et al. 2016; Das et al. 2018). The antimicrobial solution (culture supernatant or cell extracts, volume of 20–100 μL) is poured into these holes, and the plates are then incubated under the appropriate conditions based on the indicator. The culture supernatant or cell extract preparation of the producer is necessary for the agar well diffusion method, in contrast to the "direct" and "agar plug-based diffusion" assays. As the bioactivity may be overlooked or lost, this is unfavorable for substances that cannot be extracted or are unstable under the applied conditions. Furthermore, this technique is especially useful for testing the activity of fractions that are obtained during a compound's chromatographic purification (such as preparative HPLC fractions).

Agar Disk Diffusion Assay

Agar disk diffusion is an additional screening method that can be used for culture supernatants (Ganesan et al. 2017), extracts (Saravana Kumar et al. 2014), HPLC fractions (Sharma and Manhas 2019), and purified antibiotics in solutions (Sharma and Manhas 2019; Qureshi et al. 2021). To assess the antibiotic activity, test plates inoculated with an indicator strain and filter paper disks containing the antibiotic solution are prepared. Usually, 10–100 μL are put onto a filter paper disk. The well- and agar disk diffusion method permits a concentration-dependent examination of the material (Sharma et al. 2016) and an estimation of the minimal inhibitory concentration (MIC) (Saravana Kumar et al. 2014) of pure compounds with a known concentration. Moreover, using a standard calibration curve the concentration of an antibiotic in an unknown sample can be assessed based on the size of the inhibition zones for specified concentrations of the purified antibiotic substance (Ibrahim et al. 2019). Test tubes containing liquid media setups also follow a similar method (Kavanagh 1975; Pudi et al. 2016). For instance, a medium containing the indicator strain is put into microtiter plates, and then an antibiotic solution with several different concentrations is added. Antibiotic concentrations in undefined samples can be semi-quantified using the calibration curve that is produced from measurements of absorbance or optical density. A system similar to this was recently developed for *Streptomyces fradiae* to screen for "super producers" of tylosin.

When combined, agar diffusion techniques offer easy, quick, and affordable ways to evaluate the antimicrobial properties of various extracted and/or purified compounds. When dealing with non-purified material, such as culture supernatants or extracts, it is impossible to identify the specific component causing the indicator strain's suppression in an agar diffusion assay because certain actinomycetes produce a combination of compounds that may work in concert (Challis and Hopwood 2003). The main application of agar diffusion assays is the quantification of materials or agents that have been refined and proven to contain a single compound that exhibits efficacy against the indicator strain. However, these assays are very helpful for initial screenings, particularly when the compound is uncharacterized and there are no available analytical techniques. Agar nutrient content, agar layer thickness (volume), agar uniformity, temperature, interpretation standards for inhibition zones and breakpoints, and other variables all impact the precision and repeatability of agar diffusion methods (Bonev et al. 2008). As such, assays are performed under specified conditions in order to be compared with one another or used for diagnostic purposes. For instance, the Clinical and Laboratory Standards Institute (CLSI) publishes these standards and their updates for testing against yeasts and bacteria. Lastly, controls need to be incorporated to verify the process. When the solvents (methanol or ethanol) used to dissolve the concentrated extract after evaporation also have antimicrobial activity, negative controls (solvent) should also be used to remove the possibility of misinterpretation. As a result, certain conditions must be met when conducting assays that are compared to one another or used for diagnostic purposes.

Thin-Layer Chromatography (TLC) Bioautography
Thin-layer chromatography (TLC)–bioautography combines biological activity detection with TLC's separation and analytical prowess. TLC bioautography was first popularized by Martin and associates (Consden et al. 1944) and Goodall and Levi (1946), where they combined contact bioautography with paper chromatography (PC) to analyze penicillin and amino acids, respectively. Fischer and Lautner (1961) are credited with officially introducing the term TLC. After then, TLC bioautography underwent additional development, giving rise to several varieties such as HP-TLC bioautography, D-TLC bioautography, agar diffusion or contact bioautography, direct bioautography, and agar-overlay assay (immersion bioautography) (Wang et al. 2021). In a nutshell, TLC is the process of separating components in a sample utilizing a mobile phase and a stationary phase. A thin layer of adsorbent material, such as aluminum oxide or silica gel, is applied to an inert plate surface to serve as the stationary phase (usually glass, plastic, or aluminium). Using a capillary spotter, samples are spotted onto the TLC plate in five dips per TLC plate using a 5 μL microcap. It is advised to transfer small amounts of the sample solution (spots) to one location on the TLC plate and to dry the plate in between. This is beneficial for diluted samples because it concentrates them through this process. Additionally, there should be adequate space between the plate's boundaries and the locations of the various samples. The plate is positioned vertically in an enclosed chamber containing an organic solvent once it has dried. From its initial position, the mobile phase moves in the direction of the chamber lid (capillary forces). The components are transferred in the same direction by the mobile phase, but because they have distinct affinities for the stationary and mobile phases, they halt the migration at different places on the TLC plate, resulting in varied distances for each substance (spot on the TLC plate). Their dots frequently overlap and are indistinguishable due to the chemical similarities between the highly comparable compounds in this system and their attraction to one another. If the silica gel is impregnated with a fluorescent dye, the plate should be assessed in both normal and UV light. Based on the chemical properties of the substances, a variety of commercially available visualization reagents can be employed (Aszalos and Frost 1975; Skorupa and Gierak 2011). The distance traveled by the compound (spot on the plate) divided by the solvent front's travel distance is known as the Rf value.

Contact Bioautography
The identification of in situ biological activity is the next phase of TLC bioautography. Transferring the chemical to an agar test plate that has been inoculated with an indicator strain can accomplish this (contact bioautography). Consequently, the test agar plate is covered with the TLC side that has the chemicals separated in a thin layer of adsorbent material. A few minutes or hours later, the single spots are scraped off the layer (powder is transferred onto the agar plate). After the necessary incubation period, an inhibition zone will become visible if the compound is active against the indicator (Grzelak et al. 2016; Barale et al. 2022).

Direct Bioautography

Direct bioautography provides an additional and quicker means of detecting the activity. In this case, the TLC plate containing the possible bioactive compounds is submerged in or sprayed with an indicator strain suspension, and it is then incubated under conditions that promote the strain's growth (Dewanjee et al. 2015). The TLC plate is sprayed with tetrazolium salts to assess the antibacterial activity. The regions where the strain is growing change color as a result of the production of dehydrogenases by living cells, which transform the tetrazolium salts into their vibrantly coloured formazan derivatives. Parts of the TLC containing an antibacterial ingredient, on the other hand, stay clear and stand out clearly against the background (Choma and Grzelak 2011; Choma and Jesionek 2015).

Agar Overlay Bioassay

Agar diffusion and direct bioautography are combined in the agar overlay bioassay. The TLC plate is separated, and then the inoculated agar medium is placed on top of it. In order to preserve the indicator strain, it is crucial to cool the agar to about 55 °C. The chemicals diffuse during the appropriate incubation conditions (which vary based on the indicator strain), and the inhibition zone appears as a sign of their activity. Tetrazolium dye is used to stain the indicator strain, much like in direct bioautography. This allows for the assessment of the strain's development and the detection of zones of inhibition brought on by active substances.

The primary applications of TLC bioautography have been the identification of antibacterial and antifungal activities (Couillerot et al. 2014; Joseph et al. 2017; Núñez-Montero et al. 2019; Matarrita-Carranza et al. 2021). Grzelak et al. (2016), for instance, tested a number of actinomycete-derived compounds using all three techniques, that is, contact bioautography, direct bioautography, and agar-overlay assay (immersion bioautography), demonstrating a broad range of polarities for antitubercular (anti-TB) activity.

TLC bioautography is a fast, easy, sensitive, and dependable method for screening antibiotics and other chemicals without the need for complex equipment, in addition to agar diffusion methods. The separation on the TLC plate (compounds) makes the analysis of the individual spots easier. On the other hand, comparable materials could produce a mixture-containing spot. Therefore, it is helpful to combine the TLC with HPLC, liquid chromatography-mass spectrometry (LC-MS), and/or other techniques for further characterization or dereplication. The spots (compounds) acquired from the TLC are further described (e.g., spots are scraped from the plate, extracted, and the extracts are analyzed by HPLC/MS or other methods) or the samples are analyzed in parallel utilizing TLC bioautography and HPLC/MS procedures (Kreuzig 1977; Chen and Schwack 2014; Joseph et al. 2017; Heep et al. 2019).

12.2.2.2 Secondary Screening: Target-Based Assays

One of the most important phases in the development of new drugs is target identification. Different approaches to drug target characterization have been explored in the past (Farha and Brown 2016; Emmerich et al. 2021; Li et al. 2021; Leshchiner et al. 2022). These include biochemical techniques, phenotypic screening and profiling methods, and genetic approaches. Genome analysis and the identification of putative resistance genes are two computational methodologies. Naturally resistant bacterial strains and resistant strains developed in vitro after exposure to antibacterial agents are generally used for this purpose though this method is not foolproof to unambiguously identify the antibiotic targets. More focused whole-cell assays using recombinant indicator strains such as reporter systems, an overexpression mutant library for crucial genes like the ASKA library (Kitagawa et al. 2005), various biochemical screenings such as pull-down assays (Cuatrecasas et al. 1968), and in vitro assays with potential targets is usually conducted after the in silico analysis. Cell wall compartments, DNA, RNA, ribosomes, and enzymes involved in metabolic processes are frequently used as macromolecular targets in the secondary screens (Fig. 12.1).

Assays for Targeting Cell Wall

The intricate, multilayered cell envelope serves as the cell's shield, giving its stiffness and stability, and is essential to its ability to communicate with the outside world (Scheffers and Pinho 2005; Silhavy et al. 2010; Fisher and Mobashery 2020; Pasquina-Lemonche et al. 2020). Furthermore, a variety of other cellular functions, such as growth, division, and cell wall recycling, depend on the components of the cell wall (Wang and Levin 2009; Dik et al. 2018; Mayer et al. 2019; Reyes-Lamothe and Sherratt 2019). Therefore, one of the primary objectives in the field of antibiotic development is to design drugs targeting the cell wall (Schuerholz et al. 2012; Jones 2017; Müller et al. 2017). Since Gram-positive and Gram-negative bacteria have different cell wall compositions, it is possible to distinguish between the two types of bacteria by employing antibiotics that target just one of these differences.

A whole-cell test method that disrupts peptidoglycan (PG) formation, hence impeding cell wall recycling, is one instance (Barbosa et al. 2002). In this test, pretreated *E. coli* ATCC 47076 cells were fed ^{14}C-labeled UDP-N-acetylglucosamine (UDP-GlcNAc) (subjected to freezing and thawing). The direct detection of cross-linked PG, which suggested PG recycling and cell regeneration, was made easier with the use of ^{14}C-labeled UDP-GlcNAc. Antibiotics, which are inhibitors of these processes, reduced the samples' optical density (OD_{600}), and less radioactivity was found as compared to the negative control. By inhibiting the formation of radiolabeled PG with known PG inhibitors (e.g., fosfomycin, bacitracin, and flavomycin), as well as compounds that do not target the enzymes of the PG pathway as a negative control (norfloxacin, kanamycin/streptomycin, showed no effect), the assay's functionality was confirmed. However, to work with radioactivity, one typically needs requisite substrates that have been marked with radiation, the appropriate facility, and skilled personnel. This could make high-throughput screening (HTS) more difficult.

Silbert et al. (2006) devised an outstanding method for the visual and spectroscopic identification of microorganisms, particularly bacterial communications. The idea is that bacterially produced membrane-active substances interact with nanoparticles implanted in agar. To mimic a membrane, the nanoparticles are made of phospholipids and the chromatic polymer polydiacetylene (PDA). It has been shown that bacteria-produced chemicals have an impact on the PDA, causing blue-to-red changes and strong fluorescence emission (Silbert et al. 2006). This is measurable with standard HTS equipment. After an activity was found in the primary screening, the spectroscopic detection approach was used for the screening of extracts obtained from actinomycetes. Consequently, phospholipid/PDA (vesicle solution) was incubated for 1 h after the extracts had been resolved in DMSO. The assay is a very practical method since it makes quantification easier by measuring colorimetric responses. If the extracts include colored materials, this could cause issues because it could impede the detection process. Further procedures (e.g., preparative HPLC, TLC-bioautography) are advised in this situation to isolate the chemicals and retest each one individually.

Recently, an intriguing assay was developed that used the actinomycete itself to provide a screening approach rather than the metabolites that actinomycetes make (Gosschalk et al. 2020). Gosschalk and colleagues concentrated on identifying sortase enzyme inhibitors, as these enzymes make appealing therapeutic targets. These enzymes bind relevant bacterial pathogens, such as *S. aureus*, to their surface and provide virulence characteristics. The virulence factors would disappear if the sortase enzymes were blocked. *Actinomyces oris* was used to create a screening test that worked well. As this strain grows in a manner that is dependent on sortase in a lab setting, it is especially well suited to serve as a sortase inhibitor sensor. Mutants of *A. oris* were created, and a secondary screening in addition to a primary screening in which the wild type of *A. oris* was employed. It was implemented to exclude small compounds that hampered *A. oris* growth by mechanisms unrelated to sortase. An HTS was developed based on this trait and the in vitro sortase-inhibitory activity was also verified (Gosschalk et al. 2020). This approach is a straightforward and reasonably favourable option, even though sortase inhibitors must be specifically identified through primary and secondary screening.

DNA Synthesis Inhibitors Assay

Inhibiting DNA synthesis stops cells from proliferating. Since DNA gyrases (topoisomerases) play pivotal roles in transcription, recombination, and DNA replication, these are the common targets for inhibiting bacterial growth (Reece and Maxwell 1991; Chitra et al. 2020; Man et al. 2021). Numerous quinolone antibiotics that were first derived from actinomycetes, including nalidixic acid (Finegold et al. 1966), have been demonstrated to function as strong inhibitors of DNA gyrase (Bush et al. 2020; Man et al. 2021). Assays for screening DNA synthesis inhibitors include cell-based high-throughput bioluminescence screens (Hutter et al. 2004; Moir et al. 2007). Moir et al. (2007) linked a luciferase operon to a promoter that reacts to DNA damage brought on by lowered gyrase levels of *P. aeruginosa* (luciferase genes inserted in *P. aeruginosa* genome). The region encoding the pyocin gene served as the source of the promoter (PA0614). Pyocins are poisonous bacteriocins that eliminate closely related strains of *Pseudomonas*. In response to ciprofloxacin, PA0614 exhibited a reduction in GyrA levels. As a result, gyrase expression is suppressed if a substance interacting with the promoter is added to the experiment. Simultaneously, there is an upregulation of *lux* gene expression that can be detected by a luminometer. Using the recombinant strain, the transcriptional regulatory response due to the reduction of an antibacterial target (gyrase) can be assessed. Thirteen gyrase inhibitors were identified by screening of which quinolones made up ten while non-quinolone molecules made up the remaining three (mechlorethamine-, furazolidone-, and nitrofuran-like structures). The researchers can do a HTS, especially for gyrase inhibitors using the whole-cell bioluminescent test.

Various in vitro methods that screen drugs targeting the enzymes involved in DNA synthesis have also been developed. Enzymatic assays based on PCR are particularly helpful. A PCR-based assay measuring ribonucleotide reductase (RNR) activity in a microplate format was presented by Tholander and Sjöberg (2012). RNR is a rate-limiting enzyme of DNA synthesis because it catalyzes the reduction of the four ribonucleotides required for DNA synthesis to deoxyribonucleotides (Borovok et al. 2006; Torrents 2014; Ruskoski and Boal 2021). Antibiotics frequently target the RNR, but because of the difficult experimental techniques, potential inhibitors are not thoroughly investigated. The reduction of ribonucleoside-5-diphosphates (NDPs) to deoxynucleoside diphosphates (dNDP), which is catalyzed by RNR, can be measured using PCR-based assays. Consequently, the staining agent SYBR green dye (an asymmetric cyanine dye) used for staining *P. aeruginosa* nucleic acids is added to the sample and after binding to the sample, fluorescence can be detected.

The PCR yields more product and produces a brighter fluorescence signal in proportion to the RNR activity. This assay was used to assess 1364 chemicals for their ability to block *P. aeruginosa*'s class 1 RNR. Out of these compounds, 110 have demonstrated a 50% inhibition of RNR activity, which is accurate; 27 of them showed an over 90% inhibition, with IC_{50} values ranging from 30 to 200 μM. These 27 substances underwent additional testing to determine their effects on growth and proliferation as well as dose-dependent responses. Four of them have demonstrated effects on *P. aeruginosa* that are similar to those of carbenicillin and tetracycline. *Streptomyces flocculus*'s streptonigrin was one of these four highly effective inhibitors (Tholander and Sjöberg 2012). The examples of two assays in this section make it possible to test for RNR inhibitors. The first (cell-based high-throughput bioluminescence) has the benefit of filtering out those candidates that are unable to cross the cell envelope, which is where the inhibitors must pass to reach the target. The second is probably more precise, but to test it, the RNRs must be purified. In this instance, it would be intriguing to investigate whether the application of a crude extract containing the protein could streamline the process and lower expenses associated with the enzyme's purification.

Transcription and Translation Inhibitors Assays

In every living cell, the processes of transcription and translation are essential. Thus, one targeted method to combat pathogens causing severe infections is to reduce or eliminate transcription or translation in them. Furthermore, prokaryotes and eukaryotes have different molecular processes for transcription and translation which makes it more desirable that the transcription/translation in bacteria will be specifically inhibited without impairing these vital functions in eukaryotic cells—a crucial requirement for novel drugs.

Numerous assays were developed to search for novel transcription/translation inhibitors (Brandi et al. 2008; Lowell et al. 2015; Raneri et al. 2015; Osterman et al. 2016; Kavčič et al. 2020). These include in vitro protein synthesis inhibition assays (Shimizu et al. 2001; Svetlov et al. 2006), toe-printing of antibiotic-stalled ribosomes (Orelle et al. 2013), SPARK-sensitive method for monitoring peptidyl transferase activity (Polacek et al. 2002), antibiotic binding to a fluorescently labeled ribosome (Llano-Sotelo et al. 2009); biosensor assays (Galluzzi and Karp 2003), stress response assays (Bianchi and Baneyx 1999), antibiotic detection assays (Kurittu et al. 2000; Möhrle et al. 2007), attenuation-based dual fluorescent reporter assays (Osterman et al. 2012), panel of reporter strains that lack antibiotic resistance (Melamed et al. 2012), transcriptional sensors based on promoter libraries (Goh et al. 2002); fluorescent microscopy and bacterial cytological profiling (Nonejuie et al. 2013), proteomics-based methods (Bandow et al. 2003; Osterman et al. 2013), resistance-based assays to sensor the mechanism of action (Orelle et al. 2013).

In vitro transcription/translation assays are potent methods used to address many scientific queries, such as identifying a drug's target. Each element needed for transcription and translation needs to be in the sample combination in order to perform the assay. Purified enzymes are frequently substituted by cell extracts (Hoagland et al. 1958; Kim et al. 2006), such as those from *E. coli*. Additional elements are introduced to the sample to guarantee the transcription and translation of the reporter gene or protein, such as nucleotides and amino acids as substrates for transcription and translation, fructose-1,6-bisphosphate for energy sources, and a reporter system (such as pET28-eGFP). Compared to the control, samples treated with a transcription or translation inhibitor show either no signal at all or a reduced signal (such as fluorescence). Since this is linked to the production of the reporter protein (eGFP, for instance), the corresponding spectrometry method can determine transcription or translation blockage (e.g., using a fluorometer). An investigation of the efficacy of kirromycin derivatives generated by a genetically modified strain of *Streptomyces collinus* Tü 365 was conducted, for instance, using a transcription/translation test. It has previously been determined that elongation factor Tu (EF-Tu) is the target of kirromycin (Wolf et al. 1977). If substances causing an inhibitory effect are subjected to the in vitro transcription/translation experiment and the target is unknown, more characterization may be required. Certain in vitro techniques that present the appropriate target for interaction can accomplish this. An instance of this can be found in an experiment known as the real-time fluorescence polarization activity assay (FP/FA test), wherein the bacterial enzyme RNase P was employed as a target and subjected to several inhibitors (Liu et al. 2014). The mature tRNA with a 5′ end was produced by the cleavage of the 5′ leader sequence from precursor tRNAs (pre-tRNAs) catalyzed by the enzyme RNase P. The technique makes it possible to identify substances that bind to pre-tRNAs and those that impede. Neomycin B and kanamycin B, two actinomycete antibiotics, were used to verify the test and optimize it for HTS. A library of 2880 compounds was screened. A novel *Bacillus subtilis* RNase P inhibitor, iriginol hexaacetate, was discovered (Liu et al. 2014).

Bacterial Enzymes Inhibitors Assays

Enzymes that catalyze vital bacterial cell activities are frequently employed in the assays to identify possible antibiotics or their targets. To prevent toxicity, these enzymes should ideally not exist in eukaryotic cells or differ from them fundamentally. Typically, toxicity screenings using eukaryotic cell lines come after testing the bacterial target's activity and specificity. Several enzymatic assays employ traditional targets, such as enzymes involved in cell wall compartment formation and/or recycling, particularly peptidoglycan. Additional instances comprise ribonucleotide reductase (RNR), bacterial RNase P, gyrase, and other enzymes that catalyze critical stages in the manufacture of vitamins, amino acids, or coenzymes (Spry et al. 2008; Magalhães et al. 2020; Nowak et al. 2021; Panchal and Brenk 2021). In these experiments, proteins of the ribosomal subunits and formed ribosomes are finally exposed to the chemicals under test. In order to quantify the kinetics of ribosomal subunit synthesis, several of these techniques use a pulse and chase labeling strategy. Other techniques make it easier to examine ribosome reformation following the removal of antibiotics to study the impacts that follow (Champney 2008; Nikolay et al. 2015). The fact that a sizable portion of the antibiotics from actinomycetes targets the bacterial ribosome emphasizes the significance of techniques that allow for the screening as well as identification of the antimicrobial agents (Poehlsgaard and Douthwaite 2005; Yassin et al. 2005; Yassin and Mankin 2007; Blanchard et al. 2010; Champney 2020; Kavčič et al. 2020). Furthermore, histidine kinases (HK) were added to the list of targets in the hunt for antibiotics (Bem et al. 2015; Fihn and Carlson 2021; Chen et al. 2022a). Membrane receptors called HK regulate various cellular reactions (e.g., virulence, secretion systems, and antibiotic resistance). They work in signal transduction pathways with two components. Two-component systems (TCSs) comprise a response regulator (RR) and a HK. By enhancing stress responses such as the cell wall stress response, the suppression of TCS may either kill the host or increase its antibiotic sensitivity. For instance, TCS inhibition made a strain of *S. aureus* that was resistant to methicillin, vulnerable (Boyle-Vavra et al. 2006). Antibiotic drug discovery thus encourages the development of effective screening tests for TCS inhibitors, including HK. One potential technique for HK activity detection and profiling is the recently released immuno-dot blot test (Kinoshita-Kikuta et al. 2020). Assays for in vitro kinase, like autophosphorylation methods using γ-^{32}P-ATP (Marina et al. 2001; Velikova et al. 2016), are frequently employed to quantify the phosphorylated histidine kinase.

Genome Mining of Natural Products into Drug Leads

The identification of novel compounds having therapeutic values relies on genomic computational mining. Nowadays, thousands of bacterial genome sequences that include an even greater variety and quantity of secondary metabolite gene clusters can be accessed. A multitude of genome-mining techniques, along with the advancement of HTS methods and an abundance of DNA data, aid in identifying and characterizing these substances (Fig. 12.2). Traditionally, the screening of natural sources for bioactivity has served as the foundation for drug development activities. Natural compounds remain the primary source of novel drugs in modern times. For decades, researchers have exploited *Streptomyces* to find novel drugs. It was possible with the use of reverse genetics or traditional genetics techniques. A *Streptomyces* genome contains, on average, about 20 secondary metabolite gene clusters. The idea of using genetic information to identify and isolate natural products without a preexisting structure gave rise to the concept of classical genome mining (Fig. 12.2). When genome information became available, the idea of genome mining was extended to other organisms, including cyanobacteria (Calteau et al. 2014), myxobacteria (Gerth et al. 2003; Wenzel and Müller 2009), and anaerobes (Behnken and Hertweck 2012; Letzel et al. 2013). Nowadays, full culture collections are being sequenced, thousands of bacterial genomes are becoming available, and novel methods such as single-cell genomics and metagenomics produce huge amounts of data requiring processing (Fig. 12.2).

The "Atlas of Biosynthetic Gene Clusters" a part of the Joint Genome Institute's "Integrated Microbial Genomes" Platform, currently houses the largest collection of autonomously mined gene clusters (JGI IMG-ABC) (Medema et al. 2015). More than 960,000 putative gene clusters were found in JGI's massive genome and metagenome databases as of February 2016, and these entries are included in IMG-ABC. An extensive manual re-annotation of over 1300 BGCs was recently completed as part of the "Minimum Information for Biosynthetic Gene clusters" (MIBiG) standardization initiative. This results in a well-curated reference dataset (Medema et al. 2015).

The first approach was typically used to identify novel compounds with strong bioactivities from the pool of natural products (NPs) generated by an individual bacterial species or a microbial consortium. This method produced NP in the following sequential order: isolation of the target microorganism(s), microbe(s) culture under various conditions to produce NP, and chemical extraction of the targeted products. Using bioassays on bacterial cultures, extracted fractions, or purified products, bioactivity-guided screening assesses NPs' effects on various species or human disease models to ascertain whether or not they exhibit desirable bioactivities (Katz and Baltz 2016). The "top-down" approach refers to this traditional NP discovery technique since it does not need to be aware of a producer's genome or biosynthetic ability beforehand (Luo et al. 2014). Despite its many benefits, the "top-down" method has significant drawbacks that have hindered the discovery of numerous therapeutic NPs. Because many bacterial species can synthesize the same NP often results in the re-isolation of known compounds from them (Genilloud et al. 2011; Wohlleben et al. 2016).

One possible substitute for the aforementioned "top-down" method is to look for the NP biosynthesis machinery in bacterial genomes (Bachmann et al. 2014). The "bottom-up" strategy, which is based on genome mining, makes use of genetic modification tools, bioinformatics analysis, and a wealth of genomic data to find BGCs that encode new NPs (Luo et al. 2014). In their search, these data-mining techniques frequently make use of conserved protein domains and signature genes of the enzymes that are in charge of the production of various NP kinds (Table 12.1). Despite their capacity to produce a wide range of chemical families, these enzymes'

Table 12.1 Novel compounds/natural products identified by various genome-mining methodologies

Core biosynthetic enzyme-guided (BGC-directed identification)			
Search genes/gene cluster ID/MiBIG gene bank	**Novel compounds**	**Source organisms**	**References**
Type I polyketides			
PKS genes/BGC0000022	Asperfuranone	*A. nidulans*	Chiang et al. (2009)
Modular PKS genes BGC0000151	Stambomycins	*S. ambofaciens*	Laureti et al. (2011)
Trans-AT polyketide			
PKS/NRPS genes/ BGC0001111	Rhizopodin	*S. aurantiaca* Sg a15	Pistorius and Müller (2012)
Type II polyketides			
Type II gene cluster with antiSMASH 2/Genbank: KT713752	Hexaricins		Tian et al. (2016)
Type III polyketides			
Genes encoding type III PKS/Genbank: AL645882 Gene: SCO7221	Isogermicidin	*S. coelicolor* A3(2)	Song et al. (2006)
NRPS			
Conserved NRPS genes/ BGC0000308	Aureusimines	*S. aureus* and other *Staphylococcus* strains	Wyatt et al. (2010)
NRPS genes/ BGC0001208	Poeamide	*Pseudomonas poae* RE*1-1-14	Zachow et al. (2015)
Resistance gene-guided			
Resistance genes	**Natural product**	**Source organism**	**References**
Pentapeptide repeat protein (PRP) sequences	Alkylpyrone-407	*Cystobacterineae* strain MCy9487	Hug et al. (2019)
	pyxidicyclineA	*Pyxidicoccus fallax* An d48	Panter et al. (2018)
Dihydroxyacid dehydratase	Aspterric acid	*Aspergillus terreus* NIH2624	Yan et al. (2018)
Tripartite efflux system PleABC	Prosekin	*Pseudomonas prosekii* LMG 26867	Girard et al. (2022)
Lanosterol 14α-demethylase	Lanomycin	*Pyrenophora dematioidea* TTI-1096	Liu et al. (2021)
Fatty acid synthase	Thiotetroamide	*Streptomyces afghaniensis* NRRL 5621	Tang et al. (2015)
D-stereospecific peptidase	Bogorol	*Brevibacillus laterosporus* DSM 25	Li et al. (2018)

(continued)

Table 12.1 (continued)

Phylogenomics-guided			
Sequences used for phylogenetic analysis	**Natural product**	**Source organism**	**References**
Whole BGCs of different families	Aryl polyenes	*Escherichia coli* CFT073	Cimermancic et al. (2014)
"Expanded-then-recruited" enzyme Families; 3-carboxyvinyl-phosphoshikimate Transferase	Arseno-organic metabolites	*Streptomyces lividans* 66	Cruz-Morales et al. (2016)
Each shared gene found in glycopeptide Antibiotic-producing BGCs	Corbomycin	*Streptomyces* sp. WAC01529	Culp et al. (2020)
ATP-grasp ligase	MdnA7	*Cyanothece* sp. PCC 7822	Ahmed et al. (2017)
LuxR	cepacinA	*Burkholderia ambifaria* BCC0191	Mullins et al. (2019)
Whole BGCs containing tauD expansion	Detoxin S1	*Streptomyces* sp. NRRL S-325	Navarro-Muñoz et al. (2020)
Terpene synthase	Hydropyrene	*Streptomyces clavuligerus* ATCC 27064	Yamada et al. (2015)
Chain length factor (CLF) protein	Oryzanaphthopyran A	*Streptacidiphilus oryzae* CGMCC 4.2012	Chen et al. (2022b)
Structure-guided			
Targeted chemical structure	**Natural product**	**Source organism**	**References**
Cationic amino acid residues	Brevicidin	*Brevibacillus laterosporus* DSM 25	Li et al. (2018)
Chemical transformations catalyzed by cytochrome P450 on cyclodipeptides	Cyctetryptomycin B	*Saccharopolyspora hirsuta* DSM 44795	Malit et al. (2021a)
Prenyl groups on cyclodipeptides	Griseocazine D1	*Streptomyces griseocarneus* 132	Malit et al. (2022b)
Thioether bonds	Freyrasin	*Paenibacillus polymyxa* ATCC 842	Hudson et al. (2019)
Chemical transformations catalyzed by the DUF–SH didomain	Guangnanmycin	*Streptomyces* sp. CB01883	Pan et al. (2017)
Phosphonic acid	Argolaphos A	*Streptomyces monomycini* NRRLB-24309	Ju et al. (2015)

(continued)

Table 12.1 (continued)

Global genome mining for RiPPs			
Combining the structure-guided strategy with precursor peptide sequence search			
RiPP family	**Natural product**	**Source organism**	**References**
Lanthipeptide	Birimositide	*Streptomyces rimosus subsp. rimosus* WC3908	Walker et al. (2020)
Cyanobactin	Tolypamide	*Tolypothrix* sp. PCC 7601	Purushothaman et al. (2021)
Polyoxazole-thiazole-based cyclopeptide	Aurantizolicin	*Streptomyces auranticaus* JA 4570	Skinnider et al. (2016)
Thiopeptide	Saalfelduracin	*Amycolatopsis saalfeldensis* NRRL B-24474	Schwalen et al. (2018)
Thioamitides	thiovarsolinA	*Streptomyces varsoviensis* DSM 40346	Santos-Aberturas et al. (2019)
Sactipeptide	Streptosactin	*Streptococcus thermophilus* JIM 8232	Bushin et al. (2020)
Lanthipeptide	Flavucin, agalacticin, etc.	*Corynebacterium* lipophiloflavum DSM 44291	van Heel et al. (2016)

Adapted from Ziemert et al. (2016), Malit et al. (2022a)

amino acid sequences are substantially conserved, making them useful as reference or seed sequences for locating gene homologs and, consequently, novel BGCs (Ziemert et al. 2016).

Genomics for Natural Product Discovery

Numerous software applications have been developed specifically to search bacterial genomes for BGCs linked to secondary metabolites (Table 12.2). Antibiotics and Secondary Metabolite Analysis Shell (antiSMASH) is the most well-known and effective software pipeline at the moment (Medema et al. 2011). Using profile Hidden Markov Models (pHMMs) based on user-defined gene cluster rules, it locates both core and additional biosynthetic genes. The latest version of antiSMASH may be used to identify more than 70 distinct types of BGCs encoding for different chemical entities (Blin et al. 2021). By utilizing assumed collinearity and the predictions of substrate specificity of protein domains found in the polyketide synthase (PKS) and non-ribosomal peptide synthase (NRPS) modules, the software may also predict the chemical structure of a core scaffold (Medema et al. 2011).

Cluster Blast, one of the integrated sequence alignment tools in antiSMASH, facilitates the comparison of BGCs, an essential step in the dereplication process. Other genome-mining techniques for BGC identification include PRISM, which locates BGCs using pHMMs and machine learning to predict biological activity

Table 12.2 List of different databases and tools used at various stages of drug discovery and development

Databases/Tools	URL of the Web server
Biosynthetic gene cluster (BGC) identification and analysis	
antiSMASH	http://antismash.secondarymetabolites.org
ClusterFinder	https://github.com/petercim/ClusterFinder
NP searcher	http://dna.sherman.lsi.umich.edu/
MultiGeneBlast	http://multigeneblast.sourceforge.net/
SMURF	http://jcvi.org/smurf/index.php
BAGEL	http://bagel.molgenrug.nl/
ClustScan	http://bioserv.pbf.hr/cms/
NaPDoS	http://napdos.ucsd.edu
eSNaPD	http://esnapd2.rockefeller.edu/
NRPS-PKS/SBSPKS	http://www.nii.ac.in/sbspks.html
Tools to connect genomics and mass spectrometric data	
GNPS	http://gnps.ucsd.edu/
Pep2Path	http://pep2path.sourceforge.net/
RiPPQuest	http://cyclo.ucsd.edu/
CycloQuest	http://cyclo.ucsd.edu/
NRPQuest	http://cyclo.ucsd.edu/
Substrate specificity predictions for NRPS and PKS enzymes	
NRPSPredictor	http://nrps.informatik.uni-tuebingen.de
LSI predictor	http://bioserv7.bioinfo.pbf.hr/LSIpredictor/
NRPSsp	http://www.nrpssp.com/
NRPS/PKS substrate predictor	http://www.cmbi.ru.nl/NRPS-PKS-substrate-predictor/
Gene cluster analysis databases	
IMG-ABC	https://img.jgi.doe.gov/cgi-bin/abc/main.cgi
MIBiG repository	http://mibig.secondarymetabolites.org/
Do BISCUIT	http://www.bio.nite.go.jp/pks/
ClusterMine360	http://www.clustermine360.ca/
ClustScan database	http://csdb.bioserv.pbf.hr/csdb/
Chemical compound databases	
ChEBI	http://www.ebi.ac.uk/chebi/
ChEMBL	https://www.ebi.ac.uk/chembl/
KNApSAcK	http://kanaya.naist.jp/KNApSAcK/
PubChem	http://pubchem.ncbi.nlm.nih.gov/
ChemSpider	http://www.chemspider.com/
ZINC	https://zinc.docking.org/
StreptomeDB	http://www.pharmaceutical-bioinformatics.de/streptomedb/
MCULE	https://mcule.com/database/
DrugBank	https://go.drugbank.com/
NORINE	http://bioinfo.lifl.fr/norine/

(continued)

Table 12.2 (continued)

Databases/Tools	URL of the Web server
Dictionary of natural products	http://dnp.chemnetbase.com/
Drug target databases	
TargetTB	https://www.targettb.org.uk/
TuberQ	http://tuberq.proteinq.com.ar/
FindTargetsWEB	http://pseudomonas.procc.fiocruz.br/FindTargetsWEB
Protein modeling tools	
MODELLER	http://bioinf.modares.ac.ir/software/easymodel/ (easy model is a web-based modeling tool based on MODELLER)
SwissModel	https://swissmodel.expasy.org/
I-TASSER	https://zhanggroup.org/I-TASSER/
QUARK	https://zhanggroup.org/QUARK/
AlphaFold	https://alphafold.ebi.ac.uk/
Molecular docking tools	
AutoDock	https://autodock.scripps.edu/
AutoDock Vina	https://vina.scripps.edu/
CDOCKER	https://www.computabio.com/cdocker-tutorial.html
SwissDock	http://www.swissdock.ch/
MD simulation tools	
AMBER	https://ambermd.org/
CHARMM	https://www.charmm.org/
GROMOS	https://www.gromos.net/
Pharmacophore modeling tools	
PharmMapper	https://www.lilab-ecust.cn/pharmmapper/
PharmaGist	https://bio.tools/pharmagist
Tools for QSAR	
QSAR-co	https://sites.google.com/view/qsar-co
Open3DQSAR	https://bio.tools/cloud3dqsar
QSAR toolbox	https://qsartoolbox.org/
Tools for ADMET assessment	
PreADMET	https://preadmet.webservice.bmdrc.org/
ALOGPS	https://vcclab.org/lab/alogps/start.html
SwissADME	http://www.swissadme.ch/
DrugMint	https://webs.iiitd.edu.in/oscadd/drugmint/

(Skinnider et al. 2020), and MetaBGC, which locates BGCs from microbial community sequencing data (Sugimoto et al. 2019). The Secondary Metabolite Bioinformatics Portal (SMBP) provides a list of genome-mining programs along with links to databases, tools, and programs helpful for NP identification utilizing omics data (Weber and Kim 2016). The Biosynthetic Gene Clusters (BGCs) identified in bacteria, fungi, archaea, and metagenome bins comprise over 400,000 BGCs in the Integrated Microbial Genomes Atlas (IMG-ABC) (Hadjithomas et al. 2017).

The Minimum Information about a Biosynthetic Gene Cluster (MIBiG) database compiles all characterized BGCs with defined roles and the secondary metabolites they produce (Kautsar et al. 2020). This database is useful for finding BGCs in sequenced genomes which can produce similar sets of chemicals based on sequence homology, hence preventing the rediscovery of already discovered compounds. As of now, the MIBiG contains the BGC data for 1923 secondary metabolites. Large-scale BGC research can be supported by a longer number of datasets, according to a recent assessment (Chevrette et al. 2021). To manage these large datasets and effectively present the unknown NP diversity hidden in cryptic genome BGCs, downstream bioinformatics technologies must be developed (Fig. 12.3). Large-scale biological sequence data can be handled by bioinformatics tools such as cd-hit (Fu et al. 2012) and Fast Tree (Price et al. 2010). They can also be integrated for data processing and visualization into the global genome-mining pipeline. Numerous omics-based approaches (that include transcriptomics, proteomics, metabolomics, and meta-omics) have been developed to identify novel secondary metabolites (Ziemert et al. 2016; Machado et al. 2017; Palazzotto and Weber 2018). Phylogenomic analysis, enzyme-catalyzed structural modifications, and NP resistance factors have all been utilized to identify new NPs by prioritizing BGCs (Table 12.1).

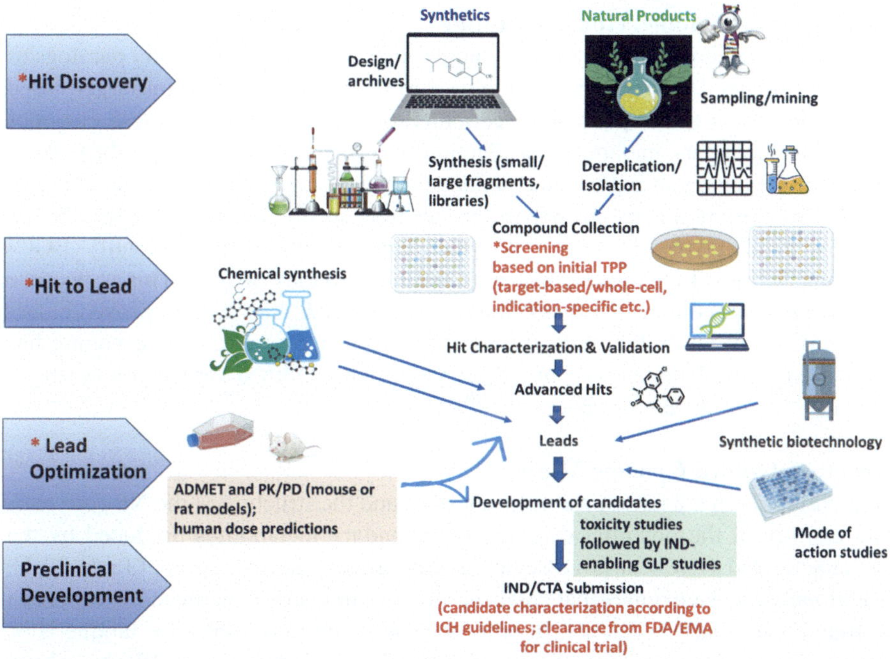

Fig. 12.3 Overview of the standard processes in the novel antibacterial drug discovery and development. The steps marked with * can be performed with various artificial intelligence and machine learning tools. [*ADMET* absorption, distribution, metabolism, excretion, and toxicity, *CTA* clinical trial application, *EMA* European medicines agency, *FDA* U.S. Food and Drug Administration, *GLP* good laboratory practice, *ICH* International Council for Harmonisation of technical requirements for pharmaceuticals for human use, *IND* investigational new drug, *PK/PD* pharmacokinetics/pharmacodynamics, *TPP* target product profile]

Resistance Gene-Guided Genome Mining

A method for doing global genome mining to find new bioactive compounds is to focus on resistance determinants linked with biosynthesis. Some BGCs include genes involved in developing immunity against the deleterious consequences of the metabolite in the producer organism, even though most BGCs primarily contain genes necessary for producing the desired secondary metabolite (O'Neill et al. 2019). Encoding enzymes that confer immunity or resistance to the producer, these genes—also known as resistance genes—typically function in three different ways. During the process of product detoxification, the protein attaches itself to the generated secondary metabolite and, quite frequently, also changes it. This stops the metabolite from attaching to its intended target in a producer strain. An NP can also be quickly removed from a cellular space by an enzyme, or it can be bound to a transporter. Last but not least, a BGC may include an enzyme that catalyzes the modification of the target to make it resistant to a generated NP or a duplicate-resistant variant of the NP's target (O'Neill et al. 2019). Resistance-guided genome mining has been used to identify a number of bioactive chemicals and their mechanisms of action.

Phylogenomics-Guided Genome Mining

Enzymes engaged in synthesizing secondary metabolites are essentially paralogs of enzymes involved in primary metabolite formation, as the resistant-target-based mining discussed in the preceding section (Baltz 2019). Similarly, these enzymes catalyzing secondary metabolite production have the ability to change throughout time, which may result in the catalysis of new chemical reactions. The general process of this evolution is the growth of the substrate specificity of the enzyme (Cruz-Morales et al. 2016). Divergent enzymes can be employed to identify BGCs encoding for novel chemical alterations and bioactivities of NPs. This is because selective pressure directs these enzymes to create metabolites with improved bioactivities (Culp et al. 2020). With the use of phylogenetic trees, this reasoning has been used to find novel NPs by identifying BGCs with divergent biosynthetic enzymes that produce new chemicals (Table 12.1).

Structure-Guided Genome Mining

Predictions about the additional tailoring steps and the structure of the NP core scaffold can reveal the potential diversity of secondary metabolites produced by the uncharacterized BGCs in large-scale genome-mining investigations (Table 12.1). Additionally, new enzymes that catalyze hitherto unheard-of chemical changes can be found using the predictions. In order to search and rank BGCs for unique NPs, enzymes that are known to catalyze intriguing chemical reactions on NPs have been employed (Rütten et al. 2022).

Global Genome Mining for Ribosomally Synthesized and Posttranslationally Modified Peptides (RiPPs)

BGCs, the gene clusters that make up ribosomally synthesized and posttranslationally modified peptides (RiPPs), encode precursor peptides and the tailoring enzymes that modify these precursors, catalyzing posttranslational modifications (PTMs) to form mature peptide structures (Arnison et al. 2013; Montalbán-López et al. 2021). In contrast to polyketide synthetases and non-ribosomal peptides, RiPPs lack genetic traits such as globally conserved enzyme domains (Santos-Aberturas et al. 2019; Kloosterman et al. 2020). However, RiPP classes or families, such as lanthipeptide, sactipeptide, thiopeptide, and lasso peptide, have core enzymes that are specific to their class and precursor peptide sequence motifs (Montalbán-López et al. 2021). The ribosomal origin of their precursor peptides makes it easier to reliably anticipate their core structure based only on gene sequence, which is useful for genome-mining investigations of RiPPs (Zhang et al. 2018; Malit et al. 2021b). The strategy involved in structure-guided genome mining is the same where the use of the core tailoring enzymes of known RiPP classes is explored for novel RiPPs. However, due to their typically short and varied lengths, RiPP precursor peptides present a hurdle. Therefore, RiPP precursors are frequently left unannotated in BLAST. Because of this, several genome-mining software programs and tools, such as RODEO (Tietz et al. 2017), RiPPER (Santos-Aberturas et al. 2019), and antiSMASH (Medema et al. 2011; Blin et al. 2021), have been created to find these brief precursor sequences (Table 12.2).

Genome-mining techniques have rapidly advanced over the last 10 years, which is facilitated by the rapid growth of genome-sequencing technologies. They are now an integral part of drug discovery initiatives. Different mining methods and methodologies have been developed to identify, dereplicate, and prioritize secondary metabolite gene clusters based on the interests and emphasis of researchers (Fig. 12.2). Wet lab experiment guidance requires a steady stream of development of efficient computational tools and algorithms due to the overwhelming amount of currently available data and will continue to be so in the future. Moreover, developments in metagenomics along with other "omics" technologies, systems biology, and synthetic biology approaches would drive automated HTS to discover novel compounds more efficiently in the future. An overview of the standard processes in the novel antibacterial drug discovery and development is provided in Fig. 12.3.

12.3 Omics and Bioinformatics in Antibacterial Drug Discovery

The conventional drug discovery and development process is a decade-long endeavor involving an investment of billions of dollars that often results in failures. The cost of a single drug from discovery to reaching the market is approximately \$2 billion. Moreover, from a hit of 10,000 compounds, the chance of finding a single drug that can successfully pass the clinical trial phases is less than 10% (Miethke et al. 2021; Shaker et al. 2021). Large pharmaceutical companies have been discouraged from investing in antibiotic development due to the difficult prospect of high attrition rates in drug discovery and development. This has significantly reduced the discovery of novel antibiotic classes in the preceding decades. The last novel class of antibiotics was discovered in the 1980s (Miethke et al. 2021). The reasons for such a significant challenge include the time-consuming and labour-intensive in vitro and in vivo experiments involved in testing the drug's efficacy. Specifically, the drug candidates need to be assessed for their absorption, distribution, metabolism, excretion, and toxicity often referred to as ADMET in the pharmaceutical language which requires huge cost and time (Shaker et al. 2021). The ADMET assessment requires the use of animal models in the preclinical phases before proceeding with the clinical trials. Though new in vitro models for ADMET assessment are constantly being developed, for example, organ-on-chip, still time remains a formidable factor. Though the automated HTS for drug candidate identification has reduced the time required for this process to some extent, still, the complexity involved in the process is a roadblock to the rapid identification process. The advancement of bioinformatics in general, and genomics in particular, over the last two to three decades has provided opportunities to reduce time, labor, and cost in the drug development process (Malathi and Ramaiah 2018; Torres and De La Fuente-Nunez 2019; Behling et al. 2023).

The groundbreaking development in the field of omics in the recent past has revolutionized the process of drug discovery. Specifically, computer-aided drug discovery (CADD) has significantly reduced the time, cost, and labor in the process. The standard workflow in the CADD involves the identification of drug targets, virtual screening of chemical (small compound) libraries to find a lead compound, further optimization of the lead, and in silico toxicity assessment (Fig. 12.4) (Malathi and Ramaiah 2018; Shaker et al. 2021; Zhang et al. 2022). Though the in vitro and in vivo assessments are still required for the complete evaluation of the drug candidate, the CADD essentially reduces the number of compounds that require preclinical assessment reducing the overall failure rates (Shaker et al. 2021). In the subsequent sections, we have highlighted the role of omics methods in the CADD processes.

12.3.1 Computer-Aided Drug Discovery (CADD) Process

CADD is a broad term to describe the methods to discover and develop novel drug candidates (Fig. 12.4). CADD generally consists of two approaches, namely structure-based drug discovery (SBDD) and ligand-based drug discovery (LBDD) (Malathi and Ramaiah 2018; Torres and De La Fuente-Nunez 2019; Shaker et al. 2021). The information available about the drug target determines which approach is best. The SBDD is the preferred approach if the structural details of the drug targets can be obtained through nuclear magnetic resonance (NMR) or X-ray crystallography methods. However, in the absence of experimental structural data, in silico prediction methods involving homology modeling and ab initio modeling can be used. If any of the above methods cannot be used, LBDD is the method of choice (Torres and De La Fuente-Nunez 2019; Shaker et al. 2021). The overall workflow of drug discovery using these approaches is described below and presented in Fig. 12.4.

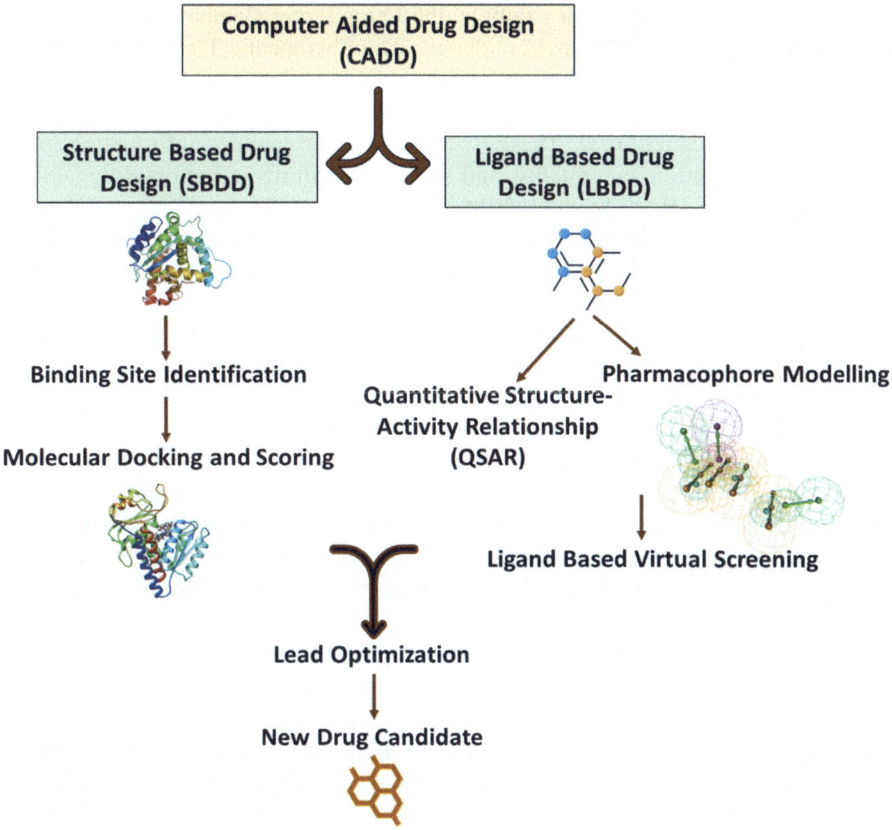

Fig. 12.4 The standard workflow in a computer-aided drug design (CADD) process (see the text for details)

12.3.1.1 Identification of Drug Target in Silico

An antibiotic target in the bacterial system is often a protein that is generally involved in some physiological processes of the cell. Finding a suitable target whose modulation can result in the control of bacterial growth is often tricky and involves a host of experimental techniques described previously. However, this process can be streamlined using in silico methods that can circumvent the use of in vitro experiments at the initial stages. To be an ideal drug target it should be able to bind to a drug-like molecule, must be essential to the organism, and should be specifically present in the pathogen and not in the host (Serral et al. 2021). Considering the above characteristics, a target protein should be "druggable" meaning a drug must be able to bind to the target resulting in desirable inhibition in its activity. Therefore, structurally the target should have a binding pocket where the candidate drug must bind specifically (Serral et al. 2021). With the availability of genomic information about bacterial pathogens in public databases, such drug targets can be predicted using various computational tools (Table 12.2). These tools focus on specific characteristics of the target, for example, the Drug Target Database uses a reverse docking process to select the target molecule. The Therapeutic Target Database is a repository of already known drug targets in the bacterial system. Target databases are also available for specific pathogens such as TuberQ and TargetTB (against *Mycobacterium tuberculosis*). These databases use proteomics information, protein-protein interaction, essentiality, and structural features to predict druggability (Raman et al. 2008; Radusky et al. 2014; Serral et al. 2021). Another example of a database that uses metabolic network information to predict drug targets is Find Targets WEB (Merigueti et al. 2019). With the recent advancement in artificial intelligence (AI) and machine learning (ML) technologies, drug target prediction has seen a significant overhaul and many of these databases are integrating such applications for drug target identification.

12.3.1.2 Structure-Based Drug Design (SBDD)

As mentioned previously the structure-based drug design (SBDD) approach takes into account the binding affinity of a ligand with the target protein considering its structural features. In SBDD primarily molecular dynamics (MD) simulations, molecular docking, and structure-based virtual screening (VS) are used to assess the binding efficiency of the drug candidate with the target protein (Shaker et al. 2021; Zhang et al. 2022).

The SBDD approach starts with finding a high-resolution structure of the target protein from the Protein Data Bank (PDB) (Rose et al. 2017). If the protein structure is not yet available, the same can be predicted using bioinformatics tools such as homology modeling and ab initio modeling. The premise of homology modeling is that proteins with high sequence similarity will eventually acquire a similar three-dimensional structural characteristic. There are several protein modeling tools available for these purposes, the most notable ones are MODELLER (Webb and Sali 2021), Swiss Model (Waterhouse et al. 2018), and Phyre2 (Kelley et al. 2015), etc. (Table 12.2). These tools select an appropriate template based on amino acid sequence similarity on which the target structure would be modeled. However, in situations where an appropriate template is not available, the ab initio protein structure prediction is considered suitable that try to build protein structures based on minimum energy conformations that are thermodynamically stable (Malathi and Ramaiah 2018; Shaker et al. 2021; Zhang et al. 2022). However, using this approach small protein (<100 amino acids) structures can only be predicted. Examples of web servers that use the ab initio structure prediction approach are I-TASSER (Yang and Zhang 2015) and QUARK (Xu et al. 2011). Recently AI-based techniques such as AlphaFold developed by DeepMind can predict 3D structures of any proteins based on their amino acid sequence with considerable accuracy (Jumper et al. 2021).

Once the target structure has been predicted, the drug-binding site needs to be identified in it. Experimental data such as crystallization with ligands are generally used to determine the binding site, otherwise in silico methods are available for predicting such sites (Shaker et al. 2021; Zhang et al. 2022). Following the target identification, molecular docking is employed to determine the ligands with the highest binding affinity for the target. The foundation of molecular docking algorithms is the prediction of a ligand's preferred orientation within a target's binding site (Torres and De La Fuente-Nunez 2019; Shaker et al. 2021).

The ligands used for docking analysis are generally selected from libraries of compounds available in several databases such as PubChem, ZINC, Drug Bank, and MCULE (Table 12.2) (Zhang et al. 2022). Among the available compounds, suitable ones are selected based on Lipinski's "Rule of Five" which considers a molecule to be bioactive if its molecular weight is <500 Da, hydrogen bond donors are not greater than 5, hydrogen bond acceptors are not greater than 10, octanol-water partition coefficient (logP) is not greater than 5, and topological polar surface area is less than 140 (Lipinski 2004).

Molecular docking is generally used in tandem with virtual screening (VS) of the ligands for streamlining the assessment of ligand–target interactions and ranking the ligands as per their binding affinity to the target (Fig. 12.4) (Zhang et al. 2022). Several molecular docking tools are currently available, the most notable of which are AutoDock, AutoDock Vina, CDOCKER, and SwissDock (Table 12.2) (Wu et al. 2003; Morris et al. 2009; Trott and Olson 2010; Grosdidier et al. 2011). The molecular docking methods can be categorized as (a) rigid docking where flexibility is not allowed between the target and the ligand, which is generally used in the initial screening stages to quickly identify a suitable ligand; (b) semi-flexible docking where the ligand is flexible and the target is rigid; and (c) flexible docking where both the target and the ligand are flexible. The flexible docking methods are generally used to refine and optimize the initial docking results (Shaker et al. 2021; Zhang et al. 2022). The molecular docking accuracy is generally based on the scoring functions that take into account the binding affinity between the target and the ligand and ultimately influence the accurate identification of the drug candidate (Malathi and Ramaiah 2018; Zhang et al. 2022).

In order to optimize the binding of the drug candidates with the target protein molecular dynamics (MD) simulation approaches are used. As protein molecules are flexible which influences its binding with the ligand, it becomes important to accurately predict the dynamics of target–ligand interaction in the 3D space. Accurate prediction can further help in optimizing the drug candidate. Therefore, MD simulations are an integral part of the in silico drug development process (Fig. 12.4). Typically, MD simulation studies are conducted on the best-docked complexes to confirm their binding and stability (Shaker et al. 2021; Zhang et al. 2022). Additionally, it aids in the discovery of extra binding sites in the target molecules, which helps in the development of more potent drug candidates (Grant et al. 2011; Shaker et al. 2021). MD simulations generally take into account the movement of the individual atoms as per Newton's laws of motion and are generally simulated using the available tools such as AMBER, CHARMM, and GROMOS simulation packages (Pearlman et al. 1995; Christen et al. 2005; Jo et al. 2008). Finally, using the MD simulations a few drug candidates can be assessed which will guide the further improvement of the ligand (Fig. 12.4).

12.3.1.3 Ligand-Based Drug Design (LBDD)

The LBDD approach is generally used when information about the drug target is not available. Therefore, knowledge of active drug compounds becomes handy in developing similar candidates in the LBDD approach. A similar drug candidate can be found based on specific characteristic features, such as aromaticity, hydrophobicity, anion and cation residues, and hydrogen bond acceptors and donors, if a known drug is already available against a given target (Fig. 12.4) (Shaker et al. 2021; Zhang et al. 2022). This is based on the assumption that compounds having similar properties would work in a similar fashion and show comparable biological activity (Martin et al. 2002; Shaker et al. 2021). Once a compound like an active drug is identified, it can be further improved by designing suitable analogs. The most commonly used methods for designing suitable analogs in the LBDD approach are pharmacophore modeling and QSAR (Fig. 12.4).

Pharmacophores are characteristic features of a compound that are essential for it to be considered as a drug molecule. It consists of specific electronic and steric features. In pharmacophore modeling these characteristic features are queried against a library of compounds to find the most suitable ones, specifically with energetically stable conformations (Fig. 12.4). Additionally, the identified compounds are stacked and overlapped to reveal functional groups that are similar and shared by the active ligands (Shaker et al. 2021; Zhang et al. 2022). For pharmacophore modeling several bioinformatics tools are available such as Ligand Scout (Wolber and Langer 2005), Pharm Mapper (Liu et al. 2010), and Pharma Gist (Schneidman-Duhovny et al. 2008) to name a few (Table 12.2).

The physicochemical and structural characteristics of compounds are typically correlated with their biological activity using quantitative structure-activity relationships (QSAR) modeling. QSAR is widely utilized in the lead discovery process because it is capable of ranking possible drug candidates based on their bioactivities (Zhang et al. 2022). For developing the QSAR models a set of compounds (generally more than 20) with known biological activities is chosen as a training set. Once the QSAR model is developed, the test set of compounds is compared against it for evaluation of their drug potential and optimization of the lead compound (Shaker et al. 2021; Zhang et al. 2022). There are several online tools available for QSAR modeling, the most notable ones are QSAR-Co (Ambure et al. 2019), Open3DQSAR (Tosco and Balle 2011), and QSAR Toolbox (Dimitrov et al. 2016) (Table 12.2).

12.3.1.4 ADMET Assessment

Once a drug candidate has been identified using any of the drug discovery methods described above, its pharmacokinetics properties, specifically its ADMET (for absorption, distribution, metabolism, excretion, and toxicity) characteristics need to be assessed. Over 50% of drug candidates are withdrawn from further development at this critical stage of the drug development process because of unfavorable ADMET concerns (Kennedy 1997; Shaker et al. 2021). Conventionally, drug ADMET is assessed using various in vitro and in vivo models including cell lines, small animal models, etc. which are very costly, laborious, and time-consuming. However, with the recent developments in the area of AI and machine learning, the ADMET can be predicted using computational tools. Even though the in vivo assessments cannot be completely bypassed, they can significantly reduce the time and labor in such processes. Several machine learning (ML)-based models have been developed that can predict the pharmacokinetic properties of the identified compounds, such as PreADMET (Viana Nunes et al. 2020), ALOGPS (Tetko and Tanchuk 2002), Swiss ADME (Daina et al. 2017), and Drug Mint (Dhanda et al. 2013) (Table 12.2).

In the preceding decades, the remarkable advancements in the omics biosciences along with computational biological tools have revolutionized the process of drug discovery and development. It has not only reduced the time required to discover a hit compound and lead optimization, but it has also reduced the number of hit molecules that are required to be used in HTS. Advancements in the areas of virtual screening methods and MD simulation algorithms have revolutionized every step in the drug development process. Moreover, the revolution in AI-based ML technologies has further enhanced the efficiency of the drug development process and is described in the following section.

12.4 Artificial Intelligence (AI)-Based Technology and Machine Learning (ML) Tools in AMR Research and Drug Development

Artificial intelligence (AI)-based technology has invaded almost every sphere of human lives in recent times. With the immense possibilities of AI technology, it has brought about a paradigm shift in the way drug discovery and development research are progressing. With the availability of a huge amount of microbial genomic information whose analysis is beyond human capabilities, the AI has been able to decipher patterns from such data. Recently, AI-based methods have been used to survey the metagenomics data to identify antibiotic resistance genes (ARGs) (Wani et al. 2022), forecast disease outbreaks (Jiménez et al. 2020), and even detection of outbreak transmission clusters (Casto et al. 2019). Moreover, AI-based methods have been successfully used to predict novel ARGs from antibiotic-resistance databases such as CARD and ResFinder (Behling et al. 2023). The machine-learning model AlphaFold has revolutionized the way protein structures have been predicted (Jumper et al. 2021). AlphaFold has been used in developing designer peptides with antimicrobial properties as well as predicting protein structural features that would inhibit resistance development (Gligorijević et al. 2021; Bolatchiev et al. 2022). From a treatment point of view AI-based methods have also been used to develop antibiotic combination therapies (Smith et al. 2020). While artificial intelligence (AI) has impacted all facets of antimicrobial resistance (AMR) research, we have highlighted its contribution to the identification and creation of new antimicrobial agents in the sections that follow (Fig. 12.5).

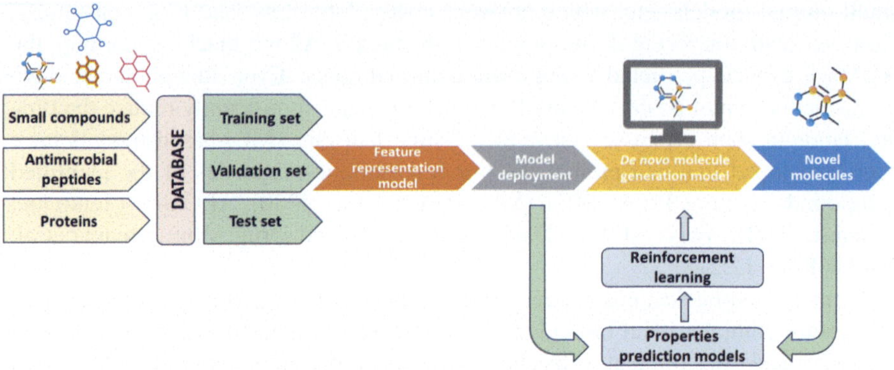

Fig. 12.5 Standard workflow in the de novo drug design using artificial intelligence and machine learning tools (see the text for details)

12.4.1 Drug Designing Using Artificial Intelligence (AI)

AI tools primarily use the information that currently exists, based on which they can establish patterns and then extrapolate that to create novel things. In this regard, it is estimated that more than 10^{30} to 10^{60} chemicals having drug-like characteristics exist in the chemical space (Melo et al. 2021; Zhang et al. 2022). AI models can be used to analyze this enormous number of chemical compounds and determine the chemical patterns within the molecule that can be utilized to create drug molecules (Fig. 12.5). Machine learning (ML) which concerns training, learning, problem-solving, and reasoning by computers to decipher complex systems, is a sub-branch of AI that has immense potential in de novo drug designing (Melo et al. 2021; Jukič and Bren 2022). The use of ML-based tools has sped up the drug discovery process in recent times. These tools have reduced the gap between finding the chemicals and their associated drug-like characteristics (Fig. 12.5).

Drug designing based on ML considers a few important steps sequentially, such as mining of chemical compounds data from public databases, filtering of data to select chemicals with drug-like characteristics, use of feature representation methods based on either simplified molecular input line entry system (SMILES) or graphs for developing learning models, which is subsequently used for de novo chemical generation based on the learned model (Li et al. 2020; Pandey et al. 2022; Zhang et al. 2022). Once the de novo molecules generative models are developed, they can be further optimized using reinforcement learning and property prediction models (Fig. 12.5). These generative approaches are effective in designing LBDD- and SBDD-based drug molecules. As the designed drug molecules must be assessed for their binding with the drug targets, the structural information of both the drug and the target must be extracted during the generative model training. Fundamentally, two approaches are used for drug designing using ML tools: (a) structure-oriented drug design and (b) ligand-oriented drug design, which are analogous to the SBDD and LBDD approaches in the CADD methods (Zhang et al. 2022).

12.4.1.1 Structure-Oriented Drug Design

In this approach, the novel compounds are generated based on the structural features of the target molecules, which are primarily structural proteins on the cell membrane, enzymes, receptors, etc. The structural features of the targets are extracted from the training set using deep generative models with which the ligand would have a stable interaction. Further, docking scores would produce a rank of the compounds based on their interaction strength and guide the design of novel compounds (Zhang et al. 2022).

12.4.1.2 Ligand-Oriented Drug Design

When a set of compounds are known to bind to a specific protein target, similar compounds with analogous binding affinity can be developed using this approach (Wang et al. 2022; Zhang et al. 2022). This approach focuses on the specific characteristics of molecules as well as adding new desirable properties (such as binding affinity, solubility, ADMET, logP, etc.) that would increase its efficacy. Different ML models can be used to optimize these properties by learning their chemical spaces. This results in the generation of novel molecules within the framework of the desired property space (Thomas et al. 2021; Zhang et al. 2022). Though the overall principle of drug design seems simple, the actual synthesis of compounds having a host of desirable features is challenging (Fig. 12.5).

12.4.2 Machine Learning Tools in Predicting Antimicrobial Activity

Apart from designing novel drugs using ML tools, the antimicrobial activity of already existing drug molecules can also be predicted. Drug repurposing is an important area where active research is going on to evaluate drug molecules for their antimicrobial properties. In this regard, combinations of different neural models have been used to create a representation of chemical compounds which are further assessed for their antimicrobial properties (Stokes et al. 2020; Melo et al. 2021). Deep neural network models have been used to predict antimicrobial properties of random amino acid sequences using sets of peptide descriptors (Fjell et al. 2009; Yan et al. 2020) which are available in a web server. As antimicrobial peptides (AMPs) are increasingly being considered to control drug-resistant bacterial pathogens, ML tools have a great potential to develop such peptides. Using ML tools in an artificial evolution framework, Yoshida et al. (2018) have identified 44 novel AMPs with improved potency (160-fold) in comparison to its ancestral 13-mer linear natural AMP. During these artificial evolution experiments, the conformations of the peptides were changed to select the best ones highlighting the potential of such experimental methods to accelerate the drug discovery process using a limited number of steps (Yoshida et al. 2018).

12.4.3 Machine Learning Tools in Predicting Drug-Likeness

Drugs developed using the traditional approaches have a very high failure rate during the preclinical stages. Specifically, many of the drug candidates have adverse ADMET during the pharmacokinetics assessment leading them to be discarded from further development. This bottleneck in drug development has been circumvented using ML tools to some extent. The ML algorithms have been developed to discriminate characteristic chemical spaces that have therapeutic potential from others lacking such properties (Melo et al. 2021). Further ML-based ligand-target binding affinity prediction tools have accelerated HTS, and lead optimization protocols to identify drug candidates with suitable properties (D'Souza et al. 2020). ML models have also been used to predict the pharmacokinetics, ADMET, solubility, and stability of AMPs during their design and development increasing the application potential of such technologies in the near future (Torres and De La Fuente-Nunez 2019; Melo et al. 2021; Jukič and Bren 2022; Behling et al. 2023).

12.5 Concluding Remarks

With the rising incidence of antibiotic resistance with the diminishing novel antibiotic discovery pipeline, the global health-care sector is witnessing a challenge of unprecedented scale. Indiscriminate use of broad-spectrum antibiotics in the clinic, animal farming, and agriculture though responsible for the current situation, the recent COVID-19 pandemic has further aggravated it. Moreover, the diminishing return on investment of billions of dollars in antibiotic discovery and development has discouraged the big pharma companies from novel antibiotic development. In this context, the decreasing sequencing cost, the advancement in single-cell genomics, and the development of novel sequencing platforms, such as PacBio and Nanopore, can help in identifying hitherto unknown metabolic capabilities in the bacterial world. This has promising consequences in the field of novel drugs as well as drug target discovery. Furthermore, the recent advancements in the fields of computational biology aided with artificial intelligence and machine learning tools show a promising future for drug discovery in general and antibiotics in particular. These tools have the potential to circumvent several steps in the initial stages of drug development, saving time, cost, and labor. This may help the big pharmaceutical companies to return to the antibiotic development field in the near future. As antibiotic resistance is a global problem, its mitigation also requires concerted efforts from all the stakeholders. As has been said earlier in the context of the COVID pandemic "no one is safe until everyone is," which can also be used for infections caused by drug-resistant pathogens.

References

Ahmed MN, Reyna-González E, Schmid B, Wiebach V, Süssmuth RD, Dittmann E et al (2017) Phylogenomic analysis of the Microviridin biosynthetic pathway coupled with targeted chemo-enzymatic synthesis yields potent protease inhibitors. ACS Chem Biol 12:1538–1546. https://doi.org/10.1021/acschembio.7b00124

Ambure P, Halder AK, González Díaz H, Cordeiro MNDS (2019) QSAR-co: an open-source software for developing robust multitasking or multitarget classification-based QSAR models. J Chem Inf Model 59:2538–2544. https://doi.org/10.1021/acs.jcim.9b00295

Arnison PG, Bibb MJ, Bierbaum G, Bowers AA, Bugni TS, Bulaj G et al (2013) Ribosomally synthesized and post-translationally modified peptide natural products: overview and recommendations for a universal nomenclature. Nat Prod Rep 30:108–160. https://doi.org/10.1039/c2np20085f

Aszalos A, Frost D (1975) Thin-layer chromatography of antibiotics. In: Methods in enzymology antibiotics. Academic Press, pp 172–213. https://doi.org/10.1016/0076-6879(75)43084-1

Bachmann BO, Van Lanen SG, Baltz RH (2014) Microbial genome mining for accelerated natural products discovery: is a renaissance in the making? J Ind Microbiol Biotechnol 41:175–184. https://doi.org/10.1007/s10295-013-1389-9

Ballestero JGA, Garcia JM, Bollela VR, Ruffino-Netto A, Dalcolmo MMP, Moncaio ACS et al (2020) Management of multidrug-resistant tuberculosis: main recommendations of the Brazilian guidelines. J Bras Pneumol 46:e20190290. https://doi.org/10.36416/1806-3756/e20190290

Baltz RH (2019) Natural product drug discovery in the genomic era: realities, conjectures, misconceptions, and opportunities. J Ind Microbiol Biotechnol 46:281–299. https://doi.org/10.1007/s10295-018-2115-4

Bandow JE, Brötz H, Leichert LIO, Labischinski H, Hecker M (2003) Proteomic approach to understanding antibiotic action. Antimicrob Agents Chemother 47:948–955. https://doi.org/10.1128/AAC.47.3.948-955.2003

Barale SS, Ghane SG, Sonawane KD (2022) Purification and characterization of antibacterial surfactin isoforms produced by Bacillus velezensis SK. AMB Express 12:7. https://doi.org/10.1186/s13568-022-01348-3

Barbosa MDFS, Yang G, Fang J, Kurilla MG, Pompliano DL (2002) Development of a whole-cell assay for peptidoglycan biosynthesis inhibitors. Antimicrob Agents Chemother 46:943–946. https://doi.org/10.1128/AAC.46.4.943-946.2002

Behling AH, Wilson BC, Ho D, Virta M, O'Sullivan JM, Vatanen T (2023) Addressing antibiotic resistance: computational answers to a biological problem? Curr Opin Microbiol 74:102305. https://doi.org/10.1016/j.mib.2023.102305

Behnken S, Hertweck C (2012) Anaerobic bacteria as producers of antibiotics. Appl Microbiol Biotechnol 96:61–67. https://doi.org/10.1007/s00253-012-4285-8

Bem AE, Velikova N, Pellicer MT, van Baarlen P, Marina A, Wells JM (2015) Bacterial histidine kinases as novel antibacterial drug targets. ACS Chem Biol 10:213–224. https://doi.org/10.1021/cb5007135

Bianchi AA, Baneyx F (1999) Stress responses as a tool to detect and characterize the mode of action of antibacterial agents. Appl Environ Microbiol 65:5023–5027. https://doi.org/10.1128/AEM.65.11.5023-5027.1999

Blair JMA, Webber MA, Baylay AJ, Ogbolu DO, Piddock LJV (2015) Molecular mechanisms of antibiotic resistance. Nat Rev Microbiol 13:42–51. https://doi.org/10.1038/nrmicro3380

Blanchard SC, Cooperman BS, Wilson DN (2010) Probing translation with small-molecule inhibitors. Chem Biol 17:633–645. https://doi.org/10.1016/j.chembiol.2010.06.003

Blin K, Shaw S, Kloosterman AM, Charlop-Powers Z, van Wezel GP, Medema MH et al (2021) antiSMASH 6.0: improving cluster detection and comparison capabilities. Nucleic Acids Res 49:W29–W35. https://doi.org/10.1093/nar/gkab335

Bolatchiev A, Baturin V, Shchetinin E, Bolatchieva E (2022) Novel antimicrobial peptides designed using a recurrent neural network reduce mortality in experimental sepsis. Antibiotics 11:411. https://doi.org/10.3390/antibiotics11030411

Bonev B, Hooper J, Parisot J (2008) Principles of assessing bacterial susceptibility to antibiotics using the agar diffusion method. J Antimicrob Chemother 61:1295–1301. https://doi.org/10.1093/jac/dkn090

Borovok I, Gorovitz B, Schreiber R, Aharonowitz Y, Cohen G (2006) Coenzyme B12 controls transcription of the Streptomyces class Ia ribonucleotide reductase nrdABS operon via a riboswitch mechanism. J Bacteriol 188:2512–2520. https://doi.org/10.1128/JB.188.7.2512-2520.2006

Boyle-Vavra S, Yin S, Daum RS (2006) The VraS/VraR two-component regulatory system required for oxacillin resistance in community-acquired methicillin-resistant Staphylococcus aureus. FEMS Microbiol Lett 262:163–171. https://doi.org/10.1111/j.1574-6968.2006.00384.x

Brandi L, Dresios J, Gualerzi CO (2008) Assays for the identification of inhibitors targeting specific translational steps. Methods Mol Med 142:87–105. https://doi.org/10.1007/978-1-59745-246-5_8

Bush NG, Diez-Santos I, Abbott LR, Maxwell A (2020) Quinolones: mechanism, lethality and their contributions to antibiotic resistance. Molecules 25:5662. https://doi.org/10.3390/molecules25235662

Bushin LB, Covington BC, Rued BE, Federle MJ, Seyedsayamdost MR (2020) Discovery and biosynthesis of Streptosactin, a Sactipeptide with an alternative topology encoded by commensal bacteria in the human microbiome. J Am Chem Soc 142:16265–16275. https://doi.org/10.1021/jacs.0c05546

Calteau A, Fewer DP, Latifi A, Coursin T, Laurent T, Jokela J et al (2014) Phylum-wide comparative genomics unravel the diversity of secondary metabolism in cyanobacteria. BMC Genomics 15:977. https://doi.org/10.1186/1471-2164-15-977

Casto AM, Adler AL, Makhsous N, Crawford K, Qin X, Kuypers JM et al (2019) Prospective, real-time metagenomic sequencing during norovirus outbreak reveals discrete transmission clusters. Clin Infect Dis 69:941–948. https://doi.org/10.1093/cid/ciy1020

Challis GL, Hopwood DA (2003) Synergy and contingency as driving forces for the evolution of multiple secondary metabolite production by Streptomyces species. Proc Natl Acad Sci 100:14555–14561. https://doi.org/10.1073/pnas.1934677100

Champney WS (2008) Three methods to assay inhibitors of ribosomal subunit assembly. Methods Mol Med 142:63–73. https://doi.org/10.1007/978-1-59745-246-5_6

Champney WS (2020) Antibiotics targeting bacterial ribosomal subunit biogenesis. J Antimicrob Chemother 75:787–806. https://doi.org/10.1093/jac/dkz544

Charusanti P, Fong NL, Nagarajan H, Pereira AR, Li HJ, Abate EA et al (2012) Exploiting adaptive laboratory evolution of Streptomyces clavuligerus for antibiotic discovery and overproduction. PLoS One 7:e33727. https://doi.org/10.1371/journal.pone.0033727

Chaudhary HS, Yadav J, Shrivastava AR, Singh S, Singh AK, Gopalan N (2013) Antibacterial activity of actinomycetes isolated from different soil samples of Sheopur (a city of Central India). J Adv Pharm Technol Res 4:118–123. https://doi.org/10.4103/2231-4040.111528

Chen Y, Schwack W (2014) High-performance thin-layer chromatography screening of multi class antibiotics in animal food by bioluminescent bioautography and electrospray ionization mass spectrometry. J Chromatogr A 1356:249–257. https://doi.org/10.1016/j.chroma.2014.06.043

Chen H, Yu C, Wu H, Li G, Li C, Hong W et al (2022a) Recent advances in histidine kinase-targeted antimicrobial agents. Front Chem 10:866392. https://doi.org/10.3389/fchem.2022.866392

Chen S, Zhang C, Zhang L (2022b) Investigation of the molecular landscape of bacterial aromatic polyketides by global analysis of type II polyketide synthases. Angew Chem Int Ed 61:e202202286. https://doi.org/10.1002/anie.202202286

Chevrette MG, Gavrilidou A, Mantri S, Selem-Mojica N, Ziemert N, Barona-Gómez F (2021) The confluence of big data and evolutionary genome mining for the discovery of natural products. Nat Prod Rep 38:2024–2040. https://doi.org/10.1039/d1np00013f

Chiang Y-M, Szewczyk E, Davidson AD, Keller N, Oakley BR, Wang CCC (2009) A gene cluster containing two fungal polyketide synthases encodes the biosynthetic pathway for a

polyketide, Asperfuranone, in aspergillus nidulans. J Am Chem Soc 131:2965–2970. https://doi.org/10.1021/ja8088185
Chitra SR, Ramalakshmi N, Arunkumar S, Manimegalai P (2020) A comprehensive review on DNA gyrase inhibitors. Infect Disord Drug Targets 20:765–777
Choma IM, Grzelak EM (2011) Bioautography detection in thin-layer chromatography. J Chromatogr A 1218:2684–2691. https://doi.org/10.1016/j.chroma.2010.12.069
Choma IM, Jesionek W (2015) TLC-direct bioautography as a high throughput method for detection of antimicrobials in plants. Chromatography 2:225–238. https://doi.org/10.3390/chromatography2020225
Christen M, Hünenberger PH, Bakowies D, Baron R, Bürgi R, Geerke DP et al (2005) The GROMOS software for biomolecular simulation: GROMOS05. J Comput Chem 26:1719–1751. https://doi.org/10.1002/jcc.20303
Cimermancic P, Medema MH, Claesen J, Kurita K, Wieland Brown LC, Mavrommatis K et al (2014) Insights into secondary metabolism from a global analysis of prokaryotic biosynthetic gene clusters. Cell 158:412–421. https://doi.org/10.1016/j.cell.2014.06.034
Consden R, Gordon AH, Martin AJ (1944) Qualitative analysis of proteins: a partition chromatographic method using paper. Biochem J 38:224–232. https://doi.org/10.1042/bj0380224
Cook MA, Wright GD (2022) The past, present, and future of antibiotics. Sci Transl Med 14:eabo7793. https://doi.org/10.1126/scitranslmed.abo7793
Couillerot O, Loqman S, Toribio A, Hubert J, Gandner L, Nuzillard J-M et al (2014) Purification of antibiotics from the biocontrol agent Streptomyces anulatus S37 by centrifugal partition chromatography. J Chromatogr B Analyt Technol Biomed Life Sci 944:30–34. https://doi.org/10.1016/j.jchromb.2013.11.008
Cruz-Morales P, Kopp JF, Martínez-Guerrero C, Yáñez-Guerra LA, Selem-Mojica N, Ramos-Aboites H et al (2016) Phylogenomic analysis of natural products biosynthetic gene clusters allows discovery of Arseno-organic metabolites in model Streptomycetes. Genome Biol Evol 8:1906–1916. https://doi.org/10.1093/gbe/evw125
Cuatrecasas P, Wilchek M, Anfinsen CB (1968) Selective enzyme purification by affinity chromatography. Proc Natl Acad Sci USA 61:636–643. https://doi.org/10.1073/pnas.61.2.636
Culp EJ, Waglechner N, Wang W, Fiebig-Comyn AA, Hsu Y-P, Koteva K et al (2020) Evolution-guided discovery of antibiotics that inhibit peptidoglycan remodelling. Nature 578:582–587. https://doi.org/10.1038/s41586-020-1990-9
D'Souza S, Prema KV, Balaji S (2020) Machine learning models for drug–target interactions: current knowledge and future directions. Drug Discov Today 25:748–756. https://doi.org/10.1016/j.drudis.2020.03.003
Daina A, Michielin O, Zoete V (2017) SwissADME: a free web tool to evaluate pharmacokinetics, drug-likeness and medicinal chemistry friendliness of small molecules. Sci Rep 7:42717. https://doi.org/10.1038/srep42717
Darby EM, Trampari E, Siasat P, Gaya MS, Alav I, Webber MA et al (2023) Molecular mechanisms of antibiotic resistance revisited. Nat Rev Microbiol 21:280–295. https://doi.org/10.1038/s41579-022-00820-y
Das R, Romi W, Das R, Sharma HK, Thakur D (2018) Antimicrobial potentiality of actinobacteria isolated from two microbiologically unexplored forest ecosystems of Northeast India. BMC Microbiol 18:71. https://doi.org/10.1186/s12866-018-1215-7
de Kraker MEA, Wolkewitz M, Davey PG, Koller W, Berger J, Nagler J et al (2011) Burden of antimicrobial resistance in European hospitals: excess mortality and length of hospital stay associated with bloodstream infections due to Escherichia coli resistant to third-generation cephalosporins. J Antimicrob Chemother 66:398–407. https://doi.org/10.1093/jac/dkq412
Dewanjee S, Gangopadhyay M, Bhattacharya N, Khanra R, Dua TK (2015) Bioautography and its scope in the field of natural product chemistry. J Pharm Anal 5:75–84. https://doi.org/10.1016/j.jpha.2014.06.002
Dhanda SK, Singla D, Mondal AK, Raghava GP (2013) DrugMint: a webserver for predicting and designing of drug-like molecules. Biol Direct 8:28. https://doi.org/10.1186/1745-6150-8-28

Dik DA, Fisher JF, Mobashery S (2018) Cell-wall recycling of the gram-negative bacteria and the nexus to antibiotic resistance. Chem Rev 118:5952–5984. https://doi.org/10.1021/acs.chemrev.8b00277

Dimitrov SD, Diderich R, Sobanski T, Pavlov TS, Chankov GV, Chapkanov AS et al (2016) QSAR toolbox – workflow and major functionalities. SAR QSAR Environ Res 27:203–219. https://doi.org/10.1080/1062936X.2015.1136680

Dodds DR (2017) Antibiotic resistance: a current epilogue. Biochem Pharmacol 134:139–146. https://doi.org/10.1016/j.bcp.2016.12.005

Economou V, Gousia P (2015) Agriculture and food animals as a source of antimicrobial-resistant bacteria. IDR 8:49–61. https://doi.org/10.2147/IDR.S55778

Elleuch L, Shaaban M, Smaoui S, Mellouli L, Karray-Rebai I, Fourati-Ben Fguira L et al (2010) Bioactive secondary metabolites from a new terrestrial Streptomyces sp. TN262. Appl Biochem Biotechnol 162:579–593. https://doi.org/10.1007/s12010-009-8808-4

Emmerich CH, Gamboa LM, Hofmann MCJ, Bonin-Andresen M, Arbach O, Schendel P et al (2021) Improving target assessment in biomedical research: the GOT-IT recommendations. Nat Rev Drug Discov 20:64–81. https://doi.org/10.1038/s41573-020-0087-3

Farha MA, Brown ED (2016) Strategies for target identification of antimicrobial natural products. Nat Prod Rep 33:668–680. https://doi.org/10.1039/c5np00127g

Fihn CA, Carlson EE (2021) Targeting a highly conserved domain in bacterial histidine kinases to generate inhibitors with broad spectrum activity. Curr Opin Microbiol 61:107–114. https://doi.org/10.1016/j.mib.2021.03.007

Finegold SM, Miller LG, Posnick D, Patterson DK, Davis A (1966) Nalidixic acid: clinical and laboratory studies. Antimicrob Agents Chemother (Bethesda) 6:189–197

Fischer R, Lautner H (1961) On the paper chromatographic detection of penicillin preparations. Arch Pharm Ber Dtsch Pharm Ges 294(66):1–7. https://doi.org/10.1002/ardp.19612940102

Fisher JF, Mobashery S (2020) Constructing and deconstructing the bacterial cell wall. Protein Sci 29:629–646. https://doi.org/10.1002/pro.3737

Fjell CD, Jenssen H, Hilpert K, Cheung WA, Panté N, Hancock REW et al (2009) Identification of novel antibacterial peptides by Chemoinformatics and machine learning. J Med Chem 52:2006–2015. https://doi.org/10.1021/jm8015365

Fu L, Niu B, Zhu Z, Wu S, Li W (2012) CD-HIT: accelerated for clustering the next-generation sequencing data. Bioinformatics 28:3150–3152. https://doi.org/10.1093/bioinformatics/bts565

Galluzzi L, Karp M (2003) Amplified detection of transcriptional and translational inhibitors in bioluminescent Escherichia coli K-12. J Biomol Screen 8:340–346. https://doi.org/10.1177/1087057103008003012

Ganesan P, Reegan AD, David RHA, Gandhi MR, Paulraj MG, Al-Dhabi NA et al (2017) Antimicrobial activity of some actinomycetes from Western Ghats of Tamil Nadu, India. Alexandria J Med 53:101–110. https://doi.org/10.1016/j.ajme.2016.03.004

Genilloud O, González I, Salazar O, Martín J, Tormo JR, Vicente F (2011) Current approaches to exploit actinomycetes as a source of novel natural products. J Ind Microbiol Biotechnol 38:375–389. https://doi.org/10.1007/s10295-010-0882-7

Gerth K, Pradella S, Perlova O, Beyer S, Müller R (2003) Myxobacteria: proficient producers of novel natural products with various biological activities – past and future biotechnological aspects with the focus on the genus Sorangium. J Biotechnol 106:233–253. https://doi.org/10.1016/j.jbiotec.2003.07.015

Girard L, Geudens N, Pauwels B, Höfte M, Martins JC, De Mot R (2022) Transporter gene-mediated typing for detection and genome mining of lipopeptide-producing Pseudomonas. Appl Environ Microbiol 88:e01869–e01821. https://doi.org/10.1128/AEM.01869-21

Gligorijević V, Renfrew PD, Kosciolek T, Leman JK, Berenberg D, Vatanen T et al (2021) Structure-based protein function prediction using graph convolutional networks. Nat Commun 12:3168. https://doi.org/10.1038/s41467-021-23303-9

Goh E-B, Yim G, Tsui W, McClure J, Surette MG, Davies J (2002) Transcriptional modulation of bacterial gene expression by subinhibitory concentrations of antibiotics. Proc Natl Acad Sci USA 99:17025–17030. https://doi.org/10.1073/pnas.252607699

Goodall RR, Levi AA (1946) A microchromatographic method for the detection and approximate determination of the different penicillins in a mixture. Nature 158:675. https://doi.org/10.1038/158675a0

Gosschalk JE, Chang C, Sue CK, Siegel SD, Wu C, Kattke MD et al (2020) A cell-based screen in Actinomyces oris to identify Sortase inhibitors. Sci Rep 10:8520. https://doi.org/10.1038/s41598-020-65256-x

Grant BJ, Lukman S, Hocker HJ, Sayyah J, Brown JH, McCammon JA et al (2011) Novel allosteric sites on Ras for lead generation. PLoS One 6:e25711. https://doi.org/10.1371/journal.pone.0025711

Grosdidier A, Zoete V, Michielin O (2011) SwissDock, a protein-small molecule docking web service based on EADock DSS. Nucleic Acids Res 39:W270–W277. https://doi.org/10.1093/nar/gkr366

Grzelak EM, Hwang C, Cai G, Nam J-W, Choules MP, Gao W et al (2016) Bioautography with TLC-MS/NMR for rapid discovery of anti-tuberculosis Lead compounds from natural sources. ACS Infect Dis 2:294–301. https://doi.org/10.1021/acsinfecdis.5b00150

Hadjithomas M, Chen I-MA, Chu K, Huang J, Ratner A, Palaniappan K et al (2017) IMG-ABC: new features for bacterial secondary metabolism analysis and targeted biosynthetic gene cluster discovery in thousands of microbial genomes. Nucleic Acids Res 45:D560–D565. https://doi.org/10.1093/nar/gkw1103

Heep J, Tuchecker PHK, Gebhardt CR, Dürr M (2019) Combination of thin-layer chromatography and mass spectrometry using cluster-induced desorption/ionization. ACS Omega 4:22426–22430. https://doi.org/10.1021/acsomega.9b03060

Hoagland MB, Stephenson ML, Scott JF, Hecht LI, Zamecnik PC (1958) A soluble ribonucleic acid intermediate in protein synthesis. J Biol Chem 231:241–257

Hudson MA, Lockless SW (2022) Elucidating the mechanisms of action of antimicrobial agents. MBio 13:e02240–e02221. https://doi.org/10.1128/mbio.02240-21

Hudson GA, Burkhart BJ, DiCaprio AJ, Schwalen CJ, Kille B, Pogorelov TV et al (2019) Bioinformatic mapping of radical S-Adenosylmethionine-dependent Ribosomally synthesized and post-translationally modified peptides identifies new Cα, Cβ, and Cγ-linked Thioether-containing peptides. J Am Chem Soc 141:8228–8238. https://doi.org/10.1021/jacs.9b01519

Hug JJ, Panter F, Krug D, Müller R (2019) Genome mining reveals uncommon alkylpyrones as type III PKS products from myxobacteria. J Ind Microbiol Biotechnol 46:319–334. https://doi.org/10.1007/s10295-018-2105-6

Hutter B, Fischer C, Jacobi A, Schaab C, Loferer H (2004) Panel of Bacillus subtilis reporter strains indicative of various modes of action. Antimicrob Agents Chemother 48:2588–2594. https://doi.org/10.1128/AAC.48.7.2588-2594.2004

Ibrahim AA, El-Housseiny GS, Aboshanab KM, Yassien MA, Hassouna NA (2019) Paromomycin production from Streptomyces rimosus NRRL 2455: statistical optimization and new synergistic antibiotic combinations against multidrug resistant pathogens. BMC Microbiol 19:18. https://doi.org/10.1186/s12866-019-1390-1

Jiménez F, Palma J, Sánchez G, Marín D, Francisco Palacios MD, Lucía López MD (2020) Feature selection based multivariate time series forecasting: an application to antibiotic resistance outbreaks prediction. Artif Intell Med 104:101818. https://doi.org/10.1016/j.artmed.2020.101818

Jo S, Kim T, Iyer VG, Im W (2008) CHARMM-GUI: a web-based graphical user interface for CHARMM. J Comput Chem 29:1859–1865. https://doi.org/10.1002/jcc.20945

Jones S (2017) Permeability rules for antibiotic design. Nat Biotechnol 35:639. https://doi.org/10.1038/nbt.3919

Joseph F-JRS, Iniyan AM, Vincent SGP (2017) HR-LC-MS based analysis of two antibacterial metabolites from a marine sponge symbiont Streptomyces pharmamarensis ICN40. Microb Pathog 111:450–457. https://doi.org/10.1016/j.micpath.2017.09.033

Ju K-S, Gao J, Doroghazi JR, Wang K-KA, Thibodeaux CJ, Li S et al (2015) Discovery of phosphonic acid natural products by mining the genomes of 10,000 actinomycetes. Proc Natl Acad Sci 112:12175–12180. https://doi.org/10.1073/pnas.1500873112

Jukič M, Bren U (2022) Machine learning in antibacterial drug design. Front Pharmacol 13:864412. https://doi.org/10.3389/fphar.2022.864412

Jumper J, Evans R, Pritzel A, Green T, Figurnov M, Ronneberger O et al (2021) Highly accurate protein structure prediction with AlphaFold. Nature 596:583–589. https://doi.org/10.1038/s41586-021-03819-2

Katz L, Baltz RH (2016) Natural product discovery: past, present, and future. J Ind Microbiol Biotechnol 43:155–176. https://doi.org/10.1007/s10295-015-1723-5

Kautsar SA, Blin K, Shaw S, Navarro-Muñoz JC, Terlouw BR, van der Hooft JJJ et al (2020) MIBiG 2.0: a repository for biosynthetic gene clusters of known function. Nucleic Acids Res 48:D454–D458. https://doi.org/10.1093/nar/gkz882

Kavanagh F (1975) Microbiological diffusion assay II: design and applications. J Pharm Sci 64:1224–1229. https://doi.org/10.1002/jps.2600640723

Kavčič B, Tkačik G, Bollenbach T (2020) Mechanisms of drug interactions between translation-inhibiting antibiotics. Nat Commun 11:4013. https://doi.org/10.1038/s41467-020-17734-z

Kelley LA, Mezulis S, Yates CM, Wass MN, Sternberg MJE (2015) The Phyre2 web portal for protein modeling, prediction and analysis. Nat Protoc 10:845–858. https://doi.org/10.1038/nprot.2015.053

Kennedy T (1997) Managing the drug discovery/development interface. Drug Discov Today 2:436–444. https://doi.org/10.1016/S1359-6446(97)01099-4

Kim T-W, Keum J-W, Oh I-S, Choi C-Y, Park C-G, Kim D-M (2006) Simple procedures for the construction of a robust and cost-effective cell-free protein synthesis system. J Biotechnol 126:554–561. https://doi.org/10.1016/j.jbiotec.2006.05.014

Kinoshita-Kikuta E, Maruta S, Eguchi Y, Igarashi M, Okajima T, Utsumi R et al (2020) An immuno-dot blot assay for screening histidine kinase inhibitors. Anal Biochem 600:113765. https://doi.org/10.1016/j.ab.2020.113765

Kitagawa M, Ara T, Arifuzzaman M, Ioka-Nakamichi T, Inamoto E, Toyonaga H et al (2005) Complete set of ORF clones of Escherichia coli ASKA library (a complete set of E. coli K-12 ORF archive): unique resources for biological research. DNA Res 12:291–299. https://doi.org/10.1093/dnares/dsi012

Kloosterman AM, Shelton KE, van Wezel GP, Medema MH, Mitchell DA (2020) RRE-finder: a genome-mining tool for class-independent RiPP discovery. mSystems 5:e00267–e00220. https://doi.org/10.1128/mSystems.00267-20

Kohanski MA, Dwyer DJ, Collins JJ (2010) How antibiotics kill bacteria: from targets to networks. Nat Rev Microbiol 8:423–435. https://doi.org/10.1038/nrmicro2333

Kreuzig F (1977) Application of quantitative high-performance thin-layer chromatography in the antibiotic industry. J Chromatogr 142:441–447. https://doi.org/10.1016/s0021-9673(01)92057-6

Kurittu J, Karp M, Korpela M (2000) Detection of tetracyclines with luminescent bacterial strains. Luminescence 15:291–297

Laureti L, Song L, Huang S, Corre C, Leblond P, Challis GL et al (2011) Identification of a bioactive 51-membered macrolide complex by activation of a silent polyketide synthase in Streptomyces ambofaciens. Proc Natl Acad Sci 108:6258–6263. https://doi.org/10.1073/pnas.1019077108

Leshchiner D, Rosconi F, Sundaresh B, Rudmann E, Ramirez LMN, Nishimoto AT et al (2022) A genome-wide atlas of antibiotic susceptibility targets and pathways to tolerance. Nat Commun 13:3165. https://doi.org/10.1038/s41467-022-30967-4

Letzel A-C, Pidot SJ, Hertweck C (2013) A genomic approach to the cryptic secondary metabolome of the anaerobic world. Nat Prod Rep 30:392–428. https://doi.org/10.1039/C2NP20103H

Lewis K (2013) Platforms for antibiotic discovery. Nat Rev Drug Discov 12:371–387. https://doi.org/10.1038/nrd3975

Li Y-X, Zhong Z, Hou P, Zhang W-P, Qian P-Y (2018) Resistance to nonribosomal peptide antibiotics mediated by D-stereospecific peptidases. Nat Chem Biol 14:381–387. https://doi.org/10.1038/s41589-018-0009-4

Li Y, Hu J, Wang Y, Zhou J, Zhang L, Liu Z (2020) DeepScaffold: a comprehensive tool for scaffold-based De novo drug discovery using deep learning. J Chem Inf Model 60:77–91. https://doi.org/10.1021/acs.jcim.9b00727
Li G, Peng X, Guo Y, Gong S, Cao S, Qiu F (2021) Currently available strategies for target identification of bioactive natural products. Front Chem 9:761609. https://doi.org/10.3389/fchem.2021.761609
Lipinski CA (2004) Lead- and drug-like compounds: the rule-of-five revolution. Drug Discov Today Technol 1:337–341. https://doi.org/10.1016/j.ddtec.2004.11.007
Liu X, Ouyang S, Yu B, Liu Y, Huang K, Gong J et al (2010) PharmMapper server: a web server for potential drug target identification using pharmacophore mapping approach. Nucleic Acids Res 38:W609–W614. https://doi.org/10.1093/nar/gkq300
Liu X, Chen Y, Fierke CA (2014) A real-time fluorescence polarization activity assay to screen for inhibitors of bacterial ribonuclease P. Nucleic Acids Res 42:e159. https://doi.org/10.1093/nar/gku850
Liu N, Abramyan ED, Cheng W, Perlatti B, Harvey CJB, Bills GF et al (2021) Targeted genome mining reveals the biosynthetic gene clusters of natural product CYP51 inhibitors. J Am Chem Soc 143:6043–6047. https://doi.org/10.1021/jacs.1c01516
Llano-Sotelo B, Hickerson RP, Lancaster L, Noller HF, Mankin AS (2009) Fluorescently labeled ribosomes as a tool for analyzing antibiotic binding. RNA 15:1597–1604. https://doi.org/10.1261/rna.1681609
Lowell AN, Santoro N, Swaney SM, McQuade TJ, Schultz PJ, Larsen MJ et al (2015) Microscale adaptation of in vitro transcription/translation for high-throughput screening of natural product extract libraries. Chem Biol Drug Des 86:1331–1338. https://doi.org/10.1111/cbdd.12614
Luo Y, Cobb RE, Zhao H (2014) Recent advances in natural product discovery. Curr Opin Biotechnol 30:230–237. https://doi.org/10.1016/j.copbio.2014.09.002
Lushniak BD (2014) Antibiotic resistance: a public health crisis. Public Health Rep 129:314–316. https://doi.org/10.1177/003335491412900402
Machado H, Tuttle RN, Jensen PR (2017) Omics-based natural product discovery and the lexicon of genome mining. Curr Opin Microbiol 39:136–142. https://doi.org/10.1016/j.mib.2017.10.025
Magalhães J, Franko N, Raboni S, Annunziato G, Tammela P, Bruno A et al (2020) Inhibition of nonessential bacterial targets: discovery of a novel serine O-acetyltransferase inhibitor. ACS Med Chem Lett 11:790–797. https://doi.org/10.1021/acsmedchemlett.9b00627
Malathi K, Ramaiah S (2018) Bioinformatics approaches for new drug discovery: a review. Biotechnol Genet Eng Rev 34:243–260. https://doi.org/10.1080/02648725.2018.1502984
Malit JJL, Liu W, Cheng A, Saha S, Liu L-L, Qian P-Y (2021a) Global genome mining reveals a cytochrome P450-catalyzed cyclization of Crownlike Cyclodipeptides with neuroprotective activity. Org Lett 23:6601–6605. https://doi.org/10.1021/acs.orglett.1c01022
Malit JJL, Wu C, Liu L-L, Qian P-Y (2021b) Global genome mining reveals the distribution of diverse Thioamidated RiPP biosynthesis gene clusters. Front Microbiol 12:635389. https://doi.org/10.3389/fmicb.2021.635389
Malit JJL, Leung HYC, Qian P-Y (2022a) Targeted large-scale genome mining and candidate prioritization for natural product discovery. Mar Drugs 20:398. https://doi.org/10.3390/md20060398
Malit JJL, Wu C, Tian X, Liu W, Huang D, Sung HH-Y et al (2022b) Griseocazines: neuroprotective Multiprenylated Cyclodipeptides identified through targeted genome mining. Org Lett 24:2967–2972. https://doi.org/10.1021/acs.orglett.2c00745
Man R-J, Zhang X-P, Yang Y-S, Jiang A-Q, Zhu H-L (2021) Recent Progress in small molecular inhibitors of DNA gyrase. Curr Med Chem 28:5808–5830
Marina A, Mott C, Auyzenberg A, Hendrickson WA, Waldburger CD (2001) Structural and mutational analysis of the PhoQ histidine kinase catalytic domain. Insight into the reaction mechanism. J Biol Chem 276:41182–41190. https://doi.org/10.1074/jbc.M106080200
Martens E, Demain AL (2017) The antibiotic resistance crisis, with a focus on the United States. J Antibiot (Tokyo) 70:520–526. https://doi.org/10.1038/ja.2017.30

Martin YC, Kofron JL, Traphagen LM (2002) Do structurally similar molecules have similar biological activity? J Med Chem 45:4350–4358. https://doi.org/10.1021/jm020155c
Matarrita-Carranza B, Murillo-Cruz C, Avendaño R, Ríos MI, Chavarría M, Gómez-Calvo ML et al (2021) Streptomyces sp. M54: an actinobacteria associated with a neotropical social wasp with high potential for antibiotic production. Antonie Van Leeuwenhoek 114:379–398. https://doi.org/10.1007/s10482-021-01520-y
Mayer C, Kluj RM, Mühleck M, Walter A, Unsleber S, Hottmann I et al (2019) Bacteria's different ways to recycle their own cell wall. Int J Med Microbiol 309:151326. https://doi.org/10.1016/j.ijmm.2019.06.006
Mazza G (1983) Rapid assay for detection of microorganisms producing DNA-damaging metabolites. Appl Environ Microbiol 45:1949–1952. https://doi.org/10.1128/aem.45.6.1949-1952.1983
Medema MH, Blin K, Cimermancic P, de Jager V, Zakrzewski P, Fischbach MA et al (2011) antiSMASH: rapid identification, annotation and analysis of secondary metabolite biosynthesis gene clusters in bacterial and fungal genome sequences. Nucleic Acids Res 39:W339–W346. https://doi.org/10.1093/nar/gkr466
Medema MH, Kottmann R, Yilmaz P, Cummings M, Biggins JB, Blin K et al (2015) Minimum information about a biosynthetic gene cluster. Nat Chem Biol 11:625–631. https://doi.org/10.1038/nchembio.1890
Melamed S, Lalush C, Elad T, Yagur-Kroll S, Belkin S, Pedahzur R (2012) A bacterial reporter panel for the detection and classification of antibiotic substances. Microb Biotechnol 5:536–548. https://doi.org/10.1111/j.1751-7915.2012.00333.x
Melo MCR, Maasch JRMA, de la Fuente-Nunez C (2021) Accelerating antibiotic discovery through artificial intelligence. Commun Biol 4:1050. https://doi.org/10.1038/s42003-021-02586-0
Merigueti TC, Carneiro MW, Carvalho-Assef APD, Silva-Jr FP, da Silva FAB (2019) FindTargetsWEB: a user-friendly tool for identification of potential therapeutic targets in metabolic networks of bacteria. Front Genet 10:633. https://doi.org/10.3389/fgene.2019.00633
Miethke M, Pieroni M, Weber T, Brönstrup M, Hammann P, Halby L et al (2021) Towards the sustainable discovery and development of new antibiotics. Nat Rev Chem 5:726–749. https://doi.org/10.1038/s41570-021-00313-1
Migliori GB, Tiberi S, Zumla A, Petersen E, Chakaya JM, Wejse C et al (2020) MDR/XDR-TB management of patients and contacts: challenges facing the new decade. The 2020 clinical update by the global tuberculosis network. Int J Infect Dis 92:S15–S25. https://doi.org/10.1016/j.ijid.2020.01.042
Möhrle V, Stadler M, Eberz G (2007) Biosensor-guided screening for macrolides. Anal Bioanal Chem 388:1117–1125. https://doi.org/10.1007/s00216-007-1300-5
Moir DT, Di M, Opperman T, Schweizer HP, Bowlin TL (2007) A high-throughput, homogeneous, bioluminescent assay for Pseudomonas aeruginosa gyrase inhibitors and other DNA-damaging agents. J Biomol Screen 12:855–864. https://doi.org/10.1177/1087057107304729
Moloney MG (2016) Natural products as a source for novel antibiotics. Trends Pharmacol Sci 37:689–701. https://doi.org/10.1016/j.tips.2016.05.001
Montalbán-López M, Scott TA, Ramesh S, Rahman IR, van Heel AJ, Viel JH et al (2021) New developments in RiPP discovery, enzymology and engineering. Nat Prod Rep 38:130–239. https://doi.org/10.1039/d0np00027b
Morris GM, Huey R, Lindstrom W, Sanner MF, Belew RK, Goodsell DS et al (2009) AutoDock4 and AutoDockTools4: automated docking with selective receptor flexibility. J Comput Chem 30:2785–2791. https://doi.org/10.1002/jcc.21256
Müller A, Klöckner A, Schneider T (2017) Targeting a cell wall biosynthesis hot spot. Nat Prod Rep 34:909–932. https://doi.org/10.1039/c7np00012j
Mullins AJ, Murray JAH, Bull MJ, Jenner M, Jones C, Webster G et al (2019) Genome mining identifies cepacin as a plant-protective metabolite of the biopesticidal bacterium Burkholderia ambifaria. Nat Microbiol 4:996–1005. https://doi.org/10.1038/s41564-019-0383-z
Munita JM, Arias CA (2016) Mechanisms of antibiotic resistance. Microbiol Spectr 4. https://doi.org/10.1128/microbiolspec.VMBF-0016-2015

Murray CJ, Ikuta KS, Sharara F, Swetschinski L, Robles Aguilar G, Gray A et al (2022) Global burden of bacterial antimicrobial resistance in 2019: a systematic analysis. Lancet S0140673621027240:629. https://doi.org/10.1016/S0140-6736(21)02724-0

Navarro-Muñoz JC, Selem-Mojica N, Mullowney MW, Kautsar SA, Tryon JH, Parkinson EI et al (2020) A computational framework to explore large-scale biosynthetic diversity. Nat Chem Biol 16:60–68. https://doi.org/10.1038/s41589-019-0400-9

Nikolay R, Schloemer R, Mueller S, Deuerling E (2015) Fluorescence-based monitoring of ribosome assembly landscapes. BMC Mol Biol 16:3. https://doi.org/10.1186/s12867-015-0031-y

Nonejuie P, Burkart M, Pogliano K, Pogliano J (2013) Bacterial cytological profiling rapidly identifies the cellular pathways targeted by antibacterial molecules. Proc Natl Acad Sci USA 110:16169–16174. https://doi.org/10.1073/pnas.1311066110

Nowak MG, Skwarecki AS, Milewska MJ (2021) Amino acid based antimicrobial agents – synthesis and properties. ChemMedChem 16:3513–3544. https://doi.org/10.1002/cmdc.202100503

Nunes AMV, de Andrade FDCP, Filgueiras LA, de Carvalho Maia OA, Cunha RL, Rodezno SV, Maia Filho ALM, de Amorim Carvalho FA, Braz DC, Mendes AN (2020) preADMET analysis and clinical aspects of dogs treated with the Organotellurium compound RF07: a possible control for canine visceral leishmaniasis? Environ Toxicol Pharmacol 80:103470. https://doi.org/10.1016/j.etap.2020.103470

Núñez-Montero K, Lamilla C, Abanto M, Maruyama F, Jorquera MA, Santos A et al (2019) Antarctic Streptomyces fildesensis So13.3 strain as a promising source for antimicrobials discovery. Sci Rep 9:7488. https://doi.org/10.1038/s41598-019-43960-7

O'Neill J (2016) An audience with Jim O'Neill. Nat Rev Drug Discov 15:526–526. https://doi.org/10.1038/nrd.2016.160

O'Neill EC, Schorn M, Larson CB, Millán-Aguiñaga N (2019) Targeted antibiotic discovery through biosynthesis-associated resistance determinants: target directed genome mining. Crit Rev Microbiol 45:255–277. https://doi.org/10.1080/1040841X.2019.1590307

Orelle C, Carlson S, Kaushal B, Almutairi MM, Liu H, Ochabowicz A et al (2013) Tools for characterizing bacterial protein synthesis inhibitors. Antimicrob Agents Chemother 57:5994–6004. https://doi.org/10.1128/AAC.01673-13

Ortlieb N, Klenk E, Kulik A, Niedermeyer THJ (2021) Development of an agar-plug cultivation system for bioactivity assays of actinomycete strain collections. PLoS One 16:e0258934. https://doi.org/10.1371/journal.pone.0258934

Osterman IA, Prokhorova IV, Sysoev VO, Boykova YV, Efremenkova OV, Svetlov MS et al (2012) Attenuation-based dual-fluorescent-protein reporter for screening translation inhibitors. Antimicrob Agents Chemother 56:1774–1783. https://doi.org/10.1128/AAC.05395-11

Osterman IA, Ustinov AV, Evdokimov DV, Korshun VA, Sergiev PV, Serebryakova MV et al (2013) A nascent proteome study combining click chemistry with 2DE. Proteomics 13:17–21. https://doi.org/10.1002/pmic.201200393

Osterman IA, Bogdanov AA, Dontsova OA, Sergiev PV (2016) Techniques for screening translation inhibitors. Antibiotics (Basel) 5:22. https://doi.org/10.3390/antibiotics5030022

Palazzotto E, Weber T (2018) Omics and multi-omics approaches to study the biosynthesis of secondary metabolites in microorganisms. Curr Opin Microbiol 45:109–116. https://doi.org/10.1016/j.mib.2018.03.004

Pan G, Xu Z, Guo Z et al (2017) Discovery of the leinamycin family of natural products by mining actinobacterial genomes. Proc Natl Acad Sci 114:E11131–E11140. https://doi.org/10.1073/pnas.1716245115

Panchal V, Brenk R (2021) Riboswitches as drug targets for antibiotics. Antibiotics (Basel) 10:45. https://doi.org/10.3390/antibiotics10010045

Pandey M, Fernandez M, Gentile F, Isayev O, Tropsha A, Stern AC et al (2022) The transformational role of GPU computing and deep learning in drug discovery. Nat Mach Intell 4:211–221. https://doi.org/10.1038/s42256-022-00463-x

Panter F, Krug D, Baumann S, Müller R (2018) Self-resistance guided genome mining uncovers new topoisomerase inhibitors from myxobacteria. Chem Sci 9:4898–4908. https://doi.org/10.1039/C8SC01325J

Pasquina-Lemonche L, Burns J, Turner RD, Kumar S, Tank R, Mullin N et al (2020) The architecture of the gram-positive bacterial cell wall. Nature 582:294–297. https://doi.org/10.1038/s41586-020-2236-6

Pearlman DA, Case DA, Caldwell JW, Ross WS, Cheatham TE, DeBolt S et al (1995) AMBER, a package of computer programs for applying molecular mechanics, normal mode analysis, molecular dynamics and free energy calculations to simulate the structural and energetic properties of molecules. Comput Phys Commun 91:1–41. https://doi.org/10.1016/0010-4655(95)00041-D

Piddock LJV (2017) Understanding drug resistance will improve the treatment of bacterial infections. Nat Rev Microbiol 15:639–640. https://doi.org/10.1038/nrmicro.2017.121

Pistorius D, Müller R (2012) Discovery of the Rhizopodin biosynthetic gene cluster in Stigmatella aurantiaca sg a15 by genome mining. Chembiochem 13:416–426. https://doi.org/10.1002/cbic.201100575

Poehlsgaard J, Douthwaite S (2005) The bacterial ribosome as a target for antibiotics. Nat Rev Microbiol 3:870–881. https://doi.org/10.1038/nrmicro1265

Polacek N, Swaney S, Shinabarger D, Mankin AS (2002) SPARK – a novel method to monitor ribosomal peptidyl transferase activity. Biochemistry 41:11602–11610. https://doi.org/10.1021/bi026040s

Price MN, Dehal PS, Arkin AP (2010) FastTree 2 – approximately maximum-likelihood trees for large alignments. PLoS One 5:e9490. https://doi.org/10.1371/journal.pone.0009490

Pudi N, Varikuti GD, Badana AK, Gavara MM, Kumari S, Malla R (2016) Studies on optimization of growth parameters for enhanced production of antibiotic alkaloids by isolated marine Actinomycetes. J App Pharm Sci 6:181–188. https://doi.org/10.7324/JAPS.2016.601025

Purushothaman M, Sarkar S, Morita M, Gugger M, Schmidt EW, Morinaka BI (2021) Genome-mining-based discovery of the cyclic peptide Tolypamide and TolF, a Ser/Thr forward O-Prenyltransferase. Angew Chem Int Ed 60:8460–8465. https://doi.org/10.1002/anie.202015975

Qureshi KA, Bholay AD, Rai PK, Mohammed HA, Khan RA, Azam F et al (2021) Isolation, characterization, anti-MRSA evaluation, and in-silico multi-target anti-microbial validations of actinomycin X2 and actinomycin D produced by novel Streptomyces smyrnaeus UKAQ_23. Sci Rep 11:14539. https://doi.org/10.1038/s41598-021-93285-7

Radusky L, Defelipe LA, Lanzarotti E, Luque J, Barril X, Marti MA et al (2014) TuberQ: a mycobacterium tuberculosis protein druggability database. Database (Oxford) 2014:bau035. https://doi.org/10.1093/database/bau035

Raman K, Yeturu K, Chandra N (2008) targetTB: a target identification pipeline for mycobacterium tuberculosis through an interactome, reactome and genome-scale structural analysis. BMC Syst Biol 2:109. https://doi.org/10.1186/1752-0509-2-109

Raneri M, Sciandrone B, Briani F (2015) A whole-cell assay for specific inhibitors of translation initiation in bacteria. J Biomol Screen 20:627–633. https://doi.org/10.1177/1087057114566376

Reece RJ, Maxwell A (1991) DNA gyrase: structure and function. Crit Rev Biochem Mol Biol 26:335–375. https://doi.org/10.3109/10409239109114072

Reyes-Lamothe R, Sherratt DJ (2019) The bacterial cell cycle, chromosome inheritance and cell growth. Nat Rev Microbiol 17:467–478. https://doi.org/10.1038/s41579-019-0212-7

Rose PW, Prlić A, Altunkaya A, Bi C, Bradley AR, Christie CH et al (2017) The RCSB protein data bank: integrative view of protein, gene and 3D structural information. Nucleic Acids Res 45:D271–D281. https://doi.org/10.1093/nar/gkw1000

Ruskoski TB, Boal AK (2021) The periodic table of ribonucleotide reductases. J Biol Chem 297:101137. https://doi.org/10.1016/j.jbc.2021.101137

Rütten A, Kirchner T, Musiol-Kroll EM (2022) Overview on strategies and assays for antibiotic discovery. Pharmaceuticals 15:1302. https://doi.org/10.3390/ph15101302

Saleem HGM, Aftab U, Sajid I, Abbas Z, Sabri AN (2015) Effect of crude extracts of selected actinomycetes on biofilm formation of a. schindleri, M. aci, and B. cereus. J Basic Microbiol 55:645–651. https://doi.org/10.1002/jobm.201400358

Santos-Aberturas J, Chandra G, Frattaruolo L, Lacret R, Pham TH, Vior NM et al (2019) Uncovering the unexplored diversity of thioamidated ribosomal peptides in Actinobacteria

using the RiPPER genome mining tool. Nucleic Acids Res 47:4624–4637. https://doi.org/10.1093/nar/gkz192
Saravana Kumar P, Duraipandiyan V, Ignacimuthu S (2014) Isolation, screening and partial purification of antimicrobial antibiotics from soil Streptomyces sp. SCA 7. Kaohsiung J Med Sci 30:435–446. https://doi.org/10.1016/j.kjms.2014.05.006
Sarmiento-Vizcaíno A, Martín J, Reyes F, García LA, Blanco G (2021) Bioactive natural products in Actinobacteria isolated in rainwater from storm clouds transported by Western winds in Spain. Front Microbiol 12:773095. https://doi.org/10.3389/fmicb.2021.773095
Scheffers D-J, Pinho MG (2005) Bacterial cell wall synthesis: new insights from localization studies. Microbiol Mol Biol Rev 69:585–607. https://doi.org/10.1128/MMBR.69.4.585-607.2005
Schneidman-Duhovny D, Dror O, Inbar Y, Nussinov R, Wolfson HJ (2008) PharmaGist: a webserver for ligand-based pharmacophore detection. Nucleic Acids Res 36:W223–W228. https://doi.org/10.1093/nar/gkn187
Schuerholz T, Dömming S, Hornef M, Dupont A, Kowalski I, Kaconis Y et al (2012) Bacterial cell wall compounds as promising targets of antimicrobial agents II. Immunological and clinical aspects. Curr Drug Targets 13:1131–1137. https://doi.org/10.2174/138945012802002438
Schwalen CJ, Hudson GA, Kille B, Mitchell DA (2018) Bioinformatic expansion and discovery of Thiopeptide antibiotics. J Am Chem Soc 140:9494–9501. https://doi.org/10.1021/jacs.8b03896
Serral F, Castello FA, Sosa EJ, Pardo AM, Palumbo MC, Modenutti C et al (2021) From genome to drugs: new approaches in antimicrobial discovery. Front Pharmacol 12:647060. https://doi.org/10.3389/fphar.2021.647060
Shaker B, Ahmad S, Lee J, Jung C, Na D (2021) In silico methods and tools for drug discovery. Comput Biol Med 137:104851. https://doi.org/10.1016/j.compbiomed.2021.104851
Sharma M, Manhas RK (2019) Purification and characterization of actinomycins from Streptomyces strain M7 active against methicillin resistant Staphylococcus aureus and vancomycin resistant enterococcus. BMC Microbiol 19:44. https://doi.org/10.1186/s12866-019-1405-y
Sharma P, Kalita MC, Thakur D (2016) Broad Spectrum antimicrobial activity of Forest-derived soil Actinomycete, Nocardia sp. PB-52. Front Microbiol 7:347. https://doi.org/10.3389/fmicb.2016.00347
Shimizu Y, Inoue A, Tomari Y, Suzuki T, Yokogawa T, Nishikawa K et al (2001) Cell-free translation reconstituted with purified components. Nat Biotechnol 19:751–755. https://doi.org/10.1038/90802
Silbert L, Ben Shlush I, Israel E, Porgador A, Kolusheva S, Jelinek R (2006) Rapid chromatic detection of bacteria by use of a new biomimetic polymer sensor. Appl Environ Microbiol 72:7339–7344. https://doi.org/10.1128/AEM.01324-06
Silhavy TJ, Kahne D, Walker S (2010) The bacterial cell envelope. Cold Spring Harb Perspect Biol 2:a000414. https://doi.org/10.1101/cshperspect.a000414
Singh V, Haque S, Singh H, Verma J, Vibha K, Singh R et al (2016) Isolation, screening, and identification of novel isolates of Actinomycetes from India for antimicrobial applications. Front Microbiol 7:1921. https://doi.org/10.3389/fmicb.2016.01921
Skinnider MA, Johnston CW, Edgar RE, Dejong CA, Merwin NJ, Rees PN et al (2016) Genomic charting of ribosomally synthesized natural product chemical space facilitates targeted mining. Proc Natl Acad Sci 113:E6343–E6351. https://doi.org/10.1073/pnas.1609014113
Skinnider MA, Johnston CW, Gunabalasingam M, Merwin NJ, Kieliszek AM, MacLellan RJ et al (2020) Comprehensive prediction of secondary metabolite structure and biological activity from microbial genome sequences. Nat Commun 11:6058. https://doi.org/10.1038/s41467-020-19986-1
Skorupa A, Gierak A (2011) Detection and visualization methods used in thin-layer chromatography. JPC-J Planar Chromat 24:274–280. https://doi.org/10.1556/JPC.24.2011.4.1
Smith NM, Lenhard JR, Boissonneault KR, Landersdorfer CB, Bulitta JB, Holden PN et al (2020) Using machine learning to optimize antibiotic combinations: dosing strategies for meropenem and polymyxin B against carbapenem-resistant Acinetobacter baumannii. Clin Microbiol Infect 26:1207–1213. https://doi.org/10.1016/j.cmi.2020.02.004

Song L, Barona-Gomez F, Corre C, Xiang L, Udwary DW, Austin MB et al (2006) Type III polyketide synthase β-Ketoacyl-ACP starter unit and Ethylmalonyl-CoA extender unit selectivity discovered by Streptomyces coelicolor genome mining. J Am Chem Soc 128:14754–14755. https://doi.org/10.1021/ja065247w

Souza MJ, Bittencourt CF, da Se Souza Filho P (2004) Microbiological assay for enrofloxacin injection. Int J Pharm 271:287–291. https://doi.org/10.1016/j.ijpharm.2003.11.012

Spry C, Kirk K, Saliba KJ (2008) Coenzyme a biosynthesis: an antimicrobial drug target. FEMS Microbiol Rev 32:56–106. https://doi.org/10.1111/j.1574-6976.2007.00093.x

Stokes JM, Yang K, Swanson K, Jin W, Cubillos-Ruiz A, Donghia NM et al (2020) A deep learning approach to antibiotic discovery. Cell 180:688–702.e13. https://doi.org/10.1016/j.cell.2020.01.021

Subramaniam G, Girish M (2020) Antibiotic resistance – a cause for reemergence of infections. Indian J Pediatr 87:937–944. https://doi.org/10.1007/s12098-019-03180-3

Sugimoto Y, Camacho FR, Wang S, Chankhamjon P, Odabas A, Biswas A et al (2019) A metagenomic strategy for harnessing the chemical repertoire of the human microbiome. Science 366:eaax9176. https://doi.org/10.1126/science.aax9176

Svetlov MS, Kommer A, Kolb VA, Spirin AS (2006) Effective cotranslational folding of firefly luciferase without chaperones of the Hsp70 family. Protein Sci 15:242–247. https://doi.org/10.1110/ps.051752506

Tacconelli E, Carrara E, Savoldi A, Harbarth S, Mendelson M, Monnet DL et al (2018) Discovery, research, and development of new antibiotics: the WHO priority list of antibiotic-resistant bacteria and tuberculosis. Lancet Infect Dis 18:318–327. https://doi.org/10.1016/S1473-3099(17)30753-3

Tang X, Li J, Millán-Aguiñaga N, Zhang JJ, O'Neill EC, Ugalde JA et al (2015) Identification of Thiotetronic acid antibiotic biosynthetic pathways by target-directed genome mining. ACS Chem Biol 10:2841–2849. https://doi.org/10.1021/acschembio.5b00658

Tetko IV, Tanchuk VY (2002) Application of associative neural networks for prediction of Lipophilicity in ALOGPS 2.1 program. J Chem Inf Comput Sci 42:1136–1145. https://doi.org/10.1021/ci025515j

Theuretzbacher U (2011) Resistance drives antibacterial drug development. Curr Opin Pharmacol 11:433–438. https://doi.org/10.1016/j.coph.2011.07.008

Tholander F, Sjöberg B-M (2012) Discovery of antimicrobial ribonucleotide reductase inhibitors by screening in microwell format. Proc Natl Acad Sci USA 109:9798–9803. https://doi.org/10.1073/pnas.1113051109

Thomas M, Smith RT, O'Boyle NM, de Graaf C, Bender A (2021) Comparison of structure- and ligand-based scoring functions for deep generative models: a GPCR case study. J Chem 13:39. https://doi.org/10.1186/s13321-021-00516-0

Tian J, Chen H, Guo Z, Liu N, Li J, Huang Y et al (2016) Discovery of pentangular polyphenols hexaricins A–C from marine Streptosporangium sp. CGMCC 4.7309 by genome mining. Appl Microbiol Biotechnol 100:4189–4199. https://doi.org/10.1007/s00253-015-7248-z

Tietz JI, Schwalen CJ, Patel PS, Maxson T, Blair PM, Tai H-C et al (2017) A new genome-mining tool redefines the lasso peptide biosynthetic landscape. Nat Chem Biol 13:470–478. https://doi.org/10.1038/nchembio.2319

Torrents E (2014) Ribonucleotide reductases: essential enzymes for bacterial life. Front Cell Infect Microbiol 4:52. https://doi.org/10.3389/fcimb.2014.00052

Torres MDT, De La Fuente-Nunez C (2019) Toward computer-made artificial antibiotics. Curr Opin Microbiol 51:30–38. https://doi.org/10.1016/j.mib.2019.03.004

Tosco P, Balle T (2011) Open3DQSAR: a new open-source software aimed at high-throughput chemometric analysis of molecular interaction fields. J Mol Model 17:201–208. https://doi.org/10.1007/s00894-010-0684-x

Trott O, Olson AJ (2010) AutoDock Vina: improving the speed and accuracy of docking with a new scoring function, efficient optimization, and multithreading. J Comput Chem 31:455–461. https://doi.org/10.1002/jcc.21334

Uddin TM, Chakraborty AJ, Khusro A, Zidan BRM, Mitra S, Emran TB et al (2021) Antibiotic resistance in microbes: history, mechanisms, therapeutic strategies and future prospects. J Infect Public Health 14:1750–1766. https://doi.org/10.1016/j.jiph.2021.10.020
van Heel AJ, Kloosterman TG, Montalban-Lopez M, Deng J, Plat A, Baudu B et al (2016) Discovery, production and modification of five novel Lantibiotics using the promiscuous Nisin modification machinery. ACS Synth Biol 5:1146–1154. https://doi.org/10.1021/acssynbio.6b00033
Velikova N, Fulle S, Manso AS, Mechkarska M, Finn P, Conlon JM et al (2016) Putative histidine kinase inhibitors with antibacterial effect against multi-drug resistant clinical isolates identified by in vitro and in silico screens. Sci Rep 6:26085. https://doi.org/10.1038/srep26085
Walker MC, Eslami SM, Hetrick KJ, Ackenhusen SE, Mitchell DA, van der Donk WA (2020) Precursor peptide-targeted mining of more than one hundred thousand genomes expands the lanthipeptide natural product family. BMC Genomics 21:387. https://doi.org/10.1186/s12864-020-06785-7
Wang JD, Levin PA (2009) Metabolism, cell growth and the bacterial cell cycle. Nat Rev Microbiol 7:822–827. https://doi.org/10.1038/nrmicro2202
Wang M, Zhang Y, Wang R, Wang Z, Yang B, Kuang H (2021) An evolving technology that integrates classical methods with continuous technological developments: thin-layer chromatography bioautography. Molecules 26:4647. https://doi.org/10.3390/molecules26154647
Wang M, Sun H, Wang J, Pang J, Chai X, Xu L et al (2022) Comprehensive assessment of deep generative architectures for de novo drug design. Brief Bioinform 23:bbab544. https://doi.org/10.1093/bib/bbab544
Wani AK, Roy P, Kumar V, Mir T, ul G. (2022) Metagenomics and artificial intelligence in the context of human health. Infect Genet Evol 100:105267. https://doi.org/10.1016/j.meegid.2022.105267
Waterhouse A, Bertoni M, Bienert S, Studer G, Tauriello G, Gumienny R et al (2018) SWISS-MODEL: homology modelling of protein structures and complexes. Nucleic Acids Res 46:W296–W303. https://doi.org/10.1093/nar/gky427
Watkins RR, Bonomo RA (2016) Overview: global and local impact of antibiotic resistance. Infect Dis Clin N Am 30:313–322. https://doi.org/10.1016/j.idc.2016.02.001
Webb B, Sali A (2021) Protein structure modeling with modeller. Methods Mol Biol 2199:239–255. https://doi.org/10.1007/978-1-0716-0892-0_14
Weber T, Kim HU (2016) The secondary metabolite bioinformatics portal: computational tools to facilitate synthetic biology of secondary metabolite production. Synth Syst Biotechnol 1:69–79. https://doi.org/10.1016/j.synbio.2015.12.002
Wenzel SC, Müller R (2009) Myxobacteria – 'microbial factories' for the production of bioactive secondary metabolites. Mol BioSyst 5:567–574. https://doi.org/10.1039/B901287G
Wilson DN, Hauryliuk V, Atkinson GC, O'Neill AJ (2020) Target protection as a key antibiotic resistance mechanism. Nat Rev Microbiol 18:637–648. https://doi.org/10.1038/s41579-020-0386-z
Wohlleben W, Mast Y, Stegmann E, Ziemert N (2016) Antibiotic drug discovery. Microb Biotechnol 9:541–548. https://doi.org/10.1111/1751-7915.12388
Wolber G, Langer T (2005) LigandScout: 3-D pharmacophores derived from protein-bound ligands and their use as virtual screening filters. J Chem Inf Model 45:160–169. https://doi.org/10.1021/ci049885e
Wolf H, Chinali G, Parmeggiani A (1977) Mechanism of the inhibition of protein synthesis by kirromycin. Role of elongation factor Tu and ribosomes. Eur J Biochem 75:67–75. https://doi.org/10.1111/j.1432-1033.1977.tb11504.x
Wu G, Robertson DH, Brooks CL, Vieth M (2003) Detailed analysis of grid-based molecular docking: a case study of CDOCKER-A CHARMm-based MD docking algorithm. J Comput Chem 24:1549–1562. https://doi.org/10.1002/jcc.10306
Wyatt MA, Wang W, Roux CM, Beasley FC, Heinrichs DE, Dunman PM et al (2010) Staphylococcus aureus nonribosomal peptide secondary metabolites regulate virulence. Science 329:294–296. https://doi.org/10.1126/science.1188888
Xu D, Zhang J, Roy A, Zhang Y (2011) Automated protein structure modeling in CASP9 by I-TASSER pipeline combined with QUARK-based ab initio folding and FG-MD-based struc-

ture refinement. Proteins: Struct Funct Bioinformatics 79:147–160. https://doi.org/10.1002/prot.23111

Yamada Y, Kuzuyama T, Komatsu M, Shin-ya K, Omura S, Cane DE et al (2015) Terpene synthases are widely distributed in bacteria. Proc Natl Acad Sci 112:857–862. https://doi.org/10.1073/pnas.1422108112

Yan Y, Liu Q, Zang X, Yuan S, Bat-Erdene U, Nguyen C et al (2018) Resistance-gene-directed discovery of a natural-product herbicide with a new mode of action. Nature 559:415–418. https://doi.org/10.1038/s41586-018-0319-4

Yan J, Bhadra P, Li A, Sethiya P, Qin L, Tai HK et al (2020) Deep-AmPEP30: improve short antimicrobial peptides prediction with deep learning. Mol Ther Nucleic Acids 20:882–894. https://doi.org/10.1016/j.omtn.2020.05.006

Yang J, Zhang Y (2015) I-TASSER server: new development for protein structure and function predictions. Nucleic Acids Res 43:W174–W181. https://doi.org/10.1093/nar/gkv342

Yassin A, Mankin AS (2007) Potential new antibiotic sites in the ribosome revealed by deleterious mutations in RNA of the large ribosomal subunit. J Biol Chem 282:24329–24342. https://doi.org/10.1074/jbc.M703106200

Yassin A, Fredrick K, Mankin AS (2005) Deleterious mutations in small subunit ribosomal RNA identify functional sites and potential targets for antibiotics. Proc Natl Acad Sci USA 102:16620–16625. https://doi.org/10.1073/pnas.0508444102

Yoshida M, Hinkley T, Tsuda S, Abul-Haija YM, McBurney RT, Kulikov V et al (2018) Using evolutionary algorithms and machine learning to explore sequence space for the discovery of antimicrobial peptides. Chem 4:533–543. https://doi.org/10.1016/j.chempr.2018.01.005

Zachow C, Jahanshah G, de Bruijn I, Song C, Ianni F, Pataj Z et al (2015) The novel Lipopeptide Poaeamide of the endophyte Pseudomonas poae RE*1-1-14 is involved in pathogen suppression and root colonization. MPMI 28:800–810. https://doi.org/10.1094/MPMI-12-14-0406-R

Zhang Y, Chen M, Bruner SD, Ding Y (2018) Heterologous production of microbial Ribosomally synthesized and post-translationally modified peptides. Front Microbiol 9:1801. https://doi.org/10.3389/fmicb.2018.01801

Zhang Y, Luo M, Wu P, Wu S, Lee T-Y, Bai C (2022) Application of computational biology and artificial intelligence in drug design. IJMS 23:13568. https://doi.org/10.3390/ijms232113568

Ziemert N, Alanjary M, Weber T (2016) The evolution of genome mining in microbes – a review. Nat Prod Rep 33:988–1005. https://doi.org/10.1039/c6np00025h

The manufacturer's authorised representative in the EU is Springer
Nature Customer Service Centre GmbH, Europaplatz 3, 69115 Heidelberg,
Germany. If you have any concerns regarding our products, please
contact ProductSafety@springernature.com

Printed and bound by CPI Group (UK) Ltd, Croydon, CR0 4YY
01/10/2026
02225923-0013